Basic Structures of Modern Algebra

Mathematics and Its Applications

Volume 265

Basic Structures of Modern Algebra

by

Yuri Bahturin
Department of Geometry and Topology,
Moscow State University,
Moscow, Russia

SPRINGER-SCIENCE+BUSINESS MEDIA, B.V.

A C.I.P. Catalogue record for this book is available from the Library of Congress.

ISBN 978-90-481-4317-7 ISBN 978-94-017-0839-5 (eBook)
DOI 10.1007/978-94-017-0839-5

typeset by AMS-TEX

Printed on acid-free paper

TABLE OF CONTENTS

PREFACE

This book has developed from a series of lectures which were given by the author in mechanics–mathematics department of the Moscow State University. In 1981 the course "Additional chapters in algebra" replaced the course "General algebra" which was founded by A.G. Kurosh (1908–1971), professor and head of the department of higher algebra for a period of several decades. The material of this course formed the basis of A. G. Kurosh's well–known book "Lectures on general algebra" (Moscow,1962; 2-nd edition: Moscow, Nauka, 1973) and the book "General algebra. Lectures of 1969–1970." (Moscow, Nauka, 1974). Another book based on the course, "Elements of general algebra" (M.: Nauka, 1983) was published by L.A. Skorniakov, professor, now deceased, in the same department. It should be noted that A.G. Kurosh was not only the lecturer for the course "General algebra" but he was also the recognized leader of the scientific school of the same name. It is difficult to determine the limits of this school; however, the "Lectures…" of 1962 mentioned above contain some material which exceed these limits. Eventually this effect intensified: the lectures of the course were given by many well–known scientists, and some of them see themselves as "general algebraists". Each lecturer brought significant originality not only in presentation of the material but in the substance of the course. Therefore not all material which is now accepted as necessary for algebraic students fits within the scope of general algebra.

There is another side to this problem. The fact is that some chapters from "general algebra" are so similar (see the second of the books cited by Kurosh) that it seems no longer advisable to study even just the basics of all these chapters for all algebraists.

These arguments lead us to change the title of this course of which the main goal is to introduce beginners to the concepts, results and problems of contemporary algebra assuming knowledge of the standard theory of linear algebra and vector spaces.

The idea of this book is not only to introduce the various concepts but also to show how they work and relate. Though the presentation of the material has a somewhat of a "chunky" character: a chunk from the theory of fields, a chunk from groups, rings, Lie algebras, etc.. There are, on the other hand, themes which "roam" from one chapter to another. One of the more significant of these themes is the one of periodic groups which started with the classical work on the theory of groups by W. Burnside. Another organizing principle is construction the bridges (links): groups – graphs, groups – rings – Lie algebras, fields – skew fields – varieties of algebras, etc.. The reader should judge for himself the value of this result.

It is pleasure for me to express my sincere gratitude to the translator D. Žviré-naité and to the editor Prof. M. Hazewinkel who did an excelent job and thus saved me from a certain amount of criticism to follow.

RECOMMENDATIONS TO THE READER

The author of this book tried to make the account self contained as far as this is possible in a text which is based on a general university courses for the first two academic years. Practically, all the concepts of abstract algebra used in this course are collected in the introduction which also contains some additional background material. References to this part of the book in other chapters contain the word Introduction: a reference Introduction $m.\ n$ means that it concerns subsection n of section m from the Introduction. Various set–theoretic concepts and theorems, as well as a number of results from topology and the theory of smooth manifolds are contained in a set–theoretic appendix at the end of the book. References to this section are indicated by Set–theoretic Supplement n, where n is a number of a subsection in the appendix. References to a certain subsection in a chapter are denoted by $m.\ n.\ k$, where m is the number of a chapter, n the number of a section, k the number of a subsection. There also occur references of the form § $m.\ n$, where m is the number of a chapter, n the number of a section. A reference inside a chapter does not mention that chapter and a reference inside a section does not mention that section.

The subject and symbol indexes indicate where a concept or notation was first introduced.

The end of a proof is marked by the sign "□". If no proof is given the same sign occurs at the end of the statement. Some simple extra statements are indicated by circles: $\circ \ldots \circ$. Supplementary material is marked by stars $* \ldots *$.

INTRODUCTION

The purpose of this introduction is to explain in a condensed form the main definitions and results in group, ring and field theory that are usually presented in elementary standard courses of a study in physics or mathematics.

§ 1. Groups

1.1. Definitions. A set G with a binary associative operation $(g, h) \mapsto gh$ is called a *group* if it contains a *unit element*, i.e. an element e such that $ge = eg = g$ for any $g \in G$, and if for each $g \in G$ there exists an *inverse* g^{-1}, i.e. $gg^{-1} = g^{-1}g = e$.

If the operation is written multiplicatively, then the unit element is called an *identity* and denoted as 1.

When an additive description of the operation is used, i.e. $(g, h) \to g + h$, the unit element is called the *zero* and is denoted as 0, the inverse element to g is called the *negative* and is denoted as $-g$; thus $g + 0 = 0 + g = g$ and $g + (-g) = (-g) + g = 0$.

The unit element is unique; unique also is the inverse element for any given g.

A non-empty subset H of a group G is called a *subgroup* if for any $h_1, h_2 \in H$ one has $h_1h_2 \in H$, $h_1^{-1}, h_2^{-1} \in H$.

∘ A subset H of a group G is a subgroup iff H is a group relative to the same operation as G. ∘

A *proper* subgroup is any subgroup not equal to the whole group.

The *trivial* subgroup is the subgroup consisting only of the unit element.

Two groups are called *isomorphic* if there exists a bijection ϕ: $G_1 \to G_2$ such that the result of the operation on the images of any two elements from G_1 coincides with the image of the result of the operation on these elements. We write $G_1 \cong G_2$.

For instance, if the operation is written multiplicatively in G_1 and G_2, $x, y \in G_1$, then the statement has the form

$$\phi(xy) = \phi(x)\phi(y).$$

∘ For isomorphic groups the image of the identity element is the identity and the image of the inverse of an element is inverse to the image of the given element. ∘

If G consists of finite number of elements n, then that number n is called the *order* of the group and we denote it as $n = |G_1|$.

A group G is called *commutative* or *abelian* if for any $x, y \in G$ we have $xy = yx$.

1.2. Examples. Some important examples are:

$\mathbf{Z}$ – the *additive group of integers*;

S_n – the *symmetric group on n letters* (the permutation group on n letters); this is the group of all bijections of the set $\{1, 2, \ldots, n\}$ (the elements from that group are often known as *permutations* or *substitutions*); the operation is written as the composition or sequential application of bijections;

$GL_n(F)$ –the *general linear group*; this is the group of all non–singular matrices of order $n \geq 1$ with coefficients from the field F; the operation is written as the product of matrices;

$\mathbf{Z}_n$ – the *group of residues modulo n*, $n \geq 2$, i.e. the group of residues $\{0, 1, 2, \ldots, n-1\}$ obtained by dividing numbers by n (with remainder); the operation is defined by: $a \oplus b = c$ if c is a residue from dividing the ordinary sum $a + b$ by n;

F^+ – the *additive group of the field F*; in particular, the group $\mathbf{Q}$ is that of the rational numbers under addition, the group $\mathbf{R}$ is the group of real numbers under addition, etc.;

F^* – the *multiplicative group of non–zero elements of the field F*; in particular, $\mathbf{R}^*$ is the group of non–zero real numbers, $\mathbf{C}^*$ is the group of non–zero complex numbers, etc..

∘ The groups $\mathbf{Z}$, $\mathbf{Z}_n$, F^+, F^* are commutative, $GL_1(F) \cong F^*$, the groups S_n and $\mathbf{Z}_n$ are finite and $|S_n| = n!$, $|\mathbf{Z}_n| = n$. ∘

1.3. Cyclic groups. If g is an element from the multiplicative group G, n is a natural number, then the integer powers of the element g are defined, as follows $g^0 = e$; $g^n = \underbrace{g \dots g}_{n \text{ times}}$; $g^{-n} = \underbrace{g^{-1} \dots g^{-1}}_{n \text{ times}}$. In the case when additive notation is used, the integer multiples of the element a are defined by

$$0a = 0, \quad na = \underbrace{a + \dots + a}_{n \text{ times}}, \quad (-n)a = \underbrace{(-a) + \dots + (-a)}_{n \text{ times}}.$$

∘ The properties of integer powers and integer multiples are as usual:

$$g^n g^m = g^{n+m}, \ (g^n)^m = g^{nm}, \ na + ma = (n+m)a, \quad n(ma) = (nm)a. \circ$$

Below in this section we use multiplicative notation. The reader can easily switch to additive notation in suitable cases.

A group G consisting of powers of a fixed element g is called *cyclic*, and the element g is said to *generate* the group G (often it is said that g is a *generator*). We then write $G = \langle g \rangle$.

∘ Any subgroup H of a cyclic group $G = \langle g \rangle$ is cyclic, and the least natural power g^m that is in H can be used as its generator. ∘

Let $\{\dots, g^{-2}, g^{-1}, g^0, g^1, g^2, \dots\}$ be the set of powers of an element g from a group G. If all the elements from this set are mutually different then it is said that g is an element of *infinite order*. Otherwise $g^k = g^l$ for some $k < l$. Thus, multiplying by g^{-k}, we get $g^{l-k} = e$. In this case we say that g is an element of *finite order* and the least natural number m such that $g^m = e$ is called the *order* of the element g. We then write $m = o(g)$. If $G = \langle g \rangle$ and g is an element of order n then $G = \{e, g, g^2, \dots, g^{n-1}\}$. Thus, $G = \langle g \rangle$ is of finite order iff g is an element of finite order.

∘ A cyclic group of infinite order is isomorphic to $\mathbf{Z}$. A cyclic group of order n is isomorphic to $\mathbf{Z}_n$. ∘

Thus two cyclic groups of the same order are isomorphic. We shall also note that the group $\mathbf{Z}_n$ is isomorphic to the group U_n of complex n–th roots of unity 1. Since the only subgroups of U_n are the U_m, where m divides n, the structure of the cyclic group of order n is as follows:

∘ Any subgroup of a cyclic group of order n has order d where d is a divisor of the number n and for any divisor d of n the subgroup of order d is unique. ∘

Thus the group $\mathbf{Z}_6 = \{0,1,2,3,4,5\} = \langle 1 \rangle$ has four subgroups : $\{0\}$, $\mathbf{Z}_6$, and also $\langle 2 \rangle = \{0,2,4\}$, $\langle 3 \rangle = \{0,3\}$.

1.4. Cosets. Lagrange theorem. Let X, Y be non–empty subsets of the group G; by XY we denote the set of all products xy, where $x \in X$, $y \in Y$. If $X = \{g\}$ is a one–element set, and $Y = H$ is a subgroup of G then instead of $\{g\}H$ (or $H\{g\}$) we write gH (respectively, Hg).

The set gH is a *left coset* of the subgroup H of G with *representative* g. Similarly, Hg is a *right coset* of the subgroup H of G with representative g.

Any two left cosets g_1H and g_2H either have no common elements or coincide. The latter possibility holds iff $g_1^{-1}g_2 \in H$ (or, equivalently, $g_2^{-1}g_1 \in H$). In fact, if $x \in g_1H \cap g_2H$, then $g_1h_1 = x = g_2h_2$, where $h_1, h_2 \in H$. In this case $g_1 = g_2h_2h_1^{-1} \in g_2H$, $g_2 = g_1h_1h_2^{-1} \in g_1H$. And, of course, $g_1H = g_2H$.

If H consists of a finite number m of elements then for any $g \in G$ the class gH consists of m elements. Any element $g \in G$ lies in certain left coset (for example, in gH). We have:

$$G = \bigcup_{g \in G} gH,$$

moreover, if we select a set $\{g_i\}_{i \in \mathcal{I}}$ from the elements of the group G such that $g_iH \neq g_jH$, for $i \neq j$, and any class gH is equal to one of the g_iH then the following partition holds

$$G = \bigcup_{i \in \mathcal{I}} g_iH. \tag{1}$$

The members of the partition written above are mutually disjoint. The cardinality of the set $\mathcal{I}$ does not depend on the selected system of representatives $\{g_i\}_{i \in \mathcal{I}}$ and is called the *index* of the subgroup H in the group G. The index is denoted by the symbol $(G : H)$. This statement is obvious in a case when G is a finite group of order n. Indeed, if H is of order m then all the members of the partition (1) have the same cardinality m. Assuming that the number of cosets in (1) equals k, we get the formula $n = km$.

Lagrange theorem. *Let G be a finite group, H a subgroup. Then $|G| = (G : H)|H|$, so that the order of a group is always divisible by the order of any of its subgroups.*

Corollary 1. *Let G be a group of prime order p. Then G is a cyclic group (isomorphic to $\mathbf{Z}_p$).*

For the proof it is sufficient to note that if G is a group and g is an element in it, then the set $H = \langle g \rangle$ of all its powers is a subgroup in G. □

Corollary 2. *Let G be a finite group, g an element of finite order in G. Then its order $o(g)$ divides the order $|G|$ of the group G.*

It was already noted above that $|\langle g \rangle| = o(g)$. It remains only to apply the Lagrange theorem.

All the definitions mentioned for left cosets can be identically transfered to the case of right cosets. The bijection $gH \mapsto Hg^{-1}$ shows that the number of right cosets coincides with the number of the left ones. So the index of a subgroup in a group does not depend on whether one considers right or left cosets.

1.5. Normal subgroup. Quotient group. A subgroup H of a group G is called *normal* if for any $g \in G$ we have $gH = Hg$, i.e. for any $h \in H$ there exists an element $h_1 \in H$ such that $gh = h_1 g$. In the case of a commutative group it is evident that each subgroup is normal.

∘ Any subgroup of index 2 is normal. ∘

In the case of the group S_n ($n \geq 3$) the subgroup S_{n-1} of all permutations with fixed symbol n is not normal; indeed, if $g = (1n)$, $h = (12)$ and $gh = h_1 g$, then $h_1 = ghg^{-1} = (1n)(12)(1n)^{-1} = (12n)(1n) = (2n) \notin S_{n-1}$.

An equivalent definition of normality is related to the concept of conjugate elements. Indeed, let us call $h_1 = ghg^{-1}$ the element *conjugate* of h by $g \in G$ (or the result of a conjugation of h). The following result is evident:

∘ A subgroup H is normal in the group G iff the result of conjugation on any $h \in H$, by any $g \in G$, lies in H, i.e. $ghg^{-1} \in H$. ∘

We can also write $gHg^{-1} \subset H$.
The following are examples of normal subgroups: in the group S_n, the subgroup A_n of even permutations; in the group $GL_n(F)$, the subgroup $SL_n(F)$ of matrices with determinant 1. The group A_n is called the *alternating* group on n letters; $SL_n(F)$ is called the *special linear group*.

A group G such that any proper normal subgroup is trivial is called *simple*. It is clear that any group of prime order is simple.

∘ Other kinds of simple abelian groups do not exist. ∘

Theorem. *For any $n \geq 5$ the alternating group A_n is a (non–abelian) simple group.*

If g_1H, g_2H are two left cosets of the group G with respect to a normal subgroup H, then it is obvious that

$$(g_1H)(g_2H) = (g_1g_2)H \tag{2}$$

(see the definition of product of subsets in the beginning of Subsection 1.4; associativity of the group operation implies the associativity of this product). Moreover, if $g_1H = g_1'H$, $g_2H = g_2'H$ then

$$(g_1H)(g_2H) = (g_1'H)(g_2'H).$$

This property (well–definedness of multiplying classes) and also the evident equalities

$$H(gH) = (gH)H = gH, \quad (g^{-1}H)(gH) = (gH)(g^{-1}H) = H$$

allow us to regard the set G/H of left cosets of the group G by a normal group H as a group with associative operation (2), unit element $H = eH$ and inverse element $(gH)^{-1} = g^{-1}H$. The group obtained is called the *quotient group* of the group G by the normal subgroup H.

∘ If $G = \mathbf{Z}$, $H = n\mathbf{Z} = \{na | a \in \mathbf{Z}\}$, then $G/H = \mathbf{Z}/n\mathbf{Z} \cong \mathbf{Z}_n$. ∘

Note the isomorphisms:

$$GL_n(F)/SL_n(F) \cong F^* \quad \text{and} \quad S_n/A_n \cong \mathbf{Z}_2.$$

∘ The subgroups of the group G/H are in one–to–one correspondence with the subgroups of the group G containing H, i.e. they are of the form K/H, where $K \supset H$. ∘

1.6. Homomorphisms. Consider two groups G_1 and G_2. A *homomorphism* from G_1 into G_2 is a mapping $\phi : G_1 \to G_2$, such that $\phi(xy) = \phi(x)\phi(y)$ for any $x, y \in G_1$. Moreover, an injective homomorphism is called a *monomorphism* and a surjective homomorphism an *epimorphism*. The *kernel*, Kerϕ, of a homomorphism ϕ is the set of elements $x \in G_1$ such that $\phi(x) = e$, where e is the unit element of G_2. It is easy to verify that Kerϕ is always a normal subgroup in G_1. Taking into account that Im ϕ, the image of the homomorphism ϕ, is a subgroup in G_2, the following formula, called the *homomorphism theorem*, holds

$$G_1/\mathrm{Ker}\phi \cong \mathrm{Im}\ \phi\,.$$

This theorem can be more exactly formulated as follows:

∘ The mapping $xK \mapsto \phi(x)$, where $K = \mathrm{Ker}\,\phi$, is well defined and is an isomorphism between $G_1/\mathrm{Ker}\,\phi$ and Im ϕ. ∘

In certain arguments the following special cases are important.

Isomorphism theorem. (1) *Let G be a group, H, K normal subgroups of it, $H \supset K$. Then the set H/K of cosets of the form hK, $h \in H$, is a normal subgroup in G/K and*

$$(G/K)/(H/K) \cong G/H. \tag{3}$$

(2) *Let G be a group, H, K subgroups, where K is normal. Then HK is a subgroup in G, K is a normal subgroup in HK, $H \cap K$ is a normal subgroup in H and*

$$HK/K \cong H/H \cap K. \tag{4}$$

The proof of the first part results from the homomorphism theorem by taking $\phi(xK) = xH$; in order to prove the second part we set $\psi(hkK) = h(H \cap K)$. The further verifications are immediate. □

Thus in the group S_4 there exists the normal subgroup $V = \{e, (12)(34), (13)(24), (14)(23)\}$ (the Klein four group) and the subgroup T leaving element 4 in place. Then $T \cong S_3$; further, $S_4 = TV$, $T \cap V = \{e\}$. By applying (4), we obtain

$$S_4/V = TV/V \cong T/T \cap V = T \cong S_3.$$

1.7. Action of groups on sets. Let G be a group, M a non–empty set. We then say that G *acts* on M if for any $g \in G$ and any $x \in M$ an element $g * x \in M$ is defined such that moreover:

(a) $(g_1 g_2) * x = g_1 * (g_2 * x)$,

(b) $e * x = x$,

where $g_1, g_2 \in G$, e is the unit element in G, $x \in M$.

Some groups are in fact defined by their action on a set. Thus, S_n acts on the set $\{1, 2, \dots, n\}$. If V is some vector space then $GL(V)$ is a group of all non–singular linear operators on V; it is clear that $GL(V)$ acts on V.

Each group G acts on itself (i.e. $M = G$) in several different ways. For example, we can put $g * x = gx$, $g \in G$, $x \in M = G$ or $g \circ x = gxg^{-1}$, $g \in G$, $x \in M = G$. The latter action was mentioned in Subsection 1.5 under the name of conjugation. The first action mentioned is called the left regular action. Let us verify, for instance, the axioms of an action for conjugation:

$$(g_1 g_2) \circ x = (g_1 g_2) x (g_1 g_2)^{-1} = g_1 (g_2 x g_2^{-1}) g_1^{-1} =$$
$$= g_1 \circ (g_2 \circ x), \quad e \circ x = exe^{-1} = x.$$

The subset $G * x \subset M$ of all elements of the form $g * x$, where $g \in G$ is arbitrary and $x \in M$ is fixed, is called the *orbit* of the element x. Any two orbits $G * x_1$ and $G * x_2$ either have no common elements or coincide (by the same reasoning as in Subsection 1.4). Therefore, as in Subsection 1.4, the set M can be written in the form of a disjoint union of a certain number of orbits:

(5) $$M = \bigcup_{j \in J} G * x_j .$$

If $x \in M$, then G_x, the set of elements g from G such that $g * x = x$ is called the *stabilizer* (or *isotropy subgroup*) of the element x. If $g \in G$ then for any $h \in G_x$ we have $(gh) * x = g * x$; in addition, $g_1 G_x \neq g_2 G_x$ implies $g_1 * x \neq g_2 * x$.

Thus the following theorem holds.

Theorem. *Let a group G act on a set M. The mapping $gG_x \mapsto g * x$ defines a bijection between the set (not a quotient group) of left cosets of the group G with respect to the stabilizer subgroup G_x and the orbit $G * x$ of this element.*

An important consequence is the validity of the formula

$$|G * x| = (G : G_x).$$

In the case of a finite group we have

$$|G| = |G * x|(G : G_x).$$

The similarity of these formulas with the formula of the Lagrange theorem can be explained in the following way: if H is a subgroup in G then G acts on the set of left cosets by the left regular action, i.e. $g * (xH) = gxH$. In this case the orbit of the element H is the whole set of the left cosets, and the stabilizer, G_H, is H. Thus we have $|G| = |H||G/H|$.

If a group G acts on a set M then a homomorphism ϕ of this group into the group $S(M)$ of all the bijections of the set M is defined. Indeed, the bijection $\phi(g) \in S(M)$ associated to the element $g \in G$ is the following

$$\phi(g)(x) = g * x$$

If $M = G$ and the action is the (left or right) regular one, then the mapping is injective. Indeed, the action of the bijection $\phi(g)$ on the unity element e of the group G gives $\phi(g)(e) = g * e = g$. Hence, $\phi(G) \cong G$ holds in this case.

Cayley theorem. *Any group G is isomorphic to a subgroup of the group $S(G)$ consisting of all the bijections of the set G. If G is finite and $|G| = n$ then G is isomorphic to a subgroup of the symmetric group $S_n = S(G)$.*

In the case when the group G acts by conjugation on itself ($M = G$) the mapping ϕ above associates to the element $g \in G$ a bijection $\phi(g)$, denoted $\imath_g$, and it is called an *interior automorphism* of the group G. The mapping $g \mapsto \imath_g$ is an homomorphism into the group $S(G)$, or to be more precise, onto its subgroup Aut (G) of all automorphisms (= isomorphisms of a group into itself) of the group G. We have $\imath_g(x) = gxg^{-1}$. The image $\phi(G)$ is denoted by $\text{Int}(G)$ and is called a group of interior automorphisms of the group G. The kernel of the mapping considered consists of those elements g for which $gxg^{-1} = x$, or $gx = xg$, for any $x \in G$. It is natural to call this set the center $Z(G)$ of the group G (see Subsection 1.8). Thus we have

$$\text{Int}(G) \cong G/Z(G).$$

1.8. Conjugacy classes. Center of a group. Conjugation is an important special case of an action of a group on a set. If G acts on itself by conjugation then the orbit of an element $x \in G$ consists of all elements of the form gxg^{-1}, where $g \in G$, and is called the *conjugacy class* of the element x. The element gxg^{-1} is often denoted as x^g. The *stabilizer* G_x of the element x relative to conjugation, is the set of elements g such that $gxg^{-1} = x$ or $gx = xg$, i.e. the set of elements that commute with x. In this case G_x is more often called the *centralizer* of the element x in G and is written as $G_x = C_G(x)$. In the special case under consideration the theorem from Subsection 1.6 takes the form:

∘ The number of elements conjugate to an element x from the group G is equal to the index of the centralizer $C_G(x)$ of that element in G, i.e.

$$|x^G| = (G : C_G(x)). \quad \circ \tag{6}$$

An element z from the group G is called *central* if $zg = gz$, for any $g \in G$. The set $Z(G)$ of all central elements of a group, a subgroup , is called the *center* of the group G. Naturally, $z \in Z(G)$ iff $x^G = \{z\}$, i.e. $Z(G)$ is a union of one–element conjugacy classes. Using formula (6) and the just mentioned concept of the center of a group we can prove a theorem on the center of a finite p–group, i.e. a group of order p^m, where p is a prime number.

Theorem. *Let $|G| = p^m$, where p is a prime number. Then the center $Z(G)$ of the group G is non–trivial, i.e. it contains non–identity elements.*

The main formula for the proof follows from (5) and (6):

$$|G| = \sum_{j=1}^{t} |x_j^G| = \sum_{j=1}^{t} (G : C_G(x_j)). \tag{7}$$

If $x_1 = e, x_2, \ldots, x_s$ are the elements of the center $Z(G)$, and $x_{s+1}, \ldots, x_t$ are the remaining elements in (7), then $(G : C_G(x_j)) \neq 1$, $j = s+1, \ldots, t$. According to the Lagrange theorem, $(G : C_G(x_j)) = p^{r_j}$ is divisible by p for any j. Considering this, formula (7) reduces to the form

$$p^m = s + \sum_{j=s+1}^{t} p^{r_j},$$

where s is a non–zero element divisible by p. Hence, the center $Z(G)$ is non–trivial. □

Using this theorem and the simple observation that the quotient group of a non–commutative group by its center can not be cyclic gives the following

Corollary. *Any group of order p^2, where p is a prime number, is an abelian group.*

As to the simple observation just mentioned, let us assume that $G/Z(G) = \langle aZ(G)\rangle$ is a cyclic group. Then any element from G has the form $a^k z$, where $z \in Z(G)$; it is clear that any two such elements commute which contradicts the assumption of non–commutativity of the group G.

1.9. Subgroups of finite groups. Let G be a finite group and $|G| = p^m k$, where p is a prime number, coprime with k. Any subgroup H in G of order p^m is called a *Sylow p–subgroup* of this group. We present the three classical Sylow theorems on existence, conjugacy and the number of Sylow p–subgroups in an arbitrary group (for proofs, see Lang [1]).

First Sylow Theorem. *For any prime divisor p of the order of the finite group G, there exists a Sylow p–subgroup H.*

In order to formulate the second theorem we should note that the group G acts by conjugation not only on the set of its own elements, but also on the set of its own subgroups: $g * H = gHg^{-1}$. Thus the orbit H^G of a Sylow subgroup consists of Sylow p–subgroups. It is possible to prove that the orbit H^G of any p–Sylow subgroup H contains all Sylow subgroups of the group G. In other words the following holds:

Second Sylow theorem. *Any two Sylow p–subgroups H_1 and H_2 of a group G are conjugate, i.e. there exists an element $g \in G$ such that $H_1 = gHg^{-1}$. Moreover, for any p–subgroup K from the group G and for any Sylow subgroup H from the group G there is an element $g \in G$ such that $gKg^{-1} \subset H$.*

The stabilizer $G_H \subset G$ of a Sylow subgroup H relative to the conjugation action, obviously contains H. Thus $(G : G_H)$ divides $(G : H)$; in particular, the order of the orbit H^G, i.e. the number of Sylow p–subgroups , is coprime with p. It turns out that this statement can be strengthened.

Third Sylow theorem. *The number N_p of Sylow p–subgroups of an arbitrary finite group and 1 are congruent modulo p, i.e. N_p has the form $1+pk$, where k is a natural number.*

1.10. Commutators. Solvable groups. If a, b are two elements from the group G then the element $c = aba^{-1}b^{-1}$ is called a *commutator* and usually is denoted as $[a, b]$. The commutator $[a, b]$ is equal to one, only in the case when a, b commute, i.e. $ab = ba$. Obviously, $[b, a] = [a, b]^{-1}$, i.e. the element, inverse to the commutator is itself a commutator; however, a product of commutators is not necessarilly equal to a commutator. Thus, in general, the set of commutators of a group is not a subgroup. The smallest subgroup G' of the group G containing all commutators is called its *commutator subgroup* (*commutant*). It is clear that the commutator subgroup G' is the set of all possible products of the form $[a_1, b_1] \ldots [a_s, b_s]$, where $a_i, b_i \in G$, and s is a natural number. From the formula $(ab)^g = a^g b^g$, $a, b, g \in G$, the following is easily derived

$$[a, b]^g = [a^g, b^g], \tag{8}$$

which, as a consequence, implies the normality of the commutator subgroup G' in the group G.

Proposition 1. *The commutator subgroup G' of a group G is the smallest among the normal subgroups H of the group G for which G/H is an abelian group.*

In fact, the commutator $[xH, yH] = [x, y]H$ is trivial iff $[x, y] \in H$, i.e. G/H is abelian iff $H \supset G'$. □

The commutator subgroup of an abelian group is trivial; the commutator subgroup of the group S_n is equal to A_n, $n \geq 1$; the commutator subgroup of the group $GL_n(F)$ is $SL_n(F)$, with exception of the case $GL_2(\mathbf{Z}_2)$. Since the non–commutative group A_n, $n \geq 5$, has no non–trivial proper normal subgroups , we have $A'_n = A_n$.

An application of formula (8) shows that the second commutator subgroup G'', i.e. the commutator subgroup of the commutator subgroup of the group G, is a normal subgroup in G. The same can also be said about the k–th commutator subgroup $G^{(k)}$, i.e. the commutator subgroup of the $(k-1)$–th commutator subgroup $G^{(k-1)}$, $k \geq 2$. Thus, any group has a sequence of commutator subgroups

$$G^{(0)} \supset G^{(1)} \supset G^{(2)} \supset \ldots \supset G^{(k-1)} \supset G^{(k)} \supset \ldots \tag{9}$$

(here $G^{(0)} = G$, $G^{(1)} = G'$, $G^{(2)} = G''$,...). It is possible, as in the case $G = A_n$, $n \geq 5$, that all members of the sequence (9) coincide. However, in many important cases, for some k, we have $G^{(k)} = \{e\}$. The groups possessing this property are called *solvable* (also *soluble*). Naturally, an abelian group is solvable, and a non–abelian simple group is not solvable.

A typical example of a solvable group is the group $G = T_n(F)$ consisting of the non–singular upper triangular matrices of order n with elements from the field F. This group contains a sequence of normal subgroups U_k, $k = 0, 1, 2, \ldots, n-1$, each of them having an identity diagonal and zeros in the first k superdiagonals. Thus, for $n = 3$,

$$U_0 = \left\{ \begin{pmatrix} 1 & a & b \\ 0 & 1 & c \\ 0 & 0 & 1 \end{pmatrix} \middle| a, b, c \in F \right\}, \quad U_1 = \left\{ \begin{pmatrix} 1 & 0 & a \\ 0 & 1 & 0 \\ 0 & 0 & 1 \end{pmatrix} \middle| a \in F \right\},$$
$$U_2 = \left\{ \begin{pmatrix} 1 & 0 & 0 \\ 0 & 1 & 0 \\ 0 & 0 & 1 \end{pmatrix} \right\}.$$

The subgroup U_{n-1} is trivial. Direct calculations show that $G' \subset U_0$, and that the commutator $[a, b]$ of two matrices from U_k is in U_{k+1}. It is clear then that $G^{(n)} = \{e\}$, i.e. $T_n(F)$ is a solvable group.

"More abstract" examples of the solvables groups are obtained by using the following easily proved statement.

Proposition 2. *A subgroup H of a solvable group G is solvable. If G is solvable and ϕ: $G \to H$ is a homomorphism, then $\phi(G) = Im\ \phi$ is a solvable group. If N is a normal subgroup in G, then G is solvable iff N and G/N are solvable groups.*

The following formulas are used to prove this:

$$H^{(k)} \subset G^{(k)}, \quad \phi(G^{(k)}) \subset (\phi(G))^{(k)}, \quad (G/N)^{(k)} \subset G^{(k)}N/N.$$

Proposition 3. *Any group of order p^m, where p is a prime number, is solvable.*

For such a group G the center $Z(G)$ and the quotient groups $G/Z(G)$ are finite p–groups of strictly smaller order (see Subsection 1.8). By using induction and Proposition 2 we see that G is a solvable group. □

∗ The smallest order for an unsolvable group is equal to 60 (the group A_5). ∗

1.11. Direct product, direct sum. Let $G_1, \ldots, G_m$ be a certain collection of groups. The group G consisting of rows $(g_1, g_2, \ldots, g_m)$, where $g_i \in G_i$, $i = 1, 2, \ldots, m$, with componentwise multiplication

$$(g_1, g_2, \ldots, g_m)(g'_1, g'_2, \ldots, g'_m) = (g_1 g'_1, g_2 g'_2, \ldots, g_m g'_m),$$

is called a (exterior) *direct product* of the groups $G_1, \ldots, G_m$ and is denoted

$$G = G_1 \times \ldots \times G_m, \quad \text{or} \quad G = \prod_{i=1}^{m} G_i.$$

If $H_i = \{(e, \ldots, e, \underset{i}{x}, e, \ldots, e) | x \in G_i\}$, then H_i is a normal subgroup in G and, moreover, every element g from G can be uniquely presented in the form $g = h_1 h_2 \ldots h_m$, where $h_i \in H_i$, $i = 1, \ldots, m$. In this case it is said that G is a (interior) direct product of the groups $H_1, \ldots, H_2$. The following condition is equivalent

∘ The group G is an interior direct product of the subgroups $H_1, \ldots, H_m$ iff $G = H_1 \ldots H_m$ and $H_i \cap H_1 \ldots H_{i-1} = \{e\}$, $i = 2, \ldots, m$. ∘

The notation for an interior direct product is similar to that for the exterior one: $G = H_1 \times \ldots \times H_m$.

In the case of additive notation, one (also) speaks of the direct sum (instead of a direct product) with a corresponding change of notation $G = H_1 \oplus \ldots \oplus H_m$. Presenting a group in the form of direct product (direct sum) leads, as a rule, to the simplification of its structure. If such a factorization (decomposition) exists (with non–trivial factors), then we say that the group is decomposable (factorable). A natural example of a decomposable group is the additive group of n–dimensional vectors F^n, $n \geq 2$. The group $\mathbf{R}^n$ has an important subgroup $\mathbf{Z}^n$, its n–dimensional integral *lattice*. We have

$$F^n = \underbrace{F \oplus \ldots \oplus F}_{n}, \quad \mathbf{Z}^n = \underbrace{\mathbf{Z} \oplus \ldots \oplus \mathbf{Z}}_{n}$$

Proposition 1. *A cyclic group of order n is decomposable iff n is divisible by two different prime numbers.*

Indeed, according to Subsection 1.3, in a cyclic group of order p^n, p a prime, any two non–trivial subgroups have a non–trivial intersection; therefore such a group is indecomposable. If the order n of a cyclic group G is of the form $n = lm$, where $(l, m) = 1$, then G is the direct sum of a subgroup of order l and a subgroup of order m. □

When analyzing a direct product it is helpful to use the following facts:

Proposition 2. 1) *If G is a finite group of order n, $G = H_1 \times \ldots \times H_m$ and H_i is of order n_i, $i = 1, \ldots, m$, then $n = n_1 \ldots n_m$.*

2) *If $G = H_1 \times \ldots \times H_m$ and $g = h_1 \ldots h_m$ then g is an element of finite order iff every $h_i \in H_i$ is of finite order. In addition, $o(g) = lcm(o(h_1), \ldots, o(h_m))$, i.e. the order of the element g is the least common multiple of the orders of the elements $h_1, \ldots, h_m$.* □

The difference between exterior and interior direct sums can be illustrated by the group of residues $\mathbf{Z}_n$, where $n = p_1^{s_1} \ldots p_m^{s_m}$. An exterior factorization is

$$\mathbf{Z}_n \cong \mathbf{Z}_{p_1^{s_1}} \oplus \ldots \oplus \mathbf{Z}_{p_m^{s_m}}.$$

An interior factorization looks as follows: let $H_i = \langle q_i \rangle$, where $q_i = \frac{n}{p_i^{s_i}}$, $i = 1, \ldots, m$. Then $\mathbf{Z}_n = H_1 \oplus \ldots \oplus H_m$.

∘ It is not true that a subgroup C of a direct sum, $C \subset A \oplus B$ is a direct sum of the intersections $C \cap A$ and $C \cap B$. For instance, the subgroup $C = \mathbf{R}(e_1 + e_2)$ in the group $\mathbf{R}^2 = \mathbf{R}e_1 \oplus \mathbf{R}e_2$, is not of the form mentioned. However, if $C \supset A$ then $C = A \oplus (B \cap C)$. ∘

This remark prevents the transferring of properties of set theoretical unions to direct sums.

It is not true that the homomorphic image of the direct sum is the direct sum of the images of the direct summands. However the following theorem is valid

Theorem. *Let $G = H_1 \times \ldots \times H_m$, and let K be a normal subgroup in G such that $K = K_1 \ldots K_m$, where K_i is a subgroup in H_i, $i = 1, \ldots, m$. Then*

$$G/K \cong H_1/K_1 \times \ldots \times H_m/K_m.$$

(In additive notation: $H_1 \oplus \ldots \oplus H_m / K_1 \oplus \ldots \oplus K_m \cong H_1/K_1 \oplus \ldots \oplus H_m/K_m$.)

To prove this we apply the homomorphism theorem: if $g = h_1 \cdot \ldots \cdot h_m$, then we set

$$\phi(g) = (h_1K_1, \ldots , h_mK_m).$$

□

A special case of the theorem is important: $A \oplus B/B \cong A$.

1.12. Abelian groups with a finite number of generators. This is the most completely known class of groups. The basic tool for studying it is the concept of a direct sum.

We say that a group G (possibly commutative) is *finitely generated* if it has a finite number of elements $a_1, \ldots , a_n$ such that the least subgroup containing $a_1, \ldots , a_n$, coincides with G. Since the set of elements of the form

$$a_{i_1}^{s_1} \ldots a_{i_t}^{s_t}, \quad \text{where} \quad a_{i_j} \in \{a_1, \ldots , a_n\}, \quad s_i \in \mathbf{Z}, \tag{10}$$

is obviously a subgroup in G, it follows that a finitely generated group contains a finite set of elements $\{a_1, \ldots , a_n\}$ by means of which the remaining elements can be written in the form (10). The elements $a_1 \ldots , a_n$ are called *generating* elements of the group G. A cyclic group is a group with one generator. Any finite group is a group with a finite number of generators. The group $\mathbf{Q}$ of rational numbers, with respect to addition, is not finitely generated.

An important example of a finitely generated abelian group is the lattice $\mathbf{Z}^n$ of points with integral coordinates in real n–dimensional space. Such a group is often called a *free abelian* group of rank n and is denoted as F_n. The group F_n has n elements $x_1, \ldots , x_n$ (a basis) such that any element a from F_n can be uniquely written in the form $a = \alpha_1 x_1 + \ldots + \alpha_n x_n$, where the α_i are integers.

If A is another abelian group with n generators $a_1, \ldots , a_n$ then the mapping $\phi : F_n \to A$, defined by the formula

$$\phi(a) = \alpha_1 a_1 + \ldots + \alpha_n a_n,$$

is a homomorphism of groups, moreover

$$F_n/ \operatorname{Ker} \phi \cong A.$$

This homomorphism allows one to prove a theorem on the structure of an abelian group A with the help of elementary transformations of the lattice $\mathbf{Z}^n$. Indeed, for any subgroup K of the group F_n we can choose a basis $x'_1, \ldots , x'_n$

in F_n and a basis $y'_1, \ldots, y'_m$ in K such that $y'_i \in \langle x'_j \rangle$, $i = 1, \ldots, n$. Hence, it follows that $F_n = \langle x'_i \rangle \oplus \ldots \oplus \langle x'_n \rangle$ and $K = \langle y'_1 \rangle \oplus \ldots \oplus \langle y'_m \rangle$. In this case

$$A \cong \langle x'_1 \rangle / \langle y'_1 \rangle \oplus \ldots \oplus \langle x'_m \rangle / \langle y'_m \rangle \oplus \langle x'_{m+1} \rangle \oplus \ldots \oplus \langle x'_n \rangle.$$

If $y'_i = r_i x'_i$, $i = 1, 2, \ldots, m$, then $\langle x'_i \rangle / \langle y'_i \rangle \cong \mathbf{Z}_{r_i}$; moreover, $\langle x'_j \rangle \cong \mathbf{Z}$. Thus, the following holds.

Fundamental theorem on finitely generated abelian groups. *Any finitely generated abelian group is isomorphic to a group of the form*

$$A = \mathbf{Z}_{r_1} \oplus \ldots \oplus \mathbf{Z}_{r_m} \oplus \mathbf{Z}^{n-m},$$

i.e. it is a direct sum of cyclic, finite or infinite groups. Moreover, the numbers $r_1, \ldots, r_m$, $r_i \geq 2$, can be chosen, in a unique way, so that r_1 divides r_2, r_2 divides r_3, ..., r_{m-1} divides r_m. The number $n - m$ is also defined uniquely. □

1.13. Finite abelian groups. If A is a finite abelian group, and p is a prime divisor of its order, then the set $A(p)$ of all elements the order of which is a power of the number p, is a subgroup in A called the *p–primary component* of the group A. In fact, $A(p)$ is a Sylow p–subgroup in A (see, Subsection 1.9).

Theorem 1. *Let $p_1, \ldots, p_r$ be the set of prime divisors of the order of a finite abelian group A. Then*

$$A = A(p_1) \oplus \ldots \oplus A(p_r).$$

□

Every primary component permits further decomposition into a direct sum of cyclic subgroups each of which will not allow further decomposition into a direct sum of proper subgroups. Combining this with theorem 1 we get the important result.

Fundamental theorem on finite abelian groups. *Any finite abelian group can be presented in the form of a direct sum of primary cyclic subgroups*

$$A = \langle a_1 \rangle_{p_{i_1}^{s_1}} \oplus \langle a_2 \rangle_{p_{i_2}^{s_2}} \oplus \ldots \oplus \langle a_{p_{i_t}^{s_t}} \rangle.$$

For any such factorization the sets of the orders of these subgroups coincide up to reordering the direct sums. □

We omit the proofs of these theorems since they follow from more general rexults in Chapter 1 in the setting of finitely generated modules over a principal ideal ring.

∗ The concept of a group is a special case of the concept of a *semigroup*, i.e. of a set with one associative operation. Recently, the theory of semigroups has been developed to a considerable extent. (For a review of the current situation, see Volkov–Shevrin [1]). A semigroup with a unit element is called a *monoid*. ∗

§ 2. Rings, fields

2.1. Definitions. Let R be a non–empty set with two operations defined on it: addition $(x, y) \mapsto x + y$ and multiplication $(x, y) \mapsto xy$. We say that R is a *ring* if relative to addition, R is an abelian group, while addition and multiplication in R are linked by the *distributive laws*

$$(x + y)z = xz + yz, \quad x(y + z) = xy + xz.$$

Usually, in the study of rings, the condition of associativity of the multiplication is included in definition as well:

$$x(yz) = (xy)z.$$

But in this book we deal also with non–associative rings (see, Chapter 4). This section only concerns associative rings.

∘ In any ring the following relations hold

$$0 \cdot a = a \cdot 0 = 0, \quad a(-b) = (-a)b = -(ab),$$
$$a(b - c) = ab - ac, \quad (a - b)c = zc - bc. \;\circ$$

A subgroup S of the additive group of the ring R which is closed relative to multiplication, is called a *subring*. Thus $S \neq \emptyset$ is a subring in R iff for any $x, y \in S$ we have

$$x + y \in S; \quad -x, -y \in S; \quad xy \in S.$$

∘ A non–empty subset of a ring is a ring iff it is a ring with respect to the operations of the containing ring. ∘

Two rings R_1 and R_2 are called *isomorphic* if there exists a bijection $\phi : R_1 \to R_2$ such that for any $a, b \in R_1$ we have

$$\phi(a + b) = \phi(a) + \phi(b), \quad \phi(ab) = \phi(a)\phi(b).$$

2.2. Examples. Some of the more important examples are:

$\mathbf{Z}$ – the ring of integers with the usual operations of addition and multiplication;

$M_n(F)$ – the full matrix ring, the ring of all quadratic matrices of order n and coefficients from the field F with respect to the usual operations of addition and multiplication;

$F[X_1, \ldots, X_n]$ the ring of polynomials in n variables $X_1, \ldots, X_n$ with coefficients from the field F;

$C[a, b]$ the ring of continuous real functions on an interval $[a, b]$ with the usual operations of pointwise addition and multiplication;

F any field; for instance, $\mathbf{Q}$ – the field of rational numbers; $\mathbf{R}, \mathbf{C}$ – the fields of real and complex numbers;

$\mathbf{Z}[G]$ – the group ring of a multiplicative group G; the underlying additive group of this ring is the free abelian group with basis G, a product of two basis elements is as in G, a product of linear combinations of basis elements is defined by distributivity; for instance, if $G = \langle a \rangle$ and $a^2 = e$ then any element from $\mathbf{Z}[G]$ has the form $\alpha e + \beta a$, $\alpha, \beta \in \mathbf{Z}$, and

$$(\alpha e + \beta a)(\alpha' e + \beta' a) = (\alpha\alpha' + \beta\beta')e + (\alpha\beta' + \alpha'\beta)a.$$

As an example of a non–associative ring one can take the ring $\mathbf{R}^3$ of three–dimensional vectors with the operation of vector multiplication (addition is as usual). This ring belongs to the class of the so–called Lie rings (see Chapter 4).

2.3. Ideal, quotient ring, homomorphism. A subring $\mathcal{I}$ of a ring R is called a *left ideal* if for any $r \in R$ and $x \in \mathcal{I}$ we have $rx \in \mathcal{I}$. A *right ideal* is defined in the same way. A subring which is a left and right ideal at the same time is called a *two–sided ideal*, or simply an *ideal*. In the case of a commutative ring any one–sided ideal is also a two–side one. In the ring of matrices $M_n(F)$, $n \geq 2$, the set of matrices for which all the elements outside a fixed column are equal to 0 is a left but not a two–sided ideal.

An important example of a left ideal of the ring R is the set Rx of all elements of the form rx, where $r \in R$ is arbitrary and x is fixed. Such a left ideal is called the *principal ideal*, generated by the element x. Properties of division with remainder of integers or polynomials of one variable permit us to prove the following.

Theorem. *Any ideal of the ring* $\mathbf{Z}$ *or* $F[X]$, F *a field, is principal.* □

Let $\mathcal{I}$ be a two–sided ideal in the ring R. Then $\mathcal{I}$ is a subgroup of the additive group R. This allows to consider the quotient group $R/\mathcal{I}$. We can convert $R/\mathcal{I}$ into a quotient ring, by defining

$$(r_1 + \mathcal{I})(r_2 + \mathcal{I}) = r_1 r_2 + \mathcal{I} \tag{1}$$

(here $r + \mathcal{I}$ is the additive notation for the left coset in the group R of the subgroup $\mathcal{I}$ with representative r. Since $\mathcal{I}$ is an ideal, the multiplication (1) is well defined, i.e. it does not depend on the selected representatives in the cosets. Moreover, the axioms of a ring are true for R, it clearly follows, that they are also true for $R/\mathcal{I}$.

If $R = \mathbf{Z}$, $\mathcal{I} = n\mathbf{Z}$ (see, Subsection 1.5) then $\mathcal{I}$ is an ideal in R and this means that the group of residues $\mathbf{Z}_n$ is a ring.
In order to multiply two residues in $\mathbf{Z}_n$ they should be multiplied as integers and then the remainder for division by n should be taken. We say that $\mathbf{Z}_n$ is the *ring of residues modulo* n.

Let R_1 and R_2 be two rings. A homomorphism of the two additive groups that is multiplication–preserving between R_1 and R_2 is called a *homomorphism of rings*. Thus $\phi : R_1 \to R_2$ is a homomorphism if

$$\phi(x+y) = \phi(x) + \phi(y), \quad \phi(xy) = \phi(x)\phi(y),$$

for any $x, y \in R_1$.

The kernel, $\operatorname{Ker}\phi$, of a homomorphism ϕ (i.e. the set of elements $x \in R_1$, for which $\phi(x) = 0$) is an ideal in R_1. The homomorphism ϕ is an isomorphism iff $\operatorname{Ker}\phi = \{0\}$.

As in the case of groups there is a homomorphism theorem:

$$R_1/\operatorname{Ker}\phi \cong \operatorname{Im}\phi$$

(or, to be more precise, the theorem on homomorphism for groups together with its proof yields the theorem on homomorphism for rings). As in Subsection 1.6 one has

Isomorphism theorem. 1) *If R is a ring, S and T are ideals and $S \supset T$, then S/T is an ideal in R/T and*

$$(R/T)/(S/T) \cong R/S.$$

2) *If R is a ring, S and T are subrings, and T is an ideal in R, then $S+T$ is a subring in R, T is an ideal in $S+T$, $S \cap T$ is an ideal in S and*

$$S+T/T \cong S/S \cap T.$$

□

To conclude this subsection we should note that a ring which has no proper non–trivial ideals is called *simple*. Any simple abelian group provided with zero multiplication is an example of a simple ring. A less trivial example is a field F: if $x \neq 0$ is an element of some ideal $\mathcal{I}$, and y is arbitrary, then $y = yx^{-1}x \in \mathcal{I}$. Hence $\mathcal{I} = F$, i.e. $\mathcal{I}$ is a non–proper ideal; so F is simple.

One of the most important examples of non–commutative simple rings are the full matrix rings $M_n(F)$, where F is a field.

Theorem. *The ring $M_n(F)$, where F is a field, is simple.*

Denote by E_{ij} an elementary matrix, i.e. a matrix with a unique non-zero element which is located at (i, j) and is equal to 1. Assume that $\mathcal{I}$ is a non-zero ideal in $M_n(F)$. Let $a = \sum_{i,j} \alpha_{ij} E_{ij}$ be a non-zero element from $\mathcal{I}$ and $\alpha_{pq} \neq 0$. Then $\mathcal{I}$ contains the element

$$E_{uv} = \alpha_{pq}^{-1} E_{up} a E_{qv}.$$

This follows from the rules of multiplication of elementary matrices: $E_{ij} E_{kl} = \delta_{jk} E_{il}$. Since $\mathcal{I}$ contains all elementary matrices, $\mathcal{I} = R$. □

2.4. Divisibility in rings. A non-zero element b from the ring F *divides* an element $a \in R$ from the left, if $a = bc$, for a suitable $c \in R$. Analogously is defined divisibility from the right. If R contains an identity element for the multiplication then an element which is a left and right divisor of the identity element, is called an *invertible* element.

∘ The set R^* of invertible elements of the ring R with identity element is a multiplicative group. ∘

Thus to any ring R with 1 there is associated the group R^* of its invertible elements. In the case of a full matrix ring we have

$$(M_n(F))^* = GL_n(F).$$

A special feature of a number of rings is the existence of zero divisors; a non-zero element a from the ring R is a *zero divisor* if there exists a non-zero b such that either $ab = 0$ or $ba = 0$. For instance, zero divisors appear in any ring of matrices $M_n(F)$, $n > 1$ ($E_{11}E_{21} = 0$); in the ring of continuous functions $C[a, b]$ in interval $[a, b]$, $a < b$, etc.. On the other hand, there are no zero divisors in the ring of integers $\mathbf{Z}$, and in the ring $F[X_1, \ldots, X_n]$ of polynomials in n variables over a field F.

A ring without zero divisors is called an *integral domain.* An ideal P of a ring R is called *prime* if the quotient ring R/P is an integral domain.

As an example of an integral domain we have any *division ring*, i.e. a ring R, with 1, such that every non-zero element a has a multiplicative inverse a^{-1}. In other words, $R^* = R\backslash\{0\}$. A division ring is also called a *skew field.*

Below, we assume that the ring R is *commutative*, i.e. $xy = yx$ for any $x, y \in R$.

An element $a \neq 0$ of the ring R is called *reducible* if $a = bc$, where neither b, nor c, are invertible.
If from a representation $a = bc$ it follows that either $b \in R^*$, or $c \in R^*$, then a is called *irreducible* (*prime*). This definition completely agrees with the irreducibility of polynomials (ring $F[X_1, \dots, X_n]$) or primality of integers (ring $\mathbf{Z}$).

A commutative integral domain R is called *factorial* (or a *unique factorization ring*) if any irreversible element a in this ring can be uniquely represented in the form

$$a = p_1 p_2 \dots p_s, \tag{2}$$

where $p_1, p_2, \dots, p_s$ are irreducible (prime) elements form R; here, under uniqueness, we understand the following:
if

$$a = q_1 q_2 \dots q_t \tag{3}$$

is another such representation, then $s = t$, and after a suitable renumbering of the elements,

$$p_1 = \epsilon_1 q_1, \quad p_2 = \epsilon_2 q_2, \dots, p_s = \epsilon_s q_s,$$

where for any $i = 1, 2, \dots, s = t$, the element ϵ_i is invertible. (If two elements differ by an invertible factors then they are called associated.)

Widely known examples of factorial rings are the ring $\mathbf{Z}$ of integers and the ring $F[X]$ of polynomials of one variable X over a field F. One can prove this by using the Euclidean algorithm. In addition for the case $\mathbf{Z}$ the irreducible elements are the prime numbers and their negatives; the invertible elements are ± 1. Thus in the factorizations (2) and (3) the numbers p_i and q_i under the suitable ordering differ only by a sign factor. In the case of the ring $F[X]$ the prime elements are the irreducible polynomials and the prime factors in a factorization only differ by a scalar factor.

An example of a non factorial ring (but which is an integral domain and does have factorization into irreducible factors) is the ring $\mathbf{Z}[\sqrt{-3}]$ of complex numbers of the form $a+b\sqrt{-3}$, where a, b are integers. In this ring the element 4 factorizes into a product of irreducible factors in two essentially different ways:

$$2 \cdot 2 = 4 = (1 + \sqrt{-3})(1 - \sqrt{-3}).$$

The unique factorization property of the ring of polynomials $F[X_1, \dots, X_n]$ in any number of variables over a field F will be proved in Subsection 1.3.1.

2.5. Field. The simplest example of a factorial ring is a commutative division ring or *field*. In this case all non–zero elements are invertible.

A non–zero subring S of a field F is called a *subfield* if together with a non–zero $a \in S$ the element a^{-1} is also in S. (We should note that the requirement $0 \neq 1$ is usually included in the definition of a field.)

Obviously, a field is an integral domain. In fact, if $ab = 0$ and $a \neq 0$ then $a^{-1}(ab) = 0$ and $b = 0$, i.e. a non–zero element cannot be a zero divisor.

In the other direction, any commutative integral domain R is isomorphic to a subring of some field. Or, to be more precise, the following holds.

Theorem. *For any commutative integral domain S, there exists a field F containing S which is unique up to an isomorphism that is the identity on S, and such that any element a from F is representable in the form $a = rs^{-1}$, where $r, s \in S$.* □

Usually, the so called *field of fractions* Fr (S) of the integral domain S is chosen as such an F. The elements of the field Fr (S) are equivalence classes of fractions of the form $\frac{r}{s}$, $r, s \in S$, $s \neq 0$. The operations are

$$\frac{r_1}{s_1} + \frac{r_2}{s_2} = \frac{r_1 s_2 + r_2 s_1}{s_1 s_2}, \quad \frac{r_1}{s_1} \cdot \frac{r_2}{s_2} = \frac{r_1 r_2}{s_1 s_2}. \tag{4}$$

Here, two fractions $\frac{r_1}{s_1}$ and $\frac{r_2}{s_2}$ are considered equivalent if $r_1 s_2 = r_2 s_1$. A slight possible arbitrariness in the definition has to be checked; viz. that the substitution in sums or products in (4) of an equivalent fraction only changes the result within equivalence. This verification and the verification that Fr (S) is a field, are left to the reader (for a detailed discussion of a more general situation, see, Subsection 1.3.2).

The statement on isomorphism is trivial.

The quotient field of a ring of polynomials $F[X_1, \dots, X_n]$ is called the field of *rational fractions* in the variables $X_1, \dots, X_n$ and denoted $F(X_1, \dots, X_n)$.

Every field F has a minimal subfield, F_0, the elements of which are fractions of the form $(m1)(n1)^{-1}$, where m and n are integers and $n1 \neq 0$. If two elements of such form coincide, $(m1)(n1)^{-1} = (m_1 1)(n_1 1)^{-1}$, but the fractions $\frac{m}{n}$ and $\frac{m_1}{n_1}$ are not equal, then $mn_1 - m_1 n \neq 0$ and $(mn_1 - m_1 n)1 = 0$. In this case, i.e. when for some natural $s \neq 0$ we have $\underbrace{1 + \ldots + 1}_{s} = s1 = 0$, we say that the field F has non–zero characteristic. The smallest such s is then called the *characteristic* of the field F. If $s = tu$ is some factorization of the number s, $1 \leq t,\ u \leq s$, then $s1 = (tu)1 = (t1)(u1)$; since $s1 = 0$ and there are no zero divisors in the field, we have $t1 = 0$ or $u1 = 0$. Consequently, either $t = s$, or $u = s$. This means that $s = p$ is a prime number.

∘ The characteristic of a field is always a prime number (or zero). ∘

If the elements $(m1)(n1)^{-1}$ and $(m_1 1)(n_1 1)^{-1}$ are equal only in the case when the fractions $\frac{m}{n}$ and $\frac{m_1}{n_1}$ are equal, then $m1 = 0$ iff $m = 0$.

We say that such field F has *characteristic* 0. In this case associating the element $(m1)(n1)^{-1}$ with the fraction $\frac{m}{n}$ amounts to an isomorphism of the minimal subfield F_0 and the field $\mathbf{Q}$ of rational numbers.

The minimal subfield F_0 of a field F of characteristic p is isomorphic to $\mathbf{Z}_p$. Indeed, then $F_0 = \{(m1)(n1)^{-1} |\ 0 \leq m,\ n \leq p - 1,\ n \neq 0\}$. According to the Euclidean algorithm, if $(n, p) = 1$, then there are integers u, v such that $un + vp = 1$. So $(u1)(n1) = 1 - (v1)(p1) = 1$; $(m1)(n1)^{-1} = (m1)(u1)$, $F_0 = \{(m1) | 0 \leq m \leq p - 1\} \cong \mathbf{Z}_p$.

∘ The ring $\mathbf{Z}_n$ of residues modulo n is a field iff $n = p$ is a prime number. ∘

2.6. Adjoining a root of a polynomial to a field. A proper ideal M of a ring R is called *maximal* if for any ideal $\mathcal{I}$ of the ring R with the property $M \subset \mathcal{I} \subset R$ we have either $\mathcal{I} = M$ or $\mathcal{I} = R$.

Proposition. *The quotient ring R/M of the commutative ring R with* 1 *with respect to an ideal M is a field iff M is a maximal ideal.*

Proof. It was noted in Subsection 2.4 that there are no proper non–trivial ideals in a field. Now, if R/M is a field and $\mathcal{I}$ is an ideal with the property $R \supset \mathcal{I} \supset M$ then $\mathcal{I}/M$ is an ideal in R/M. So either $\mathcal{I}/M$ is a zero ideal and $\mathcal{I} = M$ or $\mathcal{I}/M = R/M$ and $\mathcal{I} = R$. Therefore M is a maximal ideal.

Inversily, let M be a maximal ideal and $x + M$ a non–zero element in R/M, i.e. $x \notin M$. In this case $\mathcal{I} = xR + M$ is an ideal in R strictly containing M. Thus $\mathcal{I} = R$ and, in particular, $1 \in xR + M$, i.e. $1 = xr + m$, for some $r \in R$, $m \in M$. Going over to equivalence classes we get $1 + M = (x + M)(r + M)$, so that $r + M = (x + M)^{-1}$, and R/M is a field. □

There is the classical problem in mathematics of extending numerical systems in such a way that algebraic equations can be solved. In order to solve equations of the form $mX + n = 0$, where m and n are natural numbers, rational numbers are introduced. An equation of the form $X^n = m$ requires irrational numbers and the equation $X^2 + 1 = 0$ the complex numbers.

Here we shall discuss a way of constructing a field $\bar{F}$ such that a fixed (previously given) algebraic equation with coefficients from F has a solution.

Thus, let $f(X)$ be a polynomial of non–zero degree with coefficients from a field F. Suppose, moreover, that $f(X) = 0$ has no solutions in F. Let $p(X)$ be some irreducible factor of $f(X)$ in the ring $F[X]$. Without restriction of generality, we assume $f(X) = p(X)$.

Let us consider the ring $R = F[t]$ of polynomials of the variable t and let $M = Rp(t)$ be the principal ideal generated by the polynomial $p(t)$. The ideal M is maximal. Indeed, let $R \supset \mathcal{I} \supset M$. The ideal $\mathcal{I}$, as was mentioned in Subsection 2.3, is principal, i.e. $\mathcal{I} = Rq(t)$. Since $p(t) \in \mathcal{I}$ we get $p(t) = s(t)q(t)$, for some $s(t)$. One of $q(t)$, $s(t)$ is invertible and the other one is associated to $p(t)$. If $q(t)$ is invertible then $I = Rq(t) = R$. If $q(t) = \epsilon p(t)$ is associated to $p(t)$ we have $\mathcal{I} = R\epsilon p(t) = Rp(t) = M$. Thus M is a maximal ideal.

Now, put $\bar{F} = R/M$; according to the proposition above $\bar{F}$ is a field. In addition it should be mentioned that the field F can be identified with the subfield in $\bar{F}$ consisting of cosets of polynomials of zero degree $\lambda \in F \leftrightarrow \bar{\lambda} = \lambda + M \in \bar{F}$. (We suggest the reader to verify that this mapping is an isomorphism of fields.) By this identification the ring $F[X]$ is identified with a subring of $\bar{F}[X]$ where $f(X) = \lambda_0 + \lambda_1 X + \ldots + \lambda_n X^n$ is identified with $\bar{f}(X) = (\lambda_0 + M) + (\lambda_1 + M)X + \ldots + (\lambda_n + M)X^n = \bar{\lambda}_0 + \bar{\lambda}_1 X + \ldots + \bar{\lambda}_n X^n$.

We shall show that $E = t + M \in \bar{F}$ is a root of the polynomial $\bar{f}(X)$. In fact, using the rules of operations in a quotient ring, we get

$$\bar{f}(\bar{t}) = \bar{\lambda}_0 + \bar{\lambda}_1\bar{t} + \ldots + \bar{\lambda}_n\bar{t}^n = \overline{(\lambda_0 + \lambda_1 t + \ldots + \lambda_n t^n)} =$$
$$= p(t) + M = M = \bar{0},$$

as required.

Note, that $\bar{F} = R/M$ is a vector space over the field F: if $\lambda \in F$ and $\overline{f(t)} = f(t) + M \in \bar{F}$ then $\lambda\overline{f(t)} = \overline{\lambda f(t)}$; moreover, it is obvious that all the axioms for a vector space hold.

Proposition. *Let $p(X)$ be an irreducible polynomial with coefficients from F, where deg $p(X) = n$. Then*

$$\dim_F \bar{F} = \dim_F F[t]/F[t]p(t) = n.$$

Proof. We shall show that the set of classes $\bar{1} = 1 + M$, $\bar{t} = t + M$, ..., $\bar{t}^{n-1} = t^{n-1} + M$ is a basis of the space $\bar{F} = F[t]/F[t]p(t)$, where M, as before, is $M = F[t]p(t)$. Indeed, if $f(t) \in F[t]$ then having divided $f(t)$ by $p(t)$ with residue $r(t)$: $f(t) = p(t)q(t) + r(t)$, we see that

$$\overline{f(t)} = f(t) + M = r(t) + M = \overline{r(t)}.$$

Since deg $r(t) <$ deg $p(t) = n$, we have

$$f(t) + M = (\lambda_0 + \lambda_1 + \ldots + \lambda_{n-1}t^{n-1}) + M =$$
$$= \lambda_0\bar{1} + \lambda_1\bar{t} + \ldots + \lambda_{n-1}\bar{t}^{n-1},$$

where $r(t) = \lambda_0 + \lambda_1 t + \ldots + \lambda_{n-1}t^{n-1}$. If $\mu_0\bar{1} + \mu_1\bar{t} + \ldots + \mu_{n-1}\bar{t}^{n-1} = 0$ then $q(t) = \mu_0 1 + \mu_1 t + \ldots + \mu_{n-1}t^{n-1} \in M$, i.e. we can write $q(t)$ in the form

$$q(t) = p(t)s(t).$$

Since deg $q(t) <$ deg $p(t) = n$ it follows from this that $s(t) = 0$, i.e. $\mu_0 = \mu_1 = \ldots = \mu_{n-1} = 0$, so that the system of elements under is linearly independent. □

We say that $\bar{F}$ is obtained from F by *adjoining a root* $\bar{t}$ of the polynomial $p(X)$. Adjunction of a root can be realized in a different way. For instance, let the field F be contained in a field K and α be a root of the polynomial $p(X)$ that is in K. We denote by $F(\alpha)$ the smallest subfield in K that contains both F and α. Then $\bar{F} \cong F(\alpha)$. In fact, let ϕ be the homomorphism from $F[t]$ to K which is the identity on F and which takes t into α (this means that ϕ is the evaluation of a polynomial at the point $t = \alpha$). Since $0 = p(\alpha) = \phi(p(t))$, we see that $p(t)$ and hence $M = F[t]p(t)$ are in $\operatorname{Ker}\phi$. According to the first isomorphism theorem, ϕ is a homomorphism with as image the field $\bar{F} = R/M$. Since there are no ideals in a field, we have $\bar{F} \cong \operatorname{Im}\phi$, i.e. $\operatorname{Im}\phi$ is a field. It is clear that $\operatorname{Im}\phi = F(\alpha)$, i.e. our statement is proved.

Obviously, if we took a root α of the irreducible polynomial $p(X)$ from any field that contains F, the result of adjoining it, $F(\alpha)$, would be the same and it can be obtained by using the quotient ring of polynomials by the ideal that is generated by $p(X)$. In particular, $\dim_F F(\alpha) = \deg p(X)$.

2.7. Quaternion field. A ring R which also is a vector space over a field F is called an *algebra* over F, or F–algebra, if for any $\lambda \in F$ and any $x, y \in R$ we have

$$\lambda(xy) = (\lambda x)y = x(\lambda y).$$

An example of an algebra over the field F is the full matrix ring $M_n(F)$; in fact, more often it is called the "full matrix algebra". Another example: if F and K are fields, where F is a subfield in K, then the multiplication of elements from K with elements from F turns K into an F–algebra. For instance, the field of complex numbers is a two–dimensional algebra over the field $\mathbf{R}$ of real numbers: $\mathbf{C} = \mathbf{R}(i) = \mathbf{R}1 + \mathbf{R}i$. It is clear that $\mathbf{C}$ can be also considered as an algebra over the field $\mathbf{Q}$ of rational numbers. Then $\dim_{\mathbf{Q}}\mathbf{C} = \infty$.

The algebra $\mathbf{C}$ is an example of a finite–dimensional algebra with division over the field $\mathbf{R}$ of real numbers. Another example of such an algebra is $\mathbf{R}$ itself. The third (and the last) example of such an algebra was discovered by Hamilton; the *quaternion algebra* $\mathbf{H}$ obtained by him, an division algebra, has dimension 4 over $\mathbf{R}$ and, unlike to $\mathbf{R}$ and $\mathbf{C}$, it is not commutative.

The algebra $\mathbf{H}$ can be constructed as follows. Let us denote a basis of a four dimensional space $\mathbf{H}$ over the field $\mathbf{R}$ by the symbols $\{1, i, j, k\}$ and define a multiplication of these elements, by assuming that 1 is an identity and, further

$$
\begin{aligned}
&i^2 = j^2 = k^2 = -1, \quad ij = k = -ji, \\
&jk = i = -kj, \quad ki = j = -ik.
\end{aligned} \tag{5}
$$

A direct verification shows that **H** is an associative non–commutative ring. (The work involved in this verification can be reduced by observing that the linear span in $M_2(\mathbf{C})$ of the set of matrices $\begin{pmatrix} 1 & 0 \\ 0 & 1 \end{pmatrix}$, $\begin{pmatrix} i & 0 \\ 0 & -i \end{pmatrix}$, $\begin{pmatrix} 0 & -1 \\ 1 & 0 \end{pmatrix}$, $\begin{pmatrix} 0 & -i \\ -i & 0 \end{pmatrix}$, is an algebra over **R**, and that these matrices satisfy the same relations (5).

Any quaternion, i.e. element from the algebra **H**, can be uniquely written in the form $z = \alpha_0 + \alpha_1 i + \alpha_2 j + \alpha_3 k$, where α_0, α_1, α_2, $\alpha_3 \in \mathbf{R}$. The real number $|z| = \sqrt{\alpha_0^2 + \alpha_1^2 + \alpha_2^2 + \alpha_3^2}$ is called the *norm* of the quaternion z. The norm is multiplicative, i.e.

$$|z_1 z_2| = |z_1||z_2|.$$

The quaternion $\bar{z} = \alpha_0 - \alpha_1 i - \alpha_2 j - \alpha_3 k$ is called *conjugate* to the quaternion z. There is then the easily verified formula

$$z\bar{z} = |z|^2.$$

From this the following is obtained: any non–zero quaternion z has an inverse $z^{-1} = \bar{z}/|z|^2$, i.e. the algebra **H** of quaternions is a (non–commutative or skew) field.

The main result about division algebras over the field of real numbers is

Frobenius theorem. *Any finite–dimensional associative division algebra over the field of real numbers is isomorphic to the one of the algebras* **R**, **C**, **H**.

Let us make a few preliminary remarks. If A is a finite–dimensional algebra with 1 over the field F, $a \in A$, then the system of elements $\{1, a, a^2, \dots\}$ is linearly dependent. There exists λ_0, λ_1, $\lambda_2, \dots \in F$, not all equal to zero, such that $\lambda_0 1 + \lambda_1 a + \lambda_2 a^2 + \dots + \lambda_m a^m = 0$ for suitable m. Let $f(X) = \lambda_0 1 + \lambda_1 X + \lambda_2 X^2 + \dots + \lambda_m X^m \in F[X]$; then $f(a) = 0$ and we say that $f(X)$ *annihilates* a.

Let us select a non–zero polynomial $\mu_a(X)$ of minimal degree with leading coefficient 1 such that $\mu_a(a) = 0$; we call it the *minimal polynomial* of the element a. The usual sort of arguments related to the minimal polynomial of a linear operator are completely applicable in our, more general, setting. In particular, any annihilating polynomial of the element a is divisible by $\mu_a(X)$.

Lemma. *Let A be a finite–dimensional integral algebra with* 1 *over the field* $\mathbf{C}$. *Then* $dim_{\mathbf{C}} A = 1$.

Proof. Let $a \in A$, $a \neq 0$. Let us consider the minimal polynomial $\mu_a(X)$. Over the field $\mathbf{C}$ we can write it as

$$\mu_a(X) = (X - \lambda_1)\ldots(X - \lambda_m), \quad \lambda_1, \ldots, \lambda_m \in \mathbf{C}.$$

Substituting a for X we get

$$(a - \lambda_1 \cdot 1)\ldots(a - \lambda_m \cdot 1) = 0.$$

Since A has no zero divisors, one of the factors is zero and $a = \lambda_i \cdot 1$. Thus $\dim_{\mathbf{C}} A = 1$. □

We call an element $z \in A$ *central* if $zx = xz$ for any $x \in A$.

Corollary. *Let T be a finite–dimensional integral algebra with* 1 *over* $\mathbf{R}$ *containing an element i such that $i^2 + 1 = 0$. If $a \in T$ commutes with i, i.e. $ai = ia$, then $a = \lambda \cdot 1 + \mu i$ for suitable $\lambda, \mu \in \mathbf{R}$.*

In fact, it is clear that a subring of elements of the form $\alpha 1 + \beta i$ is isomorphic to the field $\mathbf{C}$. We denote this ring as $\mathbf{C}$.
Further, the ring $A = \mathbf{C}[a] = \{\lambda_0 1 + \lambda_1 a + \ldots + \lambda_m a^m | \lambda_i \in \mathbf{C}\}$ is a $\mathbf{C}$–algebra (here we use the condition $ia = ai$). According to the lemma, $\dim_{\mathbf{C}} A = 1$, so $a \in A$ is equal to $\lambda_0 \cdot 1 = \lambda \cdot 1 + \mu i$. □

Proof of the theorem. Let D be a finite–dimensional field over $\mathbf{R}$. We shall show that if $\dim_{\mathbf{R}} D > 1$ then there exists $D \ni i$ such that $i^2 + 1 = 0$. In fact, let $a \in D \setminus R$. We consider the minimal polynomial $\mu_a(X) \in \mathbf{R}[X]$. Since D has no zero divisors, $\mu_a(X)$ is an irreducible polynomial (see the proof of lemma). Its degree can be 1 or 2. However, in the first case we get $a \in \mathbf{R}$. Therefore

$$\mu_a(X) = X^2 - \alpha X + \beta, \quad \text{where} \quad \alpha, \beta \in \mathbf{R} \text{ and } \alpha^2 - 4\beta < 0.$$

Now, put $i = \gamma a + \delta 1$, where

$$\gamma = \frac{2}{\sqrt{4\beta - \alpha^2}}, \qquad \delta = -\frac{\alpha}{\sqrt{4\beta - \alpha^2}}.$$

A direct verification shows that, indeed, $i^2 + 1 = 0$.

As in the corollary, we consider $\mathbf{C} = \mathbf{R}1 + \mathbf{R}i \subset D$. If for any $a \in D$ we have $ia = ai$ then, according to the corollary, $D = \mathbf{C}$.

Assume that D has elements that do not commute with i. We consider D as some space (not an algebra) over $\mathbf{C}$ with respect to multiplication of complex numbers on the left. Let us denote with $\mathcal{I}$ the linear operator on this complex space given by multiplication with i on the right, i.e. $\mathcal{I}(a) = ai$. It is clear that $\mathcal{I}(a + b) = \mathcal{I}(a) + \mathcal{I}(b)$; if $z \in \mathbf{C}$ then $\mathcal{I}(za) = (za)i = z(ai) = z\mathcal{I}(a)$. Further, $(\mathcal{I}^2 + 1_D)(a) = a(i^2 + 1) = 0$. So the minimal polynomial of the operator $\mathcal{I}$ is equal to $X^2 + 1$. Therefore D decomposes into the sum of proper subspaces U and V corresponding to the roots i and $-i$ of this polynomial. If $a \in U$ then $ai = \mathcal{I}(a) = ia$. According to the corollary, $a \in \mathbf{C}$. If $b \in V$ then $bi = \mathcal{I}(b) = -ib$. We fix a non–zero $j^{-1} \in V$ and let b be an arbitrary element from V. Then $(bj^{-1})i = b(j^{-1}i) = -b(ij^{-1}) = -(bi)j^{-1} = -(-ib)j^{-1} = (ib)j^{-1} = i(bj^{-1})$, i.e. $bj^{-1} \in \mathbf{C}$ and $b \in \mathbf{C}j$. We have: $D = U \oplus V = \mathbf{C} \oplus \mathbf{C}j$. The basis of the algebra D on $\mathbf{R}$ consists of $1, i, j = 1 \cdot j$, and $k = ij$. The multiplication table can be easily verified. □

§ 3. Modules and representations

3.1. Definitions. Let R be a ring, M an abelian group. We then say that M is a *left R–module* if for any $r \in R$ and any $x \in M$ an element $rx \in M$ is defined, such that in addition, the following natural axioms are fulfilled:

$$(r_1 r_2)x = r_1(r_2 x),$$

$$(r_1 + r_2)x = r_1 x + r_2 x, \quad r(x_1 + x_2) = rx_1 + rx_2,$$

$$1 \cdot x = x \quad (\text{ if in } R \text{ there is an identity } 1).$$

The concept of a right R–module is defined analogously.

One often writes $M = {}_RM$, if M is a left R– module, and $M = M_R$ if M is a right R–module.

A *submodule* N in M is a subgroup in M such that $rx \in N$ for any $r \in R$ and any $x \in N$. In this case the quotient group M/N has well defined R–module structure (*quotient module*) if we set $r(x+N) = rx + N$.

A *homomorphism* $\phi : M_1 \to M_2$ of one R–module into another is a homomorphism of abelian groups such that for any $x \in M_1$ and any $r \in R$ we have $\phi(rx) = r\phi(x)$. A bijective homomorphism of modules is called an *isomorphism*. The kernel $\operatorname{Ker}\phi$ of any homomorphism of modules, i.e. the set of elements with zero image, is an R–submodule. The homomorphism theorem for modules has the usual form:

$$M/\operatorname{Ker}\phi \cong \operatorname{Im}\phi.$$

The *annihilator* $\operatorname{Ann}_R(M)$ of a left R–module M is the set of elements $r \in R$ such that $rx = 0$ for any $x \in M$. It is clear that $\operatorname{Ann}_R(M)$ is a left ideal in R. If $\operatorname{Ann}_R(M) = 0$ then the R–module M is called *faithful*.

If an abelian group M is a left R–module and a right S–module, where R and S are two rings and for any $r \in R$, $x \in M$, $s \in S$ there holds

$$(rx)s = r(xs),$$

then we say that M is an (R,S)–bimodule. Notation: ${}_RM_S$.

The R–module M is called *simple* (or *irreducible*) if it does not contain proper non–trivial submodules.

3.2. Examples. 1) Any abelian group A is a module over the ring $\mathbf{Z}$ of integers if na is considered to be the n–multiple of the element a (see, Subsection 1.3).

2) A vector space V over the field F is an F–module.

3) A linear vector space F^n is an $M_n(F)$–module if Av is the usual product of the $n \times n$ matrix $A \in M_n(F)$ and the column vector v of length n from F^n.

4) Let V be a vector space on the field F, $\mathcal{A} : V \to V$ a linear operator. We then transform V into $F[X]$–module, by setting $f(X)v = f(\mathcal{A})(v)$, where $f(X) \in F[X]$, $v \in V$.

5) Let G be a group, V a vector space on F, and also let G act on V. We shall consider the group algebra $R = F[G]$ of the group G with coefficients from the field F. For any element $r = \sum_{g \in G} \alpha_g g$, $\alpha_g \in F$, and any $v \in V$ we set

$$rv = \sum_{g \in G} \alpha_g(g * v).$$

Then V becomes an R–module.

6) Any ring R is both a left and a right R–module over itself and also a (R, R)–bimodule. These modules have following denotions: ${}_RR$, R_R, ${}_RR_R$. The submodules of the module ${}_RR$ are the left ideals, etc.

7) If $R = \Delta$ is a skew field then a left Δ–module V is called a *left vector space* V over the skew field Δ. The theory of the left vector spaces over a skew field extends the usual theory of vector spaces, i.e. over a commutative field. For this more general case the definitions and all the properties of such concepts as linear dependence, basis, finite–dimensionalty, dimension, are transfered without changes.

∘ A space V is finite–dimensional if there exist $v_1, \ldots, v_m$ such that $V \subset \Delta v_1 + \ldots + \Delta v_m$; in this case we can find $e_1, \ldots, e_n \in V$ such that any $v \in V$ can be uniquely written in the form

$$v = \lambda_1 e_1 + \ldots + \lambda_n e_n, \qquad \text{where } \lambda_1, \ldots, \lambda_n \in \Delta;$$

such a system of vectors $\{e_1, \ldots, e_n\}$ is called a *basis*; the number of vectors in this system is called the *dimension* of the space $V : n = \dim_\Delta V$. ∘

3.3. Free module. As it was noted in example 6 any ring R is a left R–module. We shall call this module, ${}_RR$, the *free module of rank* 1. In the case when R has an identity element, the module ${}_RR$ is cyclic in the sense of the following definition. We say that an R–module M is *cyclic* if in M there is a generating element x_0, such that $M = Rx_0 = \{rx_0 | r \in R\}$. For ${}_RR$ a generating element is 1. The mapping $r \mapsto rx_0$ is a surjective homomorphism $\phi : R \to M$. The kernel $\operatorname{Ker}\phi$ is a left ideal S in R and, according to the homomorphism theorem, $R/S \cong M$. In addition, any simple R–module M is cyclic (it is generated by any of its non–zero element); thus such a module has the form $M \cong R/S$, where S is a maximal left ideal in R.

If M_1 and M_2 are two R–modules then the abelian group $M_1 \oplus M_2$ becomes an R–module, by setting

$$r(m_1, m_2) = (rm_1, rm_2), \qquad \text{where } r \in R,\ m_1 \in M_1,\ m_2 \in M_2.$$

We say that $M_1 \oplus M_2$ is the *(exterior) direct sum* of the R–modules M_1 and M_2. The *interior direct sum* is defined in the same way as for the groups (see, Subsection 1.11). We see that $M = M_1 \oplus \ldots \oplus M_s$ if in the R–module M there exists a set of submodules $M_1, \ldots, M_s$ such that

$$M = M_1 + \ldots + M_s, \quad \text{and} \quad M_i \cap (M_1 + \ldots + M_{i-1}) = 0, \quad i = 2, \ldots, s.$$

Or we can say that any element m from M is uniquely representable in the form

$$m = m_1 + m_2 + \ldots + m_s, \qquad \text{where} \quad m_i \in M_i, \quad i = 1, 2, \ldots, s.$$

If $M_i \cong N$, $i = 1, 2, \ldots, s$, then we write $M = N^s$.

The module R^n is called a *free* left R–module of rank n. If $x_i = (0, \ldots, 1, \ldots \ldots, 0)$, $i = 1, \ldots, n$ then $R = Rx_1 \oplus \ldots \oplus Rx_n$. The elements $\{x_1, \ldots, x_n\}$ form a basis in R^n in the sense that any element $a \in R^n$ is uniquely representable in the form $a = r_1x_1 + \ldots + r_nx_n$. If the module M is generated by elements $a_1, a_2, \ldots, a_n$, i.e. $M = Ra_1 + Ra_2 + \ldots + Ra_n$, then the mapping

$$r_1x_1 + r_2x_2 + \ldots + r_nx_n \mapsto r_1a_1 + r_2a_2 + \ldots + r_na_n$$

is a surjective homomorphism, and this allows us to write M in the form $M \cong R^n/P$ for a suitable submodule P. It is clear that these statements are also valid for modules that are not fnitely generated, i.e. any module M is representable in the form $M \cong F(X)/S$, where $F(X)$ is a free R–module with, possibly, an infinite basis X.

The following property of free modules is significant.

Proposition. *Assume that the left R–module M contains a free submodule N such that $M/N = F$ is a free R–module. Then $M \cong F \oplus N$ is a free R–module.*

Indeed, if $F = F(X)$ is a free module with basis X then the mapping $x \mapsto a_x$ associating to the element $x \in F$ an element $a_x \in M$ such that $a_x + N = x$, extends to a monomorphism $\psi : F \to M$. Let $F' = \psi(F)$. Then $\psi(x)+N = x$, i.e. also $\psi(p)+N = p$ for any $p \in F$. If $m+N = p \in F$ then $b = \psi(p)-m \in N$ and $a = \psi(p)+n \in \psi(F)+N$. If $\psi(p) \in N$ then $p = \psi(p)+N = 0$. We have

$M = F' \oplus N \cong N$. Finally, it is evident that the direct sum of free modules is free, concluding the proof. □

3.4. Ring of endomorphisms of a module. The set $\mathrm{End}_R(M)$ of endomorphisms (i.e. homomorphisms into itself) of an R–module is a ring in the sense of addition and composition of endomorphisms. If $M = R^n$ then $\mathrm{End}_R(M) \cong M_n(R)$. Indeed, any endomorphism $\phi : R^n \to R^n$ is defined by the images of the basis elements: $\phi(x_i) = a_{i1}x_1 + \ldots + a_{in}x_n$, $i = 1, \ldots, n$. The numbers a_{ij}, $1 \le i, j \le n$, form a matrix A_ϕ. It is easy to verify the equalities

$$A_\phi A_\psi = A_{\phi\psi}, \quad A_\phi + A_\psi = A_{\phi+\psi},$$

which establish the desired isomorphism of rings.

Another important case can be verified with a help of the well–known Schur lemma on homomorphisms of simple modules.

Schur lemma. *Let M_1 and M_2 be two simple R–modules. Then any homomorphism $\phi : M_1 \to M_2$ is either an isomorphism or the zero mapping.*

Proof. The kernel $\mathrm{Ker}\,\phi$ is an R–submodule in M_1. Because M_1 is simple either $\mathrm{Ker}\,\phi = M_1$ and then $\phi = 0$ is a zero mapping, or $\mathrm{Ker}\,\phi = \{0\}$ and then ϕ is injective. As $M_1 \neq \{0\}$ then $\mathrm{Im}\,\phi$ is a non–zero R–submodule in M_2, so $\mathrm{Im}\,\phi = M_2$. Thus ϕ is an isomorphism. □

Corollary. *The ring $\mathrm{End}_R M$ of endomorphisms of a simple R–module M is a (skew) field.*

Indeed, according to the Schur lemma, any non–zero endomorphism of the module M is an automorphism, i.e. has an inverse. □

3.5. Definition of a representation of a group. Let V be a vector space over the field F. The set $GL(V)$ of all non–singular linear operators on this space is a group. If G is a certain arbitrary group then any homomorphism $\rho : G \to GL(V)$ is called a *representation* of the group G by linear operators on the space V.

If $\rho : G \to GL(V)$ is a representation of the group G, $R = F[G]$ the group ring for G with coefficients from F, then V becomes a module in a canonical way, by putting

$$\left(\sum_{g\in G}\alpha_g g\right) v = \sum_{g\in G}\alpha_g \rho(g)(v).$$

Inversely, given an R–module V, where $R = F[G]$, then in the first place, V is a vector space over the field F. In fact, $F[G]$ contains the field F as a subring, which introduces the structure of a vector space on V. Secondly, for any $g \in G$ the mapping $\rho(g) : v \mapsto gv$ is an automorphism, and $g \mapsto \rho(g)$ is a homomorphism of groups $\rho : G \to GL(V)$. By the same token the R–module structure on V defines a representation of the group G by linear operators on the space V.

Further, we call two representations $\rho_1 : G \to GL(V_1)$, $\rho_2 : G \to GL(V_2)$ (F is given) *equivalent* if there exists a linear isomorphism $\phi : V_1 \to V_2$ of vector spaces such that for any $g \in G$ the square of mappings

$$\begin{array}{ccc} V_1 & \xrightarrow{\rho_1(g)} & V_1 \\ \phi\downarrow & & \phi\downarrow \\ V_2 & \xrightarrow{\rho_2(g)} & V_2 \end{array}$$

is *commutative*, i.e. the result of the composition of mappings from V_1 to V_2 does not depend on the path followed:

(1) $$\phi\rho_1(g) = \rho_2(g)\phi.$$

From formula (1) it is clear that in fact ϕ is an isomorphism of R–modules, where $R = F[G]$, i.e. equivalence of representations is the same thing as an isomorphism of the corresponding modules.

If $U \subset V$ is a subspace in V and it is invariant under G via the representation $\rho : G \to GL(V)$, then the restriction $\sigma(g)$ for any operator $\rho(g)$, $g \in G$, on U is a non–singular linear operator on U, i.e. $\sigma(g) \in GL(U)$. In this case the representation $\sigma : G \to GL(U)$ which associates the linear operator $\sigma(g)$ to the element $g \in G$ is called a *subrepresentation* of the representation ρ; we write $\sigma \subset \rho$. Under the same conditions, on the quotient space V/U, the operator $\rho(g)$ defines a non–singular operator $\tau(g)$ such that $\tau(g)(v + U) = \rho(g)(v) + U$. Well definedness of the operator $\tau(g)$ is implied by the $\rho(g)$–invariance of the subspace U. The representation $\tau : G \to GL(V/U)$ is called the *quotient representation* of the representation ρ by the representation $\sigma : \tau = \rho/\sigma$.

The representation $\rho : G \to GL(V)$ is called *irreducible* if any subspace in V which is invariant under all the operators of the set $\rho(G)$ (or simply: invariant under G), coincides with $\{0\}$ or with V.

We should denote that subrepresentations of the representation $\rho : G \to GL(V)$ are in natural bijective correspondence with R–submodules of the module V, where $R = F[G]$. The same can be said about the relation between quotient representations and quotient modules.

If $\rho : G \to GL(V)$ is some representation of the group G by linear operators on the space V, $\dim V = n$, then, having fixed a basis $\{e_1, \dots, e_n\}$ of the space V, we associate a matrix $\bar{\rho}(g)$ from $GL_n(F)$ with every operator $\rho(g)$, $g \in G$. The result is a homomorphism $\bar{\rho} : G \to GL_n(F)$ that is called a *matrix representation* associated with the representation ρ. For many technical statements concerning a representation ρ it is convenient to formulate and prove them by using a matrix representation, which is associated with it. Allowing ourselves some carelessness in language we often identify ρ and $\bar{\rho}$.

3.6. Examples. 1) The *natural* representation. Assume that the group G is a subgroup of the general linear group $GL(V)$ of all non–singular linear operators from some vector space V over a field F. Then the identity mapping on the group G is a representation of this group by linear operators on the space V.

2) *Permutation* representation. Assume that G is a subgroup of the symmetric group S_n, $V = \langle e_1, \dots, e_n \rangle$ is an n–dimensional linear space on the field F, $\sigma \in S_n$, $\mathcal{A}_\sigma : V \to V$ is the linear operator such that $\mathcal{A}_\sigma(e_i) = e_{\sigma(i)}$, where $\sigma(i)$ is the result of applying the permutation σ to i, $i = 1, 2, \dots, n$. Then $\mathcal{A}_\sigma \mathcal{A}_\tau = \mathcal{A}_{\sigma\tau}$ and the association of the operator $\mathcal{A}_\sigma$ to any element $\sigma \in G \subset S_n$ is a representation of the group G by linear operators on the space V.

3) If $G = \langle a \rangle$ is a cyclic group of order n and ϵ is any complex root of unity of order n then the mapping $\rho(a^s) = \epsilon^s$, $s = 0, 1, \dots, n-1$, is a homomorphism $\rho : G \to \mathbf{C}^* \cong GL_1(\mathbf{C})$ which represents the group G by operators on the one–dimensional space $\mathbf{C}$.

3.7. Completely reducible representations. If $\rho_1 : G \to GL(V_1)$, $\rho_2 : G \to GL(V_2)$ are two representations of a group by linear operators on spaces over the same field F then the representation $\rho = \rho_1 \oplus \rho_2 : G \to GL(V_1 \oplus V_2)$

defined by formula

$$(\rho_1 + \rho_2)(g)(v_1 + v_2) = (\rho_1(g)v_1,\ \rho_2(g)v_2),$$

is called the (*exterior*) *direct sum* of the representations ρ_1 and ρ_2. It is said that the representation ρ is decomposable into a direct sum of subrepresentations ρ_1 and ρ_2 if $V = V_1 \oplus V_2$, where V_1 and V_2 are the spaces of the representations ρ_1 and ρ_2 (they lie in V). This is an interior direct sum. Naturally, such an interior sum is equivalent to the exterior direct sum. Therefore in both cases we write $\rho = \rho_1 \oplus \rho_2$. It is clear that a direct sum of several representations can be considered.

We call the representation $\rho = \rho_1 \oplus \rho_2 \oplus \ldots \oplus \rho_n$ *completely reducible* if all $\rho_1, \rho_2, \ldots, \rho_n$ are irreducible. Equivalent to the latter definition is the following: ρ is completely reducible iff for any subrepresentation $\sigma \subset \rho$ there exists a subrepresentation $\tau \subset \rho$ such that $\rho = \sigma \oplus \tau$.

It is clear that the statement mentioned can be reformulated in terms of modules. A module M over a ring R is called *semisimple* if M is the sum of all simple submodules contained in it (or a direct sum of some family of simple submodules). It is the same to say that for any $M_1 \subset M$ there exists a submodule M_2 such that $M = M_1 \oplus M_2$. (This statement will be proved in Subsection 2.1.1, therefore we skip the proof of these equivalent statements about representations.)

3.8. Maschke theorem. *Let G be a finite group, the order n of which is coprime with the characteristic p of the base field F. Then any finite-dimensional representation of the group G is completely reducible.*

The main instrument in the proof of this result is a special operator which is constructed in the following way. Let ρ_1 be a subrepresentation in ρ, $V_1 \subset V$ the corresponding spaces of representations. We write $V = V_1 \oplus W$, where W is any complementary subspace of V_1 in V. We denote the projection of the space V on V_1 along W by π. Let us modify π, by defining

$$\widetilde{\pi} = \frac{1}{n} \sum_{g \in G} \rho(g) \pi \rho(g)^{-1}.$$

Then $\widetilde{\pi}|_{V_1} = \pi$. It is easy to note that $\widetilde{\pi}^2 = \widetilde{\pi}$ and that $\widetilde{\pi}\rho(g) = \rho(g)\widetilde{\pi}$ for any $g \in G$. It follows from this that $V_2 = \operatorname{Ker} \widetilde{\pi}$ is invariant under $\rho(G)$ and, besides, $V = V_1 \oplus V_2$. In this way it is shown that the representation ρ is completely reducible. □

3.9. Regular representation. It is known (see, Subsection 3.2.) that any ring R is a left module over itself. If $R = F[G]$ is the group ring of the group G then the representation of the group G which called the left *regular* representation, is the one corresponding to the module mentioned. Since G is a basis for $F[G]$, the representation $\rho : G \to GL(F[G])$ is defined by the formula $\rho(g)(h) = gh$, $g \in G$, $h \in G \subset F[G]$.

We know (Subsection 3.3) that any cyclic module, including also any simple module V, is a quotient module of a free cyclic one, i.e. $F[G]$. We have $V \cong F[G]/K$, where K is some submodule. If the order $n = |G|$ is mutually prime with a characteristic of the field F then the module $F[G]$ is completely reducible, i.e. $F[G] = V' \oplus K$, where V' is some R–submodule. It is clear that $F[G]/K \cong V'$. This means that $V' \cong V$. Translating this in terms of representations we get:

Proposition. *Any irreducible representation of a finite group G the order of which is mutually prime with the characteristic of the field F, is equivalent to a subrepresentation of the regular representation of this group.* □

Corollary. *The number of the pairwise non–equivalent irreducible representations of a finite group G is finite.*

Proof. Let us write the regular representation ρ in the form of a sum of irreducible subrepresentations $\rho = \rho_1 \oplus \ldots \oplus \rho_s$ and, correspondingly, write the space $F[G]$ of the regular representation in the form of a sum of spaces $V_1, \ldots, V_s$. Let σ be an irreducible representation. Then $\sigma \subset \rho$ and $W \subset V_1 \oplus \ldots \oplus V_s$, where W is the space of the representation σ. Let $\pi_i : W \to V_i$ be the projection of the subspace W on the i–th component of the direct sum, $i = 1, \ldots, s$. If $\pi_i = 0$ for all i, then $W = 0$. If, say, $\pi_1 \neq 0$ then π_1 is a non-zero homomorphism of the simple $F[G]$–module W into the simple $F[G]$–module V_i. According to the Schur lemma, π_1 is an isomorphism and σ is equivalent to ρ_1. Thus any irreducible representation is equivalent to one of $\rho_1, \ldots, \rho_s$. □

3.10. Character of a representation. Let V be the space of a finite-dimensional representation ρ of the group G over a field F. We define the character of the representation as the function $\chi : G \to F$ such that

$$\chi(g) = \mathrm{tr}\ \rho(g).$$

∗ We should recall that the trace tr $\mathcal{A}$ of the linear operator $\mathcal{A} : V \to V$ is a sum of the diagonal elements of its matrix in some (in fact, in any) basis of the space V. If V is presented in the form of direct sum of $\mathcal{A}$-invariant subspaces $V = U_1 \oplus U_2$, then $\mathrm{tr}\ \mathcal{A} = \mathrm{tr}\ \mathcal{A}_1 + \mathrm{tr}\ \mathcal{A}_2$, where $\mathcal{A}_i = \mathcal{A}|_{U_i}$ is the restriction of the operator $\mathcal{A}$ to U_i, $i = 1, 2$. ∗

The character of a representation is an important invariant.

Proposition. *Let $\rho_1 : G \to GL(V_1)$, $\rho_2 : G \to GL(V_2)$ are two equivalent representations of the group G, and let χ_1, χ_2 be their characters. Then $\chi_1 = \chi_2$.*

Proof. Let $\phi : V_1 \to V_2$ be an isomorphism of the linear spaces V_1 and V_2, such that $\phi\rho_1(g) = \rho_2(g)\phi$ for any $g \in G$. If $\{e_1, \dots, e_n\}$ is a basis in V_1 then $\{\phi(e_1), \dots, \phi(e_n)\}$ is a basis in V_2. The elements a_{ij} of the matrix of the operator $\rho_1(g)$ are determined by the decomposition

$$\rho_1(g)(e_j) = \sum_{i=1}^{n} a_{ij} e_i.$$

By applying the representation ϕ to both sides of the equation and using the isomorphism property of this representation, we obtain

$$\rho_2(g)(\phi(e_j)) = \sum_{i=1}^{n} a_{ij} \phi(e_j).$$

From this it is obvious that the matrices of $\rho_1(g)$ and $\rho_2(g)$ for suitable bases coincide. Thus $\chi_1(g) = \chi_2(g)$, where χ_1 and χ_2 are the characters of the representations ρ_1 and ρ_2, respectively. □

3.11. Scalar product of characters. Let G be a finite group the order of which is mutually prime with the characteristic of the field F, and let χ_1, χ_2

be two characters of the group G. We shall define a *scalar product* $(\chi_1|\chi_2)$ by setting

$$(\chi_1|\chi_2) = \frac{1}{n}\sum_{g\in G}\chi_1(g)\chi_2(g^{-1}). \tag{2}$$

The following properties are obvious: $(\chi_1 + \chi_1'|\chi_2) = (\chi_1|\chi_2) + (\chi_1'|\chi_2)$ and $(\chi_1|\chi_2 + \chi_2') = (\chi_1|\chi_2) + (\chi_1|\chi_2')$. Since g^{-1} together with g runs through the entire group G, $(\chi_1|\chi_2) = (\chi_2|\chi_1)$. If m is a natural number, and χ is the character of the representation ρ then $m\chi$ is a character of the representation $m\rho = \underbrace{\rho \oplus \ldots \oplus \rho}_{m}$. We have

$$m(\chi_1|\chi_2) = (m\chi_1|\chi_2) = (\chi_1|m\chi_2).$$

Among the more important properties of the scalar product of characters there are the orthogonality relations.

Theorem. *Let χ_1 and χ_2 be the characters of irreducible representations ρ_1 and ρ_2 of the group G, the order of which is mutually prime with the characteristic of the algebraically closed field F. If ρ_1 is equivalent to ρ_2 then $(\chi_1|\chi_2) = 1$. Otherwise $(\chi_1|\chi_2) = 0$.*

Proof. Let V_1 and V_2 be the spaces of the representations ρ_1 and ρ_2, and let $\sigma : V_1 \to V_2$ be any linear mapping. Then, analogously to the arguments used in proof of the Maschke theorem, we construct the operator

$$\widetilde{\sigma} = \frac{1}{n}\sum_{g\in G}\rho_2(g)\sigma\rho_1(g^{-1}).$$

The operator $\widetilde{\sigma}$ is a homomorphism of the $F[G]$–modules from V_1 into V_2. Indeed, for any $h \in G$

$$
\begin{aligned}
\widetilde{\sigma}\rho_1(h) &= \frac{1}{n}\sum_{g\in G}\rho_2(g)\sigma\rho_1(g^{-1})\rho_1(h) = \\
&= \frac{1}{n}\sum_{g\in G}\rho_2(g)\sigma\rho_1(g^{-1}h) = \\
&= \rho_2(h)\frac{1}{n}\sum_{g\in G}\rho_2(h^{-1})\rho_2(g)\sigma\rho_1((h^{-1}g)^{-1}) = \\
&= \rho_2(h)\frac{1}{n}\sum_{g\in G}\rho_2(h^{-1}g)\sigma\rho_1((h^{-1}g)^{-1}) = \\
&= \rho_2(h)\frac{1}{n}\sum_{k\in G}\rho_2(k)\sigma\rho_1(k^{-1}) = \rho_2(h)\widetilde{\sigma}.
\end{aligned}
$$

Here $k = h^{-1}g$ together with g runs over the entire group G.

According to the Schur theorem, if ρ_1 is not equivalent to ρ_2 then $\widetilde{\sigma}$ is the zero mapping. Writing down the operator equality

$$0 = \frac{1}{n}\sum_{g\in G}\rho_2(g)\sigma\rho_1(g)^{-1}$$

in matrix form with respect to some basis we obtain

$$0 = \sum_{k,l,g}\rho_2^{ik}(g)\sigma_1^{kl}\rho_1^{lj}(g^{-1})$$

for any i, j.

Now, take σ to be equal to the operator $\mathcal{E}_{ij}$, the matrix of which with respect to the given basis has a 1 only at (i, j) and all other elements are zero; we then obtain

$$0 = \sum_{g\in G}\rho_2^{ii}(g)\rho_1^{jj}(g^{-1}).$$

Summing over all possible i, j we get

$$0 = \frac{1}{n} \sum_{i,j,g} \rho_2^{ii}(g) \rho_1^{jj}(g^{-1}) =$$
$$= \frac{1}{n} \sum_{g \in G} \chi_2(g) \chi_1(g^{-1}) = (\chi_2 | \chi_1).$$

Thus the second statement of the theorem is proved.

This time, let ρ_1 be equivalent to ρ_2. Since the characters of equivalent representations are equal we can assume that $\rho = \rho_1 = \rho_2$, $V = V_1 = V_2$. In this case

$$\widetilde{\sigma} = \frac{\operatorname{tr} \sigma}{\dim V} 1_V, \tag{3}$$

where 1_V is the identity operator on V.

Indeed, let λ be one of the eigenvalues of the operator $\widetilde{\sigma}$. Then it is evident that the operator $\sigma_0 = \widetilde{\sigma} - \lambda \cdot 1_V$ also, as does σ, possesses the property $\rho(g)\sigma_0 = \sigma_0 \rho(g)$. Therefore either σ_0 is a zero operator or σ_0 is an automorphism. The latter is impossible as $\sigma_0 = \widetilde{\sigma} - \lambda \cdot 1_V$ is a singular operator. Hence $\sigma_0 = 0$, and $\widetilde{\sigma} = \lambda \cdot 1_V$. It is clear that then $\widetilde{\sigma} = \frac{\operatorname{tr} \widetilde{\sigma}}{\dim V} 1_V$. However, $\operatorname{tr} \widetilde{\sigma} = \frac{1}{n} \sum_{g \in G} \operatorname{tr} (\rho(g) \sigma \rho(g)^{-1}) = \frac{1}{n} \sum_{g \in G} \operatorname{tr} \sigma = \operatorname{tr} \sigma$; i.e. the formula (3) holds.

Thus

$$\frac{\operatorname{tr} \sigma}{\dim V} = \frac{1}{n} \sum_{g \in G} \rho(g) \sigma \rho(g)^{-1}.$$

By taking for σ the mapping with matrix, which has 1 only at (k, l) and zero at all other spots, we obtain

$$\frac{\delta_{ij}}{\dim V} = \frac{1}{n} \sum_{g \in G} \rho^{ik}(g) \rho^{lj}(g^{-1}).$$

Having taken $i = k$, $j = l$ and having summed with respect to i, j, we obtain

$$\Big(\sum_{i,j} \delta_{ij} \Big) \frac{1}{\dim V} = \frac{1}{n} \sum_{g \in G} \sum_{i,j} \rho^{ii}(g) \rho^{jj}(g^{-1}),$$

i.e.

$$1 = \frac{1}{n} \sum_{g \in G} \chi(g)\chi(g^{-1}) = (\chi|\chi).$$

This compeltely proves the first statement of the theorem. □

Corollary 1. *Every finite–dimensional representation is defined by its character.*

Proof. Let ρ, the representation under consideration, be of the form $\rho = n_1\rho_1 \oplus \ldots \oplus n_s\rho_s$, where the $\rho_1, \ldots, \rho_s$ are the irreducible representations of the group G, then $\chi = n_1\chi_1 + \ldots + n_s\chi_s$. Further

$$(\chi|\chi_i) = (n_1\chi_1 + \ldots + n_s\chi_s|\chi_i) = n_1(\chi|\chi_1) + \ldots + n_i(\chi|\chi_i) + \ldots + n_s(\chi|\chi_s).$$

By using the orthogonality relations we obtain $(\chi|\chi_i) = n_i$, and hence $\rho = (\chi|\chi_1)\rho + \ldots + (\chi|\chi_s)\rho_s$. □

Corollary 2. *Let ρ be the regular representation of the group G, and let $\rho_1, \ldots, \rho_s$ be all irreducible representations of this group. Then*

$$\rho = \sum_{i=1}^{s} (\dim \rho_i)\rho_i.$$

Proof. If χ is the character of the regular representation of the group G of order n then, evidently,

$$\chi(g) = \begin{cases} 0, & \text{when } g \neq e, \\ n & \text{when } g = e. \end{cases}$$

In this case

$$(\chi_i|\chi) = \frac{1}{n} \sum_{g \in G} \chi_i(g)\chi(g^{-1}) = \frac{1}{n}(\dim \rho_i)n = \dim \rho_i$$

□

Corollary 3. *Let $n_1, \ldots, n_s$ be the dimensions of all the irreducible representations of the finite group G the order of which is not divisible by the characteristic of the ground field F. Then*

$$n = \sum_{i=1}^{s} n_i^2.$$

Proof. For the regular character χ of the group G, according to Corollary 2, we have

$$\chi = \sum_{i=1}^{s} n_i \chi_i.$$

Computing values at the point e which is the unit of the group G, we obtain

$$n = \sum_{i=1}^{s} n_i^2.$$

□

3.12. One–dimensional representations. Let $\rho : G \to GL(V)$ be a one–dimensional representation of the group G over a field F. Then it is obvious that the group $GL(V)$ is isomorphic to F^*, which is the multiplicative group of the field F. The group F^* is commutative, therefore $\operatorname{Ker}\rho$ contains the commutator subgroup G' of the group G. According to the isomorphism theorem for groups (see, Subsection 1.6) the representation ρ defines a representation $\bar{\rho} : G/G' \to GL(V) \cong F^*$. Conversely, if $\sigma : G/G' \to F^*$ is a representation and $\rho : G \to F^*$ is defined by the formula $\rho(g) = \sigma(gG')$, then it is easy to verify that $\sigma = \bar{\rho}$, moreover, ρ is a one–dimensional representation of the group G. Thus the following holds.

Lemma. *The one–dimensional representations of an arbitrary group are in bijective correspondence with the one–dimensional representations of the group G/G', which is the quotient group of the group G by its commutator subgroup.* □

The answer to the question whether according to the lemma all representations of the group G/G' occur, is positive.

Proposition. *Any irreducible representation of a finite abelian group G over an algebraically closed field is one–dimensional.*

Proof. Let $\rho : G \to GL(V)$ be a representation; let $\mathcal{A} = \rho(g_1), \dots, \rho(g_n) = \mathcal{A}_n$ be all the operators corresponding to the elements of the group G defined by the representation ρ. We have $\mathcal{A}_i\mathcal{A}_j = \mathcal{A}_j\mathcal{A}_i$ (because G is abelian). Applying induction on n we prove, that $\mathcal{A}_1, \dots, \mathcal{A}_n$ have common eigenvector. When $n = 1$, this follows from the algebraic closedness of the field F. Let λ be an eigenvalue of the operator $\mathcal{A}_n$, V_λ the corresponding proper subspace. If $v \in V_\lambda$, then

$$\mathcal{A}_n(\mathcal{A}_i(v)) = \mathcal{A}_i\mathcal{A}_n(v) = \mathcal{A}_i(\lambda v) = \lambda\mathcal{A}_i(v),$$

i.e. the subspace V_λ is invariant relatively to the operators $\mathcal{A}_1, \dots, \mathcal{A}_{n-1}$. By induction there is in V_λ an eigenvector for $\mathcal{A}_1|_{V_\lambda}, \dots, \mathcal{A}_{n-1}|_{V_\lambda}$. It is clear that v is the eigenvector sought. Thus, in V there is an invariant one–dimesional subspace $\langle v\rangle$ with respect to $\mathcal{A}_1, \dots, \mathcal{A}_n$. Because of irreducibility we have $V = \langle v\rangle$, and thus the proposition is proved. □

Corollary. *The number of pairwise non–isomorphic irreducible representations of an abelian group G of order n is equal to n.*

This statement follows directly from proposition and corollary 3, Subsection 3.11. □

3.13. Length of a module. The concept of dimension of vector spaces cannot be carried over literally to the case of modules over arbitrary rings mostly because of the fact that it is possible to have $\lambda x = 0$, for a non–zero λ from the ring R and a non–zero x from the R–module M. We should note that the concept of dimension in vector spaces can be reformulated in the following way. The space V is finite–dimensional if the number of terms in any strictly decreasing chain of subspaces

$$V = V_0 \supset V_1 \supset \dots \supset V_k \supset \dots$$

is finite. Of course, if $\dim V = n$ then the maximal number of non–zero subspaces in such a chain is equal to n (as V_i , one could take the linear span: $\langle e_{i+1}, \dots, e_n\rangle$, $i = 0, 1, \dots, n$, where $\{e_1, \dots, e_n\}$ is a basis in V). In this chain $\dim V_i/V_{i+1} = 1$, i.e. the chain cannot be refined.

We say that an R–module M has *finite length* n if it possesses a non–refinable chain of submodules of the form

$$M = M_0 \supset M_1 \supset \ldots \supset M_n = \{0\}.$$

With a help of a rather general statement, it can be proved that this is well defined.

Jordan – Hölder theorem. *Let M be a left R–module with two non–refinable series of submodules*

$$M = M_0 \supset M_1 \supset \ldots \supset M_s = \{0\}, \tag{4}$$

$$N = N_0 \supset N_1 \supset \ldots \supset N_r = \{0\}. \tag{5}$$

Then $s = r$ and for any number i, $1 \leq i \leq s$, there exists a j, $1 \leq j \leq r$, such that $M_{i-1}/M_i \cong N_{j-1}/N_j$. The corresponding map $i \mapsto j$ is bijective.

Proof. Let us apply induction with respect to s starting with $s = 0$. If $s \geq 1$ we shall consider the quotient module $M' = M/M_s$. It possesses two series of submodules

$$M' = M'_0 \supset M'_1 \supset \ldots \supset M'_{s-1} = \{0\} \tag{4'}$$

$$N' = N'_0 \supset N'_1 \supset \ldots \supset N'_r = \{0\}, \tag{5'}$$

where $M'_i = M_i/M_s$, $i = 0, 1, \ldots, s-1$ and $N'_j = N_j + M_s/M_s$. The quotients of the first series are $M'_{i-1}/M'_i \cong M_{i-1}/M_i$, $i = 1, \ldots, s-1$. It also follows that the series (4′) cannot be refined. Further, $N'_{j-1}/N'_j \cong N_{j-1}+M_s/N_j+M_s$. By using the isomorphism theorem we obtain

$$N'_{j-1}/N'_j \cong (N_{j-1} + (N_j + M_s))/(N_j + M_s) \cong$$
$$\cong N_{j-1}/N_{j-1} \cap (N_j + M_s) = N_{j-1}/N_j + N_{j-1} \cap M_s.$$

The module $N_j + N_{j-1} \cap M_s$ is intermediate between N_{j-1} and N_j in the series (5). If $N_{j-1} \cap M_s \subset N_j$, then $N'_{j-1}/N'_j \cong N_{j-1}/N_j$. Otherwise, i.e. when $N_{j-1} \cap M_s \not\subset N_j$, we have $N'_{j-1}/N'_j = 0$. However $N_{j-1} \cap M_s \not\subset N_j$ is possible only when $N_j \not\supset M_s$ and $N_{j-1} \supset M_s$. It is clear that this condition uniquely determines $j = j_0$. We set $s \mapsto j_0$. Then the series (5′) has the same quotients as the series (5), with the exception of N'_{j_0-1}/N'_{j_0}, which is equal to zero. We have $N_{j_0-1}/N_{j_0} \cong M_s$. By applying induction we get $s-1 = r-1$, i.e. it is proved that $s = r$. In addition, the quotients of the series (4′), equal to M'_{i-1}/M'_i, $i = 1, \dots, s-1$, are in bijective correspondence with the factors from the series (5′), that are equal to $N'_{j-1}/N'_j \cong N_{j-1}/N_j$, $j \neq j_0$. This proves the theorem. □

Literature. Kostrikin [1], Lang [1], Kurosh [1].

CHAPTER I

COMMUTATIVE ALGEBRA

§ 1. Algebraic and transcendental extension

1.1. Concept of finite extension. Let K be a field, F a subfield. We then say that K is an *extension* of the field F. We can see K as a vector space over F by associating to a scalar $\lambda \in F$ and a vector $x \in K$ the vector λx, which is equal to the usual product of elements λ, $x \in K$. If the dimension $\dim_F K$ of this vector space is finite then K is called a *finite* extension of the field F, and $\dim_F K$ is called the degree of this extension (often it is denoted as $(K : F)$ or $[K : F]$). For instance, the field of complex numbers $\mathbf{C}$ has degree 2 over the field of real numbers $\mathbf{R}$ and infinite degree (or to be more exact, no finite degree) over the field of rational numbers $\mathbf{Q}$. An important property of finite extensions is expressed by the following lemma.

Lemma. *Let $F \subset K \subset L$ be the successive extensions of the field F. If $\dim_F K = l < \infty$, $\dim_K L = m < \infty$ then $\dim_F L = n < \infty$, where more precisely, $n = lm$.*

Proof. Let us consider a basis $\mathcal{X} = \{\xi_1, \dots, \xi_l\}$ for K over F and $\mathcal{H} = \{\eta_1, \dots$ $\dots, \eta_m\}$ for L over K. We shall show that the set Θ of all pairwise products of the form $\xi_i \eta_j$, $i = 1, \dots, l$, $j = 1, \dots, m$, is a basis for L over F. Indeed, if $x \in L$, then $x = \lambda_1 \eta_1 + \dots + \lambda_m \eta_m$, where $\lambda_1, \dots, \lambda_m \in K$. Further, $\lambda_i = \mu_{i1} \xi_1 + \dots + \mu_{il} \xi_l$, with $\mu_{ij} \in F$, $i = 1, \dots, l$, $j = 1, \dots, m$. From this $x = \sum_{i=1}^{l} \sum_{j=1}^{m} \mu_{ij} (\xi_i \eta_j)$. Thus L is the linear span of Θ over F. If we assume that Θ is linearly dependent over F then there is a non–trivial linear combination $\sum_{i,j} \lambda_{ij} \xi_i \eta_j = 0$. Using the distributivity of multiplication we obtain $\sum_{i=1}^{l} (\sum_{j=1}^{m} \lambda_{ij} \xi_i) \eta_j = 0$. In virtue of the fact that $\mathcal{H}$ is a basis for L over K we can conclude that the coefficients of the last linear combination are equal to zero, i.e. $\sum_{i=1}^{l} \lambda_{ij} \xi_i = 0$, $j = 1, \dots, m$. Since Ξ is a basis for K over F, $\lambda_{ij} = 0$, for all i, j. This contradicts our assumption. □

1.2. Examples. Cyclotomic extension. In the Introduction, § 2, the procedure of adjoining a root α of an irreducible polynomial $p(X) \in F[X]$ to a field F was considered. As a result of this procedure we obtain a field $F(\alpha)$ which is isomorphic to the quotient ring $F[X]$ by the principal ideal $\mathcal{I}$ which is generated by the polynomial $p(X)$. A basis of the field $F(\alpha)$ over F is the set $\{1, \alpha, \dots, \alpha^{n-1}\}$, where n is the degree of the polynomial $p(X)$. Every element from $F(\alpha)$ is uniquely representable in the form $f(\alpha)$, where $f(X)$ is a polynomial in $F[X]$ of degree not higher than $n-1$. It can also be said that $F(\alpha)$ is generated over F by the element α or that $F(\alpha)$ is the smallest subfield (in $F(\alpha)$) containing F and α. In general, if $\alpha_1, \alpha_2, \dots, \alpha_s$ are the elements of some extension K of the field F then $F(\alpha_1, \alpha_2, \dots, \alpha_s)$ denotes the smallest subfield in K containing F and the elements $\alpha_1, \alpha_2, \dots, \alpha_s$. We cannot state that any element from $F(\alpha_1, \alpha_2, \dots, \alpha_s)$ is representable in the form of polynomial in $\alpha_1, \alpha_2, \dots, \alpha_s$ with coefficients from F. For instance, the field $\mathbf{Q}(T)$ of rational fractions of the variable T with coefficients from $\mathbf{Q}$ is generated by the element T as a field, though it is clear that not every rational fraction is equal to a polynomial. The field $\mathbf{Q}(T)$ is an example of an infinite extension.

A necessary and sufficient condition for the finiteness of an extension

$$F(\alpha_1, \alpha_2, \dots, \alpha_s)$$

is the fact that every element $\alpha_1, \alpha_2, \dots, \alpha_s$ is a root of a suitable polynomial with coefficients from F. The necessity is evident from: if

$$\dim_F F(\alpha_1, \alpha_2, \dots, \alpha_s) = m, \quad \alpha \in F(\alpha_1, \alpha_2, \dots, \alpha_s),$$

then the system $1, \alpha, a^2, \dots, \alpha^m$ is linearly dependent over F, i.e. $q(\alpha) = 0$, where $q(X) = \lambda_0 + \lambda_1 X + \dots + \lambda_m X^m$, moreover, $\lambda_0, \lambda_1, \dots, \lambda_m$ are not all equal to zero. Inversely, let us assume that every α_i is a root of some polynomial $q_i(X) \in F[X]$. Applying induction on s and noting that the statement obviously holds for $s = 0$, we assume that the extension $F' = F(\alpha_1, \dots, \alpha_{s-1})$ is finite. Since $F[X] \subset F'[X]$, we can consider that $\alpha_s = \alpha$ is a root of a polynomial of degree m from $F'[X]$. Let $p(X)$ be an irreducible polynomial of degree n from $F'[X]$ with root α. Then the mapping $X \mapsto \alpha$ from $F'[X]$ into $F'(\alpha)$ extends to a surjective homomorphism ϕ of these rings. The kernel of this homomorphism is the ideal $\mathcal{I} = (p(X))$. Therefore $F'(\alpha) \cong F'[X]/\mathcal{I}$. As we have seen in the Introduction, § 2, in this case $F'(\alpha)$ has degree n over F'. Using the lemma from Subsection 1.1 gives us the finiteness of the field $F(\alpha_1, \dots, \alpha_s)$ over F.

As an example of a finite extension we can take a cyclotomic extension $\mathbf{Q}(\zeta)$ of the field of rational numbers, i.e. an extension obtained by adjoining a primitive n–th root ζ of unity to the field $\mathbf{Q}$. Let us present $\mathbf{Q}(\zeta)$ in the form $\mathbf{Q}[X]/(c_n(X))$, where $c_n(X)$ is an irreducible polynomial with leading coefficient 1. Since ζ is a root of the polynomial $X^n - 1$, $c_n(X)$ is a divisor of the polynomial $X^n - 1$. By factorizing $X^n - 1$ into irreducible factors with leading coefficient 1 over the field $\mathbf{Q}$, for every n–th root of 1 we obtain exactly one factor, of which it is a root. We show that all primitive n–th roots, and only they, are roots of the same irreducible integral polynomial

(1) $$c_n(X) = \prod_{\epsilon}(X - \epsilon),$$

where ϵ runs through all primitive n–th roots of 1. To see this, induction can be used taking into account that $c_1(X) = X - 1$.

Further we denote by $c'_n(X)$ polynomial on the right side of equality (1). We have (by induction) $c'_d(X) = c_d(X)$ for $d < n$. From the equality

$$X^n - 1 = \prod_{\epsilon}(X - \epsilon) = \prod_{d|n} c'_d(X) = c'_n(X) \prod_{\substack{d|n \\ d \neq n}} c_d(X)$$

(where ϵ runs through the set of all n–th roots of 1) it follows that $c'_n(X)$ is a polynomial with integral coefficients (for more details, see Chapter 1,§ 3, Gauss lemma) and $c_n(X)$ divides $c'_n(X)$. We shall prove that $c'_n(X)$ is irreducible over $\mathbf{Q}$, i.e. that $c'_n(X) = c_n(X)$. For this, take a prime number p, which is coprime with n. Then ζ^p together with ζ is a primitive n–th root of 1, and any primitive n-th root of 1 can be obtained from ζ by successive raising to the power p. Hence it is sufficient to prove that $c_n(\zeta^p) = 0$. Write

(2) $$X^n - 1 = c_n(X)a(X).$$

If $c_n(\zeta^p) \neq 0$ then $a(\zeta^p) = 0$. According to the already mentioned Gauss lemma, $a(X) \in \mathbf{Z}[X]$, and the equality (2) can be considered in $\mathbf{Z}[X]$. By replacing every coefficient λ with its remainder $\bar{\lambda}$ modulo p we get the equality

(2′) $$X^n - \bar{1} = \bar{c}_n(X)\bar{a}(X)$$

in $\mathbf{Z}_p[X]$. Since $(u + v)^p = u^p + v^p$ and $\lambda^p = \lambda$ in $\mathbf{Z}_p[X]$, where $\lambda \in \mathbf{Z}_p$, we have $\bar{a}(\zeta)^p = \bar{a}(\zeta^p) = 0$, i.e. ζ is a root of the polynomial $\bar{a}(X)$. But

$\bar{c}_n(\zeta) = 0$, i.e. ζ is a multiple root of $X^n - \bar{1}$. However the derivative $\bar{n}X^{n-1}$ of the polynomial $X^n - \bar{1}$ is coprime with it, and so we reach a contradiction. (It should be noted that in the final step we have carried out our computations in the field $\mathbf{Z}_p(\zeta)$ obtained by adjoining the root ζ of the polynomial $\bar{c}_n(X)$ to the field $\mathbf{Z}_p$. Formally we discussed the sequence of homomorphisms $\mathbf{Z}[X] \to \mathbf{Z}_p[X] \to \mathbf{Z}_p(\zeta)$.) So any primitive n–th root of unity is a root of the polynomial $c_n(X)$, i.e. its degree is equal to the quantity of numbers that are less than n and coprime with n. This is the *Euler function* $\phi(n)$. The following formula for $\phi(n)$ is well-known: if $n = p_1^{s_1}p_2^{s_2}\dots p_t^{s_t}$, where $p_1, p_2, \dots, p_t$ are pairwise different prime numbers, then

$$\phi(n) = n(1 - \frac{1}{p_1})(1 - \frac{1}{p_2})\dots(1 - \frac{1}{p_s}) = (p_1^{s_1} - p_1^{s_1-1})\dots(p_t^{s_t} - p_t^{s_t-1}).$$

Returning to the cyclotomic field $\mathbf{Q}(\zeta)$, we get

$$[\mathbf{Q}(\zeta) : \mathbf{Q}] = \phi(n).$$

The polynomial $c_n(X)$ is called the n–th cyclotomic polynomial (i.e. the polynomial of dividing the circle into n equal parts). We have

$$c_1(X) = X - 1, \;\; c_2(X) = X + 1, \;\; c_3(X) = X^2 + X + 1,$$
$$c_4(X) = X^2 + 1, \;\; c_5(X) = X^4 + X^3 + X^2 + X + 1, \;\; c_6(X) = X^2 - X + 1$$

1.3. Algebraic extensions. An element α of an extension K of a field F is called *algebraic* (over F) if there exists a non–zero polynomial $f(X) \in F[X]$ such that $f(\alpha) = 0$. An extension K of a field F is called *algebraic* if all its elements are algebraic.

Proposition. *A finitely generated extension $K = F(\alpha_1, \dots, \alpha_s)$ is algebraic iff the elements $\alpha_1, \dots, \alpha_s$ are algebraic. Moreover, if the given condition is fulfilled, K is finite.*

Proof. Actually in the first part of Subsection 1.2 it was shown that an element of a finite extension is algebraic. On the other hand, there was shown also that the extension generated by the finite number of algebraic elements is finite and, consequently, algebraic. Hence our proposition follows. □

Thus any finite extension is algebraic. It is clear that the opposite is not true. As an example we can take a collection $\bar{\mathbf{Q}}$ of all complex numbers that are algebraic over $\mathbf{Q}$. The set $\bar{\mathbf{Q}}$ is a field. Indeed, if $\alpha, \beta \in \bar{\mathbf{Q}}$ then $\alpha + \beta$, $\alpha\beta$ and $\alpha^{-1}\beta^{-1}$ lie in $\mathbf{Q}(\alpha, \beta)$. According to the proposition, the latter field is algebraic, i.e. the elements written down lie in $\bar{\mathbf{Q}}$. The field $\bar{\mathbf{Q}}$ is called the field of algebraic numbers. The degree $[\bar{\mathbf{Q}} : \bar{\mathbf{Q}}]$ is infinite, since, for instance, $\mathbf{Q}(\zeta) \subset \bar{\mathbf{Q}}$ and $[\mathbf{Q}(\zeta) : \mathbf{Q}] = p - 1$, where ζ is a primitive root of unity of prime degree p.

While studying algebraic extensions it is useful to know the following simple statement.

Lemma. *Let $F \subset K \subset L$ be a sequence of extensions of the field F. If the extensions $F \subset K$ and $K \subset L$ are algebraic then $F \subset L$ is also an algebraic extension.*

Proof. Consider an element $\xi \in L$. Then ξ is a root of non–zero polynomial $f(X)$ with coefficients from $K[X]$ of the form

$$f(X) = \alpha_0 + \alpha_1 X + \ldots + \alpha_m X^m,$$

where $\alpha_0, \alpha_1, \ldots, \alpha_m \in K$. Let us consider the subfield $K' \subset K$ generated by these elements. It is sufficient to prove that $K'(\xi)$ is a finite extension of the field F. However, $K'(\xi) = F(\alpha_1, \alpha_2, \ldots, \alpha_m, \xi)$, and our statement follows from proposition 1.3. □

From this lemma there follows an important property of the field $\bar{\mathbf{Q}}$ making it similar to the field $\mathbf{C}$ of complex numbers. Namely, any equation $f(X) = 0$, where $f(X)$ is a polynomial of non–zero degree with coefficients from $\bar{\mathbf{Q}}$, has a solution in $\bar{\mathbf{Q}}$. Indeed, if α is a complex root of such an equation then the extension $\bar{\mathbf{Q}} \subset \bar{\mathbf{Q}}(\alpha)$ is algebraic. So $\bar{\mathbf{Q}}(\alpha)$ is also algebraic; so α is algebraic over $\mathbf{Q}$, i.e. $\bar{\mathbf{Q}}(\alpha) = \bar{\mathbf{Q}}$ and $\alpha \in \bar{\mathbf{Q}}$.

1.4. Algebraically closed field. Let us assume that the field F is such that any polynomial of non–zero degree with coefficients from F has a root in F. In this case we say that F is an *algebraically closed field*. The fields $\bar{\mathbf{Q}}$ and $\mathbf{C}$ are examples of algebraically closed fields. An extension $\bar{F}$ of the field F is called an *algebraic closure* of the field F if it is algebraically closed and algebraic over F. For instance, $\bar{\mathbf{Q}}$ is an algebraic closure for $\mathbf{Q}$ and $\mathbf{C}$ is not an algebraic closure of $\mathbf{Q}$ (for that matter also not of $\bar{\mathbf{Q}}$).

In order to formulate the following important theorem we define an *isomorphism of extensions* K' and K'' of the field F as an isomorphism of fields: $\phi : K' \to K''$, such that $\phi(x) = x$ for any $x \in F$.

Theorem. *Every field F has an algebraic closure which is unique within isomorphism of extensions of F.*

Proof. If F is imbedded in any algebraically closed field K, then the set $\bar{F}$ of all algebraic elements of the field K is the sought for algebraic closure. The proof of this fact does not differ from the corresponding arguments from Subsection 1.3 in the case of the field $\mathbf{Q}$. Thus it is sufficient to imbed F in some algebraically closed field. This field is obtained by taking the union (more exactly, a direct limit, see Section 12 of the Set–theoretic supplement) of an increasing sequence of fields

$$F = K_0 \subset K_1 \subset \ldots \subset K_i \subset K_{i+1} \subset \ldots, \tag{3}$$

where any polynomial with coefficients from K_i has a root in K_{i+1}.

If such chain is constructed then any polynomial f with coefficients from the "union" $\widetilde{F}$ is in fact a polynomial with coefficients from K_i for a suitable $i \geq 0$. Its root in K_{i+1} lies also in $\widetilde{F}$, i.e. $\widetilde{F}$ is algebraically closed. In order to carry out the following step in the construction, we consider the ring of polynomials R over a field K in the set of variables X_f, where f is an irreducible polynomial of degree not less than 2 and with coefficients from K. Let $\mathcal{I}$ be the ideal in R generated by all polynomials of the form $f(X_f)$. If $\mathcal{I} = R$ then for some $f_1, \ldots, f_m$

$$1 = f_1(X_{f_1})g_1 + \ldots + f_m(X_{f_m})g_m. \tag{4}$$

Let K' be a finite extension of the field K where all $f_1, \ldots, f_m$ have roots $\alpha_1, \ldots, \alpha_m$ (see, Subsection 1.2). By applying the homomorphism ϕ from R to K' under which X_{f_i} goes to α_i, $i = 1, \ldots, m$, to both sides of the equation (4) we obtain $1 = 0$. Thus $\mathcal{I} \neq R$ and there exist a maximal ideal M in R different from R, such that $M \supset \mathcal{I}$. The quotient ring $L = R/M$ is a field. This field in a natural way contains the field K (see, Subsection 1.2). It is clear that the element (coset) $X_f + M$ is a root of the polynomial f. The pair $K \subset L$ can be taken as an elementary step of the chain (3).

$*$ Some remarks on the character of the "unions" in (3). Let $\gamma_i : K_i \to K_{i+1}$, $i = 1, 2, 3, \ldots$, be the mapping in (3). We shall consider the set P of all the sequences $\{x_i\}$, $i = 0, 1, 2, 3, \ldots$, such that $x_i \in K_i$, and such that there exists $n \geq 0$ such that for any $m \geq n$ we have $\gamma_m(x_m) = x_{m+1}$. Two sequences from P are called *equivalent* if they coincide up to a finite subsequence. $\widetilde{F}$ itself is the set of classes of equivalent sequences. Operations on classes are defined with a help of representatives. A simple check shows that the operations are well defined. Each member from the chain (3) lies in $\widetilde{F}$ and is imbedded in every following member of the chain: if $x \in K_i$ then x is identified with the sequence $\{x_n\}$, where $x_0 = x_1 = \ldots = x_{i-1} = 0$, $x_i = x$, $x_{n+1} = \gamma_n(x_n)$, $n \geq i$. $*$

Let C_1 and C_2 be two algebraic closures of the field F. It is useful to note the following fact which plays an important role in the proof of the uniqueness of algebraic closure.

Lemma. *Let K be an algebraic extension of the field F, Q an algebraically closed field. Then any imbedding $\phi : F \to Q$ extends to an imbedding $\bar{\phi} : K \to Q$ for which $\bar{\phi}|_F = \phi$.*

Proof. We shall use the Zorn lemma. Let Π be the partially ordered set of pairs (E, ψ), where E is a subfield in K containing F, and ψ its imbedding in Q over F, i.e. $\psi|_F = \phi$. We put that $(E, \psi) \leq (E', \psi')$ if $E \subset E'$ and $\psi'|_E = \psi$. The set Π is non–empty as the pair (F, ϕ) is an element. Let Σ be a chain in Π, $S = \bigcup_{E \in \Sigma} E$ the union of the members of the chain. We define the imbedding $\chi : S \to Q$ by setting that $\chi(x) = \psi(x)$, if $x \in E$, where E is the first component of a pair (E, ψ) from the chain Σ. The definition is correct because Σ is a chain. We thus see that for any chain of the set Π there exists a maximal element for the chain. According to the Zorn lemma a maximal element also exists in Π. Let the first component of this element be the subfield H. Assume that H does not coincide with K and let the imbedding of the field H in Q over F: $\tau : H \to Q$ be the second component of the maximal element under consideration. Let us select an element $\alpha \in K$ which does not lie in H. Using a polynomial of minimal degree from $H[X]$ a root of which is α, we see that α is a root of an irreducible polynomial $p(X)$ with coefficients from H. We shall regard the polynomial $p^{\tau}(X)$ with coefficients from Q which are equal to the images of coefficients of the polynomial $p(X)$ under the imbedding τ. From the algebraic closure property of the field Q follows that there exists an element ζ such that $p^{\tau}(\zeta) = 0$. Any element x from $H' = H(\alpha)$ can be presented uniquely in the form $x = c_0 + c_1\alpha + \ldots + c_{m-1}\alpha^{m-1}$, $c_i \in H$, where $m = \deg p(X)$. We define $\tau' : H' \to Q$, by setting $\tau'(x) = \tau(c_0) + \tau(c_1)\zeta + \ldots + \tau(c_{m-1})\zeta^{m-1}$. It is clear that τ' is an imbedding such that $\tau'|_H = \tau$. Since $\tau|_F = \phi$, we see that $(H', \tau') \in \Pi$ and $(H, \tau) \leq (H', \tau')$ and this contradicts the maximality of the pair (H, τ) in Π. □

An application of the lemma above gives an imbedding $\lambda : C_1 \to C_2$ over F. It image $\lambda(C_1)$ is an algebraically closed subfield in C_2. However C_2 is an algebraic extension of the field F, so also of the field $\lambda(C_1)$. Since C_2 is algebraically closed, we have $\lambda(C_1) = C_2$. So $\lambda : C_1 \to C_2$ is the desired isomorphism, and the uniqueness of the algebraic closure of a field is proved. □

1.5. Transcendental extensions. An example of a non–algebraic extension is the field $F(T)$ of rational fractions (rational functions) of one variable T with coefficients from a field F. Here the algebraic elements over F are only those from the field F. Such extensions are called *purely transcendental.* Generally, an element α of an extension K of the field F is called *transcendental* over F if it is not algebraic over F. Further, a set $\{\alpha_1, \ldots, \alpha_n\}$ from K is *algebraically independent* if any polynomial $f(X_1, \ldots, X_n)$ with coefficients from F vanishing under substitution of $\alpha_1, \ldots, \alpha_n$ for $X_1, \ldots, X_n$, is identically the zero polynomial. If S is some (not necessary finite) subset in K then we call it *algebraically independent* over F under the condition that any finite subset of S is algebraically independent. By the Zorn lemma any extension contains a maximal algebraically independent subset B. Any such subset is called a *transcendence basis* of the field K over the subfield F (or of the extension $F \subset K$). Let $F(B)$ be the subfield in K generated by the set B. Then K is an algebraic extension of the field $F(B)$. Indeed, if $a \in K$ then the set $\{B, a\}$ is algebraically dependent, i.e. for some finite subset, necessarily of the form $\{b_1, \ldots, b_n, a\}$, there exists a non–zero polynomial $f(X_1, \ldots, X_n, X_{n+1})$ with coefficients from F such that $f(b_1, \ldots, b_n, a) = 0$. From the last equality follows that a is a root of a polynomial with coefficients from $F(b_1, \ldots, b_n) \subset F(B)$ which proves our statement.

Theorem. *The cardinalities of any two transcendence bases are the same.*

Proof. Let B and B' be the two transcendence bases for K over F. We shall prove the theorem for the case that one of the bases, say B, is finite. Also we shall replace the elements of one basis by the elements of another. Let $B = \{b_1, b_2, \dots, b_n\}$. We consider an arbitrary element b'_1 from B'. Then the set $\{B, b'\}$ is algebraically dependent. Let us consider a non–zero polynomial $f(X_1, X_2, \dots, X_n, X_{n+1})$ such that

$$f_1(b_1, b_2, \dots, b_n, b'_1) = 0 \tag{5}$$

Then this polynomial non–trivially contains at least one element form the set B, say b_1. In this case the set $B_1 = \{b'_1, b_2, \dots, b_n\}$ is a basis of transcendence for K because the equality (5) implies that b_1 is algebraic over $F(B_1)$. Now let us take $b'_2 \in B'$ different from b'_1. The set $B_1 = \{B_1, b'_2\}$ is algebraically dependent, i.e. there exists a non–zero polynomial $f_2(X_1, X_2, \dots, X_n, X_{n+1})$ such that $f_2(b'_1, b_2, \dots, b_n, b'_2) = 0$. Because of the algebraic independence of the set $\{b'_1, b'_2\}$ the latter equality non–trivially contains some element from the set $\{b_2, \dots, b_n\}$, say b_2. In this case $B_2 = \{b'_1, b'_2, b_3, \dots, b_n\}$ is a basis of transcendence for K over F. By continuing this procedure we get one of the two possibilities: either $|B'| \le |B|$ or B' is algebraic over a subset itself, $B'_n = \{b'_1, \dots, b'_n\}$. As the latter is impossible, we get that $|B'| \le |B|$. By replacing B with B', we get $|B| \le |B'|$. Thus $|B'| = |B|$, which proves the theorem in the considered case. □

The cardinality of any transcendence basis of the field K over a subfield F is called the degree of transcendence and is denoted $\operatorname{tr deg}_F K$. From the theorems in this section it follows that any extension $F \subset K$ can be represented in the form $F \subset K' \subset K$, where $F \subset K'$ is purely transcendent and $K' \subset K$ is an algebraic extension, where $K' = F(B)$, B being any basis of transcendence for K over F.

∗ The transcendence of certain complex numbers over the field of rational numbers as a rule is difficult to prove. It is known that the numbers e, π, are transcedent, as are any numbers of the form α^β, where α is algebraic different from $0, 1$, and β is algebraic irrational number (see, Feldman [1]).

The structure of purely transcendent extensions is not nearly so simple as it might appear at first glance. For instance, almost for a century the following problem (Luroth problem) remained open: let K be a subextension of degree m of the field of rational functions $F(T_1, \dots, T_n)$; is it true that there exist algebraically independent $u_1, \dots, u_m$ such that $K = F(u_1, \dots, u_m)$. The positive answer for $n = 1, 2$ was obtained in the seventies and eighties of the last century and the negative answer for $n \geq 3$ only in the beginning of the seventies of this century. *

§ 2. Galois theory

For many important cases the structure of finite extensions is completely described in terms of finite groups. This description allows us to solve many problems of classical mathematics. To that kind of problems also belong such problems as the angle trisection (it is required to divide any angle in three equal parts with a help of a rule and divider), the doubling of a cube (with a help of the same means it is required to construct an edge of the cube of double volume compared with the volume of a cube with the given edge). Though the most important application (and also the most stimulating motivation) is the problem of solvability of equations in radicals. The problem asks whether it is possible to write down the roots of an algebraic equation starting from its coefficients by using only the four arithmetic operations and taking roots of an arbitrary natural degree. Classical formulas for solving equations give positive answer to this problem only in the case when the degree of an equation is less or equal to 4. More than three hundred years of attempts to find formulas for solving equations of higher degree had passed until N. H. Abel found an example of the fifth–degree equation with rational coefficients which is unsolvable in radicals. The true cause of this phenomenon was found by E. Galois by associating a finite group with every equation. Galois theory states: an equation is solvable in radicals iff its group (the so–called Galois group) is solvable, i.e. possesses a finite sequence of normal subgroups for which the sequential factors are abelian. This solution in principle, and also the simple remark that the symmetric group S_n, $n \geq 5$, is unsolvable reduces the problem to the one of constructing an equation which is unsolvable in radicals to the technical (but not at all simple!) problem of constructing an equation, the group of which is isomorphic to the group S_5. The "simplest" example of such equation is the equation $X^5 - X - 1 = 0$ (see, Skornyakov [1]).

2.1. Galois extension. As we know from the lemma of Subsection 1.4, any imbedding of a field F into its algebraic closure $\bar{F}$ extends to an imbedding into $\bar{F}$ of any algebraic extension K of this field. If K is generated by a root α of a polynomial f of degree n, which is irreducible over F, then one tends to think that the number of different imbeddings is equal to n. Indeed, for each imbedding of this kind the image of the element α can be set equal to any root of the polynomial f in $\bar{F}$. If f is a field of characterictic zero, $n = \deg f > 0$, then $f' \neq 0$, $(f, f') = 1$ and f has no multiple roots. However, if char $F = p > 0$ then it is possible that $f' = 0$, and it can occur that an irreducible polynomial of degree n has less than n roots in $\bar{F}$ and the number of imbeddings of the field K in $\bar{F}$ can turn out to be less than n. By picking out the "best" case we state that an extension $F \subset K$ of degree n is *separable* if the number of different imbeddings K in $\bar{F}$ leaving the field F in place, is equal to n.

$*$ An example of a non–separable extension can be obtained by taking for F the field $\mathbf{Z}_2(T)$ of rational fractions of a variable T over the field of residues modulo 2. For K one can take the extension of degree 2 over F generated by the element $\sqrt{T}$. Its minimal polynomial is equal to $X^2 + T$. Of course, this polynomial has no other roots, so the extension $K = F(\sqrt{T})$ is non–separable. The given extension is purely non–separable in the sense that that there is only one imbedding in $\bar{F}$. $*$

A natural way of constructing extensions is by adjoining roots of irreducible polynomials. One should not believe that by adjoining one of the roots of irreducible polynomial the others are also adjoined. For instance, adjoining the fourth power root of 2 to the field $\mathbf{Q}$ of rational numbers we get a field consisting entirely of real numbers, but the polynomial $X^4 - 2$ has two purely imaginary roots which, therefore, do not occur in $\mathbf{Q}\sqrt[4]{2}$. The case when together with any root of irreducible polynomial f over F in the extension K of the field F any other root of this polynomial is also in K, is specially noted by introducing the concept of a normal extension. Thus a *normal* extension is given by the set of all roots of some system of polynomials. In the case of a finite extension this is equivalent to the fact that the extension is generated by a set of all roots of a finite system of polynomials. Taking a product of the polynomials of this system we obtain that the extension is generated by the set of roots of one unique polynomial $f(X)$. We also say that a *splitting field* of the polynomial $f(X)$ is given. If a normal extension K of the field F in some way is imbedded into $\bar{F}$ over F then any other imbedding of the field K in $\bar{F}$ takes it into itself, i.e. is an automorphism of the extension K.

The last observation brings us to the definition of Galois group Gal (K/F) of an extension $F \subset K$, as a group of automorphisms of the field K which leave elements from F fixed. We should note that if the extension is purely non–separable then its Galois group consists of the identity only, and in the case of non–normal extension it has but a weak connection with the extensions of the field K in the algebraic closure $\bar{F}$ of the base field F. The Galois group is most closely connected with the structure of an extension when the extension is normal and separable. Extensions with these two properties are called *Galois extensions.*

2.2. Basic theorem of Galois theory. As preparation some results of independent interest are needed.

Theorem (on primitive elements). *Any finite separable extension can be generated by one element.*

Proof. Let K be such an extension of the field F. Because of its finiteness and the lemma of Subsection 1.1, there are $\alpha_1, \alpha_2, \ldots, \alpha_s$ such that $K = F(\alpha_1, \alpha_2, \ldots, \alpha_s)$. Let us carry out induction on s starting with the obvious case $s = 1$. For $s > 1$ and because of the induction assumption we have $K = F(\alpha, \beta)$, where $F(\alpha) = F(\alpha_1, \ldots, \alpha_{s-1})$ and $\beta = \alpha_s$. We shall prove that if F is infinite a number $\theta \in F$ for which $K = F(\alpha + \theta\beta)$, can be found. For this purpose we consider all imbeddings $\sigma_1, \ldots, \sigma_m$ of the field K in the algebraic closure $\bar{F}$ of the field F. Because of separability of the extension the number of imbeddings is equal to the degree of the extension $F \subset K$. In the product

$$f(X) = \prod_{i \neq j} (\sigma_i(\alpha) + X\sigma_i(\beta) - \sigma_j(\alpha) - X\sigma_j(\beta))$$

every factor is a non–zero polynomial of degree 0 or 1. Therefore $f(X)$ is a non–zero polynomial. If the field F is infinite then there can be found $\theta \in F$ such that $f(\theta) \neq 0$. In this case all elements $\sigma_i(\alpha+\theta\beta)$, $i = 1, \ldots, m$, differ and the degree of the field $F(\alpha + \theta\beta)$ is not less than m, i.e. $F(\alpha, \beta) = F(\alpha + \theta\beta)$. The case of a finite field F is based on another idea. Since a finite extension of a finite field is also finite we have only to apply the fact (justified in

Subsection 2.3 below) that the multiplicative group K^* of a finite field is cyclic. A generating element of this group can be taken as the sought for element which generates the field K over F. □

A generating element such as in the previous theorem is called a primitive element.

Basic theorem. *Let K be an extension of Galois field F, G the group of this extension. For any subgroup H of the group G we denote by K^H the subfield in K consisting of the elements that are fixed under automorphisms from H. Then the mapping $\phi : H \mapsto K^H$ is a bijection of the set of subgroups in G to the set of subextensions of $F \subset K$.*

Proof. It is necessary to prove that from $H_1 \neq H_2$ follows $K^{H_1} \neq K^{H_2}$ and that for any field L, $F \subset L \subset K$, there exists a subgroup H such that $L = K^H$. This can be shown by defining an inverse mapping $\psi : L \mapsto S(L)$ from the set of subextensions into the set of subgroups, by setting $S(L) = \{g \in G \,|\, g(x) = x\}$ for any $x \in L$. We should note that if $H_1 \subset H_2$ then $K^{H_1} \supset K^{H_2}$, but if $L_1 \subset L_2$ then $S(L_1) \supset S(L_2)$. This implies $K^{S(L)} \supset L$ and $S(K^H) \supset H$. Let us assume that $H_1 = S(K^H)$ does not coincide with H for some subgroup H. Then $H_1 \supset H$ and $K^{H_1} \subset K^H$ according to what was mentioned above. However $K^H \subset K$ is a Galois extension with Galois group H. In this case $\dim_{K^H} K = |H|$. The latter equality follows from normality and the theorem on primitive elements. Thus, $|H_1| > |H_2|$, and $\dim_{K^{H_1}} K \leq \dim_{K^H} K$ which contradicts the last equality. So $H_1 = H$, i.e. $\psi\phi = 1$.

Analogously, let $L_1 = K^{S(L)}$. As before $S(L_1) \supset S(L)$. If $L_1 \supset L$ then as $S(L)$ is a Galois group for $K \supset L$, we have $\dim_L K = |S(L)|$. Further

$$|S(L)| = \dim_L K > \dim_{L_1} K = |S(L_1)|,$$

i.e. $|S(L_1)| < |S(L_2)|$, a contradiciton. So also $\phi\psi = 1$. Hence ϕ, ψ are mutualy inverse mappings. □

$*$ As we have seen, H is the Galois group of the extension $K \supset K^H$, and $S(L)$ is the Galois group of $K \supset L$. If H is a normal subgroup in G then $K^H \supset F$ is a normal extension of the field F (consequently, a Galois extension) and $\mathrm{Gal}(K^H/F) \cong G/H$. Inversily, if L/F is a Galois extension then $S(L)$ is normal in G and $\mathrm{Gal}(L/F) \cong G/S(L)$. Indeed, if τ is some imbedding $K^H \to \bar{F}$ then by taking an extension τ' of the automoprhism τ to an imbedding of the field K in $\bar{F}$, $x \in K^H$, $\nu \in H$, we obtain

$$\nu\tau(x) = \tau'\tau'^{-1}\nu\tau'(x) = \tau'(x) = \tau(x),$$

because $\tau'^{-1}\nu\tau' \in H$. Therefore $\tau(x) \in K^H$, i.e. $K^H \supset F$ is a normal extension. If $\mu \in \mathrm{Gal}\,(K/F)$, then all elements from the class μ^H act on K^H/F in the same way. If $\mu_1\mu_2^{-1} \notin H$ then $\mu_1\mu_2^{-1}(x) \neq x$ for some $x \in K^H$. So by defining $(\mu H)(x) = \mu(x)$ we obtain an isomorphism of the group G/H with the group $\mathrm{Gal}\,(K^H/F)$. $*$

Example. Let **Q** be the field of rational numbers, ζ a primitive n–th root of unity. We shall show that $\mathbf{Q}(\zeta)$ is a Galois extension with an abelian group of automorphisms. Indeed since the characteristic of the basis field is equal to zero, the extension is separable. Further, any root of the equation $X^n - 1 = 0$ is a power of a fixed primitive root and from that it is clear that $\mathbf{Q}(\zeta)$ is a splitting field of $X^n - 1$. Therefore $\mathbf{Q}(\zeta)$ is normal, i.e. it really is a Galois extension of the field **Q**. If p is an arbitrary number coprime with n and ϕ_p is the automorphism of the field $\mathbf{Q}(\zeta)$ extending the mapping $\zeta \mapsto \zeta^p$, then the powers of suitable automorphisms ϕ_p for various p transfers ζ to any given primitive n–th root of 1. So the products of these automorphisms exhaust the Galois group $\mathrm{Gal}\,(\mathbf{Q}(\zeta)/\mathbf{Q})$. Since they commute, the group is abelian.

To conclude this section we present a "group–theoretic" proof of the basic theorem of algebra. It is a consequence of the basic theorem of Galois theory. The ony "non–algebraic" part in the theorem is the statement about the existence of a real root of a real polynomial of odd degree.

Corollary. *The field* **C** *of complex numbers is algebraically closed.*

Proof. The field **C** is an extension of degree 2 over the field **R**. Let α be a root of an irreducible polynomial with complex coefficients. If $[\mathbf{C}(\alpha) : \mathbf{C}] \leq 2$ then the formulas for the solution of a quadratic equation show that $\alpha \in \mathbf{C}$.[1] This points out that no finite extension of the field **C** has a subextension of degree 2. Let Q be a finite extension of the field **C**. We consider it as an

[1]One checks that there is a $p+iq \in \mathbf{C}$ such that $(p+iq)^2 = a+bi$ (using the formulas for solving quadratic equations).

extension of the field $\mathbf{R}$ and imbed it into a finite Galois extension $P \supset \mathbf{R}$. Let $G = G(P/\mathbf{R})$ be Galois group of this extension. According to the theorem on primitive elements $P = \mathbf{R}(\beta)$, where β is a root of an irreducible polynomial over $\mathbf{R}$ of degree equal to the order of the group $|G|$. If $|G|$ is odd then β is a root of an irreducible polynomial over $\mathbf{R}$ of odd degree. From the remark above the formulation of the corollary it follows that we can take $\beta \in \mathbf{R}$, i.e. $|G| = 1$, which cannot occur.

Thus $|G|$ is even. Let H be the Sylow 2–subgroup in G (see, Introduction 1.9). Then according to the basic theorem $\dim_{\mathbf{R}} P^H = [G : H]$ is an odd number. From that $P^H = \mathbf{R}$, i.e. $H = \{G\}$ and $|G| = [P : \mathbf{R}] = 2^t$ for suitable t. The field P is a Galois extension of the field $\mathbf{C}$ of degree equal to 2^u, $u = t - 1$. Let $\widetilde{G} = \mathrm{Gal}\,(P/\mathbf{C})$. In the group $\widetilde{G}$ of order 2^u, if $u > 0$ there exists a subgroup K of order 2^{u-1} (this is seen using induction from the theorem on non–triviality of the center of finite p–group (see, Introduction 1.8)). We then have $[P^K : \mathbf{C}] = 2$, but it is impossible. Therefore, $|\widetilde{G}| = 1$, and P and also Q coincide with $\mathbf{C}$, i.e. $\mathbf{C}$ is algebraically closed. □

2.3. Finite fields. Let us fix a prime number p. We denote by $F(p)$ the field of residues $\mathbf{Z}_p$ modulo p and by $F(p^\infty)$ the algebraic closure of this field of residues. Any finite subfield K in $F(p^\infty)$ is a vector space of a certain dimension n over $F(p)$, i.e. it consists of $q = p^n$ elements. Its multiplicative group K^* is a finite abelian group. According to the basic theorem on finite abelian groups, there exists in K^* an element of greatest order m, i.e. of an order that is divisible by the order of any other element. In other words, the elements of the group K^* satisfy the equation $X^m = 1$. This equation cannot have more than m solutions, i.e. $m = |K^*|$ and, hence, K^* is the cyclic group generated by the element a. It is evident that K coincides with the set of solutions of the equation $X^{p^n} - X = 0$, which we denote by $F(p^n)$. In the opposite direction, for any n the set $F(p^n)$ is a subfield in $F(p^\infty)$; indeed, if $u, v \in F(p^n)$, then

$$(u+v)^{p^n} = u^{p^n} + v^{p^n} = u + v, \quad (uv)^{p^n} = u^{p^n} v^{p^n} = uv,$$
$$(u^{-1})^{p^n} = (u^{p^n})^{-1} = u^{-1}.$$

Since the polynomial $X^{p^n} - X$ has no multiple roots, the field $F(p^n)$ consists of p^n elements. So any finite subfield in $F(p^n)$ is a subfield of the form $F(p^n) = \{u | u^{p^n} = u\}$. Further, let $m \leq n$ and $n = ml + r$. If $v \in F(p^m)$ then

$$v^{p^n} = ((\ldots((v^{p^m})^{p^m})\ldots)^{p^m})^{p^r} = v^{p^r}.$$

It is clear that if $r = 0$, i.e. m divides n, then $F(p^m) \subset F(p^n)$. If $r \neq 0$ and v is a generator of the group of non–zero elements in $F(p^m)$, then $v^{p^r} \neq v$. Thus, $F(p^m) \not\subset F(p^n)$.

So $F(p^n)$ is the splitting field of a separable polynomial $X^{p^n} - X$, i.e. it is a Galois extension of the field $F(p)$. According to the basic theorem of Galois theory (see, Subsection 2.2) the lattice of subgroups of the Galois group G is anti–isomorphic to the lattice of subfields in $F(p^n)$ or by the preceding, to the lattice of natural divisors of a number n. Moreover, G contains the cyclic subgroup $\langle\phi\rangle$ generated by the Frobenius automorphism $\phi : u \longrightarrow u^p$, $u \in F(p^n)$. The lattice of subgroups in $\langle\phi\rangle$ is also antiisomorphic to the lattice of the divisors of a number n. From this, naturally, $G = \langle\phi\rangle$ is a cyclic group of order n.

If L is an arbitrary finite field, p its characteristic, then L contains a subfield isomorphic to $F(p)$ (see, Introduction 2). Let us imbed L into $F(p^\infty)$. The previous arguments show that the following theorem holds.

Theorem. *Any finite field is isomorphic to a field of the form $F(p^n)$. The field $F(p^m)$ is isomorphic to a subfield of $F(p^n)$ iff m divides n. The group of automorphisms of the field $F(p^n)$ over $F(p)$ is a cyclic group of order n generated by the Frobenius automorphism $\phi : u \mapsto u^p$, where p^n is the order of the field $F(p^n)$.* □

Commutativity of a finite field. In connection with the description of finite fields it is natural to ask about the possibility of reconstructing this result for the case of skew fields. We shall demonstrate that the "problem of skew finite fields" does not exist. So the following holds.

Wedderburn theorem. *Any finite field is commutative.*

Proof. Let K be some finite skew field consisting of s elements. We denote by Z the center of the field K. It is clear that Z is a subfield of K; indeed, if $0 \neq z \in Z$, $a \in K$,

$$z^{-1}a = (a^{-1}z)^{-1} = (za^{-1})^{-1} = az^{-1},$$

i.e. also $z^{-1} \in Z$. Let p be the characteristic of the field Z, $q = p^t$ the number of elements of the center Z. The field K is a vector space over Z; let $n = \dim_Z K$. The principal ingredient in the proof is the coset formula for the group K^* of non–zero elements of the field K (see, Introduction 1.7). We note that for any $a \in K$ the centralizer C_a is a subfield in K containing Z. If $a \in K^* \setminus Z$ then $C_a \neq K$, therefore the number of elements conjugate with a in K^* is equal to $(q^n - 1)/(q^d - 1)$, where $d = \dim_Z C_a$; moreover $d|n$. On the other hand the number of elements in Z^* is equal to $q - 1$.

The result is

$$(1) \qquad q^n - 1 = (q-1) + \sum_d \frac{q^n - 1}{q^d - 1}.$$

In this formula the summation extends over proper divisors of the number n; also equal summands can occur. Our aim is to show that the relation (1) cannot be fulfilled for any $n > 1$. To prove this properties of cyclotomic polynomials, see Subsection 1.2, will be applied. Among these properties there is the fact that each polynomial $c_m(X)$, the product of the $(X - \xi_m)$, where ξ_m runs over the primitive m–th roots of unity, is integer, and moreover,

$$(2) \qquad X^d - 1 = \prod_{\delta | d} c_\delta(X).$$

Thus the fraction $(X^n - 1)/(X^d - 1)$, where $d|n$, also is an integer polynomial and if $d < n$ this polynomial is divisible by $c_n(X)$ without remainder. Changing X to the integer q we get that $c_n(q)$ is an integer dividing $q^n - 1$ and $(q^n - 1)/(q^d - 1)$ for any proper divisor d of the number n. Applying this to formula (2) we get that $c_n(q)$ is a divisor of the number $q - 1$. In order to obtain a contradiction we estimate the value of the number $c_n(q)$. The number under consideration is a product of numbers of the form $q - \epsilon$, where ϵ is a primitive n–th root of unity. From the point of view of the complex plane the modulus of the number $q - \epsilon$ is the distance between the real point $q \geq 2$ and the point ϵ on the unity circle. It is clear that $|q - \epsilon| \geq q - 1$; moreover, equality is possible only when $\epsilon = 1$. However this possibility is unrealizable if $n > 1$. Therefore, $|c_n(q)| = \prod_\epsilon |q - \epsilon| > (q-1)^n$, and a contradiction is obtained.

Thus we have $n = 1$, i.e. that $K = Z$, so that the field K is commutative. □

∗ There exist also other conditions under which a field is commutative. There is the well known Jacobson theorem according to which a field K for which K^* is a periodic group is commutative. ∗

2.4. Solvability in radicals. Let F be a field of characteristic zero, $f(X)$ a polynomial with coefficients from F. We denote by K a splitting field of $f(X)$. Since the condition of separability for fields of zero characteristic is automatically fulfilled, K is a Galois extension of the field F. We denote by G the group Gal (K/F). Our aim is to prove that G is a solvable group if the equation $f(X) = 0$ is solvable in radicals, i.e. that G is then a sequential extension of abelian groups.

As was mentioned above, the solutions of the equation under consideration are obtainable with the help of the four arithmetic operations and of taking roots of natural degree. Indeed, there can be chosen one natural number m so that it is possible to take only m–th roots. Suppose L is obtained from F by adjoining a primitive m–th root of unity ζ. Then as follows from the arguments in Subsection 2.2, the group Gal (L/F) is abelian. We denote by M the field of factorization (splitting feld) of $f(X)$ and $X^m - 1$. Let $\bar{G} =$ Gal (M/F). In the virtue of the remark from Subsection 2.2 the group G is a quotient group of the group $\bar{G}$. On the other hand, $\bar{G}$ is an extension of the group Gal (L/F) with kernel Gal (M/L). So in order to prove the solvability of the group G it is sufficient to show the solvability of the group Gal (M/L). However the extension $M \supset L$ can be presented as a sequential extension (tower), the levels of which are obtained by adjoining a root of an equation of the form $X^m - \beta$, where β lies in the previous level. Let $M \supset M' \supset L' \supset L$ be such an elementary level extension, i.e. $M' = L'(\alpha)$, where $\alpha^m = \beta$, $\beta \in L'$. Note that M' contains all m–th roots $\xi_1, \ldots, \xi_m$ of unity. Since $(\alpha\xi_i)^m = \alpha^m = \beta$, it is evident that M' contains all roots of the equation under consideration. Thus M' is a field of factorization of the equation $X^m - \beta = 0$ and is a Galois extension. Let $\sigma \in$ Gal (M'/L'). Then $\sigma(\alpha)$ is a root of the equation $X^m - \beta = 0$. Therefore $\sigma(\alpha)\alpha^{-1}$ is an m–th root of unity. The mapping $\sigma \mapsto \sigma(\alpha)\alpha^{-1}$ is an injective homomorphism from G to the cyclic group of m–th roots of unity. So Gal (M'/L') is a cyclic group and Gal (M/L) is solvable.

In order to find an example of an equation with unsolvable Galois group we consider the field $K = F(T_1, T_2, \ldots, T_n)$ of rational functions over the field F of n algebraically independent elements $T_1, T_2, \ldots, T_n$. The group $G = S_n$ acts on K by the permutation of these elements. Of course, the elementary symmetric polynomials $s_1, s_2, \ldots, s_n$ lie in the fixed field K^G. In particular, $F(s_1, s_2, \ldots, s_n) \subset K^G$. However K is a field of factorization of a polynomial of degree n and which is of the form

$$f(X) = (X - T_1)(X - T_2)\ldots(X - T_n)$$

with coefficients from $F(s_1, s_2, \dots, s_n)$.
So the degree of the field K over $F(s_1, s_2, \dots, s_n)$ cannot exceed $n!$. As S_n acts faithfully on K, the degree of the field K over K^G is equal to $|G| = n!$. Thus $K^G = F(s_1, s_2, \dots, s_n)$, and

$$\mathrm{Gal}(F(T_1, \dots, T_n)/F(s_1, s_2, \dots, s_n)) \cong S_n.$$

In particular, the equation $f(X) = 0$ with coefficients from $F(s_1, s_2, \dots, s_n)$ is unsolvable in radicals if $n \geq 5$. We should mention that if $n < 5$ solvability occurs, and when $n = 2$, we have

$$T_1 = \frac{s_1 + \sqrt{s_1^2 - 4s_2}}{2}, \quad T_2 = \frac{s_1 - \sqrt{s_1^2 - 4s_2}}{2}.$$

It is more complicated to construct an example of an equation with symmetric Galois group over the field of rational numbers (see, Skornyakov [1]). And it is truely difficult to construct an extension of the field of rational numbers with given Galois group. This problem, the so called *inverse problem of Galois theory* is not completely solved so far.

§ 3. Affine rings

3.1. Rings of polynomials. The main objects of discussion in this section are finitely generated commutative rings and algebras, also called *affine* rings and algebras. Every ring of this kind is a homomorphic image of an integer ring of polynomials $\mathbf{Z}[X_1, \dots, X_n]$ in some finite number of variables. Respectively, a finitely generated algebra over a field F is a quotient algebra of a ring of polynomials $F[X_1, \dots, X_n]$ with coefficients from F. It is clear that the structure of rings of polynomials is important in the analysis of affine rings.

We call a commutative ring R *noetherian* if any of its ideals $\mathcal{I}$ can be generated by a finite number of elements, i.e. if fgor an ideal $\mathcal{I}$ there exist $e_1, e_2, \dots, e_m \in R$ such that any element x from $\mathcal{I}$ is representable in the form $x = r_1e_1 + r_2e_2 + \dots + r_me_m$, where $r_1, r_2, \dots, r_m$ are certain elements from R. An equivalent condition is that any increasing chain of ideals $\mathcal{I}_1 \subset \mathcal{I}_2 \subset \dots \subset \mathcal{I}_k \subset \mathcal{I}_{k+1} \subset \dots$ of the ring R stabilizes at some finite stage, i.e. for some k we have $\mathcal{I}_k = \mathcal{I}_{k+1} = \dots$.

Indeed, if the condition of stabilization is not satisfied for some chain $\mathcal{I}_1 \subset \mathcal{I}_2 \subset \dots \subset \mathcal{I}_k \subset \mathcal{I}_{k+1} \subset \dots$, then the ideal $\mathcal{I}$ which is equal to the union of all members from the chain, cannot be finitely generated. Otherwise its (finite) system of generators would lie in some member from the chain with finite index, for instance k. Then $\mathcal{I} = \mathcal{I}_k$ and consequently $\mathcal{I}_k = \mathcal{I}_{k+1} = \dots$, which is a contradiction.

Inversily, assume that the condition of finite generation of ideals is not satisfied for some ideal $\mathcal{I}$. Then, for any finite set M_n of elements $\{x_1, \dots, x_n\} \subset \mathcal{I}$ there is an element $x_{n+1} \in \mathcal{I}$ that does not lie in the ideal generated by this set. Let $M_{n+1} = M_n \cup \{x_{n+1}\}$. If $\mathcal{I}_n$ is the ideal generated by the set M_n, then all inclusions in the chain

$$\{0\} = \mathcal{I}_0 \subset \mathcal{I}_1 \subset \dots \subset \mathcal{I}_k \subset \mathcal{I}_{k+1} \dots$$

are proper, i.e. stabilization does not occur.

The basic result about noetherian rings is the following.

Hilbert basis theorem. *Let R be a noetherian commutative ring, $R_1 = R[X]$ the ring of polynomials in one variable over R. Then R_1 is a noetherian ring.*

Proof. Let $\mathcal{I}_1$ be an ideal in R_1. We shall consider the subgroup $\mathcal{I}$ in R generated by the highest non–zero coefficients of polynomials from $\mathcal{I}_1$. Then $\mathcal{I}$ is an ideal in R. Indeed, if $r \in R$, $s \in \mathcal{I}$, then R_1 contains a polynomial $f = f(X) = sX^n + s'X^{n-1} + \dots$. If $rs = 0$ then the ideal condition is satisfied, otherwise rs is the highest non–zero coefficient of the polynomial $(r \cdot 1)f \in \mathcal{I}_1$, i.e. $rs \in \mathcal{I}$. Further, $\mathcal{I}$ is an ideal in the noetherian ring R, so it is generated by some finite set of elements $r_1, \dots, r_n$. Having increased the generating set a little (if necessary) we can assume that $r_1, \dots, r_n$ are the highest coefficients of certain polynomials $f_1, \dots, f_n \in \mathcal{I}_1$. Now consider any polynomial $f \in \mathcal{I}_1$. Let r be its coefficient of highest power X^d, i.e. $f = rX^d + r'X^{d-1} + \dots$. Suppose that the degree of the polynomial f is not smaller than any of the degrees $d_1, \dots, d_n$ of the polynomials $f_1, \dots, f_n$. We represent r in the form $r = s_1 r_1 + s_2 r_2 + \dots + s_n r_n$. Then the polynomial

$$\bar{f} = s_1 X^{d-d_1} f_1 + s_2 X^{d-d_2} f_2 + \dots + s_n X^{d-d_n} f_n$$

lies in $\mathcal{I}_1$ and has the leading term $(s_1 r_1 + \dots + s_n r_n)X^d$. Therefore the polynomial $f - \bar{f}$ is in $\mathcal{I}_1$ and has a degree strictly smaller than the degree of the polynomial f. This argument shows that if d' is the maximum of the degrees of the polynomials $f_1, \dots, f_n \in \mathcal{I}_1$, then any polynomial in

$\mathcal{I}_1$ of degree not less than d' lies in the ideal generated by the elements $f_1, \ldots, f_n$. We now note that for any m the set $\mathcal{I}^{(m)}$ of the highest coefficients of polynomials of degree m situated in $\mathcal{I}_1$ is an ideal of the ring R. Having taken its finite generating set $r_1^{(m)}, \ldots, r_t^{(m)}$ and corresponding polynomials $f_1^{(m)}, \ldots, f_t^{(m)}$, we arrive at the conclusion that any polynomial of degree not higher than d' lies in the ideal generated by a finite set of polynomials $M = \{f_1^{(1)}, \ldots, f_t^{(1)}, \ldots, f_1^{(d')}, \ldots, f_t^{(d')}\}$. Accordingly, the preceding, using $M \cup \{f_1, \ldots, f_n\}$ as a generating set for $\mathcal{I}_1$, can be applied. □

Corollary 1. *The ring of polynomials in n variables over a noetherian ring is noetherian.* □

Corollary 2. *Any affine ring and any affine algebra is noetherian.*

Indeed, let A be an affine algebra over the field F and let $a_1, \ldots, a_n$ be generators of this algebra, then the mapping $\phi : F[X_1, \ldots, X_n] \to A$, under which $f(X_1, \ldots, X_n)$ goes to $f(a_1, \ldots, a_n)$, is a surjective homomorphism. For any ideal $\mathcal{I} \subset A$ its complete preimage $J = \phi^{-1}(\mathcal{I})$ is an ideal in $F[X_1, \ldots, X_n]$. Assume that $b_1, \ldots, b_m$ are generating elements for J; the existence of these is guaranteed by corollary 1. Then $\phi(b_1), \ldots, \phi(b_m)$ are generators for $\mathcal{I}$, i.e. $\mathcal{I}$ is finitely generated and A is noetherian. □

The definition and properties of noetherianness presented at the beginning of this section can be literally transferred to the case of modules over a ring. We note that the module ${}_RR$, where R is some ring, is noetherian iff the ring R is a noetherian ring. It is evident that a homomorphic image and hence a quotient module of a noetherian module is noetherian. More complicated is the statement: if the module M has a noetherian submodule N such that M/N is noetherian then the module M itself is noetherian. Indeed if $M_1 \subset M_2 \subset \ldots \subset M_n \subset \ldots$ is an increasing chain of submodules in the R–module M, then

$$M_1 \cap N \subset M_2 \cap N \subset \ldots \subset M_n \cap N \subset \ldots$$
$$M_1 + N/N \subset M_2 + N/N \subset \ldots \subset M_n + N/N \subset \ldots$$

are increasing chains of submodules in N and M/N. From the fact that the latter two modules are noetherian follows that these chains of submodules in them stabilize; therefore the original chain of submodules in M also stabilizes.

Note. *A module M over a noetherian ring R is noetherian iff it is finitely generated.*

Indeed, if M is finitely generated than it is representable in the form $M = R^n/T$, where T is some submodule of the free module R^n. Since R is a noetherian R–module, with a help of induction on n we easily obtain that R^n is a noetherian R–module and then M, as a homomorphic image for R^n, is also noetherian. In the opposite direction the statement is even clearer. □

Another corollary of the Hilbert basis theorem is known as the Artin–Tate theorem. It will be used in the analysis of algebraic groups in chapter 6.

Corollary 3. *Let R, B, A be commutative rings, $R \subset B \subset A$. We assume that the ring R is noetherian, that A is a finitely generated algebra over R, and that A is a finitely generated B–module. Then B is a finitely generated R–algebra.*

Proof. Let $a_1, \ldots, a_n$ generate A as an algebra over R and $u_1, \ldots, u_m$ generate it as B–module, i.e. $A = Bu_1 + \ldots + Bu_m$. Then

$$a_i = \sum_{j=1}^{m} b_{ij} u_j \quad \text{and} \quad u_i u_j = \sum_{k=1}^{m} b_{ijk} u_k,$$

where b_{ij} and b_{ijk}, for all i, j, k, lie in B. We denote by C the subalgebra generated by all b_{ij}, b_{ijk}. Because of the fact that R is noetherian and according to the corollary 2, C is a noetherian algebra. It is clear that $Cu_1 \oplus \ldots \oplus Cu_m$ contains both R (one can take $u_1 = 1$) and all $a_1, \ldots, a_n$. Thus $A = Cu_1 \oplus \ldots \oplus Cu_m$. Further, any submodule of the C–module A is finitely generated. If $\{b_1, \ldots, b_q\}$ is a generating set for the C–module B, then the set of all b_i, b_{ij}, b_{ijk} is a generating set for B as an R–algebra. □

Another significant property of a ring of polynomials is its factoriality. We should recall that an integral ring R is called *factorial* if any element can be decomposed in to a product of irreducible elements, such that, moreover, such a factorization is unique up to associatedness (see, Introduction 2.4).

Theorem. *The ring of polynomials in one variables $R_1 = R[X]$ over a factorial ring R is itself factorial.*

Proof. Let D be the field of fractions of the ring R (see, Introduction 2.5). The ring R_1 is imbedded into $D[X]$ which is the ring of polynomials over the field D. It is well–known that such a ring is factorial. Any irreducible polynomial p of degree $n \geq 1$ is representable in the form $p = dp'$, where $d \in D$ and $p' \in R[X]$, moreover the greatest common divisor of the coefficients of the polynomial p' can be set equal to 1. In fact, p' is an irreducible polynomial in $R[X]$. Let now f be an arbitrary polynomial from $R[X]$. It is stated that if f is irreducible in $R[X]$ then it is also irreducible in $D[X]$. Indeed, if $f = f_1 f_2$ is a factorization in $D[X]$, where $\deg f_1, \deg f_2 < \deg f$, then having reduced to the common denominator we obtain $\rho f = f_1' f_2'$, where f_1', f_2' are polynomials from $R[X]$ of same degrees as f_1, f_2 and $\rho \in R$. From that we obtain a non–trivial factorization for f in the following way. We denote by $\operatorname{Cont} f$ the greatest common divisor of coefficients of the polynomial f. Our aim is to prove that $\operatorname{Cont}(f_1 f_2)$ is associated with $(\operatorname{Cont} f_1)(\operatorname{Cont} f_2)$. It is obvious that $\operatorname{Cont}(f_1 f_2)$ is divisible by $(\operatorname{Cont} f_1)$ $(\operatorname{Cont} f_2)$. In order to prove the inverse, having reduced to common denominators, we can suppose that $\operatorname{Cont} f_1 = \operatorname{Cont} f_2 = 1$. We shall prove that also $\operatorname{Cont}(f_1 f_2) = 1$. Let p be a factor of $\operatorname{Cont}(f_1 f_2)$. Let $f_1 = a_0 + a_1 X + \ldots + a_m X^m$, $f_2 = b_0 + b_1 X + \ldots + b_n X^n$. We assume that $p|a_0, \ldots, p|a_s$, $p \nmid a_{s+1}$, $p \mid b_0, \ldots, p \mid b_r$, $p \nmid b_{r+1}$, $s < m$, $r < n$. Then the coefficient of X^{s+r+2} in $f_1 f_2$ is equal to $a_{s+1} b_{r+1} + \sum\limits_{\substack{i+j=s+r+2 \\ i \neq s+1}} a_i b_j$. According to the assumption, this coefficient is divisible by p, besides every member in the sum is divisible by p. From the factoriality of the ring R we obtain that either a_{s+1} or b_{r+1} is divisible by p, i.e. a contradiction. Thus either $s = m$ or $r = n$, from which either $\operatorname{Cont} f_1$ or $\operatorname{Cont} f_2$ is divisible into p. ($\operatorname{Cont} f$ is called the *content* of the polynomial f.)

From the relation proved (it is the well–known *Gauss lemma*) follows that if

$$f = (\operatorname{Cont} f)\bar{f},\ f_1' = (\operatorname{Cont} \bar{f}_1)\bar{f}_1',\ f_2' = (\operatorname{Cont} f_2')\bar{f}_2',$$

then

$$\rho(\operatorname{Cont} f)\bar{f} = (\operatorname{Cont} f_1')(\operatorname{Cont} f_2')\bar{f}_1'\bar{f}_2'.$$

Having reduced to common denominators we obtain $\rho' \bar{f} = \bar{f}_1' \bar{f}_2'$. From what has been proved ρ' is an invertible element. Therefore $\bar{f}$ and, consequently, f are reducible in $R[X]$.

The argument just given, using the possibility of factorization into irreducible factors in $D[X]$ and R, produces a factorization into irreducible factors in $R[X]$. If now

$$\rho_1 \ldots \rho_s f_1 \ldots f_t = \sigma_1 \ldots \sigma_u g_1 \ldots g_v,$$

where ρ_i, σ_j are the irreducible factors from R, and f_k, g_l are irreducible factors in $R[X]$ (with $\operatorname{Cont} f_k = \operatorname{Cont} g_l = 1$). From the comparison of contents follows that $\rho_1 \dots \rho_s$ is associated with $\sigma_1 \dots \sigma_u$. Using factoriality of the ring R we get $s = u$ and the associatedness of the factors of the "prefixes" $\rho_1 \dots \rho_s$ and $\sigma_1 \dots \sigma_u$. Further from the factoriality of $D[X]$ we get $t = v$ and that each f_k is associated with some g_l. Reducing to a common denominator we obtain $\rho f_k = \sigma g_l$. As $\operatorname{Cont} f_k = \operatorname{Cont} g_l = 1$, ρ and σ are associated in R. Therefore f_k and g_l are associated in $R[X]$. □

Corollary. *The ring of polynomials in variables $R[X_1, \dots, X_n]$ over a factorial ring R is factorial. In particular, the ring $\mathbf{Z}[X_1, \dots, X_n]$ is factorial.* □

3.2. Ring of fractions. Let R be some not necessarily commutative ring. A subset $S \subset R$ is called *multiplicatively closed* if all elements in S are regular, i.e. are not zero divisors in R, and also for any $a, b \in S$ we have $ab \in S$. For instance, if z is a regular element of the ring R, then the set $S = \{z, z^2, \dots z^n, \dots\}$ is multiplicatively closed. Further, if P is a prime ideal in R, i.e. R/P has no zero divisors, then $S = R \setminus P$ is multiplicatively closed. In this section we shall show that for any central muliplicatively closed set S of the ring R there exists an imbedding of the ring R into a ring R' under which the images of all elements from S are invertible in R'. (An element z of the ring R is called *central* if $za = az$ for any $a \in R$.) To this end we present a construction of the *central ring of fractions* R_S with set of denominators S. This construction generalizes the construction of the field of fractions of a commutative integral ring. Indeed, the set T of formal symbols r/s, where $r \in R$, $s \in S$, is considered. Operations of summation and multiplication are introduced on T as follows

$$\begin{aligned} r_1/s_1 + r_2/s_2 &= (r_1 s_2 + r_2 s_1)/(s_1 s_2), \\ (r_1/s_1)(r_2/s_2) &= (r_1 r_2)/(s_1 s_2). \end{aligned} \tag{1}$$

Two fractions r/s and r'/s' are called *equivalent*, ($r/s \sim r'/s'$), if $rs' = r's$. Direct verification shows, that the relation $\sim$ is, in fact, a relation of equivalence. Indeed, if $r/s \sim r'/s'$ and $r'/s' \sim r''/s''$, then $rs' = r's$, $r's'' = r''s'$. By multiplying the first equality by s'' and the second one by s, we obtain $rs's'' = r's''s = r''s's$, i.e. $(rs'' - r''s)s' = 0$ and $rs'' = r''s$, since S does not contain zero divisors. Thus $r/s \sim r''/s''$ and the transitivity of the relation $\sim$ is established.

Another verification yields the fact that the change of each component or factor to an equivalent fraction in (1) brings an equivalent result. Indeed, if, for instance, $r_1/s_1 \sim r_1'/s_1'$, $r_2/s_2 \sim r_2'/s_2'$ then

$$r_1/s_1 + r_2/s_2 = (r_1s_2 + r_2s_1)/s_1s_2 \sim$$
$$\sim (r_1's_2' + r_2's_1')/s_1's_2' = r_1'/s_1' + r_2'/s_2'.$$

We denote by the symbol R_S the set of equivalence classes of fractions of the set T with operations on these classes that are defined by means of operations (1) on its representatives. The previous remarks demonstrate this is well defined. Admitting some abuse of notation we shall denote an equivalence class by any representative. It is not difficult to check that R_S is a ring with identity s_0/s_0, $s_0 \in S$; moreover, the mapping $r \mapsto rs_0/s_0$ is an isomorphic imbedding of the ring R into R_S. Under this imbedding the element $s \in S$ goes into ss_0/s_0 the inverse of which is s_0/ss_0. We shall identify elements of the form $rs_0/s_0 \in R_S$ with the elements r of the ring R; then R is identified with a subring in R_S. Any element r/s can be represented in the form

$$r/s = (rs_0/s_0)(s_0/ss_0) = (rs_0/s_0)(ss_0/s_0)^{-1} = rs^{-1},$$

where $r \in R$, $s \in S$.

The following holds

Proposition. *For any ring R and a central multiplicatively closed set (of regular elements) S of this ring, there exists, up to isomorphism, a unique ring R_S and an isomorphic imbedding of the ring R into R_S such that S is invertible in R_S and R_S consists of elements of the form rs^{-1}, where $r \in R$, $s \in S$; moreover, any homomorphism $\phi : R \to R'$ under which the images from S are invertible extends to a unique homomorphism $\phi_S : R_S \to R'$ for which $\phi_S|_R = \phi$.*

The construction of the ring R_S and the imbedding R into it are discussed above. If $\phi : R \to R'$ satisfies the condition of the proposition, then we set $\phi_S(rs^{-1}) = \phi(r)\phi(s)^{-1}$. Since ϕ_S must extend ϕ and the inverse of an element needs to go to the inverse of the image of that element, it is not possible to define ϕ_S in a different way. In addition, ϕ_S is well defined since $\phi(s)^{-1}$ exists in R' according to the condition, and if $rs^{-1} = r's'^{-1}$, then

$rs' = r's$, $\phi(r)\phi(s') = \phi(r')\phi(s)$, hence $\phi(r)\phi(s)^{-1} = \phi(r')\phi(s')^{-1}$ and, finally, $\phi_S(rs^{-1}) = \phi_S(r's'^{-1})$. □

If R is commutative and has no zero divisors, then $R^* = R \setminus \{0\}$ is multiplicatively closed; the corresponding ring of quotients $E(R)$ was considered before in Introduction 2.5 under the name of field of quotients.

If P is a prime ideal of the commutative ring R with identity then $S = R \setminus P$ satisfies the conditions of the proposition and R_S is often denoted R_P and called the *localization* of the ring R relative to P.

* The ring R_P is *local* in the sense that it contains a unique maximal ideal $P_S = PS^{-1}$. *

As an example of a local ring we can take the ring $\mathbf{Z}_{(p)}$ of all rational numbers for which the denominator in irreducible notation is not divisible by p, p a prime number. This ring is obtained by localization of the ring $\mathbf{Z}$ relative to the prime ideal $p\mathbf{Z}$. The quotient ring R/M of the local ring R by its unique maximal ideal M is called a field of residues of this ring.

§ 4. Modules over principal ideal rings

4.1. Tensor product of modules. The principal concepts of the theory of modules are presented in Introduction, § 3. We shall introduce here the important concept of the tensor product of two modules. Let M be a right R–module, N a left R–module, where R is some, not necessary commutative, ring. We consider the free abelian group, Π, with basis consisting of all possible pairs of the form (x, y), where $x \in M$, $y \in N$. We denote by P the subgroup of Π generated by the elements of the form

$$T1 : (x\lambda, y) - (x, \lambda y),$$
$$T2 : (x_1 + x_2, y) - (x_1, y) - (x_2, y),$$
$$T3 : (x, y_1 + y_2) - (x, y_1) - (x, y_2).$$

Here $\lambda \in R$, x, x_1, $x_2 \in M$, y, y_1, $y_2 \in N$.

The abelian group Π/P is called the *tensor product* of the R–modules M and N and is denoted by $M \otimes_R N$. We write $x \otimes y = (x, y) + P$. In terms of these notations each element a from $M \otimes_R N$ is representable in the form

$$a = \sum_{i=1}^{n} x_i \otimes y_i,$$

where $x_i \in M$, $y_i \in N$. From the definition and axioms $T1 - T3$ follows that

$$(x\lambda) \otimes y = x \otimes (\lambda y), \quad (x_1 + x_2) \otimes y = x_1 \otimes y + x_2 \otimes y,$$
$$x \otimes (y_1 + y_2) = x \otimes y_1 + x \otimes y_2.$$

If M is a (S, R)–bimodule, or N is a (R, T)–bimodule, then $M \otimes_R N$ can be given the structure of a left S–module or, correspondingly, a right T–module by setting

$$\mu((x, y) + P) = (\mu x, y) + P, \quad \text{or} \quad ((x, y) + P)\nu = (x, y\nu) + P.$$

Let us now suppose that R is commutative, then a right module M can be turned into a left one (and inversily) by setting $rx = xr$ for $x \in M$, $r \in R$. Using this we shall consider the tensor product $M \otimes_R N$ of two R–modules as a left and right R–module. Then the following chain of equalities holds

$$\lambda(x \otimes y) = (\lambda x) \otimes y = (x\lambda) \otimes y =$$
$$= x \otimes (\lambda y) = x \otimes (y\lambda) = (x \otimes y)\lambda.$$

The tensor product possesses the following universality property. Let $\Phi : M \times N \to L$ be a bilinear mapping of the pair of R–modules M, N into the R–module L. Then there exists a unique homomorphism of R–modules $\phi : M \otimes_R N \to L$ such that $\phi(x \otimes y) = \Phi(x, y)$. It is not difficult to prove the assertion that $M \otimes_R N$ is completely characterized by this property.

The tensor product of modules over an arbitrary commutative ring is a somewhat surprising operation. For instance, the tensor product of non–zero abelian groups (in this example we have modules over the ring $\mathbf{Z}$ of integer numbers) can be a zero group. For instance, $\mathbf{Z}_2 \otimes_{\mathbf{Z}} \mathbf{Z}_3 = 0$, since $1 \otimes 1 = (1 + 1 + 1) \otimes 1 = 3(1 \otimes 1) = 1 \otimes (3 \cdot 1) = 0$. However if M and N are free R–modules with bases $\{x_i\}_{i \in I}$, $\{y_i\}_{j \in J}$ then $M \otimes_R N$ is also a free R–module with the basis $\{x_i \otimes y_j\}_{i \in I, j \in J}$. Indeed, the fact that $M \otimes_R N$ is generated by the indicated set is evident:

$$x \otimes y = \left(\sum_i \lambda_i x_i\right) \otimes \left(\sum_j \mu_j y_j\right) = \sum_{i,j} \lambda_i \mu_i (x_i \otimes y_j).$$

Let

$$\sum_{k,l} \lambda_{k,l}(x_k \otimes y_l) = 0 \tag{1}$$

be a linear dependence relation between the elements of the hypothetical basis. For any pair (i, j) we investigate the R–bilinear mapping $\Phi_{ij} : M \times N \to R$, determined by the rule

$$\Phi_{ij}(x_k, y_l) = \delta_{ik}\delta_{jl}1.$$

This mapping extends to a homomorphism $\phi_{ij} : M \otimes N \to R$ under which $x_k \otimes y_l$ goes to zero for all k, l except for $k = i$, $l = j$. Applying ϕ_{ij} to the left side of the equality (1) we obtain

$$\phi_{ij}\left(\sum_{k,l} \lambda_{kl}(x_k \otimes y_l)\right) = \lambda_{ij} = 0,$$

proving linear independence.

If $M_1, \ldots, M_n$ is an arbitrary finite set of R–modules then it is easy to introduce by induction

$$M_1 \otimes \ldots \otimes M_n = (M_1 \otimes \ldots \otimes M_{n-1}) \otimes_R M_n.$$

In view of natural isomorphisms $\phi : (M_1 \otimes M_2) \otimes M_3 \to M_1 \otimes (M_2 \otimes M_3)$ and $\psi : M_1 \otimes M_2 \to M_2 \otimes M_1$ the tensor product up to isomorphism does not depend on the order of the parentheses and the order of multiplication. (The basic natural isomorphisms are given by $\phi((x_1 \otimes x_2) \otimes x_3) = x_1 \otimes (x_2 \otimes x_3)$, $\psi(x_1 \otimes x_2) = x_2 \otimes x_1$; this is well defined, i.e. agrees with the relations $T1$–$T3$.)

If M is a fixed R–module, $n \geq 1$, then $T_n(M) = \overbrace{M \otimes \ldots \otimes M}^{n}$ is called the *n–th tensor power* of the R–module M. To this concept is also related the

n–th symmetric power $\Sigma_n(M)$ and the n–th alternating power $\Lambda_n(M)$. The module $\Sigma_n(M)$ is defined as the quotient group of the group $T_n(M)$ by the subgroup U_n which is spanned by all possible elements of the form:

$$x_{\sigma(1)} \otimes_{\sigma(2)} \otimes \ldots \otimes_{\sigma(n)} - x_1 \otimes x_2 \otimes \ldots \otimes x_n,$$

where $x_1, x_2, \ldots, x_n \in M$, $\sigma \in S_n$.

Since U_n is a R–submodule, $\Sigma_n(M)$ naturally becomes an R–module. (The image of the element $x_1 \otimes x_2 \otimes \ldots \otimes x_n$ in $\Sigma_n(M)$ is denoted by $x_1 x_2 \ldots x_n$.)

The module $\Sigma_n(M)$ possesses the following property of universality: for any n–linear symmetric mapping $\Phi : M \times \ldots \times M \to L$, where M, L are R–modules, there exists a unique homomorphism of R–modules $\phi : \Sigma_n(M) \to L$ such that

$$\phi(x_1 x_2 \ldots x_n) = \Phi(x_1, x_2, \ldots, x_n).$$

By using similar arguments as in the case of the tensor product it easy to prove that if $\{x_i\}_{i \in \mathcal{I}}$ is a basis of a free R–module M, then $x_{i_1} x_{i_2} \ldots x_{i_n}$, $i_1 \leq i_2 \leq \ldots \leq i_n$, is a basis in $\Sigma_n(M)$.

Finally, the alternating power $\Lambda_n(M)$ is defined as the quotient module of the module $T_n(M)$ with respect to the submodule V_n which is generated as an abelian group by the elements of the form

$$x_1 \otimes \ldots \otimes x_i \otimes \ldots \otimes x_j \otimes \ldots \otimes x_n,$$

where $x_1, \ldots, x_n \in M$ are such that there are at least two equal elements, let us say, $x_i = x_j$, among the $x_1, \ldots, x_n$. Simple arguments show that in V_n lie all the elements of the form

$$x_{\sigma(1)} \otimes x_{\sigma(2)} \otimes \ldots x_{\sigma(n)} - \epsilon_\sigma x_1 \otimes x_2 \otimes \ldots \otimes x_n,$$

where $x_1, \ldots, x_n \in M$, $\sigma \in S_n$, and ϵ_σ is the sign of permutation σ. The image of the element $x_1 \otimes x_2 \otimes \ldots \otimes x_n$ is denoted by $x_1 \wedge x_2 \wedge \ldots \wedge x_n$. For any alternating mapping $\Psi : M \times \ldots \times M \to L$, where M, L are R–modules, there exists a unique homomorphism of R–modules $\psi : \Lambda_n(M) \to L$ such that $\psi(x_1 \wedge x_2 \wedge \ldots \wedge x_n) = \psi(x_1, x_2, \ldots, x_n)$.

A basis of the alternating power of the free R–module M with basis $\{x_i\}_{i \in \mathcal{I}}$ has the form

$$x_{i_1} \wedge x_{i_2} \wedge \ldots \wedge x_{i_n}, \quad \text{where} \quad i_1 < i_2 < \ldots < i_n.$$

In particular, if the rank of the module M is equal to $m < n$, then $\Lambda_n(M) = 0$. If we set that $\Lambda_0(M) = R$, then $\Lambda(M) = \Lambda_0(M) \oplus \Lambda_1(M) \oplus \ldots \oplus \Lambda_n(M) \oplus \ldots$ is called the *Grassmann algebra* of the module M. Similarly, $T(M) = T_0(M) \oplus T_1(M) \oplus \ldots \oplus T_n(M) \oplus \ldots$ is called the *tensor algebra* of M, and $\Sigma(M) = \Sigma_0(M) \oplus \Sigma_1(M) \oplus \ldots \oplus \Sigma_n(M) \oplus \ldots$ the *symmetric algebra* of M.[2]

4.2. Finitely generated modules over principal ideal rings. The reader must have noticed the similarity between modules over a ring and vector spaces over a field. From the point of view of structure this similarity is in fact superficial. For that matter, the complexity of constructing modules is completely explicable as many difficult problems are easily formulated in the language of modules. For instance, the problem of the description of invariant subspaces of one or several commuting operators reduces to the problem of the description of submodules over a ring of polynomials of one or several variables. Indeed, if V is a vector space over the field F, $A_1, \ldots, A_n$ are pairwise commuting operators acting on it, then V becomes a module over $F[X_1, \ldots, X_n]$ if for any polynomial $f(X_1, \ldots, X_n) \in F[X_1, \ldots, X_n]$ and $v \in V$ we set

$$f(X_1, \ldots, X_n)v = f(A_1, \ldots, A_n)(v).$$

Another principal example links modules with representations. Let $R = F[G]$ be the group ring of the group G over the field F, i.e. the set of finite formal sums $\sum_{g \in G} \lambda_g g$ with multiplication extending the multiplication in G and in F, and let $\rho : G \to GL(V)$ be a representation of the group G in a linear space V over F. Then V becomes an R–module if we set

$$\left(\sum_{g \in G} \lambda_g g\right) v = \sum_{g \in G} \lambda_g \rho(g)(v).$$

The structure of subrepresentations in V is equivalent to the structure of R–submodules of the module V.

The examples presented show the impossibility in principle of general profound theorems on modules. At the same time the exhaustive theory of finitely

[2]The multiplications in $T(M)$, $\Lambda(M)$, $\Sigma(M)$ are the obvious ones and are induced by $(x_1 \otimes \ldots \otimes x_r, y_1 \otimes \ldots \otimes y_s) \mapsto x_1 \otimes \ldots \otimes x_r \otimes y_1 \otimes \ldots \otimes y_s$. The symmetric algebra of a free module M over R with basis $x_1, \ldots, x_n$ naturally identifies with the algebra of polynomials $R[X_1, \ldots, X_n]$ (with x_i corresponding to X_i). Translators note.

generated abelian groups (modules over $\mathbf{Z}$) indicates that there exist classes of rings and modules over them with a nice structure theory.

One of these classes which naturally generalizes the special case mentioned, is the class of finitely generated modules over principal ideal rings.

We should recall that a commutative ring R without zero divisors is called a *principal* ideal ring if each ideal $\mathcal{I}$ in it can be generated by one element: $\mathcal{I} = Rx$ for a suitable $x \in R$. We note the following simple result.

Proposition. *A principal ideal ring is factorial.*

Proof. According to the discussion in Subsection 3.1, the ring R is noetherian, i.e. satisfies the stabilization condition for increasing chains of ideals.

The possibility of factorization of an element into a product follows from the fact that if $r = r_1 r_2$, where r_1 and r_2 are non–invertible elements, then strict inclusions of ideals $Rr \subset Rr_1$, $Rr \subset Rr_2$ occur. By applying this idea systematically and using the noetherian property of the ring R we obtain what is required. Let us suppose that

(2) $$r_1 r_2 \dots r_t = s_1 s_2 \dots s_u,$$

where r_i, s_j ($i = 1, \dots, t$, $j = 1, \dots, u$) are irreducible elements. We should mention that an irreducible element r from a principal ideal ring divides the product ab iff it divides one of multipliers. Indeed, if r does not divide a, then the ideal Rr does not contain Ra. Therefore the sum $Rr + Ra$ is an ideal strictly containing Rr. From the irreducibility of the element r follows that the ideal Rr is maximal; therefore $Rr + Ra = R$. Then, for suitable $c, d \in R$ we have $1 = cr + da$. Having multiplied both sides by b we get $b = bcr + bda$. Since ab is divisible by r we see that b is divisible by r. Hence, from (2) it follows that either s_1 or $s_2 \dots s_u$ is divisible by r_1. By using induction we see that one of s_i is divisible by r_1. In view of the irreducibility of the element s_i is clear that $s_i = \epsilon_i r_i$, where ϵ_i is an inveribl element. Having reduced both sides in (2) by r_1 (the ring R is integral) we see that an equality of similar form is obtained, therefore induction is applicable. □

Examples of principal ideal rings are fields, the ring of integers, the ring of polynomials in one variable over a field, the ring of integer Gaussian numbers $\mathbf{Z}[i]$ (i.e. of complex numbers of the form $a + bi$, where a, b are integers) and others. We should mention that a ring of polynomials in two variables is not a principal ideal ring. All the same, such a ring is noetherian and factorial as we have seen above in § 3.

Now we recall that a module M over a ring R is called *cyclic* if M has an element x for which $M = Rx$. Thus a cyclic group is the same as a cyclic module over the ring of integers. Every cyclic module is representable in the form of a quotient module of the free cyclic module, i.e. in the case of a ring of principal ideals it has the form $M = R/Rr$. Let $p \in R$ be an irreducible element. A module of the form R/Rp^t, where t is some natural number, is called *primary cyclic*.

Theorem. *Every finitely generated module M over a principal ideal ring R is representable in the form*

$$M \cong R^s \oplus R/Ra_1 \oplus \ldots \oplus R/Ra_t,$$

where

$$Ra_1 \supset Ra_2 \supset \ldots \supset Ra_t.$$

This factorization is unique in the sense of uniqueness of the rank s of the free summand R^s and the chain of ideals $Ra_1 \supset Ra_2 \supset \ldots \supset Ra_t$.

We present M in the form $M \cong F/K$, where K is some submodule of a free module $F = R^n$ of rank n.

Lemma. *The module K is free. In F a basis $u_1, \ldots, u_n$ can be selected, and in K a basis $v_1, \ldots, v_m$ $(m \leq n)$, such that, moreover, $v_i = a_i u_i$ $(i = 1, \ldots, m)$ and $a_i | a_{i+1}$ $(i = 1, \ldots, m-1)$.*

Proof. Induction with respect to n with the evident starting point $n = 1$. To prove that the module K is free we write $F = Rx_1 \oplus \ldots \oplus Rx_n$. Let $K' = K \cap (Rx_2 \oplus \ldots \oplus Rx_n)$. The module K' is free according to the induction hypothesis. We denote by S the ideal in R consisting of those elements r such that $rx_1 + y \in K$ for some $y \in Rx_2 \oplus \ldots \oplus Rx_n$. The ideal S is the image of K under the projection of F onto its first factor. Since R is a principal ideal ring, $S = Rr_0$. If $r_0 = 0$ then $K \subset K'$ and the induction works. If $r_0 \neq 0$, we consider $z_0 = r_0 x_1 + y_0 \in K$ for a suitable $y_0 \in Rx_2 \oplus \ldots \oplus Rx_n$. We claim that $K = Rz_0 \oplus K'$. The directness of the sum follows from the conditions. Further, if $z_1 \in K$ then $z_1 = r' r_0 x_1 + y$, i.e. $z_1 - r' z_0 \in K'$ and from this we get $K \subset Rz_0 + K'$. Since the direct sum of free modules is free the first statement is proved. We should note that according to the proof the rank of the submodule does not exceed the rank of the module.

To prove the second part we denote by F^* the set of R–homomorphisms from F into R. If $f \in F^*$ and K is a submodule in F then $f(K)$ is an ideal in R. By using the noetherian property of the ring R we select a maximal ideal Ra_1 from the set of ideals of the form $f(K)$, where f runs through F^*. Let $f_1 \in F^*$ be such that $f_1(K) = Ra_1$. We take $v_1 \in K$ such that $f(v_1) = a_1$. We shall prove that for any function $g \in F^*$ we have $g(v_1) \in Ra_1$. Indeed, let us assume that $g(v_1) = b_1 \notin Ra_1$. Then for a suitable $d \in R$ we obtain $Ra_1 + Rb_1 = Rd$ or in other words, there exist $r, s \in R$ for which $ra_1 + sb_1 = d$. Using the function $h = rf + sg$ we obtain $h(v_1) = d$ and $h(K) = Rd \supset Ra_1 = f(K)$, moreover $Rd \neq Ra_1$; contradiction.

Using what has been proved we can write $v_1 = r_1x_1 + r_2x_2 + \ldots + r_nx_n$. Since the coefficients $r_1, r_2, \ldots, r_n$ are the values of a linear function on the element v_1 we have $r_i = a_1r_i'$, $i = 1, \ldots, n$, and $v_1 = a_1(r_1'x_1 + r_2'x_2 + \ldots + r_n'x_n)$. Write $u_1 = r_1'x_1 + r_2'x_2 + \ldots + r_n'x_n$ to obtain an element u_1 of the module F for which $f_1(u_1) = 1$. Given such an element we can write

$$F = Ru_1 \oplus \operatorname{Ker} f_1.$$

Since $F' = \operatorname{Ker} f_1$ is a free module by the first part of the proof, a union of a basis for it and $\{u_1\}$ produces a basis for F. From this it follows that the rank of the module F' is strictly smaller than the rank of the module F and this permit us to apply induction with respect to n to F' and $K' = K \cap F'$. Let $\{u_2, \ldots, u_n\}$, $\{v_2, \ldots, v_m\}$ be bases in F' and K' such that $v_i = a_iu_i$, $i = 2, \ldots, n$ and also $Ra_2 \supset Ra_3 \supset \ldots \supset Ra_m$. We shall show that $Ra_1 \supset Ra_2$. We should note that $\{u_1, u_2, \ldots, u_n\}$ is a basis in F. If $Ra_1 \not\supset Ra_2$ then for a suitable $d \in R$ we get $Rd = Ra_1 + Ra_2$ and Rd strictly contains Ra_1. We write $d = r_1a_1 + r_2a_2$. Suppose that $h = r_1f_1 + r_2f_2$ where f_1 is as above, and f_2 is the projection onto u_2. Then $h(r_1v_1 + r_2v_2) = r_1a_1 + r_2a_2 = d$ which contradicts the assumption about the maximality of the ideal $f_1(K)$. □

Proof of the theorem. According to the lemma the submodule K can be presented in the form

$$Ra_1u_1 \oplus Ra_2u_2 \oplus \ldots \oplus Ra_mu_m,$$

where $\{u_1, u_2, \ldots, u_m, u_{m+1}, \ldots, u_n\}$ is some basis in F, i.e. $F = Ru_1 \oplus \ldots \oplus Ru_n$. Factorizing F by K we obtain

$$M \cong F/K \cong Ru_1/Ra_1u_1 \oplus \ldots \oplus Ru_m/Ra_mu_m \oplus Ru_{m+1} \oplus \ldots \oplus Ru_n.$$

Note that the mappings $1 + Ra_i \mapsto u_i + Ra_iu_i$ ($i = 1, \ldots, m$, $1 \mapsto u_j$ ($j = m+1, \ldots, n$) provide isomorphisms from R/Ra_i and R to Ru_i/Ra_iu_i and Ru_j, respectively. Thus the existence of the searched for decomposition is proved.

To prove the uniqueness we consider the exterior powers $\Lambda_1(M)$, $\Lambda_2(M), \ldots$ $\ldots, \Lambda_n(M)$ (from what was said in Subbsection 4.1 it is clear that $\Lambda_{n+1}(M) = \{0\}$). We denote by A_k the annihilator of the module $\Lambda_k(M)$ in R, i.e. the set of all elements $r \in R$ that for any $y \in \Lambda_k(M)$ we have $ry = 0$. For convenience we set $a_{m+1} = \ldots = a_n = 0$. Then $M \cong R/Ra_1 \oplus \ldots \oplus R/Ra_n$. If $e_1, \ldots, e_n$ are the images of the element 1 in $R/Ra_1, \ldots, R/Ra_n$, respectively, then any element x from M can be represented in the form $x = r_1e_1 + \ldots + r_ne_n$. If also $x = s_1e_1 + \ldots + s_ne_n$ then for any $i = 1, \ldots, n$ we have $r_i - s_i \in Ra_i$. According to Subsection 4.1 any element from $\Lambda_k(M)$ is a linear combination of elements of the form $e_{i_1} \wedge e_{i_2} \wedge \ldots \wedge e_{i_k}$, where $i_1 < i_2 < \ldots < i_k$.

Let $r \in Ra_{n-k+1}$. Then $i_1 \leq k$ and $r(e_{i_1} \wedge e_{i_2} \wedge \ldots \wedge e_{i_k}) = (re_{i_1}) \wedge e_{i_2} \wedge \ldots \wedge e_{i_k} = 0$. We shall show that if $r(e_{n-k+1} \wedge \ldots \wedge e_n) = 0$ then $r \in Ra_{n-k+1}$. Consider the alternating mapping $\Psi_k : \underbrace{M \times \ldots \times M}_{k} \to R/Ra_{n-k+1}$ defined by the formula

$$\Psi(x_1, \ldots, x_k) = \det P_k + Ra_{n-k+1},$$

where P_k is the quadratic matrix of order k consisting of projections of the vectors $x_1, \ldots, x_k$ onto the elements $e_{n-k+1}, \ldots, e_n$ of the generating set for M. From what has been said is clear that this well defined. According to the universality property of an exterior product there exists a homomorphism $\psi_k : \Lambda_k(M) \to R/Ra_{n-k+1}$ such that $\Psi_k(x_1, \ldots, x_k) = \psi_k(x_1 \wedge \ldots \wedge x_k)$. However

$$\Psi_k(re_{n-k+1}, e_{n-k+2}, \ldots, e_n) = \det P + Ra_{n-k+1} = r + Ra_{n-k+1}.$$

(Here $P = \text{diag}\,(r, 1, \ldots, 1)$. If $re_{n-k+1} \wedge \ldots \wedge e_n = 0$ then $\psi_k(re_{n-k+1} \wedge \ldots \wedge e_n) = 0$, so according to the formula above, $r \in Ra_{n-k+1}$, which was required.

The presented argument shows that the ideals $Ra_1 \supset Ra_2 \supset \ldots$ are defined uniquely as Ann $\Lambda_n(M)$, Ann $\Lambda_{n-1}(M), \ldots$. This proves uniqueness. □

We now consider the cyclic module R/Ra. If $a = p_1^{s_1}p_2^{s_2} \ldots p_l^{s_l}$ is a decomposition of an element a in to a product of irreducible factors $p_1, p_2, \ldots, p_l$

then

$$R/Ra \cong R/Rp_1^{s_1} \oplus R/Rp_2^{s_2} \oplus \ldots \oplus R/Rp_l^{s_l}.$$

Indeed, it is sufficient to set

$$\phi(r + Ra) = (r + Rp_1^{s_1}, r + Rp_2^{s_2}, \ldots, r + Rp_l^{s_l})$$

If $\phi(r + Ra) = 0$ then $r \in Rp_1^{s_1}, \ldots, r \in Rp_l^{s_l}$, i.e. $r \in Rp_1^{s_1} \ldots p_l^{s_l} = Ra$, and hence $r + Ra = 0$. In addition, the preimage of $(r_1 + Rp_1^{s_1}, \ldots, r_t + Rp_l^{s_l})$ is the class of the element

$$r = r_1 v_1 \frac{a}{p_1^{s_1}} + \ldots + r_l v_l \frac{a}{p_l^{s_l}},$$

where $v_i \in R$ is such that

$$p_i^{s_i} u_i + \frac{a}{p_i^{s_i}} v_i = 1$$

for suitable $u_i \in R$.

Writing M in the form $M = R^s \oplus R/Ra_1 \oplus \ldots \oplus R/Ra_t$, where $Ra_1 \supset Ra_2 \supset \ldots \supset Ra_y$, we see that

$$a_i = \epsilon_i p_1^{s_1} \ldots p_l^{s_{il}}$$

for suitable irreducible $p_1, \ldots p_l \in R$ and natural $s_{i1}, \ldots, s_{il}$; moreover, $s_{ij} | s_{i+1,j}$ for any $i = 1, \ldots, t-1$, $j = 1, \ldots, 1$. The result is the following

Theorem. *Any finitely generated module M over a principal ideal ring R is representable in the form of the direct sum of free cyclic submodules and of primary cyclic submodules, i.e. submodules of the form $R/Rp_j^{s_{ij}}$ for suitable irreducible $p_1, \ldots, p_l$ and natural s_{ij}. The set $\{p_j^{s_{ij}} | i = 1, \ldots, t;\ j = 1, \ldots, l\}$ is defined uniquely up to permutation of its members.* □

Examples of principal ideal rings are the ring $\mathbf{Z}$ of integer numbers and the ring $F[X]$ of polynomials of one variable X over a field F. A finitely generated module A over $\mathbf{Z}$ is simply a finitely generated abelian group. Thus, we get the well–known theorem (see, Introduction 1.12).

Corollary 1. *Any finitely generated abelian group is uniquely (up to permutation of summands) decomposable into a direct sum of a finite number of infinite cyclic and primary cyclic groups.* □

A second corollary is the theorem on the Jordan normal form of a linear operator $\mathcal{A}$ on the finite–dimensional vector space V over a field F. In order to obtain it we convert V into an $F[X]$–module by defining the product of $f(X) \in F[X]$ and $v \in V$ as $f(X)v = f(\mathcal{A})(v)$. It is evident that it is well defined. According to the general theorem $V = V_1 \oplus V_2 \oplus \ldots \oplus V_m$, where V_i is a primary cyclic $F[X]$–module, i.e. $V_i \cong F[X]/(p_i^{s_i}(X))$, where $\chi_{\mathcal{A}}(X) = p_1^{s_1}(X)p_2^{s_2}(X)\ldots p_m^{s_m}(X)$ is the characteristic polynomial for the operator $\mathcal{A}$. Let us assume that $\chi_{\mathcal{A}}(X)$ is decomposable into linear factors over F. Then $p_i(X) = X - \lambda_i$, where λ_i is a characteristic root of the operator $\mathcal{A}$, $i = 1, 2, \ldots, m$. The vector space V_i then has the dimension s_i. Let v be a vector which is the image of 1 under the homomorphism from $F[X]$ into V with kernel $F[X](X - \lambda_i)^{s_i}$. Then a basis in V consists of the vectors $v_0, v_1, \ldots, v_{s_i-1}$, where $v_{j+1} = (\mathcal{A} - \lambda_i\mathcal{E})(v_j)$, $j = 0, \ldots, s_i - 2$, because $1, X - \lambda_i, (X - \lambda_i)^2, \ldots, (X - \lambda_i)^{s_i-1}$ form a basis in $F[X]/((X - \lambda_i)^{s_i})$. In this basis $\mathcal{A}(v_j) = \lambda_i v_j + v_{j+1}$, $j = 0, 1, \ldots, s_i - 2$, and $\mathcal{A}(v_{s_i-1}) = \lambda v_{s_i-1}$. The matrix A_i of the restriction of the operator $\mathcal{A}$ to V_i for the given basis has the form

$$\begin{pmatrix} \lambda_i & 0 & 0 & \ldots & 0 & 0 \\ 1 & \lambda_i & 0 & \ldots & 0 & 0 \\ 0 & 1 & \lambda_i & \ldots & 0 & 0 \\ \ldots & \ldots & \ldots & \ldots & \ldots & \ldots \\ \ldots & \ldots & \ldots & \ldots & \ldots & \ldots \\ 0 & 0 & 0 & \ldots & \lambda_i & 0 \\ 0 & 0 & 0 & \ldots & 1 & \lambda_i \end{pmatrix},$$

i.e. the matrix is a so–called *Jordan block* (*Jordan cell*). A corresponding basis (not unique) is called a *Jordan basis*. By taking the union of Jordan basis for all the subspaces V_i, $i = 1, \ldots, m$, we obtain a *Jordan basis* for V, i.e. a basis where the matrix of the operator is the direct sum of Jordan cells. From the uniqueness theorem follows that the number of cells of given order with given root λ on the diagonal is an invariant of the linear operator $\mathcal{A}$. □

§ 5. Algebraic sets

5.1. Definition. Let F be some field, F^n the n–dimensional arithmetic space over F. A subset S in F^n is called an *algebraic* set if it consists of all the zeros of some system J of polynomials of the ring $F[X_1, \ldots, X_n]$. In other words, $\alpha = (\alpha_1, \dots, \alpha_n)$ belongs to S iff $f(\alpha) = f(\alpha_1, \dots, \alpha_n) = 0$ for any $f = f(X_1, \dots, X_n) \in J$.

Examples. 1) The sets F^n and $\emptyset$ are algebraic (for $J = \emptyset$ and $J = F[X_1, \dots \dots, X_n]$).

2) A set $S = \{\alpha\}$ consisting only of one element is algebraic ($J = \{X - \alpha_1, \dots, X - \alpha_n\}$, where $\alpha = (\alpha_1, \dots, \alpha_n)$).

3) If the $(S_\alpha)_{\alpha \in \mathcal{I}}$ are algebraic in F^n (and corresponds to the sets of polynomials $(J_\alpha)_{\alpha \in \mathcal{I}}$) then $\bigcap_{\alpha \in \mathcal{I}} S_\alpha$ is algebraic (and corresponds to the set of polynomials $\bigcup_{\alpha \in \mathcal{I}} J_\alpha$).

4) The union $S = S_1 \cup S_2$ of algebraic sets S_1 and S_2 from F^n is algebraic. Indeed, let S_i be the set of zeros of the system J_i, $i = 1, 2$. Let us form the set J of all possible products $f_1 f_2$, where $f_1 \in J_1$, $f_2 \in J_2$. It is clear that S is contained in the set S' of zeros of the system J. Let us assume that $S' \setminus S \neq \emptyset$ and $\beta \in S' \setminus S$. Since $\beta \notin S_1$, there is an $f_1^0 \in J_1$ such that $F_1^0(\beta) \neq 0$. Similarly there can be found an $f_2^0 \in J_2$ such that $f_2^0(\beta) \neq 0$. The polynomial $f = f_1^0 f_2^0$ belongs to J and $f(\beta) = f_1^0(\beta) f_2^0(\beta) \neq 0$. This gives a contradiction. Thus $S' \setminus S = \emptyset$ and S is an algebraic set.

5.2. Zariski topology. The examples presented above demonstrate that the family of algebraic sets in F^n can be considered as the system of closed sets of a topological space. The resulting topology on F^n is called the *Zariski topology*. This topology possesses some pecularities, in particular, the topology is not Hausdorff. Thus for the real line $\mathbf{R}^1$ any closed set S is finite (the set of zeros of the polynomial $f(X)$ of one variable which is equal to the greatest common divisor of polynomials of the system $\mathcal{J} \subset \mathbf{R}[X]$). Thus any open set is the whole line from which a finite number of points have been deleted. If now $\alpha \neq \beta$ are two points from $\mathbf{R}^1$ then any of their neighborhoods have nonempty intersection.

Another pecularity of Zariski topology is the noetherian property, which says that any sequence of strict inclusions

$$S_1 \supset S_2 \supset \dots \supset S_m \supset \dots \tag{1}$$

of closed subsets in F^n is finite, or which is the same, any strictly increasing chain of open subsets is finite.

In order to make clear the validity of the latter statement we define two mappings: $\mathcal{J}$ and $\mathcal{S}$. The mapping $\mathcal{J}$ associates to any subset $S \subset F^n$

the set $J = \mathcal{J}(S)$ of all polynomials from $F[X_1, \dots, X_n]$ taking the value zero in all points from M. Conversely, for any subset $J \subset F[X_1, \dots, X_n]$ we define $\mathcal{S}(J)$ as the set S of all zeros of the polynomials J. We can see that for any $S \subset F^n$ the set $\mathcal{S}(J)$ is an ideal in $F[X_1, \dots, X_n]$ and for any $J \subset F[X_1, \dots, X_n]$ the set $\mathcal{S}(J)$ is a closed subset in F^n. It is evident that if $S_1 \supset S_2$ then $\mathcal{J}(S_1) \subset \mathcal{J}(S_2)$, and if $J_1 \supset J_2$ then $\mathcal{S}(J_1) \subset \mathcal{S}(J_2)$. Taking the closure $\bar{S}$ of the set S in the Zariski topology can be written in the following way with a help of the mappings $\mathcal{S}$ and $\mathcal{J}$: $\bar{S} = \mathcal{S}(\mathcal{J}(S))$. Moreover, if S is an algebraic set defined by the system $P \subset F[X_1, \dots, X_n]$ then it is also definible by the system $\mathcal{J}(\mathcal{S}(J))$ of all polynomials that take the value zero on a set $S = \mathcal{S}(J)$ of zeros of the system J.

Returning to the noetherian property we denote that to a decreasing sequence (1) of closed sets corresponds an increasing sequence

$$\mathcal{J}(S_1) \subset \mathcal{J}(S_2) \subset \dots \subset \mathcal{J}(S_m) \subset \dots \tag{2}$$

of ideals in $F[X_1, \dots, X_n]$. According to the Hilbert basis theorem (see, Subsection 3.1) the sequence (2) is finite and so, begining with some m, we get $\mathcal{J}(S_m) = \mathcal{J}(S_{m+1}) = \dots$. According to what was said before $S_i = \bar{S}_i = \mathcal{S}(\mathcal{J}(S_i))$ for any i. Consequently, $S_m = S_{m+1} = \dots$, as is required.

5.3. Radical ideals. A more detailed study of the mappings $\mathcal{S}$ and $\mathcal{J}$ shows that the image of $\mathcal{S}$ is the set of all closed subsets in F^n. At the same time it is evident that not all ideals from $F[X_1, \dots, X_n]$ belong to the image of the mapping $\mathcal{J}$. For instance, if $n = 1$, the principal ideal (X^2) generated by the polynomial X^2 has the same set of zeros as (X). We then have $\mathcal{J}(\{0\}) = (X)$, i.e. (X^2) has no preimage. From this example it is clear that if $J = \mathcal{J}(S)$ and $f^m \in J$ then $f \in J$, i.e. any ideal of the form $\mathcal{J}(S)$ contains together with any of its own polynomials any natural degree root of that polynomial. Let us introduce an additional requirement. If J is an ideal in a commutative ring R, then $\sqrt{J}$, the *radical of the ideal* J, is a set of elements $r \in R$ such that $r^m \in J$ for a some natural m. It is clear that the radical of an ideal is itself an ideal. If $\sqrt{J} = J$, then we say that J is a *radical* ideal. The arguments given show that the image of the mapping $\mathcal{S}$ consists of radical ideals. To see that the inverse also holds, some additional information is required.

Proposition. *Let R be a commutative ring, J a radical ideal in it. Then J is equal to the intersection of some set of prime ideals J_i, $i = 1, 2, \dots$; i.e. ideals J_i such that R/J_i has no zero divisors.*

Proof. Replacing R with R/J we can assume that R has no non–zero nilpotent elements. It is necessary to prove that the intesection of the prime ideals of the ring R is equal to zero. To this end we shall show that each non–zero element a of the ring R does not belong to some prime ideal. Let $S = \{a, a^2, \dots, a^n, \dots\}$. Then according to the Zorn lemma there exists an ideal Q of the ring R which is maximal among the ideals with an empty intersection with S. We shall show that Q is a prime ideal. Let us suppose that $uv \in Q$, $u, v \notin Q$. Then the ideals $Ru + Q$, $Rv + Q$ strictly contain Q and, consequently, $a^k \in Ru + Q$, $a^l \in Rv + Q$ for suitable k, l. In this case, $a^{k+l} = a^k a^l \in Ruv + Q = Q$ which contradicts the definition of the ideal Q. □

Corollary. *For a commutative ring the intersction of all prime ideals coincides with the set of nilpotent elements. If R is noetherian then the intersection of all prime ideals is the largest nilpotent ideal of the ring.*

In order to prove the first statement it is sufficient to note that the set indicated is a radical ideal which is contained in any prime ideal. The second statement follows from the fact that any ideal in a noetherian ring is finitely generated, and an ideal in a commutative ring which is generated by finite number of nilpotent elements is nilpotent itself. □

5.4. Homomorphisms of algebras without nilpotent elements. In Subsection 1.1.4 it was mentioned that any imbedding of the field F into its algebraic closure $\bar{F}$ can be extended to an imbedding of an algebraic extension K in $\bar{F}$. In this section we shall demonstrate that the similar statement, replacing imbedding by homomorphism, holds in the case where K is a finitely generated algebra with 1 over F. This important fact will be used to prove Hilbert's theorem on zeros and to establish the one–to–one correspondence between algebraic sets and radical ideals.

A natural way to construct a homomorphism $\phi : K \to \bar{F}$ is to take a maximal ideal M in K which does not contain 1. The quotient ring $A = K/M$ is a field (see, Introduction 2.6) which satisfies all the conditions of the statement presented above, i.e. the field A is a finitely generated algebra over F with 1. If A is algebraic over F then there exists an imbedding of the field A over F in $\bar{F}$. Composition of this imbedding with the natural homomorphism from K onto $K/M = A$ will be the sought for homomorphism.

The fact that A, as above, is always algebraic is the content of an important theorem.

Theorem. *Let F be a field, $A = F[x_1, \dots, x_n]$ a finitely generated algebra over F with unity, which is a field. Then A is an algebraic extension of the field F.*

Proof. Let $a \in A$. We shall prove that a is algebraic over F. Assuming the opposite we denote by B the subring with unity which is generated by the element a over F. Then A is a left B–module (relative to multiplication by a). The idea of the proof consists in the choice of an element $b = f(a)$ such that $B[b^{-1}] \otimes_B A$ is a free $B[b^{-1}]$–module. In this case the multiplication in $B[b^{-1}]$ by an element $c = h(a)$, which is not a divisor of any power of the element b, is not surjective. Thus its action on a free module over $B[b^{-1}]$ is also not surjective. On the other hand, if $g \in B[b^{-1}]$, $m \in A$ then

$$c(g \otimes m) = gc \otimes m = g \otimes cm.$$

Since c, m are the elements of the field A, where m is arbitrary, we obtain the surjectivity of multiplication by c.

Let us consider in detail A as a B–module. Generating elements for this set can be written down in the form $x^\nu = x_1^{\nu_1} \dots x_n^{\nu_n}$, where $\nu = (\nu_1, \dots, \nu_n)$ is some set of non–negative integers. We denote a set of all such tuples by $\mathbf{N}^n$. We then introduce an ordering on $\mathbf{N}^n$, by defining that $\nu \leq \nu' = (\nu_1', \dots, \nu_n')$, if $\nu_1 \leq \nu_1', \dots, \nu_n \leq \nu_n'$.

Lemma. *Let $S \subset \mathbf{N}^n$, and let S_0 be the set of minimal elements from S. Then S_0 is finite and any element from S dominates some element from S_0.*

Proof. Let S_0 contain an infinite subset $\{s_1, s_2, \dots, s_m, \dots\}$. We select from this an infinite subset with increasing first component. Then we select from it an infinite subset of tuples of elements with increasing second component. Having repeated the procedure n times we obtain an infinite sequence of minimal elements of the set S and ordered with respect to the ordering on $\mathbf{N}^n$. This contradiction demonstrates that S_0 is finite. The second statement follows from the fact that the number of tuples which are smaller than a given one, is finite. □

Returning to the B–module A, for any $\nu \in \mathbf{N}^n$ we introduce B–submodules

$$A_\nu = \sum_{\nu' \leq \nu} Bx^{\nu'}, \quad A_\nu^- = \sum_{\nu' < \nu} Bx^{\nu'}.$$

Of course, $A = \bigcup_{\nu \in \mathbf{N}^n} A_\nu$. Let us consider the quotient module A_ν / A_ν^-. It is generated by the unique element $x_*^\nu = x^\nu + A_\nu^-$. We denote by $\mathcal{I}_\nu$ the annihilator of the element x_*^ν in B. It is evident that $\mathcal{I}_\nu \subset \mathcal{I}_{\nu'}$, $\nu \leq \nu'$. Let Λ be the set of indexes such that $I_\nu \neq 0$. According to the lemma we can find a finite set of indices $\nu_1, \dots, \nu_r \in \Lambda$, such that any non–zero $\mathcal{I}_\nu$ contains one from $\mathcal{I}_{\nu_1}, \dots, \mathcal{I}_{\nu_r}$. Since B has no zero divisors, the ideal $\mathcal{I}_{\nu_1} \cap \dots \cap \mathcal{I}_{\nu_r}$ is non–zero. Let b be a non–zero element from this ideal. We now consider the module $B[b^{-1}]$ and for any $\nu \in \mathbf{N}^n$, the ring $B[b^{-1}] \otimes_B (A_\nu / A_\nu^-)$. If $\nu \notin \Lambda$, then $A_\nu / A_\nu^- \cong B$ and $B[b^{-1}] \otimes_B (A_\nu / A_\nu^-) \cong B[b^{-1}]$. If $\nu \in \Lambda$, then

$$\begin{aligned} B[b^{-1}] \otimes_B (A_\nu / A_\nu^-) &= B[b^{-1}]b \otimes_B (A_\nu / A_\nu^-) = \\ &= B[b^{-1}] \otimes_B b(A_\nu / A_\nu^-) = 0. \end{aligned}$$

Thus for any $\nu \in \mathbf{N}^n$ the module $B[b^{-1}] \otimes_B (A_\nu / A_\nu^-)$ is free. Since an extension of a free module by a free one is free (see, Introduction 3.3), it is possible to assert that $B[b^{-1}] \otimes_B A$ is also a free $B[b^{-1}]$–module.[3] This concludes the proof of the theorem. □

Corollary. *Let K be a finitely generated commutative algebra with unity over a field F, with no nilpotent elements. Then for any $a \neq 0$ there exists a homomorphism $\phi : K \to \bar{F}$, where $\bar{F}$ is an algebraic closure of the field F such that $\phi(a) \neq 0$.*

Proof. The ideal $\{0\}$ in K is radical, thus, according to the proposition from Subsection 5.3, it is equal to the intersection of all prime ideals. In this case the element a does not lie in one of the prime ideals, let us say P_1. Under the natural homomorphism $K \to K/P_1$ the image of the element a differs from zero and as before, K/P_1 satisfies our conditions. Having replaced K by K/P_1 we can assume that K is an algebra without zero divisors. Let us imbed K into its field of fractions $F(K)$. Let $K_1 = K[a^{-1}]$. According to the theorem proved, there exists a homomorphism $\phi_1 : K_1 \to \bar{F}$. It is clear that

[3] The set of x^ν for which $\nu \notin \Lambda$, can be used as a set of free generators.

its restriction ϕ on K is the needed homomorphism for which $\phi(a) \neq 0$. The corollary is proved. □

5.5. Hilbert Nullstellensatz (Hilbert theorem on zeros). *Let F be an algebraically closed field, J a system of polynomials from $R = F[X_1, \dots, X_n]$, S the set of zeros of the system J, f a polynomial vanishing in every point $\alpha \in S$. Then there exist $f_1, \dots, f_m \in J$, $g_1, \dots, g_m \in R$ and natural number k such that*

$$f^k = f_1 g_1 + \dots + f_m g_m.$$

Using the notation from Subsection 5.2, we can formulate this in the form

$$\mathcal{J}(\mathcal{S}(J)) = \sqrt{J}$$

for any ideal J of the ring $F[X_1, \dots, X_n]$, where F is an algebraically closed field.

Proof. For the proof we shall use the notation from Subsection 5.2. Of course, $\mathcal{J}(\mathcal{S}(J)) \supset \sqrt{J}$. In order to prove the opposite inclusion it is sufficient for any polynomial f which is not in $\sqrt{J}$, to find a point $\alpha \in \mathcal{S}(J)$, such that $f(\alpha) \neq 0$. Let $\bar{f} = f + \sqrt{J} \in K = R/\sqrt{J}$. The important step in the argument is that any homomorphism $\phi : K \to F$ for which $\phi(\bar{f}) \neq 0$, defines such a point. Indeed, let $\epsilon : R = F[X_1, \dots, X_n] \to R/\sqrt{J} = K$ be the natural homomorphism. Let us consider the composition $\psi = \phi\epsilon$ and write $\alpha_i = \phi\epsilon(X_i)$. Then $\alpha = (\alpha_1, \dots, \alpha_n)$ is the point sought. Ideed,

$$\begin{aligned} f(\alpha) =& f(\alpha_1, \dots, \alpha_n) = f(\phi\epsilon(X_1), \dots, \phi\epsilon(X_n)) = \\ =& (\phi\epsilon) f(X_1, \dots, X_n) = \phi(\bar{f}) \neq 0. \end{aligned}$$

On the other hand, if $h \in J \subset \sqrt{J}$, then

$$h(\alpha_1, \dots, \alpha_n) = h(\phi\epsilon(X_1), \dots, \phi\epsilon(X_n)) = \phi(\epsilon(h(X_1, \dots, X_n)) = \phi(\bar{h}) = 0,$$

as $\bar{h} = 0$.

To conclude the proof of the theorem it remains to recall that the ideal $\sqrt{J}$ is radical, that $K = R/\sqrt{J}$ is a finitely generated algebra without nilpotent elements, and this allows us to apply the corollary from Subsection 5.4 taking the element $\bar{f}$ as a. □

Corollary. (F is algebraically closed) *The mappings S and $\mathcal{J}$ are one–to–one bijections between the set of radical ideals in $F[X_1, \dots, X_n]$ and the set of closed subsets in F^n.* □

5.6. Other pecularities of the Zariski topology. In Subsection 5.1 we have introduced the Zariski topology on F^n. Moreover, as usual, any subset $M \subset F^n$ becomes a topological space (subspace) with the induced topology. A subspace N of the topological space M is called *irreducible* if it is not the union of two proper closed subsets. It is evident that M is irreducible iff the intersection of any two open subsets in M is not empty. Finally, both the conditions mentioned are equivalent to the fact that any non–empty open subset U of the set M is dense in M. Indeed, if $\bar{U} \neq M$ then $V = M \backslash \bar{U}$ is a non–empty open subset in M such that $U \cap V = \emptyset$. We note that any irreducible set is connected.

Theorem. *A space M with the Zariski topology has only a finite number of maximal irreducible subspaces $M_1, \dots, M_n$. In addition, $M = M_1 \cup \dots \cup M_n$ and each M_i is closed. A closed subset S in F^n is irreducible iff the ideal $J = \mathcal{J}(S)$ is prime, i.e. $F[X_1, \dots, X_n]/J$ has no zero divisors. In particular, F^n itself is irreducible.*

Proof. Let $\mathcal{A}$ be the family of all finite unions of closed irreducible subsets of the space M. It is not empty, because $\emptyset \in \mathcal{A}$, and contains a union of any two of its members. If M does not lie in $\mathcal{A}$, then by virtue of the noetherian property of the Zariski topology (see, Subsection 5.2) there exists a minimal closed subset N which is not in $\mathcal{A}$. It is clear that N is a non–empty reducible set, i.e. $N = N_1 \cup N_2 \in \mathcal{A}$, where N_1 and N_2 are proper closed subsets. By the minimality of N we have N_1, $N_2 \in \mathcal{A}$. The contradiction obtained proves that $M \in \mathcal{A}$, i.e. $M = M_1 \cup \dots \cup M_n$, where every M_i, $i = 1, \dots, n$, is a closed irreducible subset. If now L is an arbitrary maximal irreducible subset, then $L = L \cap M = (L \cap M_1) \cup \dots \cup (L \cap M_n)$. By the maximality and irreducibility of L, for some i we have $L = M_i$.

Now let S be a closed subset in F^n. We shall consider the ideal $J = \mathcal{J}(S)$. If S is irreducible and J is not a prime ideal, then there are polynomials f_1, $f_2 \notin J$ for which $f_1 f_2 \in J$. Let S_i be the closed subset in S determined inside S by the equation $f_i(X_1, \dots, X_n) = 0$. Then $S = S_1 \cup S_2$. Because of irreducibility one of S_1, S_2, let us say S_1, coincides with S. However, this fact contradicts our assumption that the ideal $J = \mathcal{J}(S)$ is not prime.

Conversely, let us assume that $J = \mathcal{J}(S)$ is a prime ideal. Also assume that $S = S_1 \cup S_2$, where S_1 and S_2 are closed subsets. Then $\mathcal{J}(S_1) \supset \mathcal{J}(S) = J$ and $\mathcal{J}(S_2) \supset \mathcal{J}(S) = J$. In addition, $\mathcal{J}(S_i) \neq J$. If $f_i \in \mathcal{J}(S_i) \setminus P$, $i = 1, 2$, and $f = f_1 f_2$, then $f(\alpha) = 0$ for any point $\alpha \in S_1 \cup S_2$ so that $f_1 f_2 \in J$, which is impossible.

In order to conclude the proof of the theorem we should note that the ideal $\{0\}$ in $R = F[X_1, \dots, X_n]$ is prime as R has no zero divisors. Thus, the space F^n is irreducible. □

If M is an irreducible algebraic set in F^n, $J = \mathcal{J}(M)$ its annihilator in $R = F[X_1, \dots \dots, X_n]$, then R/J is an integral ring and, hence, there exists the field of fractions $K = \mathrm{Fr}\,(R/J)$. The transcendence degree $\mathrm{tr}\deg_F K$ is called *dimension* of the algebraic set M and is denoted by $\dim M$. It is clear that for all cases, $\dim M \leq n$, and, moreover, $\dim F^n = n$. The next statement shows that if $M \subset N$ are two irreducible algebraic sets in F^n such that $M \neq N$, then $\dim M < \dim N$. Moreover, the maximal length of the chains of irreducible subsets in F^n of the form

$$M_1 \subset M_2 \subset \dots \subset M_s$$

does not exceed n.

Proposition. *Let A be a commutative algebra over field F with m generators. Then any chain of prime ideals in A has length not larger than m. If P, Q are prime ideals in A, such that P lies in Q in a proper way, then the transcendence degree of the field of fractions Fr (A/Q) over F is strictly less than the transcendence degree of the field of fractions Fr (A/P).*

Proof. It is clear that the transcendence degree of the field Fr (A/P) does not exceed n. Thus it is sufficient to prove the second statement. Let us select a transcendence basis $\{y_1, \dots, y_t\}$ in Fr (A/Q) over F consisting of elements of the ring A/Q. For every y_i we select an x_i in A/P which has its canonical image equal to y_i. Let x_0 be a non–zero element in Q/P. It is sufficient to show that the set $\{x_0, x_1, \dots, x_t\}$ is algebraically independent over F. We select a non–zero polynomial f of lowest degree (in x_0) with coefficients from F and such that $f(x_0, \dots, x_t) = 0$. We rewrite this in the form

$$g(x_1, \dots, x_t) + h(x_0, \dots, x_t)x_0 = 0,$$

where g, h are polynomials with coefficients from F. The canonical image of the left side of the equality above in A/Q is $g(y_1, \dots, y_t)$. Thus g is the zero polynomial. Because of the primality of the ideal we have $h(x_0, \dots, x_t) = 0$, which contradicts the minimality of the degree of the polynomial f. □

§ 6. Normed fields

6.1. Definition. Let A be a ring. A mapping assigning to every element x from the ring A a real non–negative number $\| x \|$ is called *norm* on the ring A if the following conditions are fulfilled

(*i*) $\| xy \| = \| x \| \cdot \| y \|$,

(*ii*) $\| x + y \| \leq \| x \| + \| y \|$,

(*iii*) $\| x \| = 0$ iff $x = 0$.

Directly form the definition it follows that a normed ring has no zero divisors. Conversely, any ring without zero divisors can be normed, namely: one can assume $\| x \| = 1$ for all $x \neq 0$ (the trivial norm).

6.2. Examples. 1) $A = \mathbf{R}, \mathbf{Q}$, $\| x \| = |x|$ is a modulus of the number x;

2) $A = \mathbf{C}$, $\| z \| = |z|$ is the modulus of a complex number;

3) $A = \mathbf{Z}$, $\| m \| = \gamma^{\alpha}$, where $0 < \gamma < 1$, $m = p^{\alpha} m'$, p is a fixed prime number, $(m', p) = 1$. In this case the property (*ii*) can be replaced by the stronger one:

(*ii'*) $\| x + y \| \leq \max\{\| x \|, \| y \|\}$.

A norm with the properties (*i*), (*ii'*), (*iii*) is called *non–archimedean*.

The norm on the ring $\mathbf{Z}$ of example 3) is called *p–adic*. It is easily extended to the field $\mathbf{Q}$ by putting $\| 1/n \| = 1/ \| n \|$.

4) A norm on the ring of polynomials $F[X]$ can be obtained by setting $\| f(X) \| = \gamma^{\deg f(X)}$, where $0 < \gamma < 1$.

A norm turns the ring A into a topological, even a metric, space. Namely, the distance $\rho(a, b)$, where $a, b \in A$, is set equal to $\| a - b \|$.

Below, we restrict ourselves to the investigation of norms on fields.

6.3. Equivalence of norm. Two norms $\| \, \|_1$ and $\| \, \|_2$ on the field F are called equivalent if there exists a positive $c > 0$ such that for all $x \in F$ we have $\| x \|_2 = \| x \|_1^c$.

Proposition. *Two norms of the field F are equivalent iff they induce the same topology on F.*

Proof. The condition $\| x \| < 1$ is fulfilled iff $x^n \to 0$ for $n \to \infty$ in the sense of the topology which is induced by norm $\| \ \|$. From this follows that if the topologies coincide then the inequality $\| x \|_1 < 1$ is equivalent to the inequality $\| x \|_2 < 1$. Going over to inverse elements we notice that in this case $\| x \|_1 = 1$ is also equivalent to $\| x \|_2 = 1$.

Let now $y, z \in F$, $y \neq 0$, $z \neq 0$. Applying the previous arguments to $x = y^m z^n$, $(m, n \in \mathbf{Z})$ we see that the inequality

$$m \log \| y \|_1 + n \log \| z \|_1 \geq 0$$

holds only if the following inequality is valid

$$m \log \| y \|_2 + n \log \| z \|_2 \geq 0,$$

so that

$$\frac{\log \| y \|_1}{\log \| y \|_2} = \frac{\log \| z \|_1}{\log \| z \|_2}.$$

Denoting this common number by c we get $\| x \|_1 = \| x \|_2^c$.

Conversely, if $\| x \|_1 = \| x \|_2^c$ for any $x \in F$ under $c > 0$, then the sequence $\{x_n\}$ converges to $x \in F$ in the sense of metric which is defined by the second of two norms, iff $\| x_n - x \|_2 \to 0$. According to the condition also $\| x_n - x \|_1 = \| x_n - x \|_2^c \to 0$. So convergence in the sense of the first metric implies convergence in the sense of the second one. In this case the closures of any set in the sense of each topology are the same.

6.4. Completeness. A ring A is called *complete* with respect to the norm $\| \ \|$ if it is complete as a metric space with respect to the metric $\rho(x, y) = \| x - y \|$, where $x, y \in A$. This mean that for every fundamental sequence $\{x_n\}$, $n = 1, 2, \ldots$ there can be found an $a \in A$ such that $x_n \to a$ in the sense of the norm. A sequence $\{x_n\}$ is called *fundamental* (or a *Cauchy sequence*) if for any $\epsilon > 0$ an N can be found such that from $m, n > N$ it follows that $\| x_m - x_n \| < \epsilon$.

Theorem. *Any ring A with norm $\| \ \|$ can be imbedded into a complete ring $\bar{A}$ with a norm $\| \ \|$ which extends the initial norm on A in such a way that*

$\bar{A}$ is the closure of the ring A with respect to $\| \ \|$. Moreover, the ring $\bar{A}$ is unique for A (relative to $\| \ \|$) up to an isomorphism that is the identity on A. If A is a field than $\bar{A}$ is also a field.

Proof. The ring $\bar{A}$ is defined as the set of equivalence classes Cauchy sequences of elements from A. Here, two sequences $\{x_n\}$ and $\{y_n\}$ are called equivalent if the difference sequence $\{z_n\}$, $z_n = x_n - y_n$, converges to zero. We leave to the reader the straightforward exercise of checking that this relation is reflexive, symmetric and transitive, and that is compatible with the ring operations (i.e. $\{x_n\} \sim \{x'_n\}$, $\{y_n\} \sim \{y'_n\}$ implies $\{x_n + y_n\} \sim \{x'_n + y'_n\}$, $\{x_n y_n\} \sim \{x'_n y'_n\}$). Then if $\xi, \eta \in \bar{A}$ and $\xi \ni \{x_n\}$, $\eta \ni \{y_n\}$ we define $\xi + \eta$ as the class of the sequence $\{x_n + y_n\}$, and $\xi\eta$ as the class of the sequence $\{x_n y_n\}$.

We now define a norm in $\bar{A}$ by $\| \xi \| = \lim_{n\to\infty} \| x_n \|$, where $\{x_n\} \in \xi$. Then $\bar{A}$ is a complete metric space. Indeed, let $\{\xi_m\}$ be a Cauchy sequence in $\bar{A}$. Let us assume that from every ξ_m a representative $\{x_{mn}\}$, $m = 1, 2, \ldots$, is selected. We then define a sequence $\{y_n\}$ in A, by putting $y_n = x_{nn}$, $n = 1, 2, \ldots$. Then $\{y_n\}$ is Cauchy sequence in A. Indeed, let $\epsilon > 0$. There can be found a natural number N_1 such that for any $m, n > N_1$ we have $\| \xi_m - \xi_n \| < \frac{\epsilon}{3}$, i.e. $\lim_{p\to\infty} \| x_{mp} - x_{np} \| < \frac{\epsilon}{3}$. The latter means that for $p > N_2$, where N_2 is some natural number, we have $\| x_{mp} - x_{np} \| < \frac{\epsilon}{3}$. Let now $m, n > \max\{N_1, N_2\}$. Then

$$\begin{aligned} \| y_m - y_n \| = \| x_{mm} - x_{nn} \| &\leq \| x_{mm} - x_{mp} \| \\ &+ \| x_{mp} - x_{np} \| + \| x_{np} - x_{nn} \| < \frac{\epsilon}{3} + \frac{\epsilon}{3} + \frac{\epsilon}{3} = \epsilon. \end{aligned}$$

Thus $\{y_n\}$ is really a Cauchy sequence.

Moreover,

$$\| \xi_m - \eta \| = \lim_{n\to\infty} \| x_{mn} - y_n \| = \lim_{n\to\infty} \| x_{mn} - x_{nn} \| < \epsilon$$

when m is sufficiently large, i.e. $\eta = \lim_{m\to\infty} \xi_m$. This proves the completeness of the space $\bar{A}$. An imbedding of the ring A into $\bar{A}$ is obtained by identifying an element x from A with the class $\bar{x}$ of the Cauchy sequence $\{x_n\}$ for which $x_n = x$ for all $n = 1, 2, \ldots$. It is evident that the mapping $x \mapsto \bar{x}$ is an injective homomorphism of rings and this allows identify A with its image in $\bar{A}$. In addition $\| \bar{x} \| = \lim_{n\to\infty} \| x \| = \| x \|$. We should note that the class

ξ of a sequence $\{x_n\}$, $n = 1,2,\dots$ is the limit of the sequence of the classes $\{\bar{x}_m\}$, $m = 1,2,\dots$ ($x_{mn} = x_m$, $n = 1,2,\dots$). This also means that A is dense in $\bar{A}$, i.e. that $\bar{A}$ is the closure of A. Finally, let A be a field. Consider $\xi \ni \{x_n\}$, where $\xi \neq 0$. Then $\lim_{n\to\infty} \| x_n \| = \alpha \neq 0$. Now consider the sequence $\{y_n\} = \{x_n^{-1}\}$. Then $\lim_{n\to\infty} \| y_n \| = \alpha^{-1}$, i.e. $\{y_n\}$ is a Cauchy sequence. If η is the class containing $\{y_n\}$, then $\xi\eta = 1$, i.e. $\bar{A}$ is also a field.

□

6.5. Approximation. Non–equivalent norms are in fact independent. The exact meaning of this statement can be found in the following result.

Theorem (Artin – Whaples). *Let $\| \ \|_1, \dots, \| \ \|_n$ be non–equivalent non–trivial norms on the field F, and let $x_1, \dots, x_n$ be elements from F. For every $\epsilon > 0$ there exists an $y \in F$ such that $\| y - x_1 \|_1 < \epsilon, \dots, \| y - x_n \|_n < \epsilon$.*

Proof. Let us start by noting that it is sufficient to find $y_i \in F$, $i = 1, \dots, n$, such that $\| y_i \|_i > 1$ and $\| y_i \|_j < 1$, $1 \leq i \neq j \leq n$. In this case for $r \to \infty$ we obtain

$$\frac{y_i^r}{1 + y_i^r} = \frac{1}{1 + y_i^{-r}} \to \begin{cases} 1 & \text{in the sense of norm } \| \ \|_i, \\ 0 & \text{in the sense of norm } \| \ \|_j, j \neq i. \end{cases}$$

For sufficiently large r the desired element is

$$y = \sum_{i=1}^{n} \frac{x_i y_i^r}{1 + y_i^r}.$$

Because of symmetry with respect to the indexes it is sufficient to show the existence of an y_1 such that $\| y_1 \|_1 > 1$ and $\| y_1 \|_j > 1$, when $j = 2, \dots, n$. We shall apply induction on n. When $n = 2$ the fact that $\| \ \|_1$ and $\| \ \|_2$ are non–equivalent, yields an element z' such that $\| z' \|_1 < 1$, $\| z' \|_2 \geq 1$, and a z'' such that $\| z'' \|_1 \geq 1$ and $\| z \|_2'' < 1$. In this case we can take $y_1 = z''(z')^{-1}$.

Let now $n \geq 3$. According to the induction hypothesis there exists a $z \in F$, such that $\| z \|_1 > 1$, $\| z \|_j < 1$, $j = 2, \dots, n-1$. Using the case of $n = 2$ there exists also an $u \in F$ for which $\| u \|_1 > 1$ and $\| u \|_n < 1$. Then for sufficiently large natural r we take

$$y_1 = \begin{cases} z, & \text{if } \| z \|_n < 1, \\ z^r u, & \text{if } \| z \|_n = 1, \\ \frac{z^r u}{1+z^r}, & \text{if } \| z \|_n > 1. \end{cases}$$

□

6.6. Norms on the field of rational numbers. In Subsection 6.1 some examples of norms of the field **Q** of rational numbers were mentioned. Respectively, there exist two types of the completion of the field **Q**: the field **R** of real numbers and the field $\mathbf{Q}_p$ of p–adic numbers for a prime p. In this subsection it will be shown that there are no other completions of this field.

Ostrovski's theorem. *Any non–trivial norm of the field* **Q** *of rational numbers is equivalent to the usual norm which is obtained by means of the standard absolute value, or to a p–adic norm.*

Proof. Let $\| \ \|$ be some norm on the field **Q**. Two cases are possible: either there exists at least one natural $a > 1$, for which $\| a \| > 1$, or $\| n \| \leq 1$, for all natural n. Let us start with the first case. Since

$$\| n \| \leq \| 1 + \ldots + 1 \| \leq \| 1 \| + \ldots + \| 1 \| = n,$$

we can assume that $\| a \| = a^\alpha$, where the real number α satisfies condition $0 < \alpha \leq 1$.

Take an arbitrary natural number N and write it in terms of powers of a:

$$N = x_0 + x_1 a + \ldots + x_{k-1} a^{k-1},$$

where $0 \leq x_i \leq a - 1$ $(0 \leq i \leq k-1)$, $x_{k-1} \geq 1$. Then for N the inequality $a^{k-1} \leq N < a^k$ holds. By virtue of the properties of norms and the fact noted above, we obtain

$$\begin{aligned} \| N \| &\leq \| x_0 \| + \| x_1 \| \| a \| + \ldots + \| x_{k-1} \| \| a \|^{k-1} \leq \\ &\leq (a-1)(1 + a^\alpha + \ldots + a^{(k-1)\alpha}) = (a-1)\frac{a^{k\alpha} - 1}{a^\alpha - 1} < \\ &< (a-1)\frac{a^{k\alpha}}{a^\alpha - 1} = \frac{(a-1)a^\alpha}{a^\alpha - 1} a^{(k-1)\alpha} \leq \\ &\leq \frac{(a-1)a^\alpha}{a^\alpha - 1} N^\alpha = C N^\alpha, \end{aligned}$$

i.e. $\| N \| < CN^{\alpha}$, where the constant C does not depend on N. Now, replace N by N^m, with natural m, in the inequality obtained. Then

$$\| N \|^m = \| N^m \| < CN^{m\alpha},$$

whence $\| N \| < \sqrt[m]{C} N^{\alpha}$. Letting m go to infinity we get the inequality $\| N \| \leq N^{\alpha}$. Now assume that $N = a^k - b$, where $0 < b \leq a^k - a^{k-1}$. Then

$$\| a^k \| = \| (a^k - b) + b \| \leq \| a^k - b \| + \| b \| \leq \| N \| + \| b \|,$$

i.e.

$$\| N \| \geq \| a^k \| - \| b \| = a^{\alpha k} - \| b \|.$$

According to what has been proved

$$\| b \| \leq \| b \|^{\alpha} \leq (a^k - a^{k-1})^{\alpha},$$

and hence

$$\| N \| \geq a^{\alpha k} - (a^k - a^{k-1})^{\alpha} =$$
$$(1 - (1 - \frac{1}{a})^{\alpha})a^{\alpha k} = C_1 a^{\alpha k} > C_1 N^{\alpha},$$

where the constant C_1 does not depend on N. By replacing N by N^m in the last inequality, we obtain $\| N \|^m = \| N^m \| > C_1 N^{\alpha m}$ or $\| N \| > \sqrt[m]{C_1} N^{\alpha}$. By letting m go to infinity we obtain $\| N \| \geq N^{\alpha}$. Therefore, finally

$$\| N \| = N^{\alpha}.$$

Let now $x = \pm \frac{N_1}{N_2}$ be an arbitrary rational number (N_1 and N_2 are natural). Then

$$\| x \| = \| \frac{N_1}{N_2} \| = \frac{\| N_1 \|}{\| N_2 \|} = \frac{N_1^{\alpha}}{N_2^{\alpha}} = \| x \|^{\alpha}.$$

Thus it is proved that our norm is equivalent to the norm obtained by means of the standard absolute value.

Let us pass to the case where $\| n \| \leq 1$ for all natural n. If for all primes p we would have $\| p \| = 1$ then, evidently, our metric would be trivial, i.e. $\| x \| = 1$ for any $x \neq 0$.

Let the prime p be such that $\| p \| < 1$. Assume that for another prime q we also have $\| q \| < 1$. Let us select exponents k, l such that the following inequalities are fulfilled

$$\| p \|^k < \frac{1}{2}, \quad \| q \|^l < \frac{1}{2}.$$

Because the numbers p^k and q^l are mutually prime for some integers u, v we obtain $up^k + vq^l = 1$.

Then

$$1 = \| 1 \| = \| up^k + vq^l \| \leq \| u \| \| p \|^k + \| v \| \| q \|^l < \frac{1}{2} + \frac{1}{2} = 1.$$

The contradiction obtained shows that there exists only one prime p for which $\| p \| = \rho < 1$. Since $\| q \| = 1$ for the remaining prime numbers, $\| a \| = 1$ for all a which are mutually prime with p. Let $x = p^m \frac{a}{b}$ be a rational number different from zero, where, moreover, a, b are integers and mutually prime with p. Then

$$\| x \| = \| p^m \| \frac{\| a \|}{\| b \|} = \| p \|^m = \rho^m.$$

Thus our norm is equivalent to a p–adic norm on the field of rational numbers.

6.7. Field of p–adic numbers. The completion $\mathbf{Q}_p$ of the field of rational numbers relatively to the p–adic norm is called the *field of p–adic numbers.*

Proposition 1. *Every element of the field $\mathbf{Q}_p$ of p–adic numbers can be presented uniquely in the form of a convergent series*

$$a_m p^m + a_{m+1} p^{m+1} + \dots , \tag{1}$$

where $m \in \mathbf{Z}$, and the numbers $a_m, a_{m+1}, \dots$ are integers in $\{0, 1, \dots , p-1\}$.

Proof. It is clear that a series of the form (1) converges in $\mathbf{Q}_p$. In fact the sequence of its partial sums

$$S_k = a_m p^m + a_{m+1} p^{m+1} + \dots + a_{m+k} p^{m+k}$$

is a Cauchy sequence: indeed, the number $\| S_k - S_l \| \leq \left(\frac{1}{p}\right)^{m+k+l}$, $k \leq l$, can be made arbitrarily small for all sufficiently large k and l. In virtue of the completeness of the field $\mathbf{Q}_p$ the series converges to a well defined p-adic number ξ. If $a_m \neq 0$ then $\| S_k \| = \left(\frac{1}{p}\right)^m$ for any $k = 0, 1, \ldots,$ i.e. $\| \xi \| = \left(\frac{1}{p}\right)^m$. We should note that for different sets of coefficients different elements from $\mathbf{Q}_p$ are obtained; thus if

$$a'_m p^m + a'_{m+1} p^{m+1} + \ldots \tag{2}$$

presents the same number and s is the minimal integer for which $a_s \neq a'_s$, then the difference of the series (1) and (2) would have the norm $\left(\frac{1}{p}\right)^s \neq 0$.
The contradiction obtained proves the uniqueness of representation in the form (1). Now we shall demonstrate that the set T of series (1) is a complete field containing $\mathbf{Q}$. Indeed, it evidently contains all natural numbers (take the p-adic decomposition of a natural number).

All series of the form (1) have the form $p^m(b_0 + b_1 + \ldots) = p^m \lim_{n\to\infty} y_n$, where $\{y_n\}$ is a converging sequence of natural numbers which are not divisible by p. It is clear, that the product of such two numbers, i.e.

$$\left(p^m \lim_{n\to\infty} y_n\right)\left(p^s \lim_{n\to\infty} z_n\right) = p^{m+s} \lim_{n\to\infty} (y_n z_n),$$

is again a number of the same form. Further, the inverse of a natural number a which is not divisible by p, can be found in the following way. We write $au + pv = 1$ for suitable integers u, v. Then $au = 1 - pv$. In this case

$$au(1 + pv + p^2v^2 + \ldots) = 1.$$

Now in the series $1 + pv + p^2v^2 + \ldots$ write $v, v^2, \ldots$ as a p-adic sum. This permits us to rewrite the series in the form (1). Thus, the inverse a^{-1} of a natural number a has the form (1). Since any number of the form (1) is $p^m \lim_{n\to\infty} y_n$, where the y_n are natural numbers, the inverse will be of the form $p^{-m} \lim_{n\to\infty} y_n^{-1}$, i.e. of the form (1). Finally,

$$-1 = \frac{p-1}{1-p} = (p-1)(1 + p + p^2 + \ldots) = (p-1) + (p-1)p + (p-1)p^2 + \ldots$$

Thus, T is a subfield in $\mathbf{Q}_p$, containing $\mathbf{Q}$ and the limits of all elements from $\mathbf{Q}$. Since $\mathbf{Q}_p$ is a completion of $\mathbf{Q}$, the proposition is proved. □

The subring $\mathbf{Z}_p$ of numbers ξ in $\mathbf{Q}_p$ satisfying $\| \xi \| \leq 1$ is called the *ring of integer p–adic numbers* (or the *ring of p–adic integers*).

Proposition 2.. *The set* $\mathbf{Z}_p$ *is compact.*

Proof. Let $\{\mathcal{O}_\lambda\}_{\lambda\in\Lambda}$ be some family of open sets covering $\mathbf{Z}_p$. We must prove that a finite such covering can be found. Let us suppose the opposite. The ring $\mathbf{Z}_p$ is the union of the finite set of subsets $p\mathbf{Z}_p$, $1+p\mathbf{Z}_p, \dots, (p-1)+p\mathbf{Z}_p$. Then at least one from these sets, let us say $a_0+p\mathbf{Z}_p$, is not covered by any finite subfamily from $\{\mathcal{O}_\lambda\}$. Analogously, there can be found a_1, $0 \le a_1 < p$ such that $a_0+a_1p+p^2\mathbf{Z}_p$ is not covered by any finite number of subsets from $\{\mathcal{O}_\lambda\}$. Continuing this process, we find a number $a = a_0+a_1p+a_2p^2+\dots$. According to the covering condition there is an index $\lambda_0 \in \Lambda$ such that $a+p^s\mathbf{Z}_p \subset \mathcal{O}_{\lambda_0}$, i.e. $a_0+a_1p+\dots+a_{s-1}p^{s-1}+p^s\mathbf{Z}_p$ is covered only by one set from $\{\mathcal{O}_\lambda\}$. Contradiction. The proposition is proved. □

Corollary. $\mathbf{Q}_p$ *is a locally compact space.*

In order to prove this it should be noted that $\mathbf{Q}_p$ is an additive topological group (see, Subsection 2.5). This means that the neighborhoods of any point are just like those of zero. The set $\mathbf{Z}_p$ is the closure of the neighborhood of zero defined by $\| x \| < 1$. Therefore for any point a there is the closed neighborhood $a+\mathbf{Z}_p$ which is compact. This means that $\mathbf{Q}_p$ is locally compact. □

∘ We should note that the same property is valid for the field of real numbers – the completion of the field of rational numbers relative to the usual topology defined by the usual absolute value. ∘

References. Algebraic theory of numbers, Borevich–Shafarevich [1], Lang [1], van der Waerden [1].

CHAPTER II

GROUPS

§ 1. Representations of groups

1.1. Semisimple modules. Let R be not necessarily commutative ring, M a left R–module. We recall (see Introduction 3.1) that M is a *simple (irreducible)* R–module if any R–submodule either coincides with M or is equal to the zero submodule. The simple modules over the ring $\mathbf{Z}$ of integers are the cyclic groups of prime order p. The simple modules over the ring of polynomials $F[X]$ of one variable X over an algebraically closed field F are one–dimensional. If the field is not algebraically closed then every module of this kind has the form of quotient module $F[X]/(p(X))$, where the polynomial $p(X)$ is irreducible in $F[X]$. The structure of simple modules over an arbitrary principal ideal ring is similar. Finally, if $R = F[X_1, \dots , X_n]$, then, according to the Hilbert Nullstellensatz, any irreducible module over R is finite–dimensional. Indeed an irreducible module is generated by any of its non–zero elements, so it is cyclic and, therefore, is representable in the form R/P, where P is some maximal ideal. Further, as was mentioned already, R/P is a field which is finitely generated over F as an algebra. From the Hilbert Nullstellensatz follows that each element from R/P is algebraic over F, i.e. R/P is a finitely generated algebraic extension of a field. Finally, R/P is finite–dimensional over F as was required. If F is algebraically closed then any irreducible R–module is one–dimensional.

Another important case when irreducible modules are finite–dimensional is the case of G–modules over field F, when the group G is finite. We should recall that if G is a group, and $\rho : G \to GL(V)$ a representation of it by non–singular linear operators on a vector space V over a field F then, in a natural way, V becomes a module over the group algebra $F[G]$ of the group G (see, Introduction 3.5).

In this case the elements of the form gv, where $g \in G$, $v \in V$ are defined, and this allows us to consider the G–module V. We should note, that if G is a finite group, then $R = F[G]$ is a finite–dimensional vector space over F. Since any irreducible G–module M has the form R/P, M is finite–dimensional.

When analyzing G–modules we come across the concept of a semisimple module. The fact is that according to Maschke classical theorem (see, Introduction 3.8) any finite–dimensional G–module V over the field of characteristic 0 or p, where $(p, |G|) = 1$, is completely reducible, i.e. for any G–submodule $U \subset V$ there can be found a G–module W such that $V = U \oplus W$. A direct consequence is that any finite–dimensional module is decomposable into a direct sum of irreducible modules. Generalizing this concept to the case of an arbitrary ring and a module over it, we obtain the concept of a semisimple module.

Definition. The module M over the ring R is called *semisimple* if it is equal to the direct sum of some simple submodules of it.

The condition given is equivalent to what was noted above, i.e. to the existence of a complement to every submodule. In order to prove this fact we note the following evident statement: if a module (not necessarily simple) is equal to the sum of some family of irreducible submodules, then it is equal to the direct sum of some subfamily of the family considered (Zorn's lemma should be applied). Let now U be a submodule of a semisimple R–module M. We denote by W a maximal sum of simple submodules which has a zero intersection with U. Then the sum $U \oplus W$ is direct. Let $U \oplus W$ not coincide with M. Then there can be found a simple submodule S which does not lie in $U \oplus W$, i.e. which has a zero intersection with $U \oplus W$. In this case $W \oplus S$ has a zero intersection with U and this contradicts the maximality of W. So $U \oplus W = M$ as was required. Conversily, let the module M be completely reducible. According to Zorn's lemma, M contains a maximal semisimple submodule $M_x \supset Q$ for which $x \notin M_x$. We denote by N_x the submodule generated by the module M_x and the element x, i.e. $N_x = Rx + M_x$. Let us select a complement L_x to N_x in M. Then $M = L_x \oplus N_x$. We should note that according to the definition, N_x/M_x does not contain intermediate submodules. Let K_x be a complement to M_x in M, i.e. $M = K_x \oplus M_x$. We consider $N_x \cap (K_x \oplus M_x)$. Since $N_x \supset M_x$, we have

$$N_x = N_x \cap M = N_x \cap (K_x \oplus M_x) = (N_x \cap K_x) \oplus M_x.$$

Evidently, $N_x \cap K_x \cong N_x/M_x$, i.e. it is a simple module. Finally we have

$$M = L_x \oplus (N_x \cap K_x) \oplus M_x.$$

In this case $(N_x \cap K_x) \oplus Q$ is a semisimple module strictly containing Q. Contradiction. Thus, $M = Q$ is a semisimple module.

The main aim of this section is a density theorem. In order to formulate it we should recall the definition of $K = \mathrm{End}_R(M)$. The ring of endomorphisms of an R–module M is defined as follows; it is the ring of those homomorphisms of the abelian group M into itself such that $\phi(rx) = r\phi(x)$ for any $r \in R$, $x \in M$. The product of endomorphisms is equal to its composition. The sum of endomorphisms is defined "pointiwise": if $\phi, \psi \in K$ then $(\phi + \psi)(x) = \phi(x) + \psi(x)$. The ring K is also called the commutant (centralizer) of the R–module M.

We should note that M is a K–module if we define $\phi x = \phi(x)$ for any $\phi \in K$, $x \in M$. We shall consider $L = \mathrm{End}_K(M)$. If $r \in R$, then the mapping $r_M : x \mapsto rx$ is an endomorphism of the K–module M. In particular, if M is faithful R–module then the mapping $r \mapsto r_M$ is an imbedding of the ring R in $L = \mathrm{End}_K(M)$. In a number of significant cases the image of the ring R for this imbedding coincides with L.

1.2. Density theorem. *Let M be a semisimple module over R, $K = \mathrm{End}_R(M)$, $f \in \mathrm{End}_K(M)$, $x_1, \dots, x_n \in M$. Then there exists an $r \in R$ such that $f(x_i) = rx_i$ for all $i = 1, \dots, n$.*

Proof. Let us first assume that the statement has already been proved for $n = 1$. We shall consider the direct sum M^n of n copies of the module M. This module is also semisimple. The mapping $f^n : M^n \to M^n$ associating the element $(f(y_1), \dots, f(y_n))$ with any element $(y_1, \dots, y_n)$ is an endomorphism of it. According to the assumption, there exists an $r \in R$ such that $f^n(x_1, \dots, x_n) = r(x_1, \dots, x_n)$, i.e. $(f(x_1), \dots, f(x_n)) = (rx_1, \dots, rx_n)$ as was required.

When $n = 1$ it is necessary for $f \in \mathrm{End}_K(M)$, $x \in M$, to find an $r \in R$ such that $f(x) = rx$.

By the property of complete reducibility $M = Rx \oplus M'$. We denote by π the projection of the module M on Rx. Then $\pi \in K$. We have

$$f(x) = f(\pi(x)) = \pi(f(x)) = rx$$

for a suitable $r \in R$. The theorem is proved. □

This theorem has a number of important corollaries.

Corollary 1. *Let M be a finite–dimensional vector space over an algebraically closed field F, $R \subset End_F(M)$. If M is an irreducible R–module, than $R = End_F(M)$.*

According to the Schur lemma any endomorphism of an irreducible module either is equal to zero or is an automorphism. Thus $K = \mathrm{End}_R(M)$ is a skew field; moreover $K \subset \mathrm{End}_F(M)$. From the last condition it follows that any element α from K is algebraic over F. In other word, $F(\alpha)$ is an algebraic extension of the field F; hence, $F(\alpha) = F$ and $\alpha \in F$. It follows that $K = F$. Hence, $\mathrm{End}_K(M) = \mathrm{End}_F(M)$. If $e_1, \dots, e_n$ is a basis in M, f is an arbitrary linear operator then $f \in \mathrm{End}_K(M)$. According to the density theorem there exists an $r \in R$ such that $re_1 = f(e_1), \dots, re_n = f(e_n)$, i.e. $r_M = f$, so that $R = \mathrm{End}_F(M)$. □

Corollary 2. *Let M be a finite–dimensional vector space over algebraically closed field F, G a subgroup of the group $GL(M)$ of non–singular linear operators on the space M. If M is an irreducible G–module then G contains a basis of the space $End_F(M)$.*

Proof. Let us take $R = F[G]$. Then according to the previous corollary $R = \mathrm{End}_F(M)$. Since R is the linear span of the elements of the group G the group must contain a basis of the space $\mathrm{End}_F(M)$, □

Corollary 3. *Let M be a faithful irreducible module, $K = End_R(M)$. Let us assume that M is finite–dimensional as a vector space over the field K. Then $R = End_K(E)$.* □

If the condition on finite–dimensionality in the last corollary is missing, then there still remains

Corollary 4. *Let M be an irreducible R–module, and let the field $K = End_R(M)$ be a centralizer of it. For any $x_1, \dots, x_n \in M$ linearly independent over K, and any $y_1, \dots, y_n \in M$ there can be found an $r \in R$ such that*

$y_1 = rx_1, \ldots, y_n = rx_n$. □

1.3. Induced modules. Let R be some ring, S a subring of it. Then R is a (S,S)*–bimodule*, i.e. a module from both the left and the right relative to the usual multiplication with the additional condition $(s_1 r)s_2 = s_1(rs_2)$, $s_1, s_2 \in S$, $r \in R$. If N is a certain left S–module then the tensor product $M = R \otimes_S N$ is, naturally, provided with the structure of the left R–module by putting $r(r' \otimes n) = (rr') \otimes n$. This R–module M is called the *module induced from the S–module* N. In particular, if K is an extension of a field F, V is a vector space over F, then $V^K = K \otimes_F V$ is a vector space over K. When passing from the field **R** of real numbers to the field **C** of complex numbers this is called *complexification* of the space. We should note that if $\dim_F V = n$ then $\dim_K V^K = n$. The fact is that if $\{e_1, \ldots, e_n\}$ is a basis for V then $\{1 \otimes e_1, \ldots, 1 \otimes e_n\}$ is a basis for $V^K = K \otimes_F V$ (see, Subsection 1.4.1).

In the present subsection we shall be interested in a way of constructing representations of groups, or, which is the same, of G–modules, where G is some group, that are obtained by means of inducing. Let F be a base field, H a subgroup of the group G, N some H–module, i.e. a vector space over F, on which H acts by means of non–singular linear operators. Then the induced G–module $\mathrm{Ind}_H^G N$ is

$$\mathrm{Ind}_H^G N = F[G] \otimes_{F[H]} N,$$

where $F[G]$ is the group algebra of the group G and $F[H]$ the subalgebra of it with basis consisting of the elements from the subgroup H. In order to study the structure of the induced module we select a *transversal*, T, for H in G. We have in mind a set of representatives of the left cosets of the the subgroup H in G. If such selection is made then any element $g \in G$ can be the uniquely written in the form $g = th$, $h \in H$, $t \in T$. We shall write $\bar{g} = t$. Let $T = \{t_1, \ldots, t_n\}$. We then see that $F[G] = \bigoplus_{i=1}^{n} t_i F[H]$ is a free $F[H]$–module, and, hence, $\mathrm{Ind}_H^G N = \bigoplus_{i=1}^{n} t_i \otimes_{F[H]} N$. From this it follows that if $\{e_1, \ldots, e_m\}$ is a basis for N over F, then a basis of the induced module $\mathrm{Ind}_H^G N$ can be found of the form $\{t_i \otimes e_j \,|\, i = 1, \ldots, n;\ j = 1, \ldots, m\}$.

The information obtained allows to calculate the character of the induced G–module $\mathrm{Ind}_H^G N$ given the character of the H–module N.

Proposition. *Let ψ be the character of the representation of the subgroup H*

in the H–module N, $M = \mathrm{Ind}_H^G N$. Then the character χ of the representation of the group G in the G–module M is calculated according to the following formula

$$\chi(g) = \sum_{i=1}^{n} \psi_0(t_i^{-1} g t_i),$$

where $T = \{t_1, \dots, t_n\}$ is a transversal for H in G and ψ_0 is the extension of the character ψ to G defined by $\psi_0(g) = 0$, when $g \notin H$.

Proof. As was noted, $M = \bigoplus_{i=1}^{n} t_i \otimes N$. Let $g \in G$, ρ the representation in M. We now calculate tr $\rho(g)$. We have $\rho(g)(t_i \otimes e_j) = gt_i \otimes e_j$. Write $gt_i = t_k h$. then $\rho(g)(t_i \otimes e_j) = t_k \otimes he_j$. A non–zero contribution to tr $\rho(g)$ from an element of the basis $t_i \otimes e_j$ is obtained only if $k = i$. We denote the set of such i by $\mathcal{I}$; then $h = t_i^{-1} g t_i$. For such i we get

$$\rho(g)(t_i \otimes e_j) = t_i \otimes (t_i^{-1} g t_i) e_j =$$
$$= t_i \otimes \sum_{k=1}^{m} \lambda_{jk}^{(i)} e_k = \sum_{k=1}^{m} \lambda_{jk}^{(i)} (t_i \otimes e_k).$$

Here $\sum_j \lambda_{jj}^{(i)} = \mathrm{tr}\ \rho(t_i^{-1} g t_i) = \psi(t_i^{-1} g t_i)$. The summation over $i \in \mathcal{I}$ with the noted property and over j gives the trace of the operator $\rho(g)$, i.e. $\chi(g) = \sum_{i \in \mathcal{I}} \psi(t_i^{-1} g t_i)$.

However if $i \notin \mathcal{I}$ then according to the definition $\psi_0(t_i^{-1} g t_i) = 0$. Therefore the formula (1) differs from the one just obtained only by several zero summands.

1.4. Frobenius duality. Induced representations can also be defined in a non–constructive way. Indeed, if H is a subgroup of the group G then any finite–dimensional representation for G automatically is a representation for H. Let us fix an H–module N and consider G–modules V which contain an homomorphic image of the H–submodule N and as G–modules are also generated by this module. Our induced G–module $M = \mathrm{Ind}_H^G N$ possesses this property. Now M is distinguished among all the modules of this type by the fact that it "covers" all these modules, i.e. the homomorphism $\phi : N \to V$ extends uniquely to an epimorphism of the G–modules $\bar{\phi} : M \to V$. Since two modules with such a property are evidently isomorphic, in order to prove the statement just made, it is sufficient to note how to extend the identity mapping to $\mathrm{Ind}_H^G N$. It is completely natural:

$$\bar{\phi}\Big(\sum_{i,j} \mu_{ij} t_i \otimes e_j\Big) = \sum_{i,j} \mu_{ij} t_i \phi(e_j) \in V.$$

If $g \in G$, $g t_i = t_{k_i} h_i$, $h_i e_j = \sum_s \lambda_{ij}^s e_s$, then

$$\begin{aligned}
&\bar{\phi}\Big(g \sum \mu_{ij} t_i \otimes e_j\Big) = \bar{\phi}\Big(\sum \mu_{ij} t_{k_i} h_i \otimes e_j\Big) = \\
&= \bar{\phi}\Big(\sum_{i,j} \mu_{ij} t_{k_i} \otimes h_i e_j\Big) = \bar{\phi}\Big(\sum_{i,j,s} \mu_{ij} \lambda_{ij}^s t_{k_i} \otimes e_s\Big) = \\
&= \sum_{i,j,s} \mu_{ij} \lambda_{ij}^s t_{k_i} \phi(e_s) = \sum_{i,j,s} \mu_{ij} t_{k_i} \phi(h_i e_j) = \\
&= \sum_{i,j} \mu_{ij} t_{k_i} h_i \phi(e_j) = \sum_{i,j} \mu_{ij} (g t_i) \phi(e_j) = g\bar{\phi}\Big(\sum \mu_{ij} t_i \otimes e_j\Big),
\end{aligned}$$

i.e. $\bar{\phi}$ is a well defined G–homomorphism. As an example of an induced module the regular module can serve: it is isomorphic to $F[G]$ and is induced from the one–dimensional representation of the identity subgroup . It is useful to keep in mind the following property of induction: if H is a subgroup in G, K is a subgroup in H and V is some left K–module, then

$$\mathrm{Ind}_H^G(\mathrm{Ind}_K^H V) \cong \mathrm{Ind}_K^G V.$$

Indeed it is clear that the module on the right side covers the module on the left side. Further the G–module $\mathrm{Ind}_K^G V$ has the H–submodule $K[H]V$. It is covered by the H–module $\mathrm{Ind}_K^H V$. If $\phi : \mathrm{Ind}_K^H V \to K[H]V$ is the corresponding epimorphism then there exists $\bar{\phi} : \mathrm{Ind}_H^G(\mathrm{Ind}_K^H V) \to \mathrm{Ind}_K^G V$ and it is an epimorphism because $\mathrm{Ind}_K^G V$ is generated as a G–module by the space V and, a fortiori, by $K[H]V$.

The first significant result on induced modules is the so called Frobenius duality law. It is the most convenient to present its proof in the language of characters. We shall denote by $\mathrm{Res}_H^G V$ the restriction of the G–module V to H, i.e. V as an H–module. If χ is the character of an H–module V then $\mathrm{Ind}_H^G \chi$ is the character of the induced G–module $\mathrm{Ind}_H^G V$. If ψ is the character of the G–module V, then $\mathrm{Res}_H^G \psi$ is the character of the H–module $\mathrm{Res}_H^G V$.

By $(\ |\ )_G$ (respectively, $(\ |\ )_H$) we denote the scalar product of characters from the group G (respectively, H).

Theorem (Frobenius duality). *Let χ be a character of a subgroup H of the group G, ψ a character of the group G. Then*

$$(\mathrm{Ind}_H^G \chi | \psi)_G = (\chi | \mathrm{Res}_H^G \psi)_H.$$

Proof. The theorem is easily obtainable from the proposition of Subsection 1.3. According to this proposition

$$(\mathrm{Ind}_H^G \chi) = \sum_{i=1}^{n} \chi_0(t_i^{-1} g t_i),$$

where $\{t_1, \dots, t_n\}$ is a transversal of the group G with respect to the subgroup H, and χ_0 is the character χ extended by zero on $G \setminus H$. We write

$$(\mathrm{Ind}_H^G \chi | \psi)_G = \frac{1}{|G|} \sum_{g \in G} (\mathrm{Ind}_H^G \chi)(g) \psi(g^{-1}) =$$

$$= \frac{1}{|G|} \sum_{i=1}^{n} \sum_{g \in G} \chi_0(t_i^{-1} g t_i) \psi(g^{-1}) =$$

$$= \frac{1}{|G|} \sum_{i=1}^{n} \sum_{g \in G} \chi(h) \psi(t_i h^{-1} t_i^{-1}) =$$

$$= \frac{1}{|G|} \sum_{i=1}^{n} \sum_{g \in G} \chi(h) \psi(h^{-1}) =$$

$$= \frac{1}{|H|} \sum_{h \in H} \chi(h) (\mathrm{Res}_H^G \psi)(h^{-1}) = (\chi | \mathrm{Res}_H^G \psi)_H.$$

We have used the fact for a fixed t_i each element h is representable in a unique way in the form $t_i^{-1} g t_i$ as $g \mapsto t_i^{-1} g t_i$ is an internal automorphism of the group G. Also, $|G| = n|H|$. □

Corollary. *Let N be an irreducible H–module, V – an irreducible G–module. Then the multiplicity of occurence of V in $\mathrm{Ind}_H^G N$ is equal to the multiplicity of occurence of N in $\mathrm{Res}_H^G V$.*

Proof. See the remarks in Subsection 1.3. □

1.5. Proposition. *Let K be an algebraic extension of the field F, G a group, V a vector space over F provided with the structure of a G–module, $V^K = V \otimes_F K$ a vector space over K which is given the structure of a G–module by*

$$g(v \otimes \lambda) = gv \otimes \lambda$$

for any $v \in V$, $\lambda \in K$. Then:

(1) *if V^K is a semisimple G–module then V also is semisimple;*

(2) *if V is a semisimple G–module, $\dim_F V < \infty$ and the field F is perfect, then V^K is also a semisimple G–module.*

Proof. First we shall prove (1). Let W be some G–submodule in V. Then $W^K = W \otimes K$ is a G–submodule in V^K. In this case there exists a projection $p : V^K \to W^K$ which is a homomorphism of modules. Consider any F–linear mapping $h : K \to F$ under which $h(1) = 1$. Then the linear mapping

$$x \mapsto (1 \otimes h)(p(x \otimes 1))$$

of the space V onto $W \otimes F = W$ is a homomorphism of modules which is equal to the identity on W. The kernel of this homomorphism is a complement to W, i.e. V is a semisimple G–module.

We shall now prove (2). Without restriction of generality we can assume that V is a simple G–module. Let us suppose that W is a simple submodule in V^K of minimal dimensional over K. Then there can be found a finitely generated, and consequently, finite extension K' of the field F such that $W = W' \otimes_{K'} K$. Using (1) and the perfectness of the field F permits to limit ourselves to the case when K' is an Galois extension of the field F with Galois group Γ. Let us consider the G–submodule

$$U' = \sum_{\gamma \in \Gamma} (1 \otimes \gamma)(W').$$

The submodule U' is stable under the Galois group Γ; consequently, there is a submodule $U \subset V$ such that $U' = U \otimes K'$. Since V is simple, we have $U = V$ and $U' = V' = V \otimes K'$. Thus V' is a semisimple G–module, $V' = \sum_{\gamma \in \Gamma} W'_\gamma$, where $W'_\gamma = (1 \otimes \gamma)W$. Because the dimension of the module W is minimal, each G–module $W_\gamma = W'_\gamma \otimes_{K'} K$ is simple and

$$V^K = \sum_{\gamma\in\Gamma} W_\gamma$$

is a semisimple G–module. □

§ 2. Periodic groups

2.1. Introductory remarks. A group G is called *periodic* if each of its elements has finite order. The Burnside problem on periodic groups consists of the following: will a periodic group with a finite number of generating elements be finite? For instance, until now it is unknown if a group generated by two elements is finite, if each non–identity element of this group is of order 5. In 1968 S.P. Novikov and S.I. Adyan constructed examples of infinite groups $B(n,m)$ with n generators and such that the order of each elements divides m, where n, m are arbitrary natural numbers such that $n \geq 2$, and m is an odd number not less than 4381 (later on, S.I. Adyan proved that the lower bound for m can be lowered to 665). Earlier E.S. Golod constructed examples of infinite groups $G(n,p)$ with $n \geq 2$ generators and the order of each element equal to a power of a prime number $p \geq 2$. It is of interest to compare the result of S.P. Novikov and S.I. Adyan with one of A.I. Kostrikin according to which the orders of finite groups with a given number n of generators and satisfying the identity $x^p = 1$, where p is a prime number, are bounded in terms of a generating family, i.e. there exists a finite group $K(n,p)$ such that any finite group with n generators and the order of each non–identity element equal to p is isomorphic to a quotient–group of $K(n,p)$.

In this chapter our aim is to present both some positive results on the Burnside problem and to point out an example of its negative solution.

We call a group *linear* if it is isomorphic to a subgroup of the group $GL(V)$ of linear invertible operators on a finite vector space V over some field F.

2.2. Burnside's theorem. *Let G be a group of non-singular linear operators on a finite-dimensional vector space V over a field F of zero characteristic. Let us assume that for some natural m the identity $x^m = 1$ holds in the group G. Then G is a finite group.*

Proof. First we shall assume that V is an irreducible G–module (the field F

is algebraically closed). Then, according to the corollary 2, Subsection 1.2, the group G contains n^2 elements forming a basis of the vector space $\operatorname{End} V$ of all linear operators over the field F. Let us fix a basis in the space V and denote by g_{ij} the matrix element which corresponds to the linear operator $g \in G$. Let $g^{(1)}, \dots, g^{(n^2)}$ be a basis in $\operatorname{End} V$ of elements from the group G. The matrices $(g_{ij}^{(k)})$, $1 \leq i, j \leq n$, $1 \leq k \leq n^2$, are linear independent. Let g be an arbitrary element from G. Since g is a root of the polynomial $X^m - 1$, the roots of the minimal polynomial of g belong to a finite set $\epsilon_1, \dots, \epsilon_m$ of m–th roots of unity. Therefore the trace $\operatorname{tr}\ g$ of the operator g takes its value from the finite set of sums of n elements from $\{\epsilon_1, \dots, \epsilon_m\}$. Fixing g and noticing that $gg^{(k)} \in G$, we obtain a system of equations for the g_{ij} of the form

$$\sum_{i,j=1}^{n} g_{ij} g_{ji}^{(k)} = \operatorname{tr}\ gg^{(k)}, \quad k = 1, 2, \dots, n^2. \tag{1}$$

In virtue of what was noted above, the rows of coefficients are linearly independent, and the right sides take only a finite set of values. Therefore the number of sets (g_{ij}) satisfying a system of the form (1) is finite, i.e. the group G is finite.

Let us now assume that V is reducible. We then consider a non–trivial G–invariant subspace U. Let us consider the homomorphisms: $\rho : G \to GL(U)$ and $\sigma : G \to GL(V/U)$. The operator $\rho(g)$ is simply the restriction of the operator g to the invariant subspace U. The operator $\sigma(g)$ is the corresponding quotient–operator, i.e. $\sigma(g)(v+V) = gv+U$. We leave it to the reader to verify that it is well defined.

Carrying out an induction on the dimension of the space V we can assume that the groups $\rho(G)$ and $\sigma(G)$ are finite. The mapping $\chi : G \to \rho(G) \times \sigma(G)$ which is defined by

$$\chi(g) = (\rho(g), \sigma(g)),$$

is a homomorphism of groups with kernel $\operatorname{Ker}\rho \cap \operatorname{Ker}\sigma$.

Let $g \in \operatorname{Ker}\rho \cap \operatorname{Ker}\sigma$. Then the matrix of the operator g in a suitable basis is of the form

$$\begin{pmatrix} E_k & A \\ 0 & E_{n-k} \end{pmatrix},$$

where $k = \dim U$. However $g^m = 1$, and

$$\begin{pmatrix} E_k & A \\ 0 & E_{n-k} \end{pmatrix}^m = \begin{pmatrix} E_k & mA \\ 0 & E_{n-k} \end{pmatrix}.$$

So that $mA = 0$, i.e. (in virtue of the condition for the characteristic of the field!) $A = 0$. Thus $\mathrm{Ker}\,\rho \cap \mathrm{Ker}\,\sigma = \{1\}$, i.e. G is isomomorpic to a subgroup of $\rho(G) \times \sigma(G)$, so G is finite.

In order to pass to an arbitrary field F of zero characteristic we note that if $\bar{F}$ is an algebraic closure of the field F, then the group of non–singular matrices $GL(n, F)$ of order n over F (containing an isomorphic copy of our group G) is situated in $GL(n, \bar{F})$, so that the previous arguments apply. □

We should note that over the infinite field F of characteristc $p > 0$ in $GL(n, F)$, $n \geq 2$, there exist infinite subgroups satsfying the identity $x^p = 1$. For instance, such a group is the group of all matrices of the form $\begin{pmatrix} 1 & \alpha \\ 0 & 1 \end{pmatrix}$, $\alpha \in F$.

2.3. Schur's theorem. A positive answer for the "unrestricted" version of the Burnside problem for matrix groups is contained in the Schur theorem.

Theorem. *Any finitely generated group G of periodic linear operators over a field of zero characteristic is finite.*

Proof. Let $g^{(1)}, \dots, g^{(s)}$ be generating elements of the group G. We fix a basis in the space V, where G acts. Let us consider the field L obtained by adjoining to the field $\mathbf{Q}$ of rational numbers the components $g_{ij}^{(k)}$ of the matrices of these generating elements. Then the components of any matrix from G lie in L. The eigenvalues of any of these matrices from G are roots of some degree of the identity and at the same time they are roots of polynomials of degree n with coefficients from L. The main step in the proof of the theorem is to prove that there exists an m such that each such root of the identity is an m–th root. Indeed, since the matrix of any periodic linear operator is reducible to the diagonal form over an algebraic closure of the base field, we see that $g^m = 1$ for any $g \in G$. This allows us to apply the Burnside theorem (Subsection 2.2) and conclude to the finiteness of the group G.

Lemma. *Let $L = \mathbf{Q}(a_1, \dots, a_t)$ be a finitely generated extension of the field $\mathbf{Q}$, n – a natural number. There exists a natural number m such that if ζ is a root of the identity and satisfies an equation of degree n with coefficients from L, then $\zeta^m = 1$.*

Proof. Let the elements $a_1, \dots, a_k$ form a transcendence basis for L over $\mathbf{Q}$. Let $T = \mathbf{Q}(a_1, \dots, a_k)$. We shall denote by d the degree of the field L over T. Let S be a subfield of the field L that is algebraic over $\mathbf{Q}$. We shall prove that its degree does not exceed d. Indeed, consider elements $\alpha_1, \dots, \alpha_{d+1} \in S$. There can be found coefficients from T, that are not all zero, i.e. rational fractions of $a_1, \dots, a_k$ (and thus also polynomials in these variables) such that

$$P_1(a_1, \dots, a_k)\alpha_1 + \dots + P_{d+1}(a_1, \dots, a_k)\alpha_{d+1} = 0.$$

Set $R(X_1, \dots, X_k) = \sum_{i=1}^{d+1} P_i(X_1, \dots, X_k)\alpha_i$. The coefficients of the polynomial R are algebraic over $\mathbf{Q}$. Let $\sigma_1, \dots, \sigma_s$ be all the imbeddings of the field S in $\bar{S}$ over $\mathbf{Q}$. Then

$$W = \prod_{i=1}^{s} R^{\sigma_i}(X_1, \dots, X_k)$$

is a polynomial with coefficients in $\mathbf{Q}$ (= invariant elements of the field S under the action of all imbeddings). Moreover, $W(a_1, \dots, a_k) = 0$. From the algebraic independence of the elements $a_1, \dots, a_k$ it now follows that $W(X_1, \dots, X_n) = 0$, i.e. $R(X_1, \dots, X_k) = 0$. Now take a non–zero polynomial $P_i(X_1, \dots, X_k)$. In virtue of the infinity of the field $\mathbf{Q}$ there are rational numbers $r_1, \dots, r_k$ such that $P_i(r_1, \dots, r_k) \neq 0$. Then

$$\sum_{j=1}^{d+1} P_j(r_1, \dots, r_k)\alpha_j = 0$$

is the sought for relation between the α_j over $\mathbf{Q}$.

Let now ρ be a primitive root of degree q of 1 andlet it satisfy an equation of degree n with coefficients from L. Let us assume that $f(X)$ is the minimal polynomial for ρ over L. Since $f(X)$ divides $X^q - 1$, its coefficients are algebraic over $\mathbf{Q}$ (or to be more exact, lie in $\mathbf{Q}(\sqrt[q]{1})$). Now consider $r(X) = \prod_{i=1}^{s} f^{\sigma_i}(X)$, where $\sigma_1, \dots, \sigma_s$ are all the imbeddings of S in $\bar{S}$ over $\mathbf{Q}$, to obtain a polynomial with coefficients from $\mathbf{Q}$ of which ρ is a root. Thus ρ is a root of the polynomial of degree not larger than nd with coefficients from $\mathbf{Q}$. Since the degree of the field $\mathbf{Q}(\rho)$ is equal to $\phi(q)$ (see, Subsection 1.1.2) and $\phi(q) \to \infty$, as $q \to \infty$, we see that the number q is bounded from above. □

2.4. Solvable groups. One more significant case when the Burnside problem can be solved positively is in the case of solvable groups (see Introduction 1.10). Recall that a group G is solvable if it contains a finite series of subgroups

$$G = G_0 \supset G_1 \supset \ldots \supset G_l = \{e\}, \tag{2}$$

for which G_{i+1} is normal in G_i and G_i/G_{i+1} is an abelian group. It is clear that $G_i' = [G_i, G_i] \supset G_{i+1}$, so the l–th commutator subgroup of the group G is trivial. Having replaced (2) by the series of commutator subgroups we can assume that all the subgroups G_i, $i = 0, 1, \ldots, l$, are normal in G.

In the study periodic solvable groups an important role is played by the following lemma. Its formulation contains the concept of a locally finite group. With that we mean a group for which each finitely generated subgroup is finite.

Lemma. *Let the group G contain a normal subgroup K such that G/K and K are locally finite groups. Then G is a locally finite group.*

Proof. In § 3, when discussing free groups, we shall prove that a subgroup of finite index in a finitely generated group is itself a finitely generated group (corollary 3, Subsection 3.4). Consider a subgroup H in G generated by elements $h_1, \ldots, h_n$. It has a normal subgroup $L = H \cap K$; moreover, $H/L = H/H \cap K \cong HK/K$. The group HK/K is a subgroup of G/K generated by the finite set of the elements $h_1K, \ldots, h_nK$. By assumption HK/K is finite. Thus $H \cap K$ is a subgroup of finite index in H. According to what was mentioned above, the group $H \cap K$ has a finite number of generators. Since $H \cap K$ is a finitely generated subgroup in K, it is finite. As $|H| = |H/H \cap K| \cdot |H \cap K|$, we see that H is finite. □

Theorem. *A solvable periodic group is locally finite.*

Proof. Induction on the length l of solvability of the solvable group G (l is the minimal number such that there exists a series (2)). The base of the induction, the local finiteness of a periodic abelian group, follows form the structure theorem for finitely generated abelian groups (see, Introduction 1.10). Further if G' is the commutator subgroup of the group G, then G/G' and G' have length of solvability strictly less than the length of solvability of the group G. According to the induction assumption they are locally finite, and according to the lemma, G is locally finite. □

2.5. Grigorchuk's example. In this subsection an example of a finitely generated infinite 2–group will be presented, i.e. a group of which every element is of finite order which is equal to a power two. In order to construct the group we consider the set Ω of points from the interval $[0,1]$ with the exception of the subset of binary rational points, i.e. the points of the form $m/2^n$, $m, n = 1, 2, \ldots$. In the group $S(\Omega)$ of bijections of the set Ω we select bijections a, b, c, d, which are defined in the following way. We divide Ω into the union of the intervals Δ_k of the form

$$\Delta_k = \left[\frac{2^k - 1}{2^k}, \frac{2^{k+1} - 1}{2^{k+1}}\right] \bigcap \Omega,$$

where $k = 0, 1, 2, \ldots$. The picture obtained looks as follows

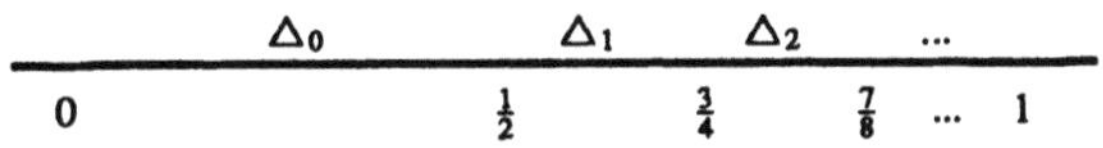

The action of a mapping ϕ on Δ_k we shall denote by ϕ_k, and the action on its complement by ϕ'_k. The transformation of an "interval" Δ which transposes the positions of its two halves, we shall denote by Π_Δ or simply Π. The identical transformation of any set will be denoted by the letter T. We shall now define the transformations a, b, c, d by setting $a = \Pi_\Omega$,

$$\begin{aligned} b_{3k} &= \Pi, & b_{3k+1} &= \Pi, & b_{3k+2} &= T, \\ c_{3k} &= \Pi, & c_{3k+1} &= T, & c_{3k+2} &= \Pi, \\ d_{3k} &= T, & d_{3k+1} &= \Pi, & d_{3k+2} &= \Pi, \end{aligned}$$

$k = 0, 1, 2, \ldots$. The elements a, b, c, d generate a group $G \subset S(\Omega)$. Of course, $a^2 = b^2 = c^2 = d^2 = e$. Further, the intervals Δ_k are invariant under b, c, d and $b_s c_s = d_s = c_s b_s$. So the subgroup K of the group G generated by the elements b, c, d, is isomorphic to the Klein four group V (see, Introduction 1.6). In the proof of the following theorem the notation $\partial_x(g)$ for the number of occurences of the letter x in a shortest writing of the element g in terms of a, b, c, d will be used; $\partial(g)$ is the total length of the word g.

Theorem. *G is a finitely generated infinite 2–group.*

Proof. We shall show that the group G is infinite. To do that consider the subgroup H in G, which leaves both halves Ω_1, Ω_2 of the set Ω in their places. The element g lies in H iff $\partial_a(g) \equiv 0(\mod 2)$. We shall consider the action of H on $\Omega_1 = [0, 1/2] \cap \Omega$. For that it is necessary to consider the component ϕ_0 of $\phi \in H$. Let $\phi(\Omega_i)$ transform Ω_i, $i = 1, 2$ in the same way as ϕ transforms Ω. Then we obtain

$$b_0 = \Pi = a(\Omega_1), \quad c_0 = \Pi = a(\Omega_1),$$
$$d_0 = T, \quad (aba)_0 = c(\Omega_1), \quad (aca)_0 = d(\Omega_1), \quad (ada)_0 = b(\Omega_1).$$

Further

$$b_0' = c(\Omega_2), \qquad c_0' = d(\Omega_2), \qquad d_0' = b(\Omega_2),$$
$$(aba)_0' = a(\Omega_2), \qquad (aca)_0' = a(\Omega_2), \qquad (ada)_0' = T.$$

Thus the image of the subgroup H under the homomorphism $\imath : \theta \to (\theta_0, \theta_0')$ from H into $S(\Omega_1) \times S(\Omega_2)$, contains the pairs (a, c), (a, d), (e, b), (c, a), (d, a), (b, e). The projection of the image on the first component contains G. Since H is a proper group in G permitting a homomorphism onto G, the group G is infinite.

Any element from G is conjugate with an element of the form

$$g = au_1au_2au_3 \dots au_m, \qquad u_i \in K.$$

An element g lies in H iff m is even. If $g \in H$ then $\imath(g) = (g_l, g_r)$, moreover $\partial(g_l) \le 2m < \partial(g)$, $\partial(g_r) \le 2m < \partial(g)$. Applying induction with respect to length we see that g_l, g_r are of finite order and so the same holds for g. If $g \notin H$ then $g^2 \in H$ and g^2 has the form

$$g^2 = au_1au_2au_3 \dots au_m au_1au_2au_3 \dots au_m.$$

If $\phi(g^2) = (\bar{g}_l, \bar{g}_r)$ then $\partial(\bar{g}_l) \le 2m$, $\partial(\bar{g}_r) \le 2m \le \partial(g)$.

Grouping the factors in g^2 according to the order: au_1a, u_2, $au_3a, \dots$, we can see that in virtue of the oddness of m we obtain the factors au_ma, u_1, $au_2a, \dots, au_{m-1}a$, u_m. If among the u_i there is a d then among the selected factors are both d and ada. However, $\imath(d) = (e, b)$, $\imath(ada) = (b, e)$. Then $\partial(\bar{g}_l) < 2m = \partial(g)$ and $\partial(\bar{g}_r) < 2m = \partial(g)$. Again we can apply the induction hypothesis. If there is no d among the u_i, but there is a c, then among the selected factors there are both c and aca. Since $\imath(c) = (a, d)$ and $\imath(aca) = (d, a)$ then either $\bar{g}_l$, $\bar{g}_r$ possess the shorter writing compared with g, or their writing contains d, and the previous reasoning is applicable. Let assume that there is no d, c in g. Since $\imath(b) = (a, c)$, $\imath(aba) = (c, a)$, we again obtain the previous case.

By similar arguments it is proved that any element from G is of order equal to a power of two. □

2.6. Example of a finitely generated infinite p–group. As was mentioned above (see, Subsection 2.1) the first example of finitely–generated infinite p–group was constructed in 1964 by E.S. Golod. Afterwards other examples possessing the same property followed. The most elegant is the example (Subsection 2.4) constructed by R.I. Grigorchuk, for $p = 2$. A similar construction for any odd p was noted by N. Gupta and Sidki based on an examination of automorphisms of trees. We shall here present this construction. Concerning graphs and trees see § 3 of this chapter.

Let p be an odd prime number and $T(0)$ the infinite regular tree of degree p (i.e. from each vertex leave p edges) with root 0. We shall denote by $T(u)$ the subtree with root $u \in T(0)$ and by $T(u, 1), \dots, T(u, p)$ the regular subtrees, the roots of which are the ends $(u, 1), \dots, (u, p)$ of edges leaving the vertex u. Each tree $T(u, i)$ is isomorphic to $T(0)$. For every vertex $u \in T(0)$ we define the automorphism $t(u) : T(u) \to T(u)$ which takes $T(u, j)$ into $T(u, j + 1)$ when $j = 1, \dots, p - 1$, and $T(u, p)$ into $T(u, 1)$.

Note that $t(u)$ is of order p.

For every vertex $u \in T(0)$ define an infinite sequence $S(u)$ of vertices by means of induction, by setting

$$S(u) = u, 1; u, 2; \dots ; u, p - 1;$$
$$S(u, p) = u, p, 1; u, p, 2; \dots ; u, p, p - 1; S(u, p, p); ldots$$

Using this sequence define an automorphism $a(u) : T(u) \to T(u)$, by $a(u) = t(u, 1)t^{-1}(u, 2)i(u, 3) \dots i(u, p-1)a(u, p)$, where t is the automorphism defined above, $i(u, j)$ is the identity automorphism of the tree $T(u, j)$, and $a(u, p)$ is the automorphism analogous to $a(u)$ but with the vertex u, p. We should note that $a(u)$ is an automorphism of order p and it leaves all the vertices $u; u, p; u, p, p; \dots$ in place.

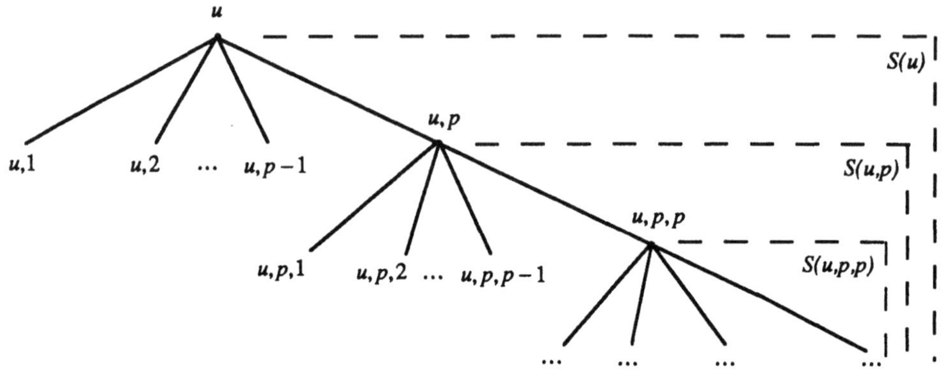

KLISE 2 (188 PSL)

Theorem. *Let p be an odd prime number and let $G(0)$ be the group generated by the elements $t(0)$ and $a(0)$ defined above. Then $G(0)$ is an infinite p–group.*

Proof. Recall that

$$a(0) = t(0,1)t^{-1}(0,2)i(0,3)\dots i(0,p-1)a(0,p) \in$$
$$\in G(0,1) \times \dots \times G(0,p),$$

where $G(u)$ is the group generated by the elements $t(u)$ and $a(u)$. For simplicity we write $a = (t, t^{-1}, i, \dots, i, a)$. For every $j = 0, \dots p-1$ the element $t^{-j}(0)a(0)t^{j}(0) = a_j$ lies in $G(0,1) \times \dots \times G(0,p)$ and can be obtained with the help of cyclic permutation of the components of the element a. We have

(4)
$$a_0 = (t, t^{-1}, i, \dots, i, a_0),$$
$$a_1 = (t^{-1}, i, i, \dots, a_0, t),$$
$$\dots\dots\dots\dots\dots\dots\dots\dots$$
$$a_{p-1} = (a_0, t, t^{-1}, \dots, i, i).$$

Let $H(0)$ be the subgroup in $G(0)$ generated by the elements $a_0, \dots, a_{p-1}$. Then $H(0)$ is a normal subgroup in $G(0)$ of index p. Since the group $G(0,j)$ is isomorphic to the group $G(0)$ and $H(0)$ is a subdirect product of the groups $G(0,j)$, we obtain that $G(0)$ is an infinite group. In order to show that $G(0)$

is a p–group, note that any element $g \in G(0)$ has the form $g = ht^j$, where $h = h(a_0, \dots, a_{p-1})$ is a word in $a_0, \dots, a_{p-1}$. We shall use induction on the number of syllables in the word under consideration, taking as syllables the powers of generators. If the number of syllables of the word g is equal to 1 then g is a p–element, i.e. an element whose order is a power of p.
Let us assume that all elements with number of syllables m are p–elements, and let g have $m+1 \geq 2$ syllables.

Case 1. $g = h(a_0, \dots, a_{p-1})t^{p-j}$, $j \in \{1, \dots, p-1\}$. The element h has m syllables, $m = \lambda(0) + \dots + \lambda(p-1)$, where $\lambda(k)$ is a number of occurences of powers of the element a_k. Further,

$$g^p = hh^{t^j} \dots h^{t^{(p-1)j}}$$

is an element from $H(0)$ with mp syllables; moreover, for any k the number of powers of the element a_k is equal to $\lambda(0) + \dots + \lambda(p-1) = m$. Expressing g^p in the form of sequences of the length p, as in (4), we see that for each j the component of the element g^p in the group $G(0, j)$ has no more than m syllables, i.e. is a p–element. Thus g is a p–element.

Case 2. $g = h(a_0, \dots, a_{p-1})$. Here h has the number of syllables $m+1 = \lambda(0) + \dots + \lambda(p-1)$. Going over, as above, to the projections on the $G(0, j)$, we see that the projections have the length $\lambda(p-j)$ or $\lambda(p-j)+1$ depending on the fact whether the component lies in $H(0, j)$ or not. If the projection has the length $\lambda(p-j)$ or $\lambda(p-j)+1 \leq m$ then, as $\lambda(p-j) \leq m$, we can apply the induction hypothesis. If the projection has the length $\lambda(p-j)+1 = m+1$, then $m = 1$, and we obtain a p–element according to the proof of case 1.

Thus it is proved that $G(0)$ is an infinite p–group. □

§ 3. Free groups and graphs

3.1. Definition of a free group. A group F is called *free* if we can find in it a generating subset X, such that for any mapping $\phi : X \to G$, where G is an arbitrary group, there is an extension to a homomorphism $\bar{\phi} : F \to G$. The condition of "extendability" means that for any $x \in X$ we have $\bar{\phi}(x) = \phi(x)$. The set X is called a *free generating set* for F or a *basis* of this group. The cardinality of the set X is called the *rank* of the group F. The rank of a free group is an invariant of it. For instance, if $|X| = r$ then $|F/F'F^2| = 2^r$ which completely defines the number r. We shall consider the set $F(X)$ consisting of the empty word e and all the words of the form $z_1 z_2 \dots z_n$, where each z_i is either an element x from X or a symbol x^{-1}, and, moreover, the letters x and x^{-1} are never adjacent. We introduce an operation of multiplication on this set by defining that the result of this operation on the words w_1 and w_2 is the word obtained by writing w_1 on the left of w_2 and by eliminating at the junction all subwords that occur of the form xx^{-1} and $x^{-1}x$, $x \in X$. The neutral element, relative to this operation, is the empty word e. The inverse to $z_1 \dots z_n$ is $z_n^{-1} \dots z_1^{-1}$. If we admit that the introduced operation is associative then it is evident that $F(X)$ becomes a free group. Indeed, if $\phi : X \to G$, then we can extend the definition of ϕ to a mapping $\widetilde{\phi} : X \cup X^{-1} \to G$, by putting $\widetilde{\phi}(x^{-1}) = (\phi(x))^{-1}$, and afterwards we take $\bar{\phi}(z_1 \dots z_n) = \widetilde{\phi}(z_1) \dots \widetilde{\phi}(z_n)$. It is clear that $\bar{\phi}$ is the desired homomorphism.

Conversely, if we assume that there exists an injective homomorphism $\bar{\phi} : F(X) \to G$, where G is a group, then from the associativity of the operation in G automatically follows the associativity of the operation in $F(X)$. Therefore a natural way to prove existence and to study free groups is to construct a model G of it. There exist a number of such models. For instance, we can consider groups of matrices, and groups of invertible elements of some other rings. In this subsection we shall approach the matter via graph theory.

3.2. Graphs and paths. A *graph* Γ is a set of the form $\Gamma = V \cup E$, where V is the set of *vertices*, and E is the set of *edges*. In addition, there are mappings $\alpha : E \to V$, $\omega : E \to V$, $\eta : E \to E$, that for any $e \in E$ we have

$$\eta^2 = 1_E, \quad \alpha(\eta(e)) = \omega(e), \quad \omega(\eta(e)) = \alpha(e). \tag{1}$$

In addition the vertex $\alpha(e)$ is called the *source* of the edge e, $\omega(e)$ – the *target* of the edge e, the edge $\eta(e)$ is called *inverse* to e and is denoted by e^{-1}. The pair $\{e, e^{-1}\}$ is called a *non–oriented edge*. It is clear that a non–empty graph has a non–empty set V.

A mapping $f : \Gamma_1 \to \Gamma_2$ is called *morphism* of the graph Γ_1 to the graph Γ_2, if $f(V_1) \subset V_2$, $f(E_1) \subset E_2$, and, moreover,

$$f(\alpha(e)) = \alpha(f(e)), \quad f(\omega(e)) = \omega(f(e)), \quad f(e^{-1}) = (f(e))^{-1}. \tag{2}$$

A sequence $p = (e_1, e_2, \dots, e_n)$ of edges of a graph Γ is called a *path* if $\omega(e_1) = \alpha(e_2)$, $\omega(e_2) = \alpha(e_3), \dots, \omega(e_{n-1}) = \alpha(e_n)$. For brevity we write $p =$

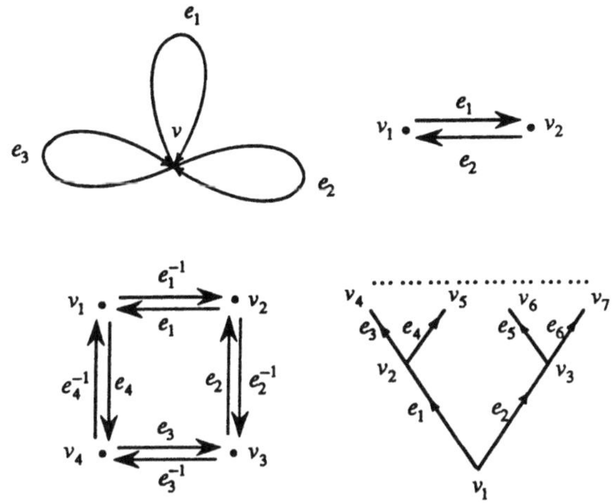

EXAMPLE OF GRAPHS KLISE 3 (192 PSL)

$e_1 e_2 \ldots e_n$. Let us define $\alpha(p) = \alpha(e_1)$, $\omega(p) = \omega(e_n)$ (the source and the target of the path p). A path p is called a *loop* at the vertex v, if $v = \alpha(e_1) = \omega(e_n)$. We shall denote by 1_v the trivial loop at the vertex v, i.e. the loop that does not contain any edges. If $p' = e_{n+1} \ldots e_m$ is another path and also $\omega(p) = \alpha(p')$, then $e_1 \ldots e_n e_{n+1} \ldots e_m$ is the path written pp' and called the *product of the paths* p and p'. The inverse to a path p is the path $p^{-1} = e_n^{-1} \ldots e_1^{-1}$. Any two loops can be multiplied and, clearly, this multiplication is associative.

Let us say that a path p is *equivalent* to a path p' (we write $p \sim p'$), if p can be obtained from p' by removing or inserting subpaths of the form ee^{-1}. It is clear that we get a relation of equivalence and such that if $p \sim p'$ and $q \sim q'$ then also $pq \sim p'q'$. The class of paths equivalent to a given path p we shall denote by the symbol $[p]$. Now, let us fix some vertex v and denote by $\pi(\Gamma, v)$ a set of classes of loops at the vertex v. In virtue of the equivalence property noted above, $\pi(\Gamma, v)$ is a group. Its neutral element is the class $1 = [1_v]$. The inverse element to $[p]$ is $[p^{-1}]$.

We shall call graph Γ *connected* if for any vertices v, w there exists a path p such that $\alpha(p) = v$, $\omega(p) = w$.

A connected graph Γ is called a *tree* if it does not contain non–neutral loops, i.e. it has no loops non–equivalent to one of the loops of the form 1_v. In any graph there is a maximal subgraph T which is a tree (apply Zorn's lemma, having noted that the union of an increasing chain of trees is again a tree). If T is a maximal subtree in a (connected) graph Γ then any vertex of the graph Γ belongs to T. Indeed, if some vertex $v \in \Gamma$ does not lie in T, then from each of the vertices of the tree T there is a path to v. Having selected among these paths a shortest one $p = e_1 \dots e_n$, $\alpha(e_1) = u \in T$, $\omega(e_n) = v$, we see that $w = \omega(e_1) \notin T$. In this case $T' = T \cup \{e_1\} \cup \{w\}$ is a tree and this contradicts the maximality of the tree T.

3.3. Theorem. *For each graph Γ and any of its vertices v the group $\pi(\Gamma, v)$ is free. If Γ is connected, then the rank of the group $\pi(\Gamma, v)$ is equal to the number $\gamma_1 - \gamma_0 + 1$, where γ_0 is the number of vertices and γ_1 is the number of (non-oriented) edges.*

Proof. It is clearly sufficient to limit ourselves to the case of a connected graph Γ. This is what we will assume. Let us select in Γ a maximal subtree T.

Introduce in $\pi(\Gamma, v)$ a set Y consisting of classes of loops of a special form. If $e \in E(\Gamma)$, $\alpha(e) = x$, $\omega(e) = y$, consider the shortest path $\overline{vx}$ connecting v, x and passing through the edges of the tree T. Put $\widetilde{e} = \overline{vx} e \overline{vy}^{-1}$. The class $[\widetilde{e}]$ of the loop $\widetilde{e}$ we shall denote by $\bar{e}$. The class $\bar{e}$ is trivial iff $e \in T$. If $\bar{e} \neq 1 = [1_v]$, then Y contains $\bar{e}$. We note that $\overline{e^{-1}}$ is in Y if $\bar{e}$ belongs to Y (and vice versa) and is an inverse element to $\bar{e}$ in $\pi(\Gamma, v)$. In order to obtain a free generating set X it is sufficient to select from each subset $\{\bar{e}, \overline{e^{-1}}\}$ one element. We shall show that X really is a free generating set for $\pi(\Gamma, v)$. According to what was noted in Subsection 3.1, it is sufficient to show that X generates $\pi(\Gamma, v)$ and that if $w = \bar{e}_1 \dots \bar{e}_n$ and $\bar{e}_i \bar{e}_{i+1} \neq 1$ for all $i = 1, 2, \dots, n-1$ then $w \neq 1$.

Any element from $\pi(\Gamma, v)$ is the class of a path $p = e_1 \dots e_n$, moreover $\alpha(e_1) = \omega(e_n) = v$. We note that

$$p \sim \overline{v\alpha(e_1)} e_1 \overline{v\omega(e_1)}^{-1} \cdot \overline{v\alpha(e_2)} e_2 \overline{v\omega(e_2)}^{-1} \dots$$

$$\dots \overline{v\alpha(e_n)} e_n \overline{v\omega(e_n)}^{-1} \sim \widetilde{e}_1 \widetilde{e}_2 \dots \widetilde{e}_n. \tag{3}$$

And it follows that $[p] = \bar{e}_1 \bar{e}_2 \dots \bar{e}_n$.

Now consider a representative p of the form (3) of the class of paths $w = \bar{e}_1 \dots \bar{e}_n$ satisfying the condition $\bar{e}_i \neq 1$, $\bar{e}_i \bar{e}_{i+1} \neq 1$. For any pair e_i, e_{i+1} in the expression (3) the interval $e_i f_1 \dots f_t e_{i+1}$ is such that all $f_1, \dots, f_t \in T$ and $e_i, e_{i+1} \notin T$. Therefore this path is equivalent either to $e_i f_{j_1} \dots f_{j_s} e_{i+1}$, $s \geq 1$, or to the path $e_i e_{i+1}$. In virtue of our assumption, a further "shortening" is not possible. Thus p is equivalent to a path with the length larger than 1 and not permitting any shortening. It is easy to see that in this case $w \not\sim 1_v$. Thus the group $\pi(\Gamma, v)$ is indeed free with a free generating set.

In order to calculate the cardinality of the set X in the case of a finite graph it is necessary to calculate the number N of edges in the maximal tree T. Since this tree has γ_0 vertices and elimination of any (ending) vertex decreases of the number of edges by 1, and moreover, if $\gamma_0 = 2$, there is only one edge, we obtain $N = \gamma_0 - 1$. Therefore $|X| = \gamma_1 - \gamma_0 + 1$. □

In virtue of the theorem proved, in order to realize a free group of rank n it is sufficient to consider the group of classes of loops $\pi(\Gamma, v)$ of the graph Γ of the form

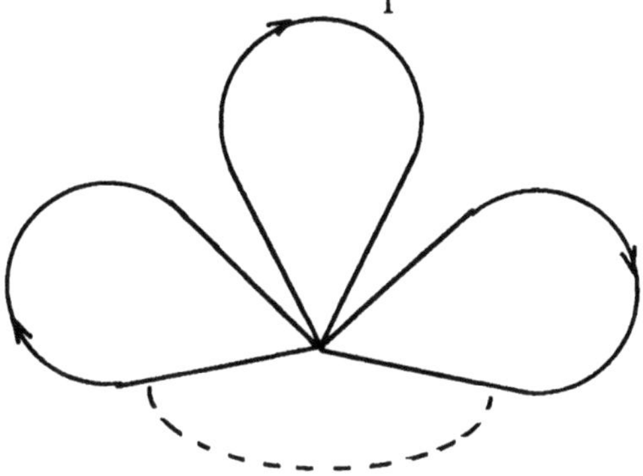

KLISE 4(PSL 196)

with one vertex v and n non–oriented edges $e_1, \dots, e_n$.

3.4. Subgroups of free groups. One of the central results in the theory of free groups is the Nielsen–Schreier theorem stating that any subgroup of a free group is free itself. The correspondence constructed above allows to give a beautiful proof of this result. The main point of this proof is the idea that if $f : \Gamma' \to \Gamma$ is a morphism of graphs, $v \in \Gamma$, $v = f(v')$, then there exist a corresponding natural homomorphism $f^* : \pi(\Gamma', v') \to \pi(\Gamma, v)$. It associates to the class of a loop $e_1 \dots e_n$ the class of the loop $f(e_1) \dots f(e_n)$.

Theorem on subgroups. *Let Γ be a connected graph, $v \in V(\Gamma)$, H a subgroup of $\pi(\Gamma, v)$. Then it is possible to construct a graph Γ' and a morphism $f : \Gamma' \to \Gamma$ such that $f(v') = v$ for some vertex v' of the graph Γ', and such that the homomorphism $f^* : \pi(\Gamma', v') \to \pi(\Gamma, v)$ is an isomorphism of the group $\pi(\Gamma', v')$ onto H.*

Corollary 1. *Any subgroup of a free group is free.* □

Proof of the theorem. Select a maximal subtree T in Γ. We shall use the notations from the proof of the previous theorem. Denote by W the set of the right cosets of the group $\pi(\Gamma, v)$ with respect to the subgroup H. Construct a new graph Γ': $V' = V \times W$, $E' = E \times W$. Every vertex has the form $v' = (v, Hg)$, every edge is of the form $e' = (e, Hg)$, $v, e \in \Gamma$, $g \in G$. Put

$$\alpha((e, Hg)) = (\alpha(e), Hg), \quad \omega((e, Hg)) = (\omega(e), Hg\bar{e}),$$
$$(e, Hg)^{-1} = (e^{-1}, Hg\bar{e}).$$

The reader can easily check that the axioms (1) for a graph hold.
For instance,

$$\alpha((e, Hg)^{-1}) = \alpha((e^{-1}, Hg\bar{e})) = (\alpha(e^{-1}), Hg\bar{e}) =$$
$$= (\omega(e), Hg\bar{e}) = \omega((e, Hg)).$$

The mapping $f : (e, Hg) \mapsto e$, $(v, Hg) \mapsto v$ is, evidently, a morphism of graphs. For v' we take (v, H). We shall find $f^*(\pi(\Gamma', v'))$. First we shall show that $f^*(\pi(\Gamma', v')) \subset H$. Since $\alpha(e_1') = (v, H)$, we see that $e_1' = (e_1, H)$; moreover $\alpha(e_1) = v$. Further, $\omega((e_1, H)) = (\omega(e_1), H\bar{e}_1)$. Since $\alpha(e_2') = \omega(e_1')$, we have $e_2' = (e_2, H\bar{e}_1)$, moreover $\alpha(e_2) = \omega(e_1)$. Continuing with this reasoning we see that $e_n' = (e_n, H\bar{e}_1\bar{e}_2 \ldots \bar{e}_{n-1})$, while $\alpha(e_n) = \omega(e_{n-1})$. However $\omega(e_n') = v'$, so that $\omega(e_n) = v$, and $H\bar{e}_1 \ldots \bar{e}_n = H$. So $e_1 \ldots e_n$ is a loop, the class, $\bar{e}_1 \ldots \bar{e}_n$, of which lies in H. Since f^* associates to the class $\bar{e}_1' \ldots \bar{e}_n'$ the class $\bar{e}_1 \ldots \bar{e}_n$, our statement about the image of the homomorphism f^* is proved. The same scheme gives a preimage under f^* for any element form H.

The injectivity of the mapping follows from the fact that the projection is taken with respect to the first component. □

Corollary 2. *Let H be a subgroup of index j of a free group of rank n. Then H is a free group of rank*

$$r(H) = (n - 1)j + 1.$$

Proof. The number of vertices in the graph Γ' is equal to $j\gamma_0$, the number of edges – $j\gamma_1$, where γ_0, γ_1 are the corresponding numbers in Γ. We have

$$r(H) = j\gamma_1 - j\gamma_0 + 1 = j(\gamma_1 - \gamma_0 + 1 - 1) = j(n-1) + 1.$$

□

Corollary 3. *A subgroup of finite index in an arbitrary finitely generated group is itself finitely generated.*

Proof. Let G be the given group with finite number of generators n, let S be a subgroup of finite index m; then we can present G in the form $G = F/R$, where F is a free group of rank n. In this case the complete preimage H in F of the group S, i.e. a set of all $a \in F$ such that $aR \in S$ is a subgroup of the same index m in F. According to the previous corollary, H is a finitely generated free group of rank $(n-1)m+1$. Now S is a homomorphic image of the group H and hence a finitely generated group; moreover, the number of generators need not exceed $(n-1)m+1$. □

§ 4. Representation of groups by generators and relations

4.1. Finitely defined groups. Let X be some non–empty set, $F(X)$ the free group with free generating set X, R a subset in $F(X)$. The pair $\langle X | R\rangle$ defines the group $F(X)/N$, where N is the normal subgroup in $F(X)$ generated by the set R. If G is an abstract group and $G \cong F(X)/N$, then the pair $\langle X|R\rangle$ is called a *presentation* of the group G in terms of the generating set X and the defining set R (or the set of defining relations $\{r = 1 | r \in R\}$. The notation $G = \langle X | R\rangle$ is used.

It is clear that the same group can have different presentations. For instance, the infinite cyclic group possesses the representations $\langle x|\emptyset\rangle$ and $\langle x, y|y = 1\rangle$. A group G which possesses a presentation $\langle X|R\rangle$, where X is finite, is a finitely generated group; if R is finite, then G is a finitely related group; if both X and R are finite, then G is called a finitely presented group.

Finitely presented groups are related to free groups of finite rank, finitely generated abelian groups and others. Any finite group G is finitely defined; indeed, one can take $X = G$, and $R = \{abc^{-1} | a, b \in G \text{ and } ab = c\}$. It is clear that this is not an economic presentation. For instance, in the case of the cyclic group of order 10 it gives 10 generators and 100 relations, while there is also the presentation $\langle x | x^{10} = 1\rangle$.

In order to pass from one presentation to another the so called Tietze transformations are applied.

Let $\langle X|R\rangle$ be an arbitrary presentation, r an element from the normal subgroup N generated by the set R. It is clear that $\langle X'|R'\rangle$, where $X' = X$ and $R' = R \cup \{r\}$, defines the same group. On the other hand, let us put $X' = X \cup \{x\}$, where $x \notin X$; if $w \in F(X)$ and $R' = R \cup \{r\}$, where $r = x^{-1}w$, then $\langle X|R\rangle$ and $\langle X'|R'\rangle$ define the same group. The transitions $\langle X|R\rangle \to \langle X'|R'\rangle$ and $\langle X'|R'\rangle \to \langle X|R\rangle$ are called *Tietze transformations.*

Proposition. *Two finite presentations $\langle X_1|R_1\rangle$ and $\langle X_2|R_2\rangle$ define the same group iff there is a finite sequence of Tietze transformations taking one to the other.*

Proof. Clearly, only necessity needs to be proved. Let $F_i = F(X_i)$ be the free group with free generating set X_i, and let N_i be a normal subgroup in F_i generated by the set R_i, $i = 1, 2$. It can be assumed that the group G and the two homomorphisms $\phi_i : F_i \to G$ with kernels N_i, $i = 1, 2$ are given. We shall suppose that $X_1 \cap X_2 = \emptyset$. Let F be the free group with generating set $X_1 \cup X_2$. Then there exists a homomorphism $\phi : F \to G$ such that $\phi|_{X_i} = \phi_i$, $i = 1, 2$.

For every $x \in X_1$ choose $w_x \in F_2$ such that $\phi(x) = \phi(w_x)$, and let S_1 be the finite set of elements of the form $s_x = x^{-1}w_x$. Analogously, we define S_2. It is possible to pass from the representation $\langle X_1|R_1\rangle$ to the representation $\langle X|R_1 \cap S_2\rangle$ by a finite sequence of Tietze transformation each of which introduces a new generator $x \in X_2$ and a new relation $x^{-1}w_x \in S_2$ expressing x in terms of X_1. It is clear that the kernel N of the mapping ϕ from F to G is the normal subgroup generated by the set $R_2 \cup S_1 \subset N$. Similarly, $\phi(R_2 \cup S_1) = \{1\}$, i.e. $R_2 \cup S_1 \subset N$. By a finite sequence of Tietze transformations the elements from $R_2 \cup S_1$ can be adjoined one by one to $R_1 \cup S_2$ so as to obtain a presentation of G of the form $\langle X|R\rangle = \langle X|R_1 \cup R_2 \cup S_1 \cup S_2\rangle$. Thus, a transition from $\langle X_1|R_1\rangle$ to $\langle X|R\rangle$ is possible by a finite sequence of Tietze transformations. Using the symmetry of the procedure we can pass from $\langle X_2|R_2\rangle$ to $\langle X|R\rangle$ in a similar way. From that it follows that a transition from $\langle X_1|R_1\rangle$ to $\langle X_2|R_2\rangle$ is also possible. □

4.2. Two–dimensional complexes. Under a *one–dimensional complex* we understand a non–oriented graph (see, § 3). A *two–dimensional complex* C

consists of a one–dimensional complex C^1 (the 1–*skeleton* of the complex C), of a set F of two–dimensional cells or *faces*, and of two mappings $\partial : F \to C$ and $\eta_2 : F \to F$. It is required that $\partial(D)$, the boundary of the cell D, is a cycle in C^1 and that $\partial(D^{-1})$, the boundary of the inverse cell $D^{-1} = \eta_2(D)$ is a path inverse to $\partial(D)$.

A vertex v is on the cell D if v is the origin of some edge from ∂D. A path in a two–dimensional complex C is a path in its 1–skeleton. Two paths p, p' are called *equivalent* ($p \sim p'$) in C if it is possible to pass from one path to another by deletion or insertion of the paths of the form ee^{-1}, where e is an edge in C^1, or of the form q, where q is a contour starting in any vertex of the face D (i.e. any loop in the cycle ∂D). The relation $\sim$ is a congruence on the set $\Pi(C)$ of the paths of the complex C, and this allows to consider the set of classes of paths $\pi(C)$ with the (natural) multiplication of these classes. The set of classes of loops $\pi(C, v)$ at the vertex v is a subgroup in $\pi(C)$. The group $\pi(C, v)$ is a homomorphic image of the group $\pi(C^1, v)$ which is the fundamental group of the one–dimensional complex C^1 (see, § 3). We know from § 3 that the group $\pi(C^1, v)$ is free. Passing to the fundamental groups $\pi(C, v)$ of the two–dimensional complexes allows to obtain arbitrary groups defined in terms of generators and defining relations.

We shall call a surjective mapping $f : C' \to C$ of a two–dimensional complex C, where C' is another two–dimensional complex, a *covering*, if it is compatible with the incidence relations, taking boundaries and taking inveres of cells. Moreover, its restriction to the set of edges and cells incident to any fixed vertex, should be one–to–one.

Lemma. *If $f : C' \to C$ is a covering, and v' is a vertex of the complex C, then f induces a monomorphism f^* from $\pi(C', v')$ into $\pi(C, f(v))$.*

Proof. If p' is a loop with its origin in a vertex v' of the complex C', then, clearly, its image $p = f(p')$ is a loop in C with the beginning $v = f(v')$. Let p'_1 and p'_2 be the loops of the complex C' with origin v' and such that $p'_1 = u'q'_1v'$, $p'_2 = u'q'_2v'$, where $q'_1q'^{-1}_2$ is a contour in some boundary of a D' in C'. Then $q - 1q_2^{-1}$ is a contour in the boundary of the cell $D = f(D')$ in C, and the images of the loops, $p_1 = uq_1v$ and $p_2 = uq_2v$, are equivalent in C. Therefore, f induces a homomorphism $f^* : \pi(C', v') \to \pi(C, v)$. The covering mapping is bijective on the set of edges incident with a fixed vertex. Hence with a help of induction on the length of the reduced path p the following is deduced: if p begins in the vertex v, $f(v') = v$, then there exists a unique reduced path p'

in C', which begins in v' and for which $f(p') = p$. Let us assume that p' is a loop in C' and that $f(p')$ bounds the cell D in C. By surjectivity there exists a cell D' such that $f(D') = D$. Then f transforms the boundary of the cell D' into $f(p')$, and according to the uniqueness property, mentioned above, the loop p' bounds D'. This means that in fact f^* is a monomorphism. □

In contrast to the fundamental group of a one–dimensional complex the fundamental group, $\pi(C, v)$, of a two–dimensional complex C is not necessarily free. The following result demonstrates that any group can be obtained in this way. Let $G = \langle X | R \rangle$ be some group. Without restriction of generality one can assume that the elements of the set R are represented by the cyclically irreducible words, i.e. the length of these words cannot be reduced by conjugations. We construct a complex $K = K(X; R)$ with a unique vertex v and edges $\widetilde{x}$ and $\widetilde{x}^{-1}$ for every $x \in X$. Thus, every path in K is a loop. Further, for each $r = x_1^{e_1} \dots x_n^{e_n} \in R$, where $x_i \in X$, $e_i = \pm 1$, introduce the cell D with boundary $\widetilde{x}_1^{e_1} \dots \widetilde{x}_n^{e_n}$ (starting in v) and the cell D^{-1}.

Proposition. *If $G = \langle X | R \rangle$ and $K = K(X; R)$ is the complex introduced above; then $\pi(K, v) \cong G$.*

Proof. Let ϕ be the mapping of the free group F with the free generating set X taking $x \in X$ into the edge $\widetilde{x}$ of the graph $K(X; R)$. According to theorem 3.3 the mapping ϕ is an isomorphism of the group F onto the fundamental group $\pi(K^1, v)$ of the 1–skeleton K^1 of the graph K. As was mentioned before there exists a natural homomorphism κ from $\pi(K^1, v)$ to $\pi(K, v)$. It is evident that $\phi(R)$ is contained in the kernel of the homomorphism κ. On the other hand, from the definition it is clear that if the loops p and p' are related by $\sim$, then their equivalence classes in $\pi(K^1, v)$ are equal modulo $\phi(R)$. □

4.3. Subgroups and coverings. The techniques of § 3, used in the theorem on subgroups of free groups, in the case of two–dimensional complexes gives information about the presentations of subgroups in terms of generators and defining relations. In order to obtain these the following result is necessary:

Proposition. *Let C be a connected two–dimensional complex, v a vertex in C, H some subgroup of the group $\pi(C, v)$. Then there exists a connected two–dimensional complex C' and a covering mapping $f : C' \to C$ taking* *some vertex $v' \in C'$ into the vertex v, such that the induced isomorphism f^* takes the group $\pi(C', v')$ onto the subgroup H of the group $\pi(C, v)$.*

Proof. With the help of the theorem from Subsection 3.4 construct a covering mapping f^1 of the one–dimensional complex C'^1 on the 1–skeleton C^1 of the complex C such that f^1 maps the group $\pi(C'^1, v')$ on the preimage H' of the subgroup H in $\pi(C^1, v)$. We need to extend C'^1 to a two–dimensional complex C', and f^1 to a mapping $f : C' \to C$. If p' is a closed path in C'^1 of which the image p bounds the cell D of the complex C then we introduce a cell D' of the complex C' with boundary p' and we put $f(D') = D$. This turns f into a covering mapping from C' into C.

Let p be a loop starting in v in C representing the element h from the subgroup H of the group $\pi(C, v)$, and let $\widehat{h}$ be from the preimage H^1 of the subgroup H in $\pi(C^1, v)$. Then f^{1*} maps some element h' of the group $\pi(C'^1, v')$ to $\widehat{h}$. This means that f^* sends the image of the element h' from $\pi(C', v')$ into the element h from H. This shows that f^* maps $\pi(C', v')$ onto H. According to the lemma from Subsection 4.2 the homomorphism f^* is injective. □

4.4. Free constructions and complexes. Among the more important objects of the theory of infinite groups there are free products, free products with an amalgamated subgroup, and HNN–extensions.

If $(G_\alpha)_{\alpha \in \mathcal{I}}$ is some family of groups with presentations $\langle X_\alpha | R_\alpha \rangle$, where $X_\alpha \cap X_\beta = \emptyset$ if $\alpha \neq \beta$, then the *free product* $G = \prod^*_{\alpha \in \mathcal{I}} G_\alpha$ can be defined as the group with the presentation $\langle X | R \rangle$, where $X = \cup_{\alpha \in \mathcal{I}} X_\alpha$, $R = \cup_{\alpha \in \mathcal{I}} R_\alpha$.

The reader can easily verify that the group G is a free product of the family of the groups $(G_\alpha)_{\alpha \in \mathcal{I}}$ (the groups G_α are identified with the subgroups of the group G by means of homomorphisms $\phi : G_\alpha \to G$) iff for any family of the homomorphisms $\psi_\alpha : G_\alpha \to G'$, $\alpha \in \mathcal{I}$, there is a unique extension to a homomorphism $\psi : G \to G'$ such that $\psi\phi_\alpha = \psi_\alpha$ for all $\alpha \in \mathcal{I}$. Now let G' be the direct product $G' = \prod_{\alpha \in \mathcal{I}} G_\alpha$ for the ψ_α take the imbeddings of the groups G_α in G'; we see that $\psi\phi_\alpha$ is injective for any $\alpha \in \mathcal{I}$, i.e. that ϕ_α is injective and that G_α by means of ϕ_α can be identified with a subgroup in G.

Suppose that from each group of the family $(G_\alpha)_{\alpha \in \mathcal{I}}$ a subgroup H_α is selected, such that, moreover, for any $\alpha \neq \beta$ from $\mathcal{I}$ there exists an isomorphism $\phi_{\alpha\beta} : H_\alpha \to H_\beta$.

The free product with *amalgamated subgroup* $\prod^*_{H_\alpha = H_\beta} G_\alpha$ is defined as the quotient group of the free product with respect to the normal subgroup K generated by all elements of the form xy^{-1}, where $x \in H_\alpha$, $y \in H_\beta$ and $y = \phi_{\alpha\beta}(x)$. It is clear how to construct a presentation for such group. It is also possible to formulate the universal property defining such a group.

Finally, an *HNN–extension* of a group $G = \langle X | R \rangle$ is defined by means of a pair of subgroups H, K together with an isomorphism $\phi : H \to K$ between them. It is the group with presentation $\langle X, a | R, \phi(h) = aha^1 (h \in H) \rangle$. The term HNN comes from the names of three well–known algebraists, G. Higman, B. Neumann, and H. Neumann, who were the first to define and apply this important construction. One of the significant results obtained with a help of HNN–extensions is the possibility to imbed any denumerable group into a group with two generating elements. For another application, see Subsection 4.6, below.

An important question concerning the structure of the constructions introduced, which is natural in view of the Nielsen–Schreier theorem, can be stated as follows: is it possible that the subgroup of a free construction has a structure similar to a structure of the entire group. A most satisfactory result has been obtained by A.G. Kurosh for a subgroup of a free product. Before formulating this result we shall introduce a theorem in the language of the two–dimensional complexes necessary to prove this result.

Theorem. *Let the two–dimensional complex C be a union of connnected subcomplexes C_α, where α runs over some set of indices $\mathcal{I}$. Let v be a common vertex for all C_α, while the different complexes C_α have no other common vertices. Then $\pi(C, v) = \prod^*_{\alpha \in \mathcal{I}} \pi(C_\alpha, v)$ is the free product of the family of the groups $(\pi(C_\alpha, v))_{\alpha \in \mathcal{I}}$.*

Proof. Let us select a maximal tree T_α of the complex C^1_α. Since there is only one common vertex v, the union $T = \bigcup_{\alpha \in \mathcal{I}} T_\alpha$ is a maximal subtree in C^1. The group $\pi(C^1, v)$ is free. As was proved in Subsection 3.3, a free generating set of it consists of the classes of loops of the form $\tilde{e} = \overline{vx} e \overline{vy}^{-1}$, where $\overline{vx}$ is the path along the tree T from v to the origin x of the edge e which does not belong to the tree, $\overline{vy}$ is the analogous path from v to the end of the edge e. In virtue of the construction of the tree this free generating set Z is the non–intersecting union of the free generating sets Z_α for the subgroups $F_\alpha = \pi(C^1_\alpha, v)$.

It remains only to show that the kernel N of the natural homomorphism of the group $F = \pi(C^1, v)$ to $G - \pi(C, v)$ is the normal closure of the set R

of elements from F each of which lies in one of the subgroups F_α. But N, according to the definition, is the normal closure of the elements from $\pi(C^1, v)$ defined by loops p with origin at the vertex v and of the form $p = qsq^{-1}$, where q is a path from v to some vertex v', and s is the boundary starting in v' of some cell D of the complex C. Replacing such an element r by a conjugate we can assume that the path q is minimal. If C_α is the subcomplex containing the cell D, then the loop p entirely lies in C_α. If $p = e_1 \dots e_n$, then all e_j lie in C_α, and $r = \widetilde{e}_1 \dots \widetilde{e}_n$, moreover $\widetilde{e}_j \in Z_\alpha$ as required. So

$$\pi(C, v) \cong \langle \bigcup_{\alpha \in \mathcal{I}} Y_\alpha \mid \bigcup_{\alpha \in \mathcal{I}} R_\alpha \rangle = \prod_{\alpha \in \mathcal{I}}{}^{*} \pi(C_\alpha, v).$$

□

4.5. Theorem on subgroups of a free product. In this subsection we give a "geometrical" proof of the theorem by A.G. Kurosh on subgroups which was originally proved using a labourious combinatorial method.

Theorem (A.G. Kurosh). *Let G be the free product of the family of groups $(G_\alpha)_{\alpha \in \mathcal{I}}$, let H be a subgroup in G. Then H is a free product of a free group and of groups which are conjugate to subgroups of the factors G_α of the group G.*

Proof. We construct for each group G_α a connected complex C_α with distinguished vertex v_α such that $G_\alpha \cong \pi(C_\alpha, v_\alpha)$. Let us form the complex C as the non–intersecting union of the complexes C_α connected by edges e_α, starting in the vertex v_α and ending in a common vertex v. If $C'_\alpha = C_\alpha \cup \{e_\alpha\}$ then $\pi(C'_\alpha, v) \cong \pi(C_\alpha, v_\alpha)$. According to the previous theorem $\pi(C, v) \cong \prod^* \pi(C'_\alpha, v)$ and we can assume that $G = \pi(C, v)$ and also $G_\alpha = \pi(C'_\alpha, v)$.

According to Subsection 4.3 for any subgroup H of the group G there exists a covering mapping f of some connected complex $\widetilde{C}$ to C which induces an isomorphism f^* of the group $\widetilde{H} = \pi(\widetilde{C}, \widetilde{v})$ to H, where $f(\widetilde{v}) = v$. The components K_β of the subcomplexes $f^{-1}(C_\alpha)$ do not intersect, and $\widetilde{C}$ is a union of these components with the one–dimensional complex $L = \bigcup_{\alpha \in \mathcal{I}} f^{-1}(e_\alpha)$ for which only vertices can be common with a K_β. We select in each K_β a maximal tree T_β. We extend the trees obtained to a maximal tree T in $\widetilde{C}$. Let $\widetilde{v}$ be a vertex in $\widetilde{C}$ such that $v = f(\widetilde{v})$. As before, the group $F = \pi(\widetilde{C}^1, v)$

is free with a free generating set X consisting of the elements $x = \widetilde{e}$ which correspond the pairs of edges $\{e, e^{-1}\}$ of the complex $\widetilde{C}$ and not belonging to the tree T. Consequently, the free generating set X can be divided into the subsets X_β and X_L depending on whether the edge e belongs to the sets K_β or L.

The group $\widetilde{H}$ has the form $\widetilde{H} = F/N$, where N is the normal closure of the set of elements r defined by the loops $p = sqs^{-1}$ beginning in $\widetilde{v}$ from $\widetilde{C}$, where q is the boundary of some cell D of the complex $\widetilde{C}$. Replacing the element r by a conjugate we can assume that s lies in the tree T and thus r is a word in the letters $x = \widetilde{e}$ of the edges e, which belong to the boundary q of the cell D. Since the cell D is in some component K_β, all x belong to X_β. Therefore, N is the normal closure of the union of sets R_β corresponding to X_β. Thus, H is the free product of the groups $\widetilde{H}_\beta = \langle X_\beta | R_\beta \rangle$ and the free group $\widetilde{H}_L = \langle X_L | \emptyset \rangle$.

Now we consider the groups of the form $\widetilde{H}_\beta$. Let K_β be a component of the subcomplex $f^{-1}(C_\alpha)$, and s a path in T from $\widetilde{v}$ to a vertex $w \in K_\beta$ such that $f(w) = v$. Then $f(s)$ is a loop with origin v in C representing some element h from G. But every element $x = \widetilde{e}$ in X_j is representable by a loop $p = sqs^{-1}$ with origin $\widetilde{v}$ in $\widetilde{C}$, where q is a loop with origin w in K_β. Therefore $f(q)$ is a loop with origin v in C'_α representing some element $g \in G_\alpha$, and $f(p)$ represents the element hgh^{-1} of the group $hG_\alpha h^{-1}$. Hence it follows that f^* maps $\widetilde{H}_j$ isomorphically onto a subgroup of the group $hG_\alpha h^{-1}$. So the subgroup H isomorphic to the group $\widetilde{H}$ is the free product of subgroups $f^*(\widetilde{H}_\beta)$ from $hG_\alpha h^{-1}$ and the free group $f^*(\widetilde{H}_L)$. □

Corollary. *Let H be a torsion free subgroup in $\prod^*_{\alpha \in I} G_\alpha$, where all the G_α are periodic groups. Then H is a free group.* □

∗ **4.6. Dehn problems.** In 1911 M. Dehn formulated three problems corresponding to the presentaion of groups by generators and relations.

I. Equality problem (of words). Let an arbitrary element of the group be given as a word in the generators. To find a method to determine in a finite number of steps whether the element is equal to unity or not.

II. Conjugacy problem. Let there be given two arbitrary elements S and T

from the group. The problem is to find a method to determine whether S and T are conjugate.

III. Isomorphism problem. Two groups are given. Determine whether they are isomorphic.

A positive solution to the equality problem, i.e. a description of the method required, has been obtained for a number of special cases, for example, such a method is known if the group is abelian or nilpotent, or finite, etc.. Without imposing extra conditions a positive solution can be obtained if there is only one defining relation or if the relations are sufficiently disjoint from each other. Here we have the following in mind. If there exists a presentation of the group $\langle X|R\rangle$ then the inverse elements are added to X and subsequently to every element its cyclic permutations are added (to the relations). We shall carry out all possible reductions of mutually inverse letters. Now we assume that r_1 and r_2 are different elements of the thus obtained set of defining relations which we denote by R. We call a word b (over $X \cup X^{-1}$) a piece, if $r_1 = bc_1$ and $r_2 = bc_2$ as words. It is said that a group satisfies the condition $C'(\lambda)$ if for any $r \in R$, $r = bc$, where b is a piece, the length of the word b is strictly less than the length of the word r multiplied by λ. For instance, the fundamental group G of a closed oriented 2–manifold of genus g has a presentation

$$G = \langle a_1, b_1, \dots , a_g, b_g | a_1^{-1} b_1^{-1} a_1 b_1 \dots a_g^{-1} b_g^{-1} a_g b_g = 1\rangle .$$

It can be easily verified that this group satisfies the condition $C'(1/(4g-1))$. The groups with condition $C'(\lambda)$ with $\lambda \leq \frac{1}{5}$ have a solvable equality problem. This result is not valid for some groups for which some pieces have length equal to $\frac{1}{5}$ of word.

The first examples of the groups with an unsolvable equality problem were constructed by P. . Novikov and afterwards by W. Boone in the middle of 50–ies.[1] This work was preceded by a decade of deep developments in mathematical logic concerning the concept of unsolvability as the absence of an algorithm which produces an answer in a finite number of steps. The details of the corresponding definitions would lead us too far aside. We only note that these results on unsolvability have, for instance, the following consequence:

[1] P. S. Novikov in 1952, 1955, and independently W. W. Boone in 1958, and also J. L. Britton in 1958. (Translator's note).

an algorithm which allows to determine from a finite presentation $G = \langle X | R \rangle$ whether the group G is the identity one or not, does not exist.

§ 5. Simple groups

5.1. Introductory remarks. In this book we practically do not concern ourselves with the extensive theory of finite simple groups. The main reason is the extreme technicality and complexity in principle of this theory. In recent years a significant result in this theory was achieved: the complete classification of finite simple groups was achieved. The proof as a complete unit, supposed to take up to tens of thousands of pages, is not published yet. The list of finite simple groups (in this situation one always speaks only of non–abelian groups) contains several sets of groups like the A_n, $n \geq 5$, the alternating groups, several infinite sets of so called Lie type and a number of groups which do not belong to these sets, the so called sporadic groups.

The simplicity of the alternating groups is usually proved in a standard course of higher algebra, and we won't discuss this here. In this section we shall prove the simplicity of the groups from the series $PSL_n(F)$, where $n \geq 2$, F is any finite field except when $n = 2$, when $F \neq \mathbf{Z}_2, \mathbf{Z}_3$. The orders of these groups will also be calculated.

Before starting on the main part of this section we should mention the famous Feit–Thompson theorem which is one of the corner stones in the theory of finite groups, which solves of still another well–known problem of Burnside (see, Subsection 2.1).

∗ **Theorem.** *Any finite group of odd order is solvable.*

Corollary. *The order of a non–cyclic finite simple group is even.* ∗

5.2. The groups PSL. Transvections. Let F be an arbitrary field, V a vector space of dimension $n \geq 2$, $GL(V)$ the complete linear group of the space V, i.e. the group of all non–singular linear operators on V. The selection of a basis in V associates to every operator from $GL(V)$ a non–singular quadratic matrix, i.e. an element from the group $GL_n(F)$. The family of operators (matrices) with determinant 1 is the subgroup $SL(V)$ (respectively, $SL_n(F)$) of the group $GL(V)$ (respectively, $GL_n(F)$).

Finally, the group $PSL_n(F)$ is the quotient group of the group $SL_n(F)$ with respect to its center. While studying this group, which is called a *projecitve special group*, we shall use both geometrical language, and, in virtue of the isomorphism $SL_n \cong SL(V)$, also the language of matrices.

We call a linear operator $\sigma \in GL(V)$ a *transvection* if $\det \sigma = 1$ and $\dim(\sigma - 1_V)V = 1$. It is clear that any transvection lies in $SL(V)$. An example of a transvection is an operator, the matrix of which in some basis is the matrix of an elementary transformation of the first kind, i.e. adding to some row of a matrix another row multiplied by a non–zero number $\lambda \in F$. We introduce also the notation $\tau_{v,\rho}$, where $v \in V$, $\rho \in V^*$, for the transvection that acts according to the rule

$$\tau_{v,\rho}(x) = x + \rho(x)v. \tag{1}$$

The reader can easily verify the following properties:

1) $\tau_{v\rho} = 1_V$ iff $v = 0$ or $\rho = 0$;
2) $(\tau_{a,\rho}\tau_{b,\phi})(x) = \{x + \rho(x)a + \phi(x)b\} + \phi(x)\rho(b)a$;
3) $\sigma\tau_{a,\rho}\sigma^{-1} = \tau_{\sigma(a),\rho\sigma^{-1}}$.

Lemma 1. *The group $SL(V)$ is generated by transvections.*

Proof. We shall use matrices to write down the elements of the group $SL(V)$ with respect to some basis. Recall the well–known theorem about the reduction of a non–singular matrix to diagonal form by elementary transformations of rows of the first kind (see above). The same transformations but applied to both rows and columns of a matrix with an identity determinant permit to turn it into the identity matrix. We illustrate the steps of such a transformation for diagonal matrices of order two:

$$\begin{pmatrix} a & 0 \\ 0 & b \end{pmatrix} \mapsto \begin{pmatrix} a & 0 \\ b & b \end{pmatrix} \mapsto \begin{pmatrix} 1 & 1-a \\ b & b \end{pmatrix} \mapsto$$
$$\mapsto \begin{pmatrix} 1 & 1-a \\ 0 & ab \end{pmatrix} \mapsto \begin{pmatrix} 1 & 0 \\ 0 & ab \end{pmatrix} = \begin{pmatrix} 1 & 0 \\ 0 & 1 \end{pmatrix}.$$

Thus, for any $\sigma \in SL(V)$ we can find transvections $\tau_1, \ldots, \tau_k$, $\tau'_1, \ldots, \tau'_l$ and such that

$$1_V = \tau_1, \ldots, \tau_k\sigma\tau'_1, \ldots, \tau'_l,$$

from which $\sigma = \tau_k^{-1} \dots \tau_1^{-1} \tau_l'^{-1} \dots \tau_1'^{-1}$ which is a product of transvections, as the inverse operator of a tranvection is also a transvection. □

Lemma 2. *Any two transvections of the group $SL(V)$ are conjugate in $GL(V)$. When $n = \dim V \geq 3$ an element from $SL(V)$ can be used as the conjugating element.*

Proof. According to the definition any transvection has the form $\sigma = \tau_{v\rho}$ for suitable $v \in V$ and $\rho \in V^*$; moreover, it is possible to choose v such that $\langle v \rangle = \langle (\sigma - 1_V)V \rangle$. In this case $\rho(v) = 0$. Indeed, from the definition of a transvection follows that its characteristic polynomial has the form $\chi_\sigma(X) = \pm(X-1)^{n-1}(X-\alpha)$; further $\det \sigma = 1$. Hence $\alpha = 1$, i.e. $\chi_\sigma(X) = \pm(X-1)^n$. Thus, the minimal polynomial, $\mu_\sigma(X)$, is equal to $(X-1)^2$. Therefore $(\sigma - 1_V)(v) = 0$. Using (1) we obtain what was required.

Now let σ' be another transvection. Then we write

$$\sigma(x) = x + \rho(x)v, \quad \sigma'(x) = x + \rho'(x)v',$$

where $\rho'(v')$ is also equal to zero. We select bases $\{v = e_1, e_2, \dots, e_n\}$ and $\{v' = e_1', e_2', \dots, e_n'\}$ in the space V such that $\rho(e_1) = \dots = \rho(e_{n-1}) = 0$, $\rho(e_n) = 1$; $\rho'(e_1') = \dots = \rho'(e_{n-1}') = 0$, $\rho'(e_n') = 1$. Let $\tau \in GL(V)$ be the operator such that $\tau(e_i) = e_i'$, $i = 1, 2, \dots, n$. Then $\tau(v) = v'$, $\rho\tau^{-1} = \rho'$ and according to property 3) we obtain $\sigma' = \tau\sigma\tau^{-1}$.

When $n \geq 3$ the conjugating element $\tau \in GL(V)$ can be replaced by the element $\bar\tau$ for which, in addition, $\bar\tau(e_2) = \lambda e_2'$, where $\lambda = (\det \tau)^{-1}$. In this case $\bar\tau \in SL(V)$, and the property $\sigma' = \bar\tau\sigma\bar\tau^{-1}$ is preserved. □

Corollary. *The commutator subgroup of the group $GL_n(F)$ is equal to $SL_n(F)$ in all cases with the exeption of $GL_2(\mathbf{Z}_2)$.*

Indeed, let $f : GL_n(F) \to A$ be a homomorphism into an abelian group. Since $\sigma' = \tau\sigma\tau^{-1}$ for any transvections σ, σ' for a suitable $\tau \in GL_n(F)$, we obtain $f(\sigma') = f(\tau\sigma\tau^{-1}) = f(\tau)f(\sigma)f(\tau)^{-1} = f(\sigma)$. Further, suppose the space V is not $\mathbf{Z}_2^2$. Then it is possible to select non-zero vectors v_1, v_2, v_3 in the hypersurface with the equation $\rho(x) = 0$, $\rho \neq 0$ such that $v_3 = v_1 + v_2$. We consider the transvections $\tau_{v_1,\rho}$, $\tau_{v_2,\rho}$, $\tau_{v_3,\rho}$ (where $\rho(v_i) = 0$, $i = 1, 2, 3$). Then according to property 2) above:

(2) $$\tau_{v_3,\rho} = \tau_{v_1,\rho}\tau_{v_2,\rho}.$$

If α is a common value of all transvections under the homomorphism f then from (2) we obtain $\alpha = \alpha^2$. Thus, all the transvections go to 1 for any homomorphism of the group $SL(V)$ into an abelian group, i.e. the group $SL(V)$ generated by them lies in the commutator subgroup of the group $GL(V)$. Since $GL(V)/SL(V) \cong F^*$ is an abelian group, we also have the inverse inclusion. Thus the corollary is proved (see, Introduction 1.10). □

5.3. Simplicity of the group $\mathbf{PSL_n(F)}$. In this subsection the following theorem will be proved

Theorem. *The group $PSL_n((F)$ is simple in all cases with the exception of $PSL_2(\mathbf{Z}_2)$ and $PSL_2(\mathbf{Z}_3)$. If the field F is finite and $q = |F|$ is its order, then the order of the group $PSL_n(F)$ (which is also denoted by $L_n(q)$) is equal to*

$$\frac{1}{d} q^{\frac{n(n-1)}{2}} \prod_{i=2}^{n} (q^i - 1),$$

where d is the greatest common divisor of the numbers n, $q-1$.

Proof. Consider a non–central normal subgroup G in $SL(V)$. We need to prove that $G = SL(V)$.

Case 1: $n \geq 3$. If we succeed to find in G at least one transvection then by using lemmas 1 and 2 we obtain the required. Thus, let $\pi \in G$ be a non–central element of the group $SL(V)$. Since $SL(V)$ is generated by transvections, there is a transvection τ such that $\tau\pi\tau^{-1} \neq \pi$, i.e. $\tau\pi\tau^{-1}\pi^{-1} \neq 1_V$. The element $\sigma = (\tau\pi\tau^{-1})\pi^{-1} = \tau(\pi\tau^{-1}\pi^{-1})$ lies in G and at the same time is a product of the two transvections τ and $\tau' = \pi\tau^{-1}\pi^{-1}$. Using this and the equality $\tau_1\tau_2 - 1_V = \tau_1(\tau_2 - 1_V) + (\tau_1 - 1_V)$ we obtain

$$\dim(\sigma - 1_V)V \leq \dim(\tau_2 - 1_V)V + \dim(\tau_1 - 1_V)V.$$

We have $\det\sigma = 1$. If also $\dim(\sigma - 1_V) = 1$ then σ is a transvection. Suppose that $\dim(\sigma - 1_V)V = 2$. Let $U = (\sigma - 1_V)V$. Since $\dim U = 2$, there exists a hypersurface W (i.e. $\dim W + 1 = \dim V$), such that $U \subset W$. If $(\sigma - 1_V)W = 0$ then the rank of the operator $\sigma - 1_V$ does not exceed 1. So there is an $a \in W$ such that $\sigma a \neq a$. Let $\rho \in V^*$ be such that $\rho(W) = 0$. Then

$$\rho(\sigma^{-1}(x) - x) \in \rho(U) \subset \rho(W) = 0$$

for all $x \in V$. Hence $\rho\sigma^{-1} = \rho$. Consequently, the group G contains the element

$$\sigma(\tau_{a,\rho}\sigma^{-1}\tau_{a,\rho}^{-1}) = (\sigma\tau_{a,\rho}\sigma^{-1})\tau_{a,\rho}^{-1} = \tau_{\sigma a,\rho\sigma^{-1}}\tau_{-a,\rho} = \tau_{\sigma a - a,\rho},$$

which is a nontrivial transvection.

Case 2: $n = 2$. Since G is non–central at least one element σ from G changes some straight line. Let x be a vector collinear with this straight line. The vectors x and $\sigma(x)$ form a basis in V. The matrix of the operator σ in this basis has the form

$$\begin{pmatrix} 0 & -1 \\ 1 & \alpha \end{pmatrix} \tag{3}$$

for some $\alpha \in F$. We shall now consider matrices instead of operators. Take an arbitrary $\lambda \neq 0$ in F and $\mu \in F$. Then G contains the matrix

$$\begin{aligned} &\begin{pmatrix} \lambda^{-1} & 0 \\ 0 & \lambda \end{pmatrix}\begin{pmatrix} 0 & -1 \\ 1 & \alpha \end{pmatrix}\begin{pmatrix} \lambda^{-1} & 0 \\ 0 & \lambda \end{pmatrix}^{-1}\begin{pmatrix} 0 & -1 \\ 1 & \alpha \end{pmatrix}^{-1} = \\ &= \begin{pmatrix} \lambda^{-2} & 0 \\ \alpha(\lambda^2 - 1) & \lambda^2 \end{pmatrix} \in G. \end{aligned} \tag{4}$$

Using the property of normality of the subgroup G we obtain also the following matrix in G

$$\begin{aligned} &\begin{pmatrix} 1 & 0 \\ \mu & 1 \end{pmatrix}\begin{pmatrix} \lambda^{-2} & 0 \\ \alpha(\lambda^2 - 1) & \lambda^2 \end{pmatrix}\begin{pmatrix} 1 & 0 \\ \mu & 1 \end{pmatrix}^{-1}\begin{pmatrix} \lambda^{-2} & 0 \\ \alpha(\lambda^2 - 1) & \lambda^2 \end{pmatrix}^{-1} = \\ &= \begin{pmatrix} 1 & 0 \\ \mu(1 - \lambda^4) & 1 \end{pmatrix} \in G. \end{aligned} \tag{5}$$

Conjugating the element obtained by the matrix $\begin{pmatrix} 0 & -1 \\ 1 & 0 \end{pmatrix}$ we obtain

$$\begin{pmatrix} 1 & -\mu(1 - \lambda^4) \\ 0 & 1 \end{pmatrix} \in G. \tag{6}$$

If $F \neq \mathbf{Z}_5$ then either $|F| > 5$ or $|F| = 4$, so there exists $\lambda \in F \setminus \{0\}$ for which $\lambda^4 \neq 1$, i.e. $\lambda^4 - 1 \neq 0$. Since μ is an arbitrary element, (5) and (6) are arbitrary elementary matrices and $G = SL_2$. If $F = \mathbf{Z}_5$ and $\alpha = 0$ in (3) then conjugating (3) by the matrix $\begin{pmatrix} 1 & 3 \\ 0 & 1 \end{pmatrix}$ we obtain $\begin{pmatrix} 3 & 0 \\ 1 & 2 \end{pmatrix} \in G$. Further

$$\begin{pmatrix} 1 & 0 \\ 2 & 1 \end{pmatrix} = \begin{pmatrix} 1 & 0 \\ 1 & 1 \end{pmatrix} \begin{pmatrix} 3 & 0 \\ 1 & 2 \end{pmatrix} \begin{pmatrix} 1 & 0 \\ 1 & 1 \end{pmatrix}^{-1} \begin{pmatrix} 3 & 0 \\ 1 & 2 \end{pmatrix} \in G.$$

If $\alpha \neq 0$ then in view of (4) the matrix $\begin{pmatrix} 4 & 0 \\ 3\alpha & 4 \end{pmatrix}$ lies in G. Then its square $\begin{pmatrix} 1 & 0 \\ 4\alpha & 1 \end{pmatrix}$ also lies in G. Taking its powers we obtain that G contains all elementary matrices with zero above the diagonal. Conjugating with the matrix $\begin{pmatrix} 0 & -1 \\ 1 & 0 \end{pmatrix}$ then gives all elementary matrices with zero below the diagonal. Again $G = SL_2$.

If now $\bar{G}$ is a normal subgroup in $PSL_n(F)$ then its complete preimage in $SL_n(F)$ is a normal subgroup in that group containing its center. As we have noted such a subgroup is either central or coincides with $SL_n(F)$. Then $\bar{G}$ is either the identity or coincides with $PSL_n(F)$.

Going over to the statement about orders we should mention that the first row of an arbitrary matrix $A = GL_n(F)$, $|F| = q$, can be selected in $q^n - 1$ ways (a non–zero vector in F^n), the second row in $(q^n - q)$ ways (a vector from F^n which does not lie in the linear span of the first row), the third row in $(q^n - q^2)$ ways (a vector from F^n which does not lie in the linear span of the first two rows), etc.. For the last row only $q^n - q^{n-1}$ possibilities are left. Thus the order of the group $GL_n(F)$ is equal to

$$q^{\frac{n(n-1)}{2}} \prod_{i=1}^{n} (q^i - 1).$$

The order of the group $SL_n(F)$ is equal to the number obtained divided by the number $q - 1$ of possible nonzero values of the determinant in the field F, i.e.

$$q^{\frac{n(n-1)}{2}} \prod_{i=2}^{n} (q^i - 1).$$

If C_n is the subgroup of scalar matrices in $SL_n(F)$ then its order is equal to the number of roots of degree n in F. Since F^* is a cyclic group of order $q - 1$, this order is equal to $d = (n, q - 1)$. From here

$$|PSL_n(F)| = \frac{1}{d} q^{\frac{n(n-1)}{2}} \prod_{i=2}^{n} (q^i - 1),$$

as was required. □

Note. In the exceptional cases we have
$PSL_2(\mathbf{Z}_2) \cong S_3$, i.e. a group hat is not simple;
$|PSL_2(\mathbf{Z}_3)| = 12$, this group contains the normal subgroup

$$G = \pm\left\{ \begin{pmatrix} 1 & 0 \\ 0 & 1 \end{pmatrix}, \begin{pmatrix} -1 & 1 \\ 1 & 1 \end{pmatrix}, \begin{pmatrix} 1 & 1 \\ 1 & -1 \end{pmatrix}, \begin{pmatrix} 0 & -1 \\ 1 & 0 \end{pmatrix} \right\}$$

of order 4.

§ 6. Topological groups

6.1. Basic definitions. A group G is called a *topological group* if G is a Hausdorff topological space such that the mappings $(x, y) \mapsto xy$ and $x \mapsto x^{-1}$ are continuous mappings from $G \times G$ into G and from G into G, respectively.

The continuity requirement of the two mappings can be replaced by the continuity of the one mapping $(x, y) \mapsto xy^{-1}$ from $G \times G$ in G. In terms of neighborhoods the continuity requirement is the following: for any $x, y \in G$ and any neighborhood W of the point xy^{-1} there can be found neighborhoods U of the point x and V of the point y, such that $UV^{-1} \subset W$.

If $a \in G$ then the mapping $l(a) : x \mapsto ax$ from G in G is called the *left translation* by the element a. Analogously, the right translation $r(a)$ is defined. It is clear that the left and right translations of a group by any element are homeomorphisms of the topological space G. Application of these translations (shifts) shows that locally, i.e. in the neighborhood of any point, a topological group is structured in the same way, namely, as it is structured in the neighborhood of the identity e. Indeed, if V is a neighborhood of the point e then $l(a)V$ is a neighborhood of the point a that is homeomorphic to V.

Proposition 1. *Let G be a topological space and a group at the same time. Then G is a topological group iff:*

(*i*) *the set $\{e\}$ is closed;*

(*ii*) *for any* $a \in G$ *the translations* $l(a)$ *and* $r(a)$ *are continuous;*

(*iii*) *the mapping* $(x, y) \to xy^{-1}$ *from* $G \times G$ *into* G *is continuous in the point* (e, e).

Proof. In one direction the statement is clear. Now we shall prove that under the conditions stated, G is a Hausdorff space. Let $a \neq b$, $a, b \in G$. We then have $\{a\} = l(a)\{e\}$ and $\{b\} = l(b)\{e\}$. Since the translation is a homeomorphism, the sets $\{a\}$ and $\{b\}$ are closed. Then, of course, there exists a neighborhood U of the point b such that $U \cap \{a\} = \emptyset$. Further, $U = bV$, where V is a neighborhood of the identity e. Using (*iii*) we can find a neighborhood W of the point e such that $WW^{-1} \subset V$. Then $aW \cap bW = \emptyset$. Indeed, if $x \in aW \cap bW$ then $x = aw_1 = bw_2$ while $w_i \in W$. So $a = bw_2w^{-1} \in bWW^{-1} \subset bV$, which is impossible. Thus, G is a Hausdorff topological space. We shall prove that multiplication in G is continuous. Take $a, b \in G$. Let W be a neighborhood of the point ab^{-1}. Consider $W' = l(b)l(a)^{-1}W$. According to property (*ii*) this is a neighborhood of the point e. We select neighborhoods U and V such that $UV^{-1} \subset W'$. Then aU and bV are neighborhoods of the points a, b, and

$$(aU)(bV)^{-1} = aUV^{-1}b^{-1} \subset aW'b^{-1} = a(a^{-1}Wb)b^{-1} = W.$$

This proves the continuity of multiplication. □

Lemma. *Let* U *be a neighborhood of the identity* e *of a topological group* G. *Then there exists a symmetrical neighborhood* V *of the point* e *contained in* U, *i.e. a neighborhood such that* $V = V^{-1}$ *and* $VV = VV^{-1} \subset U$. *In this case the closure* $\bar{V}$ *of* V *also lies in* U.

Proof. Since the multiplication $G \times G \to G$ is continuous, there exist neighborhoods P, Q of the point e in G such that $PQ \subset U$. Write $W = P \cap Q$ and put $W \cap W^{-1} = V$. Then it is evident that V possesses the required properties. Let $x \in \bar{V}$. Then xV is a neighborhood of the point x and hence $V \cap xV \neq \emptyset$. Thus, for suitable v, $v_1 \in V$ we have $xv = v_1$, i.e. $x = v_1v^{-1} \in VV^{-1} \subset U$. □

Examples.

1) Groups of matrices with a topology of the space $\mathbf{R}^{n^2}$;

2) Any group with the discrete topology, i.e. the topology in which any subspace is open;

3) A direct product of topological groups with the topology of a direct product of topological spaces.

6.2. Subgroups and homogeneous spaces. A *topological subgroup* H of the topological group G is a subgroup in G with the induced topology, i.e. U is open in H, iff there exists a set V that is open in G and such that $U = V \cap H$.

We should mention that an open group H in G is always closed; indeed, for any $a \in G$ the coset aH is also an open set. Since G is the union of pairwise non–intersecting cosets, $H = G \setminus \bigcup_{a \notin H} aH$ is a closed set. It is evident that the closure of a subgroup is always a subgroup .

Theorem. *Let H be a topological subgroup in the topological group G, and $\pi : G \to G/H$, $\pi(a) = aH$, the natural projection. Then:*

(a) G/H can be given a unique structure of a topological space such that:

(i) the projection $\pi : G \to G/H$ is continuous;

(ii) if N is a topological space and the mapping $f : G/H \to N$ is such that $f\pi : G \to N$ is continuous then f is continuous.

(b) The mapping $\pi : G \to G/H$ is open, i.e. the image $\pi(U)$ of an open set $U \subset G$ is open in G/H.

(c) The quotient topology is Hausdorff iff H is a closed subset in G.

Proof. We define a topology on G/H by taking U open in G/H iff $\pi^{-1}(U)$ is open in G. From this the continuity of the mapping π and the property (ii) follow.

Let now T be any topology on G/H, T' any other topology, both with the properties (i) and (ii) and let $f : G/H \to G/H$ be the identity mapping. Then according to the properties (i) and (ii) it follows that f is continuous. Interchanging T and T' we obtain that the inverse to f (it is equal to f) is also continuous, i.e. f is a homeomorphism. Hence, $T = T'$.

Now, let U be open in G. The complete preimage of $\pi(U)$ is UH. But $UH = \bigcup_{a \in H} Ua$, as a union of open sets, is open. Now $\pi(U)$ also is open (by definition). This means that $\pi : G \to G/H$ is an open mapping.

If G/H is Hausdorff, then $\{eH\}$ is closed, and also $\pi^{-1}(\{eH\}) = H$ is closed in G, because π is continuous. Inversily, let $aH \neq bH$. The sets aH and bH are closed. A neighborhood U of the point e can be found such that

$aU \cap bH = \emptyset$. According to the lemma in Subsection 6.1, there exsists a neigborhood V of the point e such that $V^{-1}V \subset U$; so $(Va)H = V(aH)$ and $V(bH)$ are neighborhoods of aH and bH, respectively. These neighborhoods do not intersect, because if $p \in V(aH) \cap V(bH)$ then $p = vah = v_1bh_1$, for $v, v_1 \in V$, $h, h_1 \in H$. Hence $q = v_1^{-1}va = bh_1h^{-1} \in Ua \cap bH$, giving a contradiction. □

Corollary 1. *Let H be a closed normal subgroup of the topological group G, G/H the quotient group. Then with respect to the quotient–topology, G/H becomes a topological group; moreover, the projection $\pi : G \to G/H$ is an open continuous mapping.*

Proof. We consider a neighborhood W of the point $(xH)(yH)^{-1}$. Then the set $\pi^{-1}(W)$ is a neighborhood of $xy^{-1} \in G$, containing H. There are neighborhoods U and V of x, y such that $UV^{-1} \subset \pi^{-1}(W)$. Then $\pi(U)\pi(V)^{-1} \subset W$. Since π is open, the sets $\pi(U)$ and $\pi(V)$ are open and that proves the continuity of the operations in G/H. □

Corollary 2. *If H is an open normal subgroup in the topological group G, then G/H is discrete.* □

Corollary 3. *Let $f : G \to \bar{G}$ be a homomorphism of topological groups. Then f is continuous iff it is continuous in the point $e \in G$.* □

Homomorphism theorem. *Let $f : G \to \widetilde{G}$ be a continuous homomorphism of the topological groups G and $\widetilde{G}$, and $H = \{x \in G | f(x) = \widetilde{e}\}$ the kernel of f, where $\widetilde{e}$ is the identity in $\widetilde{G}$. Then:*

(*a*) *H is a closed normal subgroup and $\pi : G \to G/H$ is a continuous homomorphism;*

(*b*) *there exists a continuous homomorphism $g : G/H \to \widetilde{G}$ such that $f = g\pi$;*

(*c*) *let H and N be the closed topological subgroups in G such that $N \subset H$, then G/H is topologically isomorphic to the group $(G/N)/(H/N)$.* □

Examples. **R** with the usual metric topology, **Z** the additive subgroup of integers. Then **Z** is closed in **R** and $\mathbf{R}/\mathbf{Z} \cong \mathbf{T}^1$. For this we can use the fact that $f : \mathbf{R} \to \mathbf{T}^1$, given by $f(x) = e^{2\pi ix}$ is a continuous homomorphism with kernel $\mathbf{Z}$.

Analogously, $\mathbf{R}^n/\mathbf{Z}^n \cong \mathbf{T}^n$. The group $\mathbf{T}^n$ is called the n–dimensional torus.

6.3. Action of a group on a set. Let M be a Hausdorff topological space, G a topological group. The group G acts continuously on M if an action of G is defined on M (see, Introduction 1.7) such that $G \times M \to M : (g, x) \mapsto gx$ is a continuous mapping.

Example. If H is a closed subgroup in the topological group G, then $M = G/H$ is a Hausdorff space. For any $g \in G$ the mappig $(g, xH) \mapsto gxH$ is continuous. For any $g \in G$ the mapping $\tau(g) : M \to M$, $\tau(g)(xH) = gxH$, is a homeomorphism. The space $M = G/H$ is called a *homogeneous* space.

We say that G acts *transitively* on M if for any $x, y \in M$ there exists such $g \in G$ that $y = gx$.

Theorem. *Let M be a Hausdorff space provided with a transitive continuous action of a topological group G. Let p be a given point in M and G_p its stabilizer. Then G_p is a closed subgroup in G, and the mapping $f : G \to G_p$, $f(a) = ap$, induces a contionuous bijection $\bar{f} : G/G_p \to M$ such that $\bar{f}\pi = f$.*

Proof. Since the action is transitive, we have $Gp = M$. Further, as the action $f : G \to M$ is continuous and $\{p\}$ is closed, $G_p = f^{-1}(\{p\})$ is closed.

We have already seen in Subsection 6.2 that $\bar{f}$ is well defined and is bijective. Further, as $f = \bar{f}\pi$ and $f : G \to M$ is continuous, according to the theorem from Subsection 6.2, $\bar{f}$ is continuous. □

Corollary. *If $\bar{f} : G/G_p \to M$ is open, then $\bar{f}$ is a homeomorphism and M is a homogeneous spacce.*

Proof. If $\bar{f} : G/G_p \to M$ is open, then by continuity for $\bar{f}$, the mapping $\bar{f}^{-1}$ is also open, i.e. $\bar{f}$ is a homeomorphism. □

6.4. Connected groups. Let M be a topological space and $p \in M$. Then p lies in the unique maximal connected subset $C(p)$, the so called *connected component of the point* p. The connected component of the identity is simply called the *identity component*.

Theorem. *Let G be a topological group, G_0 its identity component. Then:*

(a) G_0 is a closed normal subgroup in G, and $C(a) = aG_0$ for any point $a \in G$.

(b) If G is locally connected (i.e. any point has a connected neighborhood), then G/G_0 is discrete.

Proof. First of all, as G_0 is the component of the identity, $C(e)$, it is clear that G_0 is a closed set. Let now $a \in G_0$. Since multiplication is continuous, $a^{-1}G_0$ is a connected set containing e. We have $a^{-1}G_0 \subset G_0$, i.e. G_0 is a subgroup. Analogously, for any $a \in G$, aG_0a^{-1} is a connected set containing e. Thus, $aG_0a^{-1} \subset G_0$, i.e. G_0 is a normal subgroup in G. Since left translation is a homeomorphism, it is clear that aG_0 is the connected component for $a \in G$.

In order to prove (b) we assume that G is locally connected and U is a connected neighborhood of the identity e in G. Since $\pi : G \to G/G_0$ is an open mapping, the set $\pi(U)$ is a neighborhood of the element eG_0 in G/G_0. At the same time, because U is connected, we have $U \subset G_0$, so that $\pi(U) = \{eG_0\}$. Hence, $\{eG_0\}$ is an open set and G/G_0 – a discrete group. □

Proposition 1. *Let G be a topological group, G_0 its identity component, and U any neighborhood of the identity in G.*

(a) If U is a symmetrical neighborhood, then $H = \bigcup_{k=1}^{\infty} U^k$ is an open and closed subgroup in G. If U is a connected neighborhood, then H is also a connected subgroup .

(b) $G_0 = \left(\bigcup\limits_{k=1}^{\infty} U^k\right) \bigcap G_0.$

(c) If G is a connected group, then $G = \bigcup_{k=1}^{\infty} U^k$. In this case any neighborhood of the identity e is a set of generators of the connected topological group as an abstract group.

Proof. (a) If U is a symmetrical neighborhood, then for $x \in U^m$ and $y \in U^n$ we have $xy \in U^{m+n}$ and $x^{-1} \in (U^{-1})^m = U^m$, so that H is a subgroup . Further, since U is an open set, $U^2 = \cup\{aU | a \in U\}$ is open and, by induction, U^k is an open set. Thus, H is open and, consequently, it is also a closed subgroup . If U is connected then every U^k is connected, so that H is a connected subgroup (use the fact that $e \in U^k$).

(b) Let V be a symmetrical neighborhood of the identity such that $V \subset U$ and let $W = V \cap G_0$. Then W is a symmetrical neighborhood of the identity

$e \in G_0$ and $H = \cup W^k \subset \cup V^k \cap G_0$. At the same time H is a non–empty, open and closed subgroup from G_0; since G_0 is connected, we have $G_0 = H$. As $\cup V^k \subset \cup U^k$, we obtain our result.

(*c*) follows from (*b*), because $G = G_0$ in this case. □

Proposition 2. *Let G be a connected topological group, H a discrete normal topological subgroup of the group G. Then $H \subset Z(G)$.*

Proof. Let $a \in H$. Then the mapping $x \mapsto xax^{-1}$ from G into H is continuous. Since G is connected, and H is discrete, the image of the mapping is a single point $p \in H$. As $p = eae^{-1}$, $xax^{-1} = a$, for any $x \in G$. Therefore, $H \subset Z(G)$. □

Proposition 3. *Let G be a topological group, H a closed topological subgroup such that H and G/H are connected. Then G is also connected.*

Proof. Assume that $G = U \cup V$, where U and V are disjoint non–empty open sets. The open mapping $\pi : G \to G/H$ maps U and V onto the open sets $U_1 = UH$ and $V_1 = VH$ in G/H; we have $G/H = U_1 \cup V_1$ so that there is an $aH \in UH \cap VH$. Since $aH \in UH$, there exists such $h \in H$ that $ah = u \in U$. Thus, $aH \cap U$ is non–empty. Analogously, $aH \cap V$ is also non–empty. Then $aH = (aH \cap U) \cup (aH \cap V)$ which is impossible, since aH is connected together with H. □

6.5. Lie groups. A group G whose underlying set has the structure of an n–dimensional real analytic manifold is called a *Lie group* if the mapping $(x, y) \mapsto xy^{-1}$ from $G \times G$ in G is a morphism of analytic manifolds. In particular, G is a topological group so everything that was mentioned above about topological groups applies to a Lie group. An example of a Lie group is the additive group $\mathbf{R}$ of real numbers, the multiplicative group $\mathbf{R}^*$ of non–zero real numbers, the groups $\mathbf{C}$ and $\mathbf{C}^*$ of complex numbers (these two groups are two–dimensional), the groups $GL_n(\mathbf{R})$ and $GL_n(\mathbf{C})$ of non–singular quadratic matrices of order n with real and complex coefficients (the dimensions of these groups are n^2 and $2n^2$, respectively). In the Set–theoretic supplement, Section 14 it is shown how an analytical structure is introduced on $GL_n(\mathbf{R})$; the verification of the fact that $(x, y) \mapsto xy^{-1}$ is a morphism presents no difficulties and is left to the reader. We should note that any discrete topological group is a zero–dimensional Lie group. Another simple observation is that a direct product of Lie groups is itself a Lie group. Therefore, $\mathbf{T}^n$ (then n–dimensional torus) is a commutative Lie group.

Finally, we note that the space $\mathbf{R}$ of real numbers with its usual structure of a manifold is a topological group with respect to the operation $(x, y) \to \sqrt[3]{x^3 + y^3}$. This group is not a Lie group as the given function is not analytic in zero.

Let G_1, G_2 be two Lie groups. A morphism $f : G_1 \to G_2$ of analytic manifolds which is a homomorphism of groups is called an *analytic* homomorphism.

A subgroup H of a group G is called a *Lie subgroup* if it is submanifold, i.e. if in a neighborhood of any of its points p this subgroup can be defined by a system of equations of the form

$$F_i(x_1, \dots, x_n) = 0, \quad i = 1, \dots, m, \tag{1}$$

where $x_1, \dots, x_n$ are local coordinates on the group G in the neighborhood of the point p. Here $F_1, \dots, F_m$ are analytical functions in the point p and

$$\operatorname{rank} \left(\frac{\partial F_i}{\partial x_j} \right)_{x=p} = m \tag{2}$$

Then there is a structure of $(n - m)$–dimensional analytical manifold induced on H, and H becomes a Lie group. If $z_1 = x_{i_1}, \dots, z_m = x_{i_m}$ are the coordinate functions of the columns constituting a non–zero minor of order m in (2) above, $z_{m+1}, \dots, z_n$ are the remaining coordinate functions, then the correspondence which assigns the set $(z_{m+1}, \dots, z_n)$ to the point q from the neighborhood of the point p in which the minor mentioned is not equal to zero, is a homeomorphism defining an $(n - m)$–dimensional chart for H around p. To prove this fact the implicit fuction theorem is applied.

Remark. *Let a subgroup H of a Lie group G be defined in some neighborhood of the identity element by a system of equations* (1) *satisfying condition* (2). *Then H is a Lie subgroup* .

Indeed, the subset $H \subset G$ is invariant with respect to the translations $l(h) : x \mapsto hx (h \in H)$ which are diffeomorphisms of the manifold G. With the help of $l(h)$ we transfer the coordinate system, which is defined in a neighborhood of the point $e \in G$, to a neighborhood of the point h. Then the subgroup H in the neighborhood of the point h will be defined in that coordinate system by the same equations which were used in the neighborhood of the identity element. □

Using what was said above some additional examples of the Lie group can be given.

1) The group $SL_n(\mathbf{R}) \subset GL_n(\mathbf{R})$ of matrices with the determinant 1. This group is defined by one equation $\det X = 1$ ($X = (x_{ij}) \in GL_n(\mathbf{R})$), moreover,

$$\left.\frac{\partial \det X}{\partial x_{11}}\right|_{X=E} = 1 \neq 0.$$

Thus, $SL_n(\mathbf{R})$ is a (n^2-1)–dimensional Lie group.

2) The orthogonal group O_n is definable in $GL_n(\mathbf{R})$ by the equations $F_{ij}(X) = \sum_{k=1}^{n} x_{jk}x_{ik} = \delta_{ij}$, where $i \leq j$ is assumed, so the number of equations is $\frac{n(n+1)}{2}$. We then have

$$\left.\frac{\partial F_{ij}(X)}{\partial x_{st}}\right|_{X=E} = \begin{cases} 2 \text{ or } 1 & \text{if } \{i,j\} \neq \{s,t\}, \\ 0 & \text{if } \{i,j\} \neq \{s,t\}. \end{cases}$$

This extracts a non–zero minor of order $\frac{n(n+1)}{2}$. Thus, O_n is a Lie group of dimension $\frac{n(n-1)}{2}$.

3) The group SO_n coincides with O_n close to E. Consequently, according to the remark above, SO_n is a Lie subgroup in $GL_n(\mathbf{R})$ of the same dimensional as O_n. One can show that SO_n is the connected component of the identity of the group O_n.

A linear representation of the Lie group G is a homomorphism of Lie groups $\rho : G \to GL(V)$. Any representation of a discrete group is a representation of it as a zero–dimensional Lie group. The mapping $t \mapsto e^{it}$ is a one–dimensional representation of the Lie group $\mathbf{R}$. The group $GL_n(\mathbf{R})$ is represented by the linear operators of the space $M_n(\mathbf{R})$, if one associates to the matrix A the mapping $X \mapsto AXA^{-1}$, $X \in M_n(\mathbf{R})$ (the adjoint representation) or the mapping $X \mapsto AX$ (regular representation). A rather important property of a linear representation of a Lie group is its definability by its tangent mapping.

We recall (see, Set–theoretic supplement, Section 15), that each morphism $\phi : M \to N$ of manifolds give rise to a family of linear mappings $T_p(\phi) : T_p(M) \to T_{\phi(p)}(N)$. In particular, if $\phi : G \to H$ is a homomorphism of Lie groups then the mapping determined by it $T_e(\phi) : T_e(G) \to T_e(H)$ is called the *tangent mapping*: it is denoted by $d\phi$.

Theorem 1. *A homomorphism of a connected Lie group G into an arbitrary Lie group H is completely determined by its tangent mapping.*

For simplicity we shall use the symbol $g\xi$ to denote the vector $T_e(l(g))\xi$ in $T_g(G)$ for $\xi \in T_e(G)$. In particular, if $G = GL(V)$, then $g\xi$ is the usual product of matrices (linear operators).

Lemma. *Let $\phi : G \to H$ be a homomorphism of the Lie groups. Then for any $g \in G$, $\xi \in T_e(G)$ we have*

$$T_g(\phi)(g\xi) = \phi(g)(d\phi)\xi.$$

The validity of the statement follows from the fact that the commutativity of the diagram of mappings

$$\begin{array}{ccc} G & \xrightarrow{l(g)} & G \\ {\scriptstyle\phi}\downarrow & & \downarrow{\scriptstyle\phi} \\ H & \xrightarrow{l(\phi(g))} & H \end{array}$$

implies the commutativity of the diagram of tangent mappings

$$\begin{array}{ccc} T_e(G) & \xrightarrow{dl(g)} & T_g(G) \\ {\scriptstyle d(\phi)}\downarrow & & \downarrow{\scriptstyle T_g(\phi)} \\ T_e((H) & \xrightarrow{dl(\phi(g))} & T_{\phi(g)}(H) \end{array}$$

Writing out the commutativity property for $T_e(G)$ we obtain the required relation. □

Proof of the theorem. Let $d\phi : T_e(G) \to T_e(H)$ be the tangent mapping of the homomorphism $\phi : G \to H$. We shall show how to recover ϕ from

$d\phi$. Since a connected topological group is generated by any of its identity neighborhoods, it is sufficient to find $\phi(g)$ for g from an arbitrarily small neighborhood of the identity of the group G. In order to find $\phi(g)$ we connect the identity with the point g by the differentiable curve $g(t)$, $0 \leq t \leq 1$, $g(0) = e$, $g(1) = g$. For any t, $0 \leq t \leq 1$, the tangent vector $g'(t) \in T_{g(t)}(G)$ can be represented in the form $g'(t) = g(t)\xi(t)$, where $\xi(t) \in T_e(G)$ (the isomorphism $dl(g) : T_e(G) \to T_g(G)$ is to be used). By the lemma $T_{g(t)}(\phi)g'(t) = \phi(g(t))(d\phi)\xi(t)$. Put $\phi(g(t)) = h(t)$, $(d\phi)\xi(t) = \eta(t)$. Then, by the definition of the tangent mapping, $T_{g(t)}(\phi)g'(t) = h'(t)$, and we can write $h'(t) = h(t)\eta(t)$. If the tangent mapping $d\phi$ is known, then $\eta(t)$ is also known, and the latter equality can be considered as a system of differential equations for the coordinates of the point $h(t)$. The solution of the system mentioned is uniquely defined by the initial condition $h(0) = e$. In particular, $h(1) = \phi(g)$ is defined uniquely, which is what was required. □

In particular, if $\rho : G \to GL(V)$ is a linear representation of a Lie group G, then the linear mapping $\phi : T_e(G) \to T_E(GL(V)) = \text{End}\,(V)$ is called its *tangent representation* .

Corollary. *A linear representation of a coherent Lie group is determined by its tangent representation.* □

For instance, if $\text{Ad} : GL(V) \to GL(\text{End}\,(V))$ is the adjoint representation for $GL(V)$ (see, Subsection 4.5.4), then the tangent representation is given by $\text{ad} : \text{End}\,(V) \to \text{End}\,(\text{End}\,(V))$. We have $(\text{Ad}\,\alpha)(\beta) = \alpha\beta\alpha^{-1}$. Let $\alpha(t)$, $0 \leq t \leq 1$, be a curve in the group $GL(V)$ such that $\alpha(0) = E$, $\alpha'(0) = \xi \in \text{End}\,(V)$. For $\eta \in \text{End}\,(V)$ we have

$$[(d\text{Ad}\,)\xi]\eta = \frac{d}{dt}[\text{Ad}\,\alpha(t)]\eta\bigg|_{t=0} = \frac{d}{dt}\alpha(t)\eta\alpha(t)^{-1}\bigg|_{t=0} = \xi\eta - \eta\xi.$$

(Here we took advantage of the fact that $\frac{d}{dt}\alpha(t)^{-1}|_{t=0} = -\xi$; this is obtainable by differentiation at $t = 0$ of the identity $\alpha(t)\alpha(t)^{-1} = E$.) Thus, it has been shown that $(\text{ad}\,\xi)\eta = \xi\eta - \eta\xi$.

If $\rho : G \to GL(V)$ is an irreducible representation of the Lie group G, and $\rho : T_e(G) \to \text{End}\,(V)$ is the corresponding tangent representation, then it is natural to expect that $d\rho$ is also irreducible. In fact, this is the case, however a stronger result is also correct.

Theorem 2. *If a representation ρ of the Lie group G is irreducible, then its tangent l representation $d\rho$ of the tangent space $T_e(G)$ is irreducible. Inversely, if G is connected and $d\rho$ is irreducible, then ρ is also irreducible.*

Proof. We shall prove that a subspace U in V, invariant with respect to ρ, is also invariant with respect to $d\rho$, and if G is connected, then the opposite is also valid.

Thus, let $U \subset V$ be invariant with respect to $\rho : G \to GL(V)$. For $\xi \in T_e(G)$ take a curve $g(t)$, $0 \leq t \leq 1$, in G and such that $g(0) = e$, $g'(0) = \xi$. Since for any $u \in U$, $\rho(g(t))(u) \in U$ holds for all t, it follows that

$$[(d\rho)\xi](u) = (\frac{d}{dt}\rho(g(t))|_{t=0})(u) = \left.\frac{d(\rho(g(t))(u))}{dt}\right|_{t=0} \in U,$$

as the passage to the limit in t does not take us out of the subspace.

Inversely, assume that U is invariant with respect to $d\rho$ and that G is connected. As before, we connect the identity e of the group G with the point $g \in G$ by the differentiable curve $g(t)$, $0 \leq t \leq 1$, $g(0) = e$, $g(1) = g$. We have $g'(t) = g(t)\xi(t)$, where $\xi(t) \in T_e(G)$. Using the lemma, we can write

$$\frac{d}{dt}\rho(g(t)) = \rho(g(t))(d\rho)\xi(t). \tag{3}$$

Choose a basis $\{e_1, \ldots, e_n\}$ in V in such a way that $U = \langle e_1, \ldots, e_m \rangle$. We denote the matrix of $\rho(g(t))$ by $A(t) = (a_{ij}(t))$, and that of $(d\rho)\xi(t)$ by $C(t) = (c_{ij}(t))$. According to the irreducibility condition, $c_{ij}(t) = 0$, when $i > m$, $j \leq m$. We shall prove that $a_{ij}(t) = 0$ for same i, j. We rewrite (3) in the matrix form $A'(t) = A(t)C(t)$. For $i > m$, $j \leq m$, we obtain

$$a'_{ij}(t) = \sum_{s=1}^{m} a_{is}(t)c_{sj}(t)$$

These equalities can be considered as a system of differential equations for the $a_{ij}(t)$ ($i > m$, $j \leq m$). Because $A(0) = E$ we have $a_{ij}(0) = 0$, for $i > m$, $j \leq m$. By the theorem on the uniqueness of solutions of the system of differential equations with given initial condition, we get $a_{ij}(t) = 0$, for all t. For $t = 1$ we obtain $\rho(g)(U) \subset U$. □

Thus, the study of irreducible representations of a connected Lie group G is reduced in a significant way to the study of the tangent representation of the space $T_e(G)$. In Ch. 4, § 5, we shall introduce an operation on $T_e(G)$, transforming this space into a Lie algebra, and we shall demonstrate that the tangent representation is a representation of that algebra.

Literature.

Automorphisms of classical groups, Adian [1], Artin [1], Vinberg [1], Gorenstein [1], Grigorchuk [1], Kostrikin [2], Curtis–Reiner [1], Lyndon–Schup [1], Hall [1], Gupta–Sidki [1], Sagle–Walde [1].

CHAPTER III

ASSOCIATIVE RINGS

§ 1. Radical

1.1. Definition of the Jacobson radical. One of the main tools in studying rings is the investigation of their modules. The most important are the simple irreducible modules. The fact is that according to the density theorem (see, Subsection 2.1.1), a ring which posseses a faithfull irreducible module, a so–called *primitive* ring, has a sufficiently determined structure which is close to the structure of an algebra of all linear operators on a linear space. It is clear that this statement acquires a precise meaning only by imposing certain definite restrictions. As an example of what is meant we offer the algebra $A(X)$ of all associative non–commuting polynomials in X with coefficients from some field F of zero characteristic. Indeed, such an algebra is primitive.

If the ring A is not primitive then it can happen that for any $a \in A$, $a \neq 0$, there is an irreducible A–module M_a, on which a acts not as zero. Such ring is called *semiprimitive* (*semisimple*). Such rings also can be completely described.

The general case is the case of the ring A, which contains elements that act as zero in any irreducible module. Then the set $J(A)$ of such elements forms a two–sided ideal which is called the *Jacobson radical* of the ring A.

Theorem. *The Jacobson radical $J(A)$ of the ring A with an identity can be characterized in the following way:*

(1) *$J(A)$ is equal to the intersection of all maximal left ideals;*

(2) *$J(A)$ is the largest two–sided ideal, such that for any $a \in J(A)$ the element $1 - a$ has a two–sided inverse;*

(3) *$J(A)$ is the maximal left ideal which is a group with respect to the operation $a \circ b = a + b - ab$.*

Proof. (1) Assume that $J(A) \not\subset \mathcal{I}$, where $\mathcal{I}$ is some maximal left ideal of the ring A. Then we consider the irreducible A–module $A/\mathcal{I}$. If $x \in J(A) \setminus \mathcal{I}$, then $x(1+\mathcal{I}) = x+\mathcal{I} \neq 0$ in $A/\mathcal{I}$. This contradiction proves that $J(A)$ lies in the intersection of all maximal left submodules of A. Inversily, assume that $x \in A$ lies in the intersection of all maximal left ideals of the ring A. If M is an arbitrary irreducible A–module then for any $m \neq 0$ from M the subgroup $Am \neq 0$ coincides with M. Therefore the mapping $\phi : a \mapsto am$ from A to M is an epimorphism of A–modules. Its kernel $\mathcal{I}$ is a maximal left ideal in virtue of the irreducibility of the module M, and $M \cong A/\mathcal{I}$. We have: $xM = x\phi(A) = \phi(xA) \subset \phi(\mathcal{I}) = \{0\}$. Hence $x \in J(A)$, and our statement is proved.

In order to prove the second statement we need a supplementary fact.

Lemma. *Let a, b be the elements from the ring A with identity* 1. *If* $1 - ab$ *is invertible from the left (from the right) in A, then* $1 - ba$ *is also invertible from the left (from the right) to A.*

Proof. Assume that the element $u \in A$ is such that $u(1 - ab) = 1$. Let $v = 1 + bua$. Then

$$v(1 - ba) = (1 + bua)(1 - ba) = 1 + bua - b(1 + uab)a = 1 + bua - bua = 1. \qquad \square$$

(2) Using the characterization (1), we note that $J(A)$ is the set of elements $x \in A$ such that for any $y \in A$ the element $1 - yx$ has a left inverse. Indeed, if $x \in J(A)$, $y \in A$, then $yx \in J(A)$. If $\mathcal{I}$ is a maximal left ideal in A then $1 - yx \notin \mathcal{I}$, hence $A(1 - yx) = A$. In particular, we can find an $u \in A$ such that $u(1 - yx) = 1$, i.e. $1 - yx$ is invertible. Inversily, let x be such that $1 - yx$ is invertible from the left for any $y \in A$. If x does not lie in some maximal left ideal $\mathcal{I}$, then $\mathcal{I} + Ax = A$, i.e. there exists $u \in \mathcal{I}$, $y \in A$, such that $u + yx = 1$. In this case the element $u \in \mathcal{I}$ has the form $1 - yx$, i.e. is invertible from the left. But then $1 \in Au \subset \mathcal{I}$, giving $\mathcal{I} = A$, a contradiction.

Thus, if $a \in J(A)$ then $1 - a$ is invertible from the left (and according to the lemma, also from the right). Let K be a two–sided ideal such that for any x the element $1 - x$ is invertible. Then $yx \in K$, i.e. $1 - yx$ is invertible, and hence $K \subset J(A)$.

The statement proved demonstrates that the Jacobson radical $J(A)$ as the intersection of the annihilators of irreducible left A–modules is equal to the intersection of the annihilators of irreducible right A–modules.

(3) The neutral element with respect to the operation $\circ$ is zero. We call an element a quasi–regular if it has a quasi–inverse b, i.e. $a \circ b = 0$. Associating to the element a the element $1-a$ takes $a \circ b$ into $(1-a)(1-b)$, and, consequently, is an isomorphism. If $\mathcal{I}$ is an ideal which is a subgroup with respect to $\circ$, then the elements of the form $1 - x$, $x \in \mathcal{I}$, form a subgroup ; hence $\mathcal{I} \subset J(R)$. The opposite is evident. □

$*$ We should note that some important left properties of a ring do not coincide with their right analogies. One such property is the property of being noetherian. In the non–commutative case there are both left– and right–noetherian rings, i.e. satisfying the maximality condition for left and right ideals, respectively. An example of the right–noetherian ring which is not left–noetherian is the subring of the ring of matrices of order two over the field $\mathbf{Q}$ of rational numbers of the following form

$$A\left\{ \begin{pmatrix} a & b \\ 0 & c \end{pmatrix} \middle| a \in \mathbf{Z}, \quad b, c \in \mathbf{Q} \right\}.$$

We should note that the theorem and its proof are presented for the case of the ring A with identity. It is also valid in a general case. The statement (3) still gives the definition of the radical without any changes in that case. $*$

Corollary. (1) *If $\mathcal{I}$ is a left (right) nil–ideal of the ring A (i.e. consisting of nilpotent elements), then $\mathcal{I} \subset J(A)$;*

(2) $J(A/J(A) = \{0\}$.

Proof. The first statement can be easily derived from the characterization of the radical, which is obtained in the proof of the statement (2) of the theorem, and by using the fact that from $x^n = 0$ follows

$$(1 - x)(1 + x + \ldots + x^{n-1}) = 1.$$

The second statement follows immediately from the statement (1) of the previous theorem. □

1.2. The Jacobson radical of an Artinian ring. It is not necessarily the case that the Jacobson radical is nilpotent. For instance, in the ring $F[[t]]$ of power series for one variable the Jacobson radical is equal to the set of series with a constant term equal to zero. It is clear that all these series are not nilpotent and even are not zero divisors.

An important case when Jacobson radical is nilpotent, is the case of *artinian* rings, i.e. of rings with the minimality condition for the left (or right) ideals. This condition is equivalent to the fact that any descending chain of left (respectively, right) ideals stabilizes, i.e. if

$$\mathcal{I}_1 \supset \mathcal{I}_2 \supset \ldots \supset \mathcal{I}_n \supset \mathcal{I}_{n+1} \supset \ldots,$$

then starting with some n we have $\mathcal{I}_n = \mathcal{I}_{n+1} = \ldots$. Examples for such rings are finite rings or finite–dimensional algebras over a field.

A subring S of a ring R is called *nilpotent* if there exists an m such that $x_1 x_2 \ldots x_m = 0$ for any $x_1, x_2, \ldots, x_m \in S$. This can be written $S^m = \{0\}$.

Theorem. *The Jacobson radical of an Artinian ring is nilpotent.*

Proof. Consider the descending chain of ideals

$$J = J^1 \supset J^2 \supset \ldots \supset J^n \supset \ldots$$

A number N can be found such that $J^N = J^{N+1} = \ldots$. assume that $J^N \neq 0$. We investigate the two–sided ideal $W = \{x \in A \mid J^N x = 0\}$. If $W \supset J^N$, then $\{0\} = J^N J^N = J^{2N} = J^N$, a contradiction. Thus, $W \not\supset J^N$. Consider now $\bar{A} = A/W$. Then $J^N + W/W$ is a non–zero ideal in $\bar{A}$. Let $\bar{S} = S/W$ be a minimal left ideal. Then $\bar{S}$ is an irreducible A–module, hence J acts as a zero on $\bar{S}$, in other words, $JS \subset W$. In particular, $J^N S \subset W$. According to the definition of the ideal W we have $J^N S = J^N(J^N S) \subset J^N W = \{0\}$. Hence $S \subset W$, a contradiction. Thus, the assumption that $J^N \neq \{0\}$ is not correct. □

Corollary. *If A is an Artinian ring, and $\mathcal{I}$ is a nil–ideal, then $\mathcal{I}$ is nilpotent.* □

1.3. Radical of a ring of matrices and of a ring of polynomials. If S is a set, then the symbol $M_n(S)$ denotes the set of quadratic matrices of order n with the elements from the set S. If A is an algebra, then $M_n(A)$ is an algebra. If B is an ideal in A, then $M_n(B)$ is an ideal in $M_n(A)$.

Theorem 1. *The radical $J(M_n(A))$ is equal to $M_n(J(A))$.*

Proof. Let $\mathcal{I}_j$ be the set of $n \times n$–matrices from $M_n(J(A))$ with zeros in all columns with exception of the j–th one. Then $\mathcal{I}_j$ is a left ideal in $M_n(A)$. We shall prove that $\mathcal{I}_j \subset J(M_n(A))$. Let $Z_j \in \mathcal{I}_j$; moreover, let the (j, j) entry of the matrix Z_j be z_j. We denote by z_j' a quasi–inverse element to z_j, and by Z_j' a matrix in $\mathcal{I}_j$ with the (j, j) entry z_j'. Then for $Z_j \circ Z_j'$ and $Z_j' \circ Z_j$ the (j, j) entry is zero and this matrix lies in $\mathcal{I}_j$. Therefore $(Z_j \circ Z_j')^2 = (Z_j' \circ Z_j)^2 = 0$. In this case the elements $Z_j \circ Z_j'$ and $Z_j' \circ Z_j$ are quasi–regular. Then Z_j is also quasi–regular. Therefore $\mathcal{I}_j \subset J(M_n(A))$. Thus, $M_n(J(A)) \subset J(M_n(A))$. The radical $J(M_n(A))$ is an ideal in $M_n(A)$ and, so it has the form $M_n(B)$, where B is an ideal in A. Let $b \in B$, and consider the matrix $bE_n = \operatorname{diag}\{b, b, \ldots, b\}$. It is in $M_n(B) = J(M_n(A))$, i.e. it is quasi–regular. It is clear that b is then also quasi–regular in A. Thus, B is a quasi–regular ideal in A, i.e. $B \subset J(A)$, so $J(M_n(A)) = M_n(B) \subset M_n(J(A))$. Consequently, $J(M_n(A)) = M_n(J(A))$. □

Let now $A[t]$ be the algebra of polynomials in a variable t with coefficients from A. The second result, on rings of polynomials, is the following.

Theorem 2. *If A has no nil–ideals which differ from zero, then $A[t]$ is a semisimple ring in the sense of Jacobson, i.e. $J(A[t]) = 0$.*

Lemma 1. *If A is a commutative ring then the polynomial $a_0 + a_1t + \ldots + a_nt^n$ is invertible in $A[t]$ iff a_0 is invertible and the remaining coefficients a_i are nilpotent.*

Proof. The invertibility of the element a_0 is evident. In addition, if A has no zero divisors, then all a_i must be zero when $i > 0$. Consequently, if P is a prime ideal in A, and $\bar{a} = a + P \in A/P$, then $\bar{a}_0 + \bar{a}_1t + \ldots + \bar{a}_nt$ is an invertible element in $(A/P)[t]$, i.e. $\bar{a}_i = 0$ when $i > 0$. Thus, each element a_i lies in every prime ideal P of the commutative ring A, and, consequently, is nilpotent (see, Subsection 1.5.3). The proof of the opposite implication is evident. □

Lemma 2. *If A is an arbitrary ring and $f(t) \in A[t]$ is a polynomial with constant term equal to zero that is quasi–invertible in $A[t]$ then the coefficients of its quasi–inverse lie in the subalgebra which is generated by the coefficients of the polynomial $f(t)$.*

Proof. Let $g(t)$ be a quasi-inverse for $f(t)$, so that

$$(1 - f(t))(1 - g(t)) = 1 = (1 - g(t))(1 - f(t)).$$

For any $n = 1, 2, \ldots$ we have

$$1 - f(t)^{n+1} = (1 - f(t))(1 + f(t) + \ldots + f(t)^n).$$

Consequently,

$$1 - g(t) = (1 - g(t))f(t)^{n+1} + 1 + f(t) + \ldots + f(t)^n$$

and

$$g(t) = -f(t) - \ldots - f(t)^n + (g(t) - 1)f(t)^{n+1}.$$

If $\deg g(t) = n$, then from this relation we have that the coefficients of the polynomial $g(t)$ are obtainable from $-f(t) - \ldots - f(t)^n$ and, hence, lie in the subalgebra which is generated by the coefficients of the polynomial $f(t)$. The lemma is proved. □

Proof of the Theorem 2. Let us assume that $J(A[t]) \neq \{0\}$ and select a non-zero element $f(t) = a_0 + a_1 t + \ldots + a_m t^m \in J(A[t])$ with a minimal number of non-zero coefficients a_i. We can assume that $a_m \neq 0$. Then $[a_i, f(t)] = a_i f(t) - f(t) a_i \in J(A[t])$. If this polynomial is non-zero then it has less non-zero coefficients than $f(t)$. Thus, $[a_i, f(t)] = 0$, i.e. $[a_i, a_j] = a_i a_j - a_j a_i = 0$ for all i, j. Thus the subalgebra B generated by the coefficients a_i is commutative. Since $t f(t) \in J(A[t])$ and has a zero constant term, it follows from lemma 2 that the quasi-inverse $g(t)$ is a polynomial from $B[t]$. Then from lemma 1 follows that a_m is a nilpotent element. However for any $a \in A$ the polynomials $a f(t)$ and $f(t) a$ lie in $J(F[t])$. Hence it is clear that the complete ideal which is generated by the element a_m is nilpotent. This contradicts the assumptions of the theorem.

The contradiction obtained concludes the proof of the theorem. □

1.4. Generalization of the concept of a radical. A.G. Kurosh and other authors have generalized the concept of radical in the following way. A class $\mathcal{C}$ of rings containing all homomorphic images of its members is called *radical* if any ring A contains a largest ideal $R(A) \in \mathcal{C}$ such that the quotient ring $A/R(A)$ does not contain non-zero ideals which belongs to the class $\mathcal{C}$. The ideal $R(A)$ is called $\mathcal{C}$–radical of the ring A. If $R(A) = \{0\}$ then the ring is called the *$\mathcal{C}$–semisimple.*

The usual Jacobson radical emerges if in the role of $\mathcal{C}$ one takes the class $\mathcal{Q}$ of quasi–regular rings, i.e. the class of rings which are groups with respect to the multiplication $a \circ b = a + b - ab$.

Other radicals are the nil–radical corresponding the class $\mathcal{N}il$ of nil–rings, i.e. rings such that $a^n = 0$ for any a for a suitable n depending on a. The radicality of the class is evident, since if $A \supset B$, and A/B and B are the nil–rings, then A is also a nil–ring. Therefore $\mathcal{N}il \subset \mathcal{Q}$, the nil–radical of any ring lies in the Jacobson radical. The ideal $\mathcal{N}il(A)$ is called the *upper nil–radical* of the ring A.

A more restricted class of rings compared with $\mathcal{N}il$ is the class $\mathcal{N}$ of nilpotent rings. Any finite ring A or finite–dimensional algebra A has a largest nilpotent ideal which can be selected as the nilpotent radical $H(A)$ of that ring; moreover, the condition $N(A/N(A)) = \{0\}$ is fulfilled. However, if the additional conditions are omitted then the statement presented is not correct. In order to obtain a radical which is close to the nilpotent one several constructions have been investigated. One of the constructions uses the fact that the class $\mathcal{LN}$ of locally nilpotent rings is radical, so each ring has a *locally nilpotent radical.*

Another construction uses a transfinite induction.

Let A be a ring. We denote by $N_0(A)$ the zero ideal and by $N_{i+1}(A)$, where $N_i(A)$ has already been defined, the complete preimage in A of the sum of all nilpotent ideals of the ring $A/N_i(A)$.

Having constructed $N_i(A)$ for all $i = 1, 2, \ldots$, we obtain the ideal $N_\omega(A) = \bigcup_{i=1}^{\infty} N_i(A)$. It can happen that $A/N_\omega(A)$ also has nilpotent ideals different from zero. Then we define $N_{\omega+1}(A)$ in the same way as above. Generally speaking, for any non–limit α, i.e. of the form $\alpha = \beta + 1$, we define $N_\alpha(A)$ as the complete preimage of all nilpotent ideals in $A/N_\beta(A)$. If α is a limit ordinal then $N_\alpha(A) = \bigcup_{\beta<\alpha} N_\beta(A)$. By the principle of transfinite induction either there exists an ordinal α such that $N_\alpha(A) = A$, and then we call the ring A *radical,* or $N_\alpha(A) = N_{\alpha+1}(A)$ for some ordinal α, and in this case $N_\alpha(A) = LR(A)$ is called the *lower nil–radical* of the ring A.

Proposition. *The lower nil–radical of a ring is a locally nilpotent ideal. It is equal to the intersection of all prime ideals of that ring.*

Proof. The first statement easily follows from the fact that an extension by a locally nilpotent ideal of a locally nilpotent quotient algebra, is itself locally nilpotent. In fact, if B and A/B are locally nilpotent then any finite set $a_1, \ldots, a_s$ generates an algebra A_1 such that $A_1 + B/B$ is nilpotent by hypothesis. Then there is an m such that $A_1^m \subset B$. The subalgebra A_1^m is generated by the finite number of products of the elements $a_1, \ldots, a_s$ of degree between m and $2m-1$. By assumption, $(A_1^m)^k = \{0\}$, for a suitable k, in other words, $A_1^{mk} = \{0\}$. Thus, A is a locally nilpotent algebra. It is easy to see by using the isomorphism $(A+B)/B \cong A/A \cap B$, that a sum of locally nilpotent ideals (even of an infinite number) is locally nilpotent. Finally, the local nilpotency of the radical $LR(A)$ is obtained from the previous remarks by a transfinite induction the ordinal α (we recall that $LR(A) = N_\alpha(A)$, for a suitable α).

Define N' as the intersection of all prime ideals of the ring A. Then $N_1(A) \subset N'$. Indeed, if $\mathcal{I}$ is a nilpotent ideal of the ring A, and P is an arbitrary prime ideal in A, then $\mathcal{I} + P/P$ is a nilpotent ideal in the prime ring A/P. Thus $\mathcal{I} + P = P$, i.e. $\mathcal{I} \subset P$. Since P is arbitrary, we have $\mathcal{I} \subset N'$. As $N_1(A)$ is a sum of nilpotent ideals, we see that $N_1(A) \subset N'$. Passing to $A/N_1(A)$ we prove that $N_\beta(A/N_1(A)) \subset N'/N_1(A)$. Then also $N_{\beta+1}(A) \subset N'$, and so $LR(A) \subset N'$.

Lemma. *A ring A has no nilpotent ideals iff the intersection of its prime ideals is equal to zero.*

Indeed, if the intersection $\cap_\alpha P_\alpha$ of all prime ideals of the ring A is equal to zero, then A is imbedded in the Cartesian product $D = \prod_\alpha (A/P_\alpha)$. In order to get such an imbedding $\imath\colon A \to D$ one associates to an element a the function $f_a \in D$ such that $f_a(\alpha) = a + P_\alpha$. The kernel of the mapping just constructed is equal to $\cap_\alpha P_\alpha$, i.e. is zero. Now project a nilpotent ideal N of the ring A to A/P_α. Since A/P_α is prime, the projection is zero. Then also $N = \{0\}$. Inversily, let Q be the intersection of all prime ideals of the ring A. We shall show that if $Q \neq \{0\}$ then A has non–zero nilpotent ideals. Indeed, in Q we construct a sequence of non–zero elements $b_1, b_2, \ldots, b_n, \ldots$ such that $b_i = b_{i-1}a_{i-1}b_{i-1}$ for suitable $a_{i-1} \in A$. For b_1 we can use any non–zero element from Q. If b_{i-1} is constructed then the ideal $Ab_{i-1}A$ is non–zero. So $(Ab_{i-1}A)^2 \neq \{0\}$. In this case $b_{i-1}Ab_{i-1} \neq \{0\}$ and for some a_{i-1} the element $b_i = b_{i-1}a_{i-1}b_{i-1} \neq 0$. Select an ideal P that is maximal for the property that $P \cap \{b_i | i = 1, 2, \ldots\} = \emptyset$. We shall show that P is prime. Indeed, if C and D are ideals such that $CD \subset P$ and $C \not\subset P$, then $C + P$ strictly contains P, i.e. $C+P$ contains b_i for some i. Analogously, $D+P$ contains some b_j if $D \not\subset P$.

When $k > i, j$ we have $b_k \in (C+P)(D+P) \subset CD+P = P$, a contradiction. Hence $C \subset P$ and P is prime. However, the intersection of all prime ideals is equal to Q and contains all $b_1, b_2, \ldots$. We arrived at a contradiction. The lemma is proved. □

It is clear that the remaining part of the theorem follows from the lemma. □

Corollary. *Let A be a noetherian ring. Then the intersection of all prime ideals is a nilpotent ideal.*

Indeed, if a ring is (left)–noetherian then its lower nil–radical is nilpotent. Indeed, $N_1(A)$ is then the sum of a finite number of nilpotent ideals, i.e. is nilpotent, let us say, $x_1 \ldots x_n = 0$ for any $x_1, \ldots, x_n \in A$. The quotient algebra $N_2(A)/N_1(A)$ is nilpotent, i.e. for any $y_1 + N_1(A), \ldots, y_m + N_1(A)$ we have $(y_1 + N_1(A)) \ldots (y_m + N_1(A)) = 0$. Consequently, $z_1 \ldots z_{mn} = 0$ for any $z_1, \ldots, z_{mn} \in N_2(A)$. Hence $N_2(A) = N_1(A)$. It is now clear that $LR(A)$ is a nilpotent ideal. □

§ 2. Classical semisimple rings

2.1. Structure of a simple ring. A ring A is called a *left artinian* if any sequence of its left ideals satisfies the minimality condition. According to the Set–theoretic supplement, § 6, this condition means that for any non–empty subset of left ideals of the ring there is a minimal ideal or that any descending chain of left ideals

$$\mathcal{I}_1 \supset \mathcal{I}_2 \supset \ldots \supset \mathcal{I}_n \supset \ldots,$$

of the ring A stabilizes, i.e. starting at some n we have $\mathcal{I}_n = \mathcal{I}_{n+1} = \ldots$.

A ring A is called *primitive* (from the left) if it possesses a faithfull irreducible (left) module.

Proposition. *A primitive artinian ring is isomorphic to a ring of matrices over a skew field.*

Proof. Let $D = \mathrm{End}_A(M)$, where M is a faithfull irreducible module of the primitive ring A under consideration. According to the Schur lemma D is a (skew) field. We investigate the algebra $E = \mathrm{End}_D(M)$ of linear operators of the right vector spaces M over the field D. We shall prove that M is finite–dimensional over D. Let, on the contrary, $v_1, v_2, \dots, v_n, \dots$ be an infinite system of linearly independent elements from M over D. Denote by $\mathcal{I}_n$ the set of elements from A that annihilate the elements $v_1, \dots, v_n$. Then $\mathcal{I}_n$ is a left ideal in A and

$$\mathcal{I}_1 \supset \mathcal{I}_2 \supset \dots \supset \mathcal{I}_n \supset \mathcal{I}_{n+1} \supset \dots \tag{1}$$

Note that for any n, we have $\mathcal{I}_n \neq \mathcal{I}_{n+1}$. Indeed, the mapping f such that $f(v_1) = \dots = f(v_n) = 0$, $f(v_{n+1}) = v_{n+1}$, extends to an element from E, i.e. to a linear operator over the field D. According to the density theorem from Subsection 2.1.1 there exists an $a \in A$ such that $av_1 = \dots av_n = 0$, $av_{n+1} = v_{n+1}$. Hence $a \in \mathcal{I}_n \setminus \mathcal{I}_{n+1}$. The existence of the chain (1) contradicts the artinian property of the ring under consideration. Thus, $\dim_D(M) = m < \infty$. Applying the density theorem once again we obtain $A = E = M_m(D)$. □

Theorem. *A simple artinian ring A is isomorphic to a ring of matrices over field.*

Proof. Consider a minimal non–zero left ideal M of the ring A. Then M is an irreducible left A–module. Its annihilator I is a two–sided ideal in A. If $\mathcal{I} = A$ then $M^2 \subset AM = IM = \{0\}$, i.e. M is a non–zero nilpotent ideal of the ring A. Such an ideal would lie in $J(A)$ which is the Jacobson radical of the ring A. A simple ring cannot be nilpotent. Thus, $J(A) = \{0\}$ and $M = \{0\}$. There remains the case when $\mathcal{I} = \{0\}$. Then M is a faithfull irreducible A–module and A is a primitive Artinian ring. It remains to apply the proposition. □

It is clear that the previous results are valid for finite rings and also for finite–dimensional algebras over field. In fact, in the latter case an ideal of the algebra is a subspace over the base field.

2.2. Structure of artinian semisimple rings. The fundamental result on semisimple artinian rings is the Wedderburn–Artin theorem.

Theorem. *A semisimple artinian ring is isomorphic to the direct sum of a finite number of matrix rings over skew fields.*

Proof. Let A be a semisimple artinian ring. The intersection of the annihilators of irreducible left A–modules is therefore equal to zero. Note that hence the intersection of some finite number of such annihilators is zero. Indeed, assumption of the opposite allows to construct the strictly descending chain of ideals

$$R_1 \supset R_1 \cap R_2 \supset R_1 \cap R_2 \cap R_3 \cap \ldots \tag{2}$$

Indeed, consider $R_1 \cap R_2 \cap \ldots \cap R_k$. According to our assumption this ideal is non–zero. If for any annihilator R_{k+1} we have $(R_1 \cap \ldots \cap R_k) \cap R_{k+1} = R_1 \cap \ldots \cap R_k$, then $R_1 \cap \ldots \cap R_k$ lies in the Jacobson radical $J(A)$, which is impossible. Thus, (2) is a strictly descending chain of ideals in A, giving a contradiction. So A contains primitive ideals $R_1, R_2, \ldots, R_n$ such that $R_1 \cap R_2 \cap \ldots \cap R_n = \{0\}$.We can assume that the number n with this property is minimal. In that case, for any $i = 1, \ldots n$, the ideal

$$S_i = \bigcap_{\substack{j=1 \\ j \neq i}}^{n} R_j$$

is not zero and $R_i \cap S_i = \{0\}$.

The ring A/R_i, $i = 1, \ldots, n$, is primitive and, consequently, according to theorem 2.1, isomorphic to a ring $M_n(D)$, i.e. simple. Thus $A = S_i + R_i$, or, to be more precise, $A = S_i \oplus R_i$. Let $U_k = R_1 \cap \ldots \cap R_k$, $k = 1, 2, \ldots, n$. We shall prove by induction on k, that $A = S_1 \oplus \ldots \oplus S_k \oplus U_k$. The case $k = 1$ has just been done. Suppose that the statement is correct when $k = m < n$. We then consider U_m. According to the isomorphism theorem for rings we have

$$U_m/U_m \cap R_{m+j} \cong U_m + R_{m+j}/R_{m+j} = R/R_{m+j}.$$

The latter equality is valid, since $U_m \supset S_{m+j}$, when $j \geq 1$. Thus the ring $U_m/U_m \cap R_{m+j}$ is simple, i.e. $U_m \cap R_{m+j}$ is a maximal two–sided ideal in U_m. Moreover,

$$\bigcap_{j=1}^{n-m} U_m \cap R_{m+j} = \{0\}, \tag{3}$$

and for any proper subset of the ideals that are involved in (3) the intersection is not zero. As before, we obtain

$$U_m = U_{m+1} \oplus \Big(\bigcap_{j=2}^{n-m} U_m \cap R_{m+j} \Big) = U_{m+1} \oplus S_{m+1}.$$

Hence $A = S_1 \oplus \ldots \oplus S_m \oplus U_{m+1}$ as desired. Taking $k = n$ we get $A = S_1 \oplus \ldots \oplus S_n$. From the equality $A = S_i \oplus R_i$, $i = 1, \ldots, n$, we see that $S_i = A_i/R_i$ is a primitive artinian ring. According to theorem 2.1 there is a field D_i such that $S_i \cong M_{m_i}(D_i)$, for suitable m_i. □

We should also note that the opposite is correct: the direct sum of the finite number of matrix rings over fields is artinian. This follows from the artinian property of the ring $M_n(D)$, where D is a field, and from the fact that a direct sum of artinian rings is artinian. Indeed, let A, B be artinian rings. Let $T_1 \supset T_2 \supset \ldots$ be a descending chain of left ideals in $A \oplus B$. Then

$$T_1 \cap A \supset T_2 \cap A \supset \ldots \quad \text{and} \quad T_1 + A/A \supset T_2 + A/A \supset \ldots$$

are descending chains of ideals in A and $A + B/A \cong B$. According to the assumption there exists a natural n (the same for both chains) such that

$$T_n \cap A = T_{n+1} \cap A = \ldots \quad \text{and} \quad T_n + A/A = T_{n+1} + A/A = \ldots$$

Then also $T_n = T_{n+1} = \ldots$, i.e. $A \oplus B$ is artinian. The artinian property of the ring $M_m(D)$ follows from the fact that the left ideals of this ring are subspaces of the left vector space $M_m(D)$ over the field D which has dimension m^2. (For this, see Introduction 3.2.)

2.3. Structure of non–semisimple artinian rings. Let us state without proof the Wedderburn–Mal'tsev theorem (cf Ch. 5, § 3). In order to formulate it we define the concept of a perfect field. A field is *perfect* if it is of characteristic zero or if it is of characteristic p and $F^p = F$.

Theorem. *Let A be a finite dimensional algebra over a perfect field F. Then $A = S \oplus J(A)$ where S is a (semisimple) subalgebra and $J(A)$ is the Jacobson radical. If $A = S' \oplus J(A)$ where S' is another subalgebra, then there is an $x \in J(A)$ such that $S' = (1 + x)^{-1} S(1 + x)$.* □

§ 3. Structure of noetherian rings

3.1. Non–commutative localization. We recall that a ring is called left noetherian if it satisfies the stabilization condition for the increasing chains of left ideals. Example of noetherian rings are finite–dimensional algebras over fields. These rings are also artinian in addition and the theorem from the previous section gives satisfactory information about the structure of these rings.

When investigating noetherian rings which are not artinian techniques related to the concept of a non–commutative rings of fractions (see, § 1.3) apppear to be highly efficient. We remind the reader that any commutative ring R without zero divisors, has a field of fractions K, i.e. a field containing R as a subring, moreover, any of its elements $x \in K$ is representable in the form $x = rs^{-1}$, where $r, s \in R$, $s \neq 0$. This result can not be extended to the case of non–commutative rings. Indeed, in 1937 A. I. Mal'tsev constructed an example of the ring A without zero divisors that is not imbeddable as a subring in any skew field.

Definition. Let A be a ring , S a multiplicatively closed subset consisting of regular elements, i.e. of non–zero elements which are not zero divisors. A ring D is called a *right ring of fractions* for A with respect to the set S, if D contains A as a subring, any element from S has an inverse in D, and any element from D is representable in the form as^{-1}, $a \in A$, $s \in S$. If S is the set of all non zero divisors then we obtain a *classical* ring of fractions.

The set S is also called a set of *denominators*.

Theorem. *Assume that S is a multiplicatively closed subset of regular elements of the ring A; suppose, moreover, that for any $s \in S$ and $a \in A$ there exist $t \in S$ and $b \in A$ such that $at = sb$. Then there exists a ring of fractions for A with set of denominators S.*

Proof. Consider the Cartesian product $A \times S$. We call the pair (a, s) equivalent to the pair (b, t), $(a, s) \sim (b, t)$, if there exist $c, d \in A$ such that $ac = bd$ and $sc = td \in S$. (Heuristically: then $d = t^{-1}sc$ and $ac = bd = bt^{-1}sc$, therefore $as^{-1} = bt^{-1}$.) We should note that if the latter condition is satisfied for some pair c, d then for any pair c', d' with the condition $sc' = td' \in S$ we have $ac' = bd'$. Indeed, let $e \in A$, $e' \in S$ are such that $(sc)e = (sc')e'$. Then $(td)e = (td')e'$. Since s, c are regular elements, we have $ce = c'e'$, $de = d'e'$.

Therefore $ac'e' = ace = bde = bd'e'$ and $ac' = bd'$, as is required. The remark proved just allows to deduce transitivity of the relation $\sim$ (reflectivity and symmetry are evident). Thus, let $(a, s) \sim (b, t)$, $(b, t) \sim (c, u)$. Then there exist $d, e, p, q \in A$ such that $sd = te$, $tp = uq$, $ad = be$, $bp = cq$. We select f, h in such way that $tef = tpq$. Then $ef = pq$ and

$$sdf = tef = tpq = uqh.$$

Using what has just been proved we obtain

$$adf = bef = bpq = cqh,$$

i.e. $(a, s) \sim (c, u)$. We denote by A_S the set of classes of equivalent pairs. We shall denote the class of the pair (a, s) by a/s.

Now we define operations on the set A_S. If a/s, $b/t \in A_S$, then there exist $c, d \in A$ such that $sc = td = e \in S$. We put

$$a/s + b/t = (ac + bd)/e.$$

In addition, there exist $c \in A$, $u \in S$ for which $bu = sc$. We put

$$(a/s)(b/t) = (ac)/(tu).$$

It is possible to write this down in a different way:

$$(a/s)(b/t) = (ac/sc)(bu/tu).$$

The definitions given need to be checked for consistency. We shall do this, for instance, for the operation of multiplication. Thus, let $(a, s) \sim (a', s')$, $(b, t) \sim (b', t')$. For some u, u', v, v' we have $(au, su) = (a'u', s'u')$, $bv, tv) = (b'v', t'v')$, where su, $tv \in S$.

It is sufficient to prove that $(a/s)/(b/t) \sim (au/su)(bv/tv)$. Let p, q be such that $sup = bvq$. Then $(au/su)(bv/tv) = aup/tvq$. So we need to prove the equivalence of the pairs (ac, tu), (aup, tvq) (c, u, as above, are such that $sc = bu$). We choose $x, y \in A$ so that $tux = tvqy$; then $ux = vqy$, hence, $bux = bvqy$. Hence $scx = supy$, then $cx = upy$, and so $acx = aupy$. This proves the desired equivalence.

The mapping $a \mapsto a/1$ is injective and compatible with the operations. Therefore, the elements of the form $a/1$ form a subring in A_S which is isomorphic to A. Further, the element $1/s$ is inverse to $s/1 \in S$. We shall write the elements from $A/1$ without their denominator. Then $1/s = s^{-1}$. Since $a/s = (a/1)(1/s)$, we see that $a/s = as^{-1}$.

We do not verify that A_S is really an associative ring. The reader can either check this immediately or refer to § 5 of this chapter where the verification is automatic with a help of the construction of complete ring of fractions.

A multiplicative subset S for which the conditions of the theorem are fulfilled, is called a *left Ore set*. If $S = A \setminus \{0\}$ is a left Ore set then we say, that the ring A satisfies the left Ore condition. If the Ore condition is fulfilled from both sides then we say, that the ring A satisfies the Ore condition.

3.2. Structure of noetherian semi–prime rings. We call a ring A *prime* if from the condition $RS = \{0\}$, where R and S are two–sided ideals in A, it follows that either R or S is the zero ideal. If the ring A has no zero divisors then, in fact, it is prime. The opposite is not correct: for instance, the ring of matrices of order $n \geq 2$ is simple, thus prime, even though it has zero divisors. Further, a prime ring may not be simple: for instance, the ring of integers $\mathbf{Z}$ is prime, but not simple. We draw reader's attention to one detail: the ring of fractions for $\mathbf{Z}$, i.e. $\mathbf{Q}$, is a simple ring. In this example the ring $\mathbf{Z}$ is also noetherian. It turns out that the example presented is very characterstic and any noetherian prime ring possesses a classical ring of quotients which is simple artinian.

In addition, we shall prove a stronger result concerning so called semi–prime rings. Let a ring A be *semi–prime* if it has no non–zero nilpotent ideals. This terminology completely agrees with the one introduced in § 2: in a semi–prime ring the intersection S of all prime ideals, i.e. such that the quotient–rings with respect to them are prime, is equal to zero. We should note that in any ring A any nilpotent ideal N is contained in every prime ideal P. Really, in the opposite case $N+P/P$ is a non–zero nilpotent ideal in A/P. If $(N+P/P)^{t+1} = \{0\}$ and $(N+P/P)^t \neq \{0\}$, then $(N+P/P)^t(N+P/P) = \{0\}$, and the factors are not zero. Thus S contains any nilpotent ideal.

We should note that a noetherian semi–prime ring even has no one–sided nil–ideals. In fact, let L be such a non–zero ideal, x an element from A with maximal right annihilator $r(x)$. We shall show that $(AxA)^2 = \{0\}$ contradicting the semi–primality. It is sufficient to show that for any $y \in A$ we have $xyx = 0$. If $yx = 0$ then this is clear. Otherwise, there can be found a $k > 1$ such that $(yx)^{k-1} \neq 0$ and $(yx)^k = 0$. Then $(yx)^{k-1} \in L$ and $r((yx)^{k-1}) \supset r(x)$, i.e. $r((yx)^{k-1}) = (r(x)$. Therefore $yx \in r(x)$, as required.

Theorem. *Let A be a left noetherian and right semi–prime ring. Then the set S of its regular elements is a set of denominators and the ring of fractions A_S is a semi–prime ring.*

The most complicated part is to prove the Ore condition holds in A with respect to S. We shall need the concept of an essential ideal: namely, a left ideal L of a ring A is called *essential* if for any non–zero left ideal $\mathcal{I}$ of the ring A the intersection $L \cap \mathcal{I}$ is not equal to zero. In all lemmas below A is a noetherian ring.

Lemma 1. *Let x be a left regular element in A. Then Ax is an essential ideal in A.*

Proof. Assume that $\mathcal{I}$ is a non–zero left ideal in A such that $\mathcal{I} \cap Ax = \{0\}$. Then for any natural n the ideal $\mathcal{I}x^n$ is non–zero. The sum $\mathcal{I}+\mathcal{I}x+\mathcal{I}x^2+\ldots+\mathcal{I}x^n+\ldots$ cannot be direct in virtue of the noetherian property of the ring A. Thus, there are not all zero $a_0, a_1, \ldots, a_n \in \mathcal{I}$ such that $a_0+a_1x+\ldots+a_nx^n = 0$. Then $a_0 \in \mathcal{I} \cap Ax = \{0\}$. Hence $(a_1 + a_2x + \ldots + a_nx^{n-1})x = 0$, i.e. also $a_1 + a_2x + \ldots + a_nx^{n-1} = 0$. Continuing this scheme we see that $a_0 = a_1 = \ldots = a_n = 0$, a contradiction. □

Let $a \in A$, $l(a)$ is the left annihilator of the element a in A, i.e. $l(a) = \{x \in A | xa = 0\}$. We note that the set Z of all elements a of the ring A, for which $l(a)$ is an essential left ideal, is a two–sided ideal of the ring A. Indeed, it is evident that the intersection of two essential ideals is essential. Further, $l(a+b) \supset l(a) \cap l(b)$, i.e. if $a, b \in Z$ then also $a+b \in Z$. If $a \in Z$, $x \in A$ then $l(ax) \supset l(a)$, i.e. $ax \in Z$. Now consider xa. If $y \in A \setminus \{0\}$ and $yx = 0$, then $(Ay)(xa) = 0$, i.e. $Ay \cap l(ax) \neq \{0\}$. If $yx \neq 0$, then there is an $z \in A$ such that $z(yx) \neq 0$ and $(zyx) \neq 0$ and $(zyx)a = 0$. Then zy is a non–zero element from $Ay \cap l(xa)$. Thus, $l(xa)$ is an essential ideal and $xa \in Z$.

An analogous argument "almost" proves the left Ore condition for A. Indeed, let $a \in A$, $s \in S$. We denote by L the left ideal in A consisting of elements t such that $ta \in As$.

We shall show that L is an essential ideal. Let $x \in A \setminus \{0\}$. If $xa = 0$, then $Ax \cap L \neq \{0\}$; if $xa \neq 0$ then $Axa \cap As \neq \{0\}$ according to lemma 1, hence there is an $z \in A$ such that $zx \neq 0$, $(zx)a \in As$, i.e. $Ax \cap L \neq \{0\}$. Thus the ideal L is essential. If we can show that $L \cap S \neq \{0\}$ then taking $t \in L \cap S$, we obtain for a suitable $b \in A$, that $ta = bs$ (the left Ore condition).

So it remains to prove that in a noetherian semi–prime ring any essential left ideal contains a regular element.

Lemma 2. *Any one–sided regular element of a ring A is (two–sided) regular.*

Proof. First we show that the ideal Z introduced below lemma 1, is zero. Let $z \in Z$, then $z^n \in Z$ and $l(z^n)$ is an essential ideal for any n such that $z^n \neq 0$. Let $z^n \neq 0$ for all n. In virtue of the noetherian property, $l(z^m) = l(z^{m+1}) = \ldots$ for suitable m. We consider the non–zero left ideal Az^m; let $y \in A$ be such that $0 \neq yz^m \in Az^m \cap l(z^m)$. Then $yz^{2m} = 0$, i.e. also $yz^m = 0$, a contradiction. Thus, for every element $z \in Z$ there exists an n such that $z^n = 0$. As was noted above the statement of the theorem, it then follows that $Z = \{0\}$.

Assume now that $x \in A$ does not divide zero from the right. Then for any ideal $\mathcal{I} \neq \{0\}$ we have $l(z) \supseteq Ax$, an essential ideal. Therefore $r(x) \subset Z = \{0\}$. Thus x is not a zero divisor from the left. □

In order to conclude the proof of the part concerning the Ore condition we investigate an essential ideal L of the ring A. As was mentioned before, not all of its elements are nilpotent. If b is not a nilpotent element in L then as was noted in the proof of the previous lemma, we have $ab^n \cap l(b^n) = \{0\}$ for some natural n. Thus, $a = b^n$ is a non–zero element in L satisfying the condition $Aa \cap l(a) = \{0\}$. Applying the method of mathematical induction, we can assume that there are non–zero $a_1, \ldots, a_n$ such that the sum $Aa_1 + \ldots + Aa_n + l(a_1) \cap \ldots \cap l(a_n) \cap L$ is direct. If the ideal $L' = l(a_1) \cap \ldots \cap l(a_n) \cap L$ is non–zero then there is an $a_{n+1} \in L' \setminus \{0\}$ for which $Aa_{n+1} \cap l(a_{n+1}) = \{0\}$. Thus the sum $Aa_1 + \ldots + Aa_n + l(a_1) \cap \ldots \cap l(a_n) \cap L$ is direct. Using the noetherian property we see that there are non–zero $a_1, \ldots, a_n \in L$ such that the sum $Aa_1 + \ldots + Aa_n$ is direct and $l(a_1) \cap \ldots \cap l(a_n) \cap L = \{0\}$. Since the ideal L is essential, we have $l(a_1) \cap \ldots \cap l(a_n) = \{0\}$. In this case $a_1 + \ldots + a_n$ is the desired regular element. Indeed, if $x(a_1 + \ldots + a_n) = 0$ then, as the sum is direct, $xa_1 = \ldots = xa_n = 0$, hence, $x \in l(a_1) \cap \ldots \cap l(a_n) = \{0\}$, as required.

In order to prove that A_S is artinian semisimple we shall first show that A_S is a completely reducible left A_S–module. For that note that any essential left ideal L of the ring A_S coincides with A_S. In fact, $L \cap A$ is an essential left ideal in A. So, there can be found an $s \in S \cap L \cap A$. For any $a \in A$ we obtain $as^{-1}s \in L$, i.e. $L \supset A$. And, finally, $L = A_S$. Now let M be a non–zero left ideal in A_S. Then according to the Zorn lemma there exists a left ideal M' maximal with respect to the condition $M \cap M' = \{0\}$. In virtue of the maximality the ideal $M \oplus M'$ is essential, i.e. $A_S = M \oplus M'$.

We shall verify that the artinian property holds. For that note that A_S is noetherian, since any left or right ideal is determined by its intersection with A. Thus, let $M_1 \subset M$, then $A = M_1 \oplus M_1'$ and $M = A \cap M = M_1 \oplus (M' \cap M) = M_1 \oplus M_1''$. If $M_2 \subset M_1$ then, similarly, $M_1 = M_2 \oplus M_2'$, etc. As a result from the presence of the strictly descending chain of ideals $M \supset M_1 \supset M_2 \supset \dots$ there follows the presence of the strictly increasing chain

$$M_1' \subset M_1' \oplus M_2' \subset M_1' \oplus M_2' \oplus M_3' \subset \dots$$

In this way we see that A_S is artinian. Since A_S is completely reducible, we see that $A_S = \bigoplus_\alpha L_\alpha$, where the ideals L_α are minimal. The intersection of the annihilators of the ideals L_α is zero, i.e. A_S is semisimple. □

Corollary. *Let A be a prime noetherian ring, and let S be its set of regular elements. Then A_S is an artinian simple ring (i.e. a ring of matrices of some order over a skew field). If A is a noetherian ring without zero divisors then it has a skew field of fractions.*

Proof. Let A be a prime ring and A_S not simple. Since A_S is a direct sum of the finite number of ideals which are simple rings, there are $\mathcal{I}$, $\mathcal{J} \neq \{0\}$ in A_S such that $\mathcal{I}\mathcal{J} = \{0\}$. But $\mathcal{I} \cap A$, $\mathcal{J} \cap A$ are non–zero, though $(\mathcal{I} \cap A)(\mathcal{J} \cap A) = \{0\}$, a contradiction.

Now, assume that A_S is not a skew field, then A_S is a ring of matrices of order higher than 1, and A_S has zero divisors $a_1 s_1^{-1}$ and $a_2 s_2^{-1}$. From $a_1 s_1^{-1} a_2 s_2^{-1} = 0$ it follows that $a_1 a_2' = 0$, where a_2' is such that $s_1' u_2 = a_2' t^{-1}$ for a suitable t. □

Examples.

1) $A = M_n(F)$, where F is a field. Then S is the set of invertible matrices and $A_S \cong A$.

∗ 2) An algebra A over the field F is called a PI–algebra if there is a non–zero polynomial of non–commutative variables $f(X_1, \dots, X_n)$ such that $f(a_1, \dots, a_n) = 0$ for any $a_1, \dots, a_n \in A$. The Jacobson radical of a finitely generated PI–algebra A is nilpotent (a difficult theorem that has been proved by the efforts of several authors). The set S of non–zero divisors of the prime

PI–ring is a set of denominators and A_S is an artinian semisimple ring with the same identity as A (see, Ch. 7). *

§ 4. Central simple algebras

4.1. Centroid of an algebra. Let A, B be some rings. We can turn $C = A \otimes_{\mathbf{Z}} B$ into a ring by setting $(a \otimes b)(a' \otimes b') = (aa') \otimes (bb')$. The reader can check without difficulty that this is well defined. The ring C thus obtained is called the *tensor product* of the rings A and B. If A, B are algebras over a field F then, similarly, $D = A \otimes_F B$ becomes an F–algebra which is the tensor product of the F–algebras A and B. If K is an extension of the base field F then $A_K = K \otimes_F A$ is an algebra over the field K. We then say, that A_K is obtained from A by *extension of the field of coefficients* (field of scalars).

Let A be a simple ring. Consider the opposite ring A^{op}, i.e. the same set A, but with the operation $(a, b) \mapsto ba$. Let R denote the tensor product $R = A \otimes_{\mathbf{Z}} A^{op}$. If $a \in A$, $b \in A^{op}$ then the mapping $(a, b) \mapsto l_a r_b$ (l_a is the multiplication by a from the left, r_b is the multiplication by b from the right) from $A \times A^{op}$ into $\mathrm{End}(A)$ extends to a homomorphism of rings $R = A \otimes A^{op} \to \mathrm{End}(A)$. Under it, A becomes an irreducible R–module. According to the Schur lemma the centralizer of the R–module A is a field $\Delta \subset \mathrm{End}(A)$. We note that Δ is commutative. Indeed, if $a_1, a_2 \in A$, $\gamma, \delta \in \Delta$ then $(a_1 a_2)(\gamma\delta) = ((a_1 a_2)\gamma)\delta = ((a_1\gamma)(a_2)\delta = (a_1\gamma)(a_2\delta)$. In addition, $(a_1 a_2)(\gamma\delta) = (a_1(a_2\gamma))\delta = (a_1\delta)(a_2\gamma)$. Transposing γ and δ we obtain $(a_1 a_2)(\delta\gamma) = (a_1\delta)(a_2\gamma)$. Hence $(a_1 a_2)(\gamma\delta - \delta\gamma) = 0$. Since $A^2 = A$, in virtue of the simplicity of A, we obtain that every $c \in A$ can be written as $c = \sum_i a_i b_i$ ($a_i, b_i \in A$). Then $c(\gamma\delta - \delta\gamma) = 0$ for any $c \in A$. So $\gamma\delta = \delta\gamma$ and Δ is commutative.

The field $\Delta = C(A)$ is called the *centroid* of the ring A. Naturally, setting $\gamma a = a\gamma$ we turn the ring A into an algebra over Δ. It is evident that if A is a ring with identity, then A as an algebra over Δ is central in the sense that its center coincides with the base field Δ.

Thus, it is totally justified to consider only central simple algebras.

4.2. Tensor product of simple algebras

Proposition. *Let A be a central simple subalgebra of the algebra E over a field K, $B = C_e(A)$ the centralizer of A in E. Then AB is the tensor product* *of algebras A and B.*

Proof. Since A, B commute pairwise, the mapping $a \otimes b \mapsto ab$, where $a \in A$, $b \in B$, is a homomorphism of K–algebras. In order to prove its injectivity it is sufficient to demonstrate that if $a_1, \ldots, a_n$ are linearly independent over K and $\sum_{i=1}^{n} a_i b_i = 0$, $b_1, \ldots, b_n \in B$, then $b_1 = \ldots = b_n = 0$.

The action of the algebra $A \otimes A^{op}$ on A extends to an action on E such that $m(xb) = (mx)b$, where $m \in A \otimes A^{op}$, $b \in B$, $x \in E$. Here $m = \sum g_i \otimes h_i$, $mx = \sum g_i x h_i$. In virtue of the density theorem and the centrality of the algebra A for any j, $1 \leq j \leq n$, there can be found an $m_j \in A \otimes A^{op}$ such that $m_j a_j = 1$, $m_j a_k = 0$, if $k \neq j$. Applying m_j to $\sum_{i=1}^{n} a_i b_i$ we obtain that $b_j = 0$. □

Theorem. *Let A be a central simple algebra over a field K, B an algebra over K. We consider A and B as subalgebras in $A \otimes B$. Then the mappings $\mathcal{I}' \mapsto \mathcal{I}' \cap B$ and $\mathcal{I} \mapsto A\mathcal{I}$ $(= \mathcal{I}A)$, where $\mathcal{I}'$ is an ideal in $A \otimes B$, and $\mathcal{I}$ is an ideal in B, are inverse bijections between the sets of ideals in $A \otimes B$ and in B. In addition B is the centralizer for A in $A \otimes B$.*

Proof. If $\mathcal{I}$ is an ideal in B, then $A\mathcal{I}$ is an ideal in $A \otimes B = AB$. Consider a basis $\{1, u_\alpha\}$ for A. Then $A\mathcal{I} = \mathcal{I}A$ is the set of all elements of the form $1 \cdot b_0 + u_{\alpha_1} b_1 + \ldots + u_{\alpha_n} b_n$, $b_j \in \mathcal{I}$. From the independence property it follows that if an element of this form lies in B, then $b_j = 0$, for all $j \neq 0$, and the element b_0 lies in $\mathcal{I}$. Let now $\mathcal{I}'$ be an ideal in AB. Then $\mathcal{I} = \mathcal{I}' \cap B$ is an ideal in B and $A\mathcal{I}$ is an ideal in AB, contained in $\mathcal{I}'$. Form $AB/\mathcal{I}'$. If $\bar{u} = u + \mathcal{I}$, then in virtue of the simplicity of the algebra A, the set $\{\bar{1}, \bar{u}_\alpha\}$ is a basis for $\bar{A}$ in $AB/\mathcal{I}'$. The elements of the algebra $AB/\mathcal{I}'$ have the form $\bar{b}_0 + \sum \bar{u}_{\alpha_j} \bar{b}_j$, $\bar{b}_j \in \bar{B}$. As in the previous statement such an element is equal to zero only in the case when $\bar{b}_0 = 0$ and all $\bar{b}_j = 0$. Hence $b_0 + \sum u_{\alpha_j} b_j \in \mathcal{I}'$ iff b_0, $b_j \in \mathcal{I} = B \cap \mathcal{I}'$. Then $\mathcal{I}' = A\mathcal{I}$. Let now $\{1, b_\beta\}$ be a basis for B. Then any element in AB can be represented in the form $u = \sum_{j=1}^{m} a_j b_{\beta_j}$, $a_j \in A$, moreover, $\sum a_j b_{\beta_j} = 0$ implies a_j, $j = 1, \ldots, m$. We compute $ua - au$, where a is an arbitrary element from A. Then we obtain $\sum (a_j a - a a_j) b_{\beta_j}$. If this element is zero then $a a_j = a_j a$, i.e. a_j is an element of the center, or $a_j \in K \subset B$. Thus $C_{AB}(A) = B$. □

Corollary. *If the algebra A is central simple over the field K, B is simple,* *then $A \otimes B$ is a simple algebra. If B is also a central simple, then $A \otimes B$ is a central simple algebra.* □

4.3. Automorphisms of simple algebras. Let A be a (left) artinian algebra and B a K–algebra that is finitely generated as a K–module, $B = \sum_{j=1}^{m} Kb_j$. Then $A \otimes_R B$ as a left A–module is generated by the elements $1 \otimes b_j$. Since A is (left) artinian it follows that $A \otimes_K B$ satisfies the minimality condition for A–submodules. It is also clear that any left ideal of the algebra $A \otimes_K B$ is also a A–submodule. Therefore $A \otimes_K B$ is left artinian.

Recall that if A is a central simple algebra over the field K, and B is simple over K, then $A \otimes_K B$ is simple over K (see, Subsection 4.2). If in this connection A is artinian and B is finite–dimensional over K, then $A \otimes_K B$ is a simple artinian algebra.

Theorem. *Let A be a central simple artinian algebra over K, and let B be a simple subalgebra in A. Then any isomorphism σ from B into A extends to an internal automorphism of the algebra A.*

Proof. Identify A with the algebra $\mathrm{End}_{\Delta}(V)$ of linear oprators of a finite–dimensional vector space V over the field Δ. Consider $\Delta \otimes_K B$. This algebra is simple artinian and V becomes a module over that algebra in two ways. For the first way we put

$$(\delta \otimes b)x = \delta b x = b\delta x.$$

The module obtained is simply noted by V. For the second way we set

$$(\delta \otimes b)x = \delta\sigma(b)x = \sigma(b)\delta x,$$

where σ is the given isomorphism from B into A. A module V_σ is obtained.

As Δ–modules, V and V_σ coincide and are equal to n copies of irreducible Δ–modules, $n = \dim_\Delta V$. Therefore both modules have the form of direct sums of a finite number of irreducible modules which are isomorphic to the same module M. Let the number of irreducible components for the module V be equal to h and for the module V_σ equal to k. Hence $h = k$, i.e. V and V_σ are isomorphic $\Delta \otimes_K B$–modules. This gives us an isomorphism of additive groups $u : V \to V$ such that $u(\delta b)x = (\delta\sigma(b))ux$. Taking $b = 1$, we obtain $u\delta x = \delta(ux)$ so that $u \in A = \mathrm{End}_\Delta(V)$. Taking $\delta = 1$ we obtain $ubx = \sigma(b)ux$ for all x. So $\sigma(b) = ubu^{-1}$, since V is a faithfull A–module. □

Corollary (Noether–Skolem theorem). *Any automorphism of a finite–dimensional simple central algebra is interior.* □

4.4. Splitting field

Definition. Let A be a finite–dimensional central simple algebra over the field K. An extension F of the field K is called a *splitting field* for A if $A_F = F \otimes_K A \cong M_n(F)$ for a suitable n.

Theorem 1. *The algebraic closure $\bar{K}$ of the field K is a splitting field. Moreover, if $A \cong M_r(\Delta)$, where Δ is a skew field and F is a maximal commutative subfield in Δ, then F is a splitting field for A.*

Proof. For any extension F of the field K the algebra $A_F = F \otimes_K A$ is central simple over F and $\dim_F A_F = \dim_K A$. According to the Wedderburn–Artin theorem (see, § 3), $A_F \cong M_n(\Delta)$ for suitable finite–dimensional skew field Δ over F and a natural number n. If $F = \bar{K}$ is algebraically closed then $\Delta = F$ and $A_{\bar{K}} = M_n(\bar{K})$. This proves the first statement.

In order to prove the second part we shall consider A as an algebra of linear operators on a finite–dimensional vector space V over the field Δ. Let F be a maximal subfield in Δ. We consider V as a vector space over F. We denote by Δ^L, F^L the subalgebras in End(V) generated by left multiplication by elements from Δ and F, respectively. Then Δ^L is the centralizer for A in End(V). The subalgebras F^L and A centralize mutually. We form their product A'. Since V is an irreducible A–module, V also is an irreducible A'–module. Let us find its centralizer Δ'. An endomorphism $\phi \in \Delta'$ commutes with A, i.e. it lies in Δ^L. It is also commutes with F^L. Thus $F^L(\phi)$ is a subfield in Δ^L containing F^L. From the maximality property we have $\phi \in F^L$. According to the density theorem A' is dense in V over F. Thus if $\dim_F V = n$, then $A' = M_n(F)$. From the proposition 1 from Subsection 4.2 we derive $A' = F^L A \cong F \otimes_K A$. So $A_F = F \otimes_K A \cong M_n(F)$, i.e. F is a splitting field. □

Corollary. *The dimension of a finite–dimensional central simple algebra is a square.* □

Theorem 2. *Let Δ be a finite–dimensional division algebra over the field K,* *then a finite extension L of the field K is a splitting field for Δ iff L is a subfield in $M_r(\Delta) = A$ and satisfies $C_A(L) = L$. If this condition is fulfilled then $[L : K] = rn$, where $n^2 = [\Delta : K]$.*

Proof. First we assume that L is a subfield in $A = M_r(\Delta)$ satisfying $C_A(L) = L$. We identify A with $\mathrm{End}_\Delta(V)$, where $\dim_\Delta V = r$. Then V is a $L \otimes_K \Delta$–module as in the proof of the theorem 1. Since A is the centralizer of V as a Δ–module and $C_A(L) = L$, we see that the centralizer of L as an $L \otimes_K \Delta$–module is L. From the simplicity of the algebra $L \otimes_K \Delta$ follows that V is a faithfull completely reducible $L \otimes_K \Delta$–module. Since its centralizer is a field, this module is simple. Moreover, we have

$$[V : K] = [V : \Delta][\Delta : K] = rn^2 = [V : L][L : K],$$

i.e. V is finitely dimensional over L. From the density theorem applied to V as an irreducible $L \otimes_K \Delta$–module, it follows that $L \otimes_K \Delta \cong M_n(L)$, where $n' = [V : L]$. Comapring dimensions we get $n' = n$. Thus L is a splitting field and $rn^2 = n[L : K]$, i.e. $[L : K] = rn$.

Inversily, assume that $L \otimes_K \Delta \cong M_n(L)$. Let V be an irreducible module for $L \otimes_K \Delta$. Since $L \otimes_K \Delta \cong M_n(L)$, V is an n–dimensional vector space over L and $L \otimes_K \Delta$ can be identified with $\mathrm{End}_L(V)$. At the same time V is a vector space over Δ and if $[V : \Delta] = r$, since L centralizes Δ, $L \subset \mathrm{End}_\Delta(V) \cong M_r(V)$. The centralizer of the field L in $\mathrm{End}_\Delta(V)$ is contained in the centralizer of $L \otimes_K \Delta$ in $\mathrm{End}_L(V)$. Since $L \otimes_K V = \mathrm{End}_L(V)$, this centralizer is equal to L. Therefore $C_{\mathrm{End}_\Delta(V)}(L) = L$. □

The basic result on splitting fields is the following.

Theorem 3. *Any finite–dimensional central simple algebra admits a separable splitting field.*

To prove this theorem two lemmas are needed.

Lemma 1. *Let Δ be a central division algebra and let it be algebraic over its base field K. If $\Delta \neq K$, then Δ contains an element which does not lie in K and is separable over K.*

Proof. In the opposite case the characteristic of the field K is equal to p and every element a from Δ satisfies a relation of the form $a^{p^s} = \alpha$, $\alpha \in K$. Since $\Delta \neq K$, there exists $d \in \Delta \backslash K$ such that $d^p \in K$. We consider the internal derivation $\mathcal{D} : x \mapsto [d, x] = dx - xd$ of the field Δ. Then $\mathcal{D} \neq 0$. It is easy to see that $\mathcal{D}^p$ is the internal derivation $x \mapsto [d^p, x] = 0$. Therefore there exists an element $a \in \Delta$ such that $b = \mathcal{D}a \neq 0$, but $\mathcal{D}^2 a = \mathcal{D}b = 0$. Put $c = ab^{-1}d$. Then $\mathcal{D}c = (\mathcal{D}a)b^{-1}d = d$, hence $dc - cd = d$ and $dcd^{-1} = c+1$. In this case the subfield $K(c)$ maps into itself under the internal automorphism $x \mapsto dxd^{-1}$; moreover, this automorphism is not the identity on $K(c)$. However a purely non–separable extension can not have non–trivial automorphisms. So we got a contradiction. (For more about derivations of algebras see in Chapter 4.) □

Corollary. *Under the conditions of lemma 1 let E be a maximal separable subfield in Δ satisfying the double centralizer condition $C_\Delta(C_\Delta(E)) = E$. Then E is a maximal subfield in Δ.*

Proof. From the double centralizer condition follows that $\Delta' = C_\Delta(E)$ is central over E. Therefore, if $\Delta' \neq E$ then Δ' contains the element c which does not lie in E and is separable over E. Then $E(c) \neq E$ and this field is separable over K, which contradicts to the condition of maximality for E. Thus $C_\Delta(E) = E$ and E is a maximal subfield in Δ. □

Lemma 2. *Let Δ be a central division algebra over K and E a finite–dimensional subalgebra in Δ. Then $C_\Delta(C_\Delta(E)) = E$*

Proof. We consider Δ as a $\Delta \otimes_K \Delta^{op}$–module in the usual way. We have the subalgebra $E \otimes \Delta^{op} \subset \Delta \otimes_K \Delta^{op}$ so that Δ is a module over this subalgebra. Since Δ is a skew field, it is an irreducible Δ^{op}–module, hence also an irreducible $E \otimes \Delta^{op}$–module. Let us show that $(\mathrm{End}_{E \otimes \Delta^{op}} \Delta)^{op} = C_\Delta(E)$. Since $d \in \Delta^{op}$ acts on $x \in \Delta$ by multiplication from the right, $(\mathrm{End}_{\Delta^{op}} \Delta)^{op} = \Delta$. Further, since $d(ex) = e(dx)$ for $e \in E$, we see that $d \in C_\Delta(E)$. The opposite is also valid, which proves our statement. According to the density theorem $E \otimes \Delta^{op}$ is dense in the algebra of linear operators on the field Δ as a vector space over $C_\Delta(E)$. We show that $[\Delta : C_\Delta(E)]_L \leq [E : K] < \infty$. First we should note that $E \otimes \Delta^{op}$ is naturally a left vector space over Δ^{op}. According to the properties of tensor products the dimension of this space is equal to $[E : K]$. Let $u_1, u_2, \dots, u_m$ be the elements from Δ and linearly independent over $C_\Delta(E)$. According to the density theorem there can be found

$l_i \in E \otimes \Delta^{op}$ such that $l_i u_j = \delta_{ij}$. So if $d \in \Delta^{op}$, then $(\sum d_i l_i)u_j = d_j$; from this and from $\sum d_i l_i = 0$ it follows that all $d_i = 0$. Therefore, the l_i are linearly independent over Δ^{op}. Hence $m \leq [E : K]$ and so $[\Delta : C_\Delta(E)] \leq [E : K]$. Now from the density theorem we obtain that $E \otimes \Delta^{op}$ can be identified with the algebra of linear operators in Δ which is a vector space over $C_\Delta(E)$. It is clear that the elements from $C_\Delta(C_\Delta(E))$ are linear operators in Δ over $C_\Delta(E)$. Thus $C_\Delta(C_\Delta(E)) \subset E \otimes \Delta^{op}$. Let now $\{u_\alpha\}$ be a basis for Δ^{op} over K such that $u_0 = 1$. Then these elements form a basis for $\Delta \otimes \Delta^{op}$ over Δ. Let $c \in C_\Delta(C_\Delta(E))$. Then $c = \sum e_\alpha u_\alpha$ where $e_\alpha \in E$. Further $c = cu_0 = \sum e_\alpha u_\alpha$. Hence $c - c_0 \in E$. Thus $C_\Delta(C_\Delta(E)) \subset E$ and since the opposite is evidently true, we have $C_\Delta(C_\Delta(E)) = E$. □

Note. From the proof it follows that$[\Delta : C_\Delta(E)]_L = [E : K]$. Similarly, $[E : K] = [\Delta : C_\Delta(E)]_R$.

Now we can prove the theorem.

Proof of the theorem. It is sufficient to show that any finite–dimensional central division algebra Δ admits a separable splitting field. In other words, that in Δ there is a maximal subfield which is separable over K. Let E be some maximal separable subfield. According to lemma 2, $E = C_\Delta(C_\Delta(E))$. It remains to apply the corollary of lemma 2. □

4.5. Similarity of central simple algebras. Let K be a field A_1, A_2 finite dimensional simple algebras over K. We call A_1 and A_2 *similar* ($A_1 \sim A_2$) if for suitable m, n the following holds: $M_m(A_1) \cong M_n(A_2)$. It is clear that this relation is reflexive and symmetric. Let further $A_1 \sim A_2$, $A_2 \sim A_3$. Then there exist m, n and k, l such that $M_m(A_1) \cong M_n(A_2)$ and $M_k(A_2) \cong M_l(A_3)$. Because $M_a(K) \otimes M_b(K) \cong M_{ab}(K)$, taking the tensor product of the first isomorphism by $M_k(K)$ and taking the tensor product of the second one by $M_n(K)$ and using the associativity of the tensor product we obtain

$$M_{mk}(A_1) \cong M_k(K) \otimes M_m(K) \otimes A_1 \cong$$
$$\cong M_k(K) \otimes M_n(K) \otimes A_2 \cong M_{kn}(A_2),$$
$$M_{nk}(A_2) \cong M_n(K) \otimes M_k(K) \otimes A_2 \cong$$
$$\cong M_n(K) \otimes M_l(K) \otimes A_3 \cong M_{nl}(A_3).$$

Thus $M_{mk}(A_1) \cong M_{nl}(A_3)$ and $A_1 \sim A_3$. So the relation $\sim$ is also transitive. The class of similar algebras containing A is denoted by $\{A\}$; the class of matrix algebras $M_n(K)$, $n = 1, 2, \ldots$, is denoted by 1. If $A_1 \sim A_2$, $B_1 \sim B_2$ then $A_1 \otimes A_2 \sim B_1 \otimes B_2$. This follows from the commutativity and associativity of tensor products of matrix algebras. Consequently, $\{A\}\{B\} = \{A \otimes B\}$ is a well defined operation on the set of classes of similar algebras. Naturally, 1 is the neutral element for this operation and the operation is associative and commutative. Finally, if A is a central simple algebra then the algebra A^{op} opposite to it is also central simple. The algebra $A \otimes A^{op}$ is simple, hence its irreducible module A is faithfull. The centralizer of this module is equal to K. Let $n = [A : K]$ then according to the density theorem, $A \otimes A^{op} \cong M_{n^2}(K) \sim 1$. Hence $\{A\}\{A^{op}\} = \{A^{op}\}\{A\} = 1$. Thus the set of classes of similar central simple algebras is an abelian group. It is called the *Brauer group* of the field K and is denoted by $\mathrm{Br}(K)$. It should be noted that every similarity class has exactly one algebra with division which is central over K. This allows us to note that the Brauer group of an algebraically closed field is trivial. From the Frobenius theorem (see, Introduction 2.7) it follows that $\mathrm{Br}(\mathbf{R}) \cong \mathbf{Z}_2$. The non–identity element of the group $\mathrm{Br}(\mathbf{R})$ is the class of the quaternion field (the field of complex numbers, $\mathbf{C}$, is not a field which is central over the subfield of real numbers $\mathbf{R}$).

4.6. Clifford algebra. Definition and universal property. Let $Q : V \to F$ be a qadratic form on the vector space V over the field F, $T(V)$ the tensor algebra of the space V, $\mathcal{I} = \mathcal{I}(Q)$ the two–sided ideal in $T(V)$ generated by the tensors of the form $x \otimes x - Q(x) \cdot 1$, $x \in V$. The quotient algebra $C(Q) = T(V)/\mathcal{I}(Q)$ is called the *Clifford algebra* of the form Q.

We denote by ρ the linear mapping $\rho : x \to x + \mathcal{I}$ from V in $C(Q)$. It should be noted that the algebra $C(Q)$ is generated by the set $\rho(V)$ and for any $x \in V$ we have

$$\rho(x)^2 = Q(x) \cdot 1;$$

from this, replacing x with $x + y$ $(x, y \in V)$, we obtain

$$\rho(x)\rho(y) + \rho(y)\rho(x) = \Phi(x, y) \cdot 1,$$

where Φ is the bilinear form which is the polarization of Q, i.e. $Q(x) = \Phi(x, x)$.

We denote by T^+ the set of tensors of even degree in $T(V)$, by T^- the set of tensors of odd degree in $T(V)$. Then $T(V) = T^+ \oplus T^-$. Since $\mathcal{I}(Q)$ is generated by tensors of even degree, $\mathcal{I}(Q) = (T^+ \cap \mathcal{I}(Q)) \oplus (T^- \cap \mathcal{I}(Q))$.

Therefore $C(Q) = C^+(Q) \oplus C^-(Q)$, where $C^+ = C^+(Q)$ is the image of T^+ and $C^-(Q)$ is the image of T^- under the natural homomorphism $\sigma : T(V) \to C(Q)$. The following inclusions are evident

$$C^+C^+ \subset C^+, \quad C^+C^- \subset C^-, \quad C^-C^+ \subset C^-, \quad C^-C^- \subset C^+;$$

in particular, C^+ is a subalgebra in $C(Q)$.

Proposition. *Let f be a linear mapping from V into an algebra A over the field F and such that $f(x)^2 = Q(x) \cdot 1$ for any $x \in V$. Then there exists a unique homomorphism $\bar{f} : C(Q) \to A$ such that $f = \bar{f}\rho$.*

If $e_1, \dots, e_n$ is a basis in V then define the homomorphism $h : T(V) \to A$, by setting

$$h(e_{i_1} \otimes \dots \otimes e_{i_m}) = f(e_{i_1}) \dots f(e_{i_m}).$$

Then $h(x \otimes x - Q(x) \cdot 1) = f(x)^2 - Q(x) \cdot 1 = 0$, i.e. h vanishes on the ideal $\mathcal{I}(Q)$ and by means of the quotient mapping it defines the desired homomorphism $\bar{f}$. □

The basic result of the present section is a theorem giving a basis of the algebra $C(Q)$ in terms of a basis of the space V. The main idea of the proof is to show that this basis does not depend on the form of Q. If this can be proved then it remains only to find a basis for the special case $Q = 0$. However in this case the algebra $C(Q)$ is the exterior algebra $\Lambda(V)$ of the space V. It is known (see, § 1.4), that this algebra has a basis of the form

$$e_{i_1} \wedge e_{i_2} \wedge \dots \wedge e_{i_m},$$

where $e_1, \dots, e_n$ is a basis of the space V and $i_1 < i_2 < \dots < i_n$.

Using the definitions we obtain a basis in $C(Q)$ consisting of the elements of the form $\rho(e_{i_1})\rho(e_{i+_2}) \dots \rho(e_{i_m})$, $i_1 < i_2 < \dots < i_m$.

In order to prove independence several lemmas concerning the structure of a tensor algebra are necessary.

4.7. Three lemmas on the tensor algebra. We denote for $x \in V$ by e_x the linear operator $e_x(u) = x \otimes u$ in the tensor algebra $T(V)$.

Lemma 1. *Let $f \in V^*$. There exists a unique linear operator $i_f : T(V) \to T(V)$ for which*

$$i_f(1) = 0, \quad i_f \circ e_x + e_x \circ i_f = f(x) \cdot 1_{T(V)} \tag{1}$$

for any $x \in V$. The mapping $f \mapsto i_f$ of the space V^* into the algebra of linear operators on the space $T(V)$ is linear. The following relations hold

$$i_f(T^n) \subset T^{n-1}, \quad (i_f)^2 = 0, \quad i_f i_g + i_g i_f = 0$$

for all $f, g \in V^*$. The operator i_f is equal to zero on the subalgebra of the algebra $T(V)$ generated by the kernel of the element f. The ideal $\mathcal{I}(Q)$ is stable under i_f and, consequently, the operator i_f defines a linear operator on the algebra $C(Q)$.

Proof. If $u \in T(V)$ then, from (1), it is necessary that

$$i_f(x \otimes u) = -x \otimes i_f(u) + f(x)u \tag{2}$$

Since the action of i_f on T_0 is known (zero), the equality (2) with the help of induction on n defines our mapping and proves its uniqueness. The linearity of the mapping $f \mapsto i_f$ follows from uniqueness and the fact that for $f = \alpha g + \beta h$ the mapping $\alpha i_g + \beta i_h$ satisfies the conditions (1), i.e. is equal to i_f. We have

$$(i_f)^2 e_x = -i_f e_x i_f + f(x) i_f = e_x (i_f)^2,$$

from which, with a help of induction on n, in virtue of the equality $(i_f)^2(1) = 0$, we obtain that $(i_f)^2 = 0$ on T_n. Having replaced f with $f + g$ we also obtain $i_f i_g + i_g i_f = 0$. Finally, from (2) it follows that the set of elements $u \in \mathcal{I}(Q)$ such that $i_f(u) \in \mathcal{I}(Q)$, is a left ideal in $T(V)$ and, moreover, if $u = (x \otimes x - Q(x) \cdot 1)v)$ ($x \in V$, $v \in T(V)$), then

$$\begin{aligned} i_f(u) =& f(x)x \otimes v - x \otimes i_f(x \otimes v) - Q(x) i_f(v) = \\ =& f(x)x \otimes v - f(x)x \otimes v + x \otimes x \otimes i_f(v) - Q(x) i_f(v) = \\ =& (x \otimes x - Q(x) \cdot 1) i_f(v), \end{aligned}$$

consequently, $\mathcal{I}(Q)$ is stable under i_f, and lemma is proved. □

If $\Phi : V \times V \to F$ is a bilinear form on V then we denote by i_x^Φ the mapping i_f corresponding to the linear form $f : y \mapsto \Phi(x, y)$ on the space V.

Lemma 2. *There exists a unique mapping λ_Φ of the algebra $T(V)$ onto itself satisfying the following conditions*

$$\lambda_\Phi(1) = 1, \qquad \lambda_\Phi e_x = (e_x + i_x^\Phi)\lambda_\Phi \quad (x \in V) \tag{3}$$

For any $f \in V^*$ the following equality holds

$$\lambda_\Phi i_f = i_f \lambda_\Phi \tag{4}$$

Proof. According to formula (3) we have

$$\lambda_\Phi(x \otimes u) = x \otimes \lambda_\Phi(u) + i_x^\Phi \lambda_\Phi(u) \quad (x \in V,\ u \in T(V)).$$

As in the previous lemma this completely defines the mapping λ_Φ.

Equality (4) is proved by induction; assume that it is valid for $\sum_{i=0}^{n-1} T_{n-1}$. Then for $x \in V$ and $u \in T_{n-1}$ we have

$$\begin{aligned}
&(\lambda_\Phi i_f)(x \otimes u) = (-\lambda_\Phi e_x i_f + f(x)\lambda_\Phi)(u) = \\
&= -(e_x + i_f^\Phi)\lambda_\Phi i_f(u) + f(x)\lambda_\Phi(u) = \\
&= -(e_x + i_x^\Phi) i_f \lambda_\Phi(u) + f(x)\lambda_\Phi(u) = \\
&=(i_f e_x \lambda_\Phi - f(x)\lambda_\Phi + i_f i_x^\Phi \lambda_\Phi + f(x)\lambda_\Phi)(u) = \\
&=(i_f(e_x + i_x^\Phi)\lambda_\Phi)(u) = (i_f \lambda_\Phi)(x \otimes u),
\end{aligned}$$

as required. □

Lemma 3. *Let Φ and Ψ be bilinear forms on V. Then $\lambda_\Phi \lambda_\Psi = \lambda_{\Phi+\Psi}$. For every bilinear form Φ on V the mapping λ_Φ is a bijection of the algebra $T(V)$ onto itself.*

Proof. We shall verify property (3) for the mapping $\lambda_\Phi\lambda_\Psi$. It is clear that $(\lambda_\Phi\lambda_\Psi)(1) = 1$; in addition,

$$\lambda_\Phi\lambda_\Psi(e_x) = \lambda_\Phi(e_x + i_x^\Psi)\lambda_\Psi =$$
$$=(e_x + i_x^\Psi + i_x^\Phi)\lambda_\Phi\lambda_\Psi = (e_x + i_x^{\Phi+\Psi})\lambda_\Phi\lambda_\Psi.$$

For $\Phi = 0$ we have $i_x^\Phi = 0$, from where $\lambda_\Phi = 1_{T(V)}$ and then it follows that $\lambda_\Phi\lambda_{-\Phi} = \lambda_{-\Phi}\lambda_\Phi = 1_{T(V)}$ for any $\Phi : V \times V \to F$. □

4.8. Basis of a Clifford algebra

Proposition. *Let Q and Q' be quadratic forms, Ψ a bilinear form on V and $Q'(x) = Q(x)+\Psi(x,x)$ for any $x \in V$. Then λ_Ψ maps the ideal $\mathcal{I}(Q')$ into the ideal $\mathcal{I}(Q)$ and defines an isomorphism (denoted by $\bar{\lambda}_\Psi$) of the space $C(Q')$ onto $C(Q)$.*

Proof. Since λ_Ψ is a bijection and $\lambda_{-\Psi}$ is a bijection inverse to it, it is sufficient to prove that the inclusion $\lambda_\Psi(\mathcal{I}(Q')) \subset \mathcal{I}(Q)$. The ideal $\mathcal{I}(Q)$ is a left ideal stable under the mapping i_x^Ψ, so the set of elements $u \in T(V)$ such that $\lambda_\Psi(u) \in \mathcal{I}(Q)$, is a left ideal. Therefore it is sufficient to show that for any $u \in T(V)$ and $x \in V$ we have

$$\lambda_\Psi(x \otimes x \otimes u - Q'(x)u) \in \mathcal{I}(Q).$$

According to formula (3) and lemma 1

$$\lambda_\Psi e_x^2 = (e_x + i_x^\Psi)^2\lambda_\Psi = (e_x^2 + \Psi(x,x)\cdot 1_{T(V)})\lambda_\Psi,$$

so that

$$\lambda_\Psi(x \otimes x \otimes u - Q'(x)u) = (e_x^2 + \Psi(x,x) - Q'(x))\cdot 1_{T(V)})\lambda_\Psi(u) =$$
$$=(x \otimes x - Q(x)) \otimes \lambda_\Psi(u) \in \mathcal{I}(Q).$$

The proposition is proved. □

Theorem. *Let the space V have the basis $(e_i)_{i\in\mathcal{I}}$, moreover, let $\mathcal{I}$ be linearly ordered. For any finite $J \subset \mathcal{I}$ we put $e_J = \rho(e_{j_1})\rho(e_{j_2})\ldots\rho(e_{j_q})$, where*

$j_1 < j_2 < \ldots < j_q$ are elements from the set J. Then a basis of the algebra $C(Q)$ is formed by all possible elements of the form e_J.

Proof. As was noted above, the exterior algebra $\Lambda(V)$ is a Clifford algebra for the form $Q = 0$. Its basis consists of the elements $\bar{e}_J = e_{j_1} \wedge e_{j_2} \wedge \ldots \wedge e_{j_q}$, $j_1 < j_2 < \ldots < j_q$, (which are the images of the elements of the form $e_{j_1} \otimes e_{j_2} \otimes \ldots \otimes e_{j_q}$ under the quotient mapping defined by the ideal $\mathcal{I}(Q)$ for $Q = 0$).

Now let the form Q be arbitrary, Φ its polarization. We consider the bilinear form Ψ of the form $\Psi(e_i, e_j) = -\Phi(e_i, e_j)$, for $i > j$, $\Phi(e_i, e_j) = 0$, for $i < j$, and $\Phi(e_i, e_j) = -Q(e_i)$. Then it is clear that $Q(x) + \Psi(x, x) = 0$ for all $x \in V$. From the previous proposition follows that the mapping $\bar{\lambda}_\Psi : \Lambda(V) \to C(Q)$ is a linear isomorphism of spaces. By induction with respect to the number of elements in J we shall prove that

$$\bar{\lambda}_\Psi(\bar{e}_j) = \rho(e_{j_1} \ldots \rho(e_{j_q}).$$

If J is empty or contains one element then the formula above is immediately clear.

For the opposite case we can write $J = \{j\} \cup K$, where j is the smallest element in J and $|K| = q - 1$. According to the formula (3) and the induction assumption we have

$$\begin{aligned} &\lambda_\Psi(e_J) = \lambda_\Psi(e_j \otimes e_K) = \\ =&e_j \otimes \lambda_\Psi(e_K) + i^\Psi_{e_j}(\lambda_\Psi(e_K)) = e_J + i^\Psi_{e_j}(e_K). \end{aligned}$$

For any $i \in K$ the element e_i is contained in the kernel of the linear form $f : y \to (e_j, y)$, which proves what is required. □

Corollary 1. *If $dim\, V = n$, then $dim\, C(Q) = 2dim\, C^+ = 2dim\, C^- = 2^n$.* □

Corollary 2. *The canonical mappings $\rho : V \to C(Q)$ and $\lambda \mapsto \lambda \cdot 1$ from F into $C(Q)$ are injective.* □

Corollary 3. *If $V = V_1 \oplus V_2$, Q_i is the restriction of the form Q to V_i, $i = 1, 2$ then the space $C(Q)$ is canonically isomorphic to $C(Q_1) \otimes C(Q_2)$.* □

4.9. Central simplicity of the Clifford algebra

Theorem. *Let the dimension $m = 2r$ of the space V be an even number, and let Q be a non–degenerate quadratic form on V. Then the algebra $C(Q)$ is a central simple algebra of dimension 2^m.*

Proof. To start we consider the case when the form Q is neutral, i.e. when Q is non–degenerate and the space V is decomposed into a direct sum of two completely singular subspaces U and W of dimension r, i.e. $Q(U) = Q(W) = \{0\}$. In the case of algebraically closed field such a decomposition exists for any non–degenerate quadratic form.

Let S be the subalgebra in $C(Q)$ generated by the space U. Then $S = \Lambda(U)$ is the Grassmann algebra (exterior algebra) of the space U. For any $u \in U$ we denote by e'_u the mapping $t \mapsto ut$ of the algebra A onto itself.

Let $\{u_1, \dots, u_r\}$ be some basis in U; we denote by $\{w_1, \dots, w_r\}$ a basis in W such that $\Phi(u_i, w_j) = \delta_{ij}$, where Φ is the symmetric bilinear form which is the polarization of Q. For $w \in W$, a linear form $w' : U \to F$ is defined by the formula $w'(u) = \Phi(u, w)$, and let i_w be the endomorphism of the algebra S obtained with a help of quotient mapping from the endomorphism $i_{w'}$ of the algebra $T(U)$ defined for the form w' in lemma 1, Subsection 4.7. Then according to formula (1) we have

$$e'_u i_w + i_w e'_u = \Phi(u, w) \cdot 1_{T(U)} \quad (u \in U,\ w \in W) \tag{5}$$

For $v = u + w \in V$ (where $u \in U$, $w \in W$) we put $s(v) = e'_u + i_w$.

It is clear that s is a linear mapping from V into $\mathrm{End}(S)$ the algebra of linear operators on the space S. From the equality

$$s(v)^2 = (e'_u + i_w)^2 = Q(u) + \Phi(u, w) = Q(v)$$

follows that s extends to a homomorphism from $C(Q)$ into $\mathrm{End}(S)$.

We shall show that this homomoprhism is surjective. From that, in view of the fact that $C(Q)$ and $\mathrm{End}(S)$ have the same dimension 2^{2r} follows that s is an isomorphism.

In fact, let $\mathcal{I}$ be the interval $[1, r]$. For any subset H in $\mathcal{I}$ we put $H' = \mathcal{I} \backslash H$ and denote by u_H (respectively, by w_H) the product of the elements u_i (respectively, w_i) for $i \in H$ in the order of increasing indexes. The elements of the form u_H form a basis for the algebra S (see, theorem 1). Finally, we put

$x_{H,K} = u_H w_{\mathcal{I}} u_{K'}$ for any two subsets H, K of the set $\mathcal{I}$. We shall prove that the elements $s(x_{H,K})$ of the image $s(C(Q))$ generate End(S). But, if $j \notin H$ then according to lemma 1, Subsection 4.7, $s(w_j)(u_H) = i_w(u_H) = 0$, because all the u_i for $i \in H$ belong to the kernel of the linear form $u \mapsto \Phi(u, w)$ on the space U; on the other hand according to formula (5) we have

$$s(w_j)(u_j u_H) = -(i_{w_j} e'_{u_j})(u_H) = \\ =\Phi(w_j, u_j)u_H - u_j s(w_j)u_H = u_H.$$

Since s is a homomorphism, it follows that for any two subsets H and K of the set $\mathcal{I}$ the following equalities hold: $s(w_K)(u_H) = 0$, if $K \not\subset H$, and $s(w_K)(u_H) = \pm u_{H\setminus K}$, if $K \subset H$. But from the definition we have that for M, $L \subset \mathcal{I}$, $s(u_M)(u_L) = u_M u_L$ and the product $u_M u_L$ is equal to zero if $M \cap L \neq \emptyset$ and otherwise is equal to $\pm u_M u_L$. Therefore for any subsets H, K, L in $\mathcal{I}$ the element $s_{H,K}(u_L) = s(u_H)s(w_{\mathcal{I}})s(u_{K'})(u_L)$ is equal to zero when $K \neq L$ and equal to $\pm u_H$, when $K = L$. This means that the elements $s(x_{H,K})$ generate End(S).

Now we return to the case of the arbitrary non–degenerate form Q on an even–dimensional space V over a field F. We extend the field F to its algebraic closure $\bar{F}$, the space V to $\bar{F} \otimes_F V$ and the form Q to $\bar{Q}$ on $\bar{V}$. It was already noted that $C(\bar{Q})$ is isomorphic to $\bar{F} \otimes_F C(Q)$. The form $\bar{Q}$ already possesses the needed properties and the algebra $\bar{F} \otimes_F C(Q)$ is the algebra of all linear operators of some space over $\bar{F}$. The center of the algebra $\bar{F} \otimes_F C(Q)$ contains $\bar{F} \otimes_F Z$, where Z is a center in $C(Q)$. Since $\bar{F} \otimes_F C(Q)$ is central, we have $Z = F$. Similarly, from the ideal $\mathcal{I}$ of the algebra $C(Q)$ we obtain the ideal $\bar{F} \otimes_F \mathcal{I}$ of the algebra $\bar{F} \otimes_F C(Q)$. Therefore $C(Q)$ is a simple algebra. The theorem is proved. □

* In virtue of the simplicity of the algebra $C(Q)$ it has only one class of non–equivalent irreducible representations. These representations are called *spinors*. The group G of invertible elements g from $C(Q)$ and such that $gVg^{-1} = V$, where $V \subset C(Q)$ is the space of definition of the form Q, is called a *Clifford group*. If $x, y \in V$, $g \in G$, then $\Phi(gxg^{-1}, gyg^{-1}) = gxg^{-1} \cdot gyg^{-1} + gyg^{-1} \cdot gxg^{-1} = g(xy + yx)g^{-1} = \Phi(x, y)$. By the same token the mapping $\phi : G \to GL(V)$ associating to the element g the mapping $x \mapsto gxg^{-1}$ is an orthogonal mapping of the space V with the form Q. Thus $\phi(G) \subset O(Q)$. The kernel of the mapping lies in the center of the algebra

$C(Q)$, i.e. is equal to F^*. The image of ϕ for even m is equal to $O(Q)$, for odd m it is equal to $SO(Q)$. *

§ 5. Complete ring of fractions

5.1. Injective modules. Let R be a some ring with 1. A left R–module Q is called *injective* if any homomorphism of a submodule $N \subset M$, M an R–module, into Q extends to a homomorphism of M into Q. It is convenient to present this property in the form of diagram

DIAGRAMA

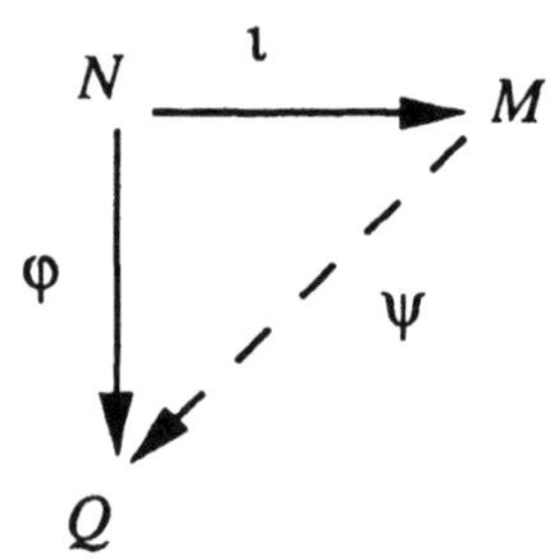

If $\imath$ is a monomorphism of R–modules, and ϕ is a homomorphism then there exists a homomorphism ψ such that the diagram is commutative, i.e. $\phi = \psi\imath$.

∘ A characteristic property of an injective module is that such a module differs by a direct summand from any module of which it is a submodule. ∘

Examples of injective modules over the ring **Z** of integers (they are also called the divisible or complete groups) are the group **Q** of rational numbers with respect to the operation of addition, the group $U(p^\infty)$ of all complex roots of 1 of power p^k, where p is a fixed prime number, $k = 1, 2, \ldots$, and those groups A where for any $a \in A$ and any natural n the equation $nx = a$ is solvable (or the equation $x^n = a$ depending on the fact if the group is written additively or multiplicatively). We shall prove that the *divisible group Q is injective*.

Suppose we have a homomorphism $\phi : N \to Q$ of a subgroup N of the group M into Q. According to the Zorn lemma there exists a maximal subgroup $\bar{N}$ to which ϕ can be extended. If $\bar{N} \neq M$ then we take $x \in M \backslash \bar{N}$ and consider the subgroup $\widetilde{N} = \langle x, \bar{N} \rangle$. If $\langle x \rangle \cap \bar{N} = \{0\}$ then $\widetilde{N} = \langle x \rangle \oplus \bar{N}$ and ϕ extends to $\widetilde{N}$, if we put $\phi(x) = 0$. Otherwise, there can be found an $x' \in \widetilde{N} \setminus \bar{N}$ and a prime number p such that $px' \in \bar{N}$. Let y be a solution in Q of the equation $py = \phi(px')$. Every element from $\langle x', \bar{N} \rangle$ can be written in a unique way in the form $ux' + n$, $n \in \bar{N}$, $u = 0, 1, \ldots, p-1$. We put $\phi(ux' + n) = uy + \phi(n)$. We thus obtain an extension of the homomorphism ϕ to the subgroup $\langle x', \bar{N} \rangle$ which is larger than $\bar{N}$. This is a contradiction. Concerning the divisibility of an injective group this appears if $\mathbf{Z}$ is taken as M and the subgroup $n\mathbf{Z}$ as N. Thus if $a \in Q$ then the mapping $n \mapsto a$ extends to a homomorphism $\phi : n\mathbf{Z} \to Q$. If $\bar{\phi} : \mathbf{Z} \to Q$ is an extension for ϕ, then $n\phi(1) = \phi(n) = a$ and thus $\phi(1)$ is a solution of the equation $nx = a$.

The argument just presented (divisibility = injectivity) can be generalized to the case of arbitrary rings with 1 and is called the *Baer criterion* of injectivity of modules.

Proposition. *Let R be some ring with identity. The left R–module Q is injective iff any R–homomorphism ϕ of an arbitrary left ideal $\mathcal{I}$ of the ring R into Q extends to a homomorphism of the R–module R in Q.*

Proof. It is clear that only the sufficiency of the condition needs to be proved. Let M be an arbitrary R–module, N a submodule, $\phi : N \to Q$ an arbitrary R–homomorphism. Let $\mathcal{S}$ be the set of pairs (S, τ), where S is a submodule in M, containing N, and τ a homomorphism of the R–module S in Q extending ϕ. Let us partly order $\mathcal{S}$, by setting $(S, \tau) \leq (S', \tau')$, if $S \subset S'$ and $\tau'|_S = \tau$. If $\mathcal{C}$ is a chain in $\mathcal{S}$ then (C, π), where $C = \bigcup_{S \in \mathcal{C}}$, $\pi|_S = \tau$, $(S, \tau) \in \mathcal{C}$, is a maximal element for the chain. According to the Zorn lemma there exists a maximal element $(\bar{N}, \bar{\phi})$ in $\mathcal{S}$. If $\bar{N} = M$ then $\bar{\phi}$ is the desired extension. We assume that $\bar{N} \neq M$ and select $x \in M \backslash \bar{N}$. If $Rx \cap \bar{N} = \{0\}$, then the homomorphism $\bar{\phi}$ is easily extended to $Rx \oplus \bar{N}$, by mapping x into 0. In the opposite case, $Rx \cap \bar{N} \neq \{0\}$. The set $\mathcal{I} = \{r | rx \in \bar{N}\}$, evidently, is a left ideal of the ring R different from R. We construct a mapping $\psi : \mathcal{I} \to Q$ by putting $\psi(r) = \bar{\phi}(rx)$. It is clear that this is a R–homomorphism. Let $\bar{\psi}$ be an extension of the homomorphism ψ to R, i.e. $\bar{\psi} : R \to Q$ and $\bar{\psi}|_{\mathcal{I}} = \psi$. We define$\widetilde{\ }$: $Rx + \bar{N} \to Q$ by setting $\widetilde{\phi}(rx + n) = \bar{\psi}(r) + \bar{\phi}(n)$. We verify that this is well defined. If $rx + n = r'x + n'$ then $(r - r')x = n' - n$, i.e. $r - r' \in \mathcal{I}$. In this case $\widetilde{\phi}((r - r')x) = \bar{\psi}(r - r') = \psi(r - r') = \bar{\phi}((r - r')x) = \bar{\phi}(n' - n) = \widetilde{\phi}(n' - n)$. It is clear that $\widetilde{\phi}$ is an R–homomorphism. Since $Rx + \bar{N}$ is strictly larger than

$\bar{N}$, we obtain a contradiction. Thus $\bar{N} = M$ and the desired extension always exists. □

5.2. Injective hull. A submodule N of a module M is called *essential* if it has a non–zero intersection with any non–zero submodule of the module M. An injective module Q is called the *injective hull* of the module M if M is isomorphic to an essential sumbmodule in Q. For instance, the group of rational numbers with respect to addition is the injective hull, or, as it is called in group theory, the *completion*, of the group $\mathbf{Z}$ of integers. In this subsection we shall prove the existence of an injective hull for any module. The first step is to prove the possibility of imbedding any module into an injective one.

To start with note that the abelian group $\mathbf{Q}/\mathbf{Z}$ is divisible as a quotient group of a divisible group. For the abelian group M we define the abelian group M^* as the group of homomorpisms $M^* = \mathrm{Hom}_{\mathbf{Z}}(M, \mathbf{Q}/\mathbf{Z})$. The group M^* is called the *group of characters* of the group M. We turn M^* into a right R–module by putting $(fr)m = f(rm)$, $r \in R$, $m \in M$. If $\phi : L \to M$ is an epimorphism of R–modules then we define $\phi^* : M^* \to L^*$ by the formula $\phi^*(f)(l) = f(\phi(l))$. The mapping ϕ^* is an injective homomorphism of R–modules. Indeed,

$$\phi^*(fr)(l) = (fr)(\phi(l) = f(r(\phi(l)) = f(\phi(rl)) = (\phi^*(f))(rl) =$$
$$=(\phi^*(f)r)(l), \quad \text{i.e. } \phi^*(fr) = \phi^*(f)r.$$

If $\phi^*(f) = 0$ then $f(\phi(l)) = 0$. Since ϕ is an epimorphism, $f(M) = \{0\}$, i.e. $f = 0$. We should also note that there exists a natural imbedding $\imath$ of the module M in M^{**}. It is given by the formula

$$(\imath(m))(f) = f(m).$$

Its injectivity means that for any $m \neq 0$ there exists a $\mathbf{Z}$–homomorphism $f : M \to \mathbf{Q}/\mathbf{Z}$ such that $f(m) \neq 0$. This follows from the fact that $\mathbf{Q}/\mathbf{Z}$ is a divisible group. Since any cyclic group, including also $\mathbf{Z}_m$, allows a non–zero homomorphism to a cyclic group of prime order p and $\frac{1}{p} + \mathbf{Z}$ is an element of order p in $\mathbf{Q}/\mathbf{Z}$ then the desired f is any homomorphism of the group M into $\mathbf{Q}/\mathbf{Z}$ extending the mapping $m \mapsto \frac{1}{p} + \mathbf{Z}$.

Before starting with the theorem on injective hulls we shall formulate and prove the key result.

Lemma. *Let F be a free right R–module, $F^* = Hom_{\mathbf{Z}}(F, \mathbf{Q}/\mathbf{Z})$ its module of characters. Then F^* is an injective left R–module.*

Proof. It is evident that a direct product of injective modules is injective. Since F is a direct sum of some number of copies of the free right cyclic R–module R, F^* is the Cartesian product of the same number of left R–modules, $R^* = \mathrm{Hom}_{\mathbf{Z}}(R, \mathbf{Q}/\mathbf{Z})$. It remains to prove that R^* is an injective left R–module. Let $\mathcal{I}$ be a left ideal in R, $\phi : \mathcal{I} \to R^*$. The $\mathbf{Z}$–homomorphism $\psi : \mathcal{I} \to \mathbf{Q}/\mathbf{Z}$ given by the formula

$$\psi(x) = \phi(x) \cdot 1,$$

in virtue of the divisibility of the group $\mathbf{Q}/\mathbf{Z}$, extends to a $\mathbf{Z}$–homomorphism $\bar{\psi} : R \to \mathbf{Q}/\mathbf{Z}$. We define $\phi : R \to F^*$ by

$$\bar{\phi}(r)(s) = \bar{\psi}(rs).$$

This is well defined as R is a free left R–ideal. Moreover, if $x \in \mathcal{I}$, $s \in R$ then

$$\bar{\phi}(xs) = \psi(xs) = \phi(xs)(1) = \phi(x)(s),$$

i.e. $\bar{\phi}$ is the desired extension of ϕ. □

Theorem. *Any (left) R–module M possesses an injective hull which is unique up to isomorphism.*

Proof. We present M^* in the form of a quotient module of a suitable free right R–module F. The module F^* is injective and M^{**} is naturally imbeded in F^* (see the beginning of this subsection) Since $M \subset M^{**}$, we finally obtain that M is imbedded in the injective module $F^* = \mathrm{Hom}_{\mathbf{Z}}(F, \mathbf{Q}/\mathbf{Z})$.

In order to form the injective hull we shall take an imbedding of the module M in an injective module Q and consider the set S of all submodules L of the module Q which contain M and in which M is an essential submodule. It is clear that the union of any chain of the set S again lies in S. Therefore according to the Zorn lemma there exists a maximal element $\mathcal{I}(M)$. It remains to show that $\mathcal{I}(M)$ is an injective module. We consider a maximal submodule M' of the module $\mathcal{I}$ which has a zero intersection with $\mathcal{I}(M)$. Then in virtue of the choice of M', $\mathcal{I}(M)+M'/M'$ is an essential submodule in $\mathcal{I}/M'$. Since

$$\mathcal{I}(M)+M'/M' \cong \mathcal{I}(M)/\mathcal{I}(M)\cap M' \cong \mathcal{I}(M),$$

we see that either $\mathcal{I}(M)+M' = \mathcal{I}$ or $P = \mathcal{I}/M'$ is an R–module where $\mathcal{I}(M)$ is a proper essential submodule.

Applying the property of injectivity of the module $\mathcal{I}$ to its submodule $\mathcal{I}(M)$ which also lies in P, we see that the identity mapping of the module $\mathcal{I}(M)$ into $\mathcal{I}$ extends to a homomorphism $\phi : P \to \mathcal{I}$. This homomorphism is injective, because if $\operatorname{Ker}\phi$ is a non–zero submodule in P then $\mathcal{I}(M)\cap\operatorname{Ker}\phi$ is non–zero, i.e. some elements from $\mathcal{I}(M)$ are mapped to 0, which is impossible. Thus $\mathcal{I} = \mathcal{I}(M)+M' = \mathcal{I}(M)\oplus M'$ is an injective module, i.e. also an injective hull. If $\mathcal{I}'$ is one more injective hull for M then the imbedding $\phi : M \to \mathcal{I}'$, as we have seen above, extends to a monomorphism $\bar{\phi} : \mathcal{I}(M) \to \mathcal{I}'$. Moreover, the image $\phi(\mathcal{I}(M)) = \mathcal{I}''$ will be a direct summand $\mathcal{I}'$, $\mathcal{I}' = \mathcal{I}'' \oplus T$. We see that $T \cap \phi(M) = \{0\}$. Since $\phi(M)$ is essential in $\mathcal{I}'$, $T = \{0\}$ and $\mathcal{I}' = \bar{\phi}(\mathcal{I}(M))$, i.e. $\mathcal{I} \cong \mathcal{I}'$. □

5.3. Complete ring of quotients. Let $\mathcal{I} = \mathcal{I}(R)$ be the injective hull of the left R–module R. We denote by $H = H(R) = \operatorname{Hom}_R(\mathcal{I},\mathcal{I})$ the ring of endomorphisms of the injective hull. The action of an element $h \in H$ on $x \in \mathcal{I}$ is denoted by xh. Then $\mathcal{I}$ is a right H–module. The centralizer of this H–module is denoted by $Q = Q(R) = \operatorname{Hom}_H(\mathcal{I},\mathcal{I})$. The ring Q is called the *complete left ring of fractions* of the ring R.

Note that R is imbedded in Q by assigning to the element $r \in R$ the mapping $x \mapsto rx$ $(x \in \mathcal{I})$. For any $x \in H$ the mapping $r \mapsto rx$ extends to the homomorphism $h \in \operatorname{Hom}_R(\mathcal{I},\mathcal{I})$. We have $1 \cdot x = x$, hence $1 \cdot H = \mathcal{I}$. Thus the mapping $h \mapsto 1 \cdot h$ is an epimorphism of the right H–module H onto the right H–module $\mathcal{I}$. Finally, the mapping $q \mapsto q(1)$ from Q into $\mathcal{I}$ is a monomorphism, as from $q(1) = 0$ it follows that $q(\mathcal{I}) = q(1 \cdot H) = q(1)H = 0$, i.e. $q = 0$.

In order to obtain a constructive definition of the complete ring of fractions we introduce the concept of a *dense submodule* D of the left H–module Q as a submodule D such that $Dh = \{0\}$ implies $Rh = \{0\}$, where $H = H(R)$. A dense left ideal of the ring F can also be considered. An intersection of two dense ideals is dense, an ideal containing a dense ideal is itself dense. A construction of the ring Q is as follows.

Theorem. *Let D run through the set of all dense ideals of the ring R. We* *consider the set of classes of equivalent homomorphisms from D into R*

$$\bigcup_D \mathrm{Hom}(D,R)/\sim,$$

where the equivalence $\sim$ is such that $f_1 \sim f_2$, if $f_1|_{D_1\cap D_2} = f_2|_{D_1\cap D_2}$. The operations $f_1+f_2 \in Hom_R(D_1\cap D_2), R)$, $f_1, f_2 \in Hom(f_2^{-1}D_1, R)$, are defined by

$$(f_1+f_2)(d) = f_1(d) + f_2(d), \quad (f_1 f_2)(d) = f_1(f_2(d)),$$

and preserve the relation of equivalence. We thus obtain a ring which is isomorphic to $Q(R)$.

Lemma 1. *If D is a dense left ideal of the ring R, then $Dq^{-1} = \{r|rq \in D\}$ for any $q \in Q$, is also a dense ideal.*

Proof. Let $h \in H$ and $(Dq^{-1})h = \{0\}$. Consider the mapping $\phi : D+Rq \to R$ defined by the equality $\phi(d+rq) = rh$. (If $d+rq = 0$ then $rq \in Q$ and, consequently, $rh = 0$. Thus ϕ is a single–valued mapping.) Extend ϕ to a mapping $h' : {}_R\mathcal{I} \to {}_R\mathcal{I}$. Then $Dh' = \{0\}$ and, consequently, $Rh' = \{0\}$. Thus $1 \cdot h = \phi(q) = qh' = q(1\cdot h') = 0$. So $1 \cdot h = \phi(q) = qh' = q(1 \cdot h') = 0$. Finally, $Rh = R(1\cdot h) = \{0\}$. □

Let further, D and G be submodules in ${}_RQ$. We denote by $(G : D)$ the set of elements $q \in Q$ such that $Dq \subset G$.

Proposition 1. *The canonical image of ${}_RQ$ in ${}_R\mathcal{I}$ consists precisely of the elements of the injective hull ${}_R\mathcal{I}$ which are annihilated by all elements from H annihilating R.*

Proof. Let $h \in H$, $Rh = \{0\}$. Then $Q(1)h = Q(1h) = \{0\}$. Thus $Q(1)$ is contained in the indicated submodule of the injective hull ${}_R\mathcal{I}$. Inversily, assume that $x \in \mathcal{I}$ possesses the property, then from $Rh = \{0\}$ follows $xh = 0$, for any $h \in H$. We can find a $q \in Q$ such that $x = q(1)$.

As we have already seen, any element from $\mathcal{I}$ is equal to $1 \cdot h$ for some $h \in H$. We consider the mapping $1 \cdot h \mapsto xh$ of the module $\mathcal{I}_H$ onto itself. (In order to convince yourself about the well definedness of this mapping assume that $1 \cdot h' = 1 \cdot h$. Then $R(h - h') = \{0\}$. Consequently, in virtue of the proposition, $x(h - h') = 0$.) Thus $q(1 \cdot h) = xh$ for some $q \in \mathrm{Hom}_H(\mathcal{I}, \mathcal{I})$. Taking that $h = 1$, we obtain the required $q(1) = x$. □

Lemma 2. *Let D be a dense submodule in ${}_RQ$. Then the modules $Hom_R(D, G)$ and $(G : D)$ are canonically isomorphic.*

Proof. A canonical isomorphism from $(G : D)$ into $\mathrm{Hom}_R(D, G)$ is, clearly, given by assigning to the element $q \in (G : D)$ the homomorphism $d \mapsto dq$ of the module D in G. It is evident that this canonical isomorphism is a monomorphism, since from the equality $Dq = \{0\}$ follows $q = 0$. (We recall that $q(1) = 1 \cdot h$ for a suitable $h \in H$ and that D is a dense submodule.) If for every $f \in \mathrm{Hom}_R(D, G)$ it is possible to find a $q \in Q$ such that $df = dq$ for all $d \in D$, then it follows that the canonical homomorphism is an isomorphism.

In order to find such a q we extend f to $h : \mathcal{I} \to \mathcal{I}$ in the sense that $d(1h) = dg(1)$ for any $d \in D$. Now consider $h' \in H$ such that $Rh' = \{0\}$. Then by proposition 1, $D(1hh') \in Q(1h') = \{0\}$. Because D is a dense submodule we have $Q(1hh') = \{0\}$. Consequently, in virtue of lemma 1, $Q(1h) \subset Q(1)$. In particular, $1h = q(1)$ for some $q \in Q$. Thus $df(1) = d(1h) = dq(1)$ and, consequently, $df = dq$ for all $d \in D$, as required. □

Proof of the theorem. If $q \in Q$ then let $D = Rq^{-1} = \{d | dq \in R\}$. In virtue of lemma 1, the set D is a dense left ideal; moreover, $q \in (R : D)$.

Therefore $Q = \bigcup_D (R : D)$. From lemma 2 follows the existence of a canonical homomorphism $(R : D) \cong \mathrm{Hom}_R(D, R)$ under which the element $q \in (R : D)$ corresponds to the homomorphism $f \in \mathrm{Hom}_R(D, R)$ such that $df = dq$, for all $d \in D$.

Let $q_i \in (R : D)$, $i = 1,2$, correspond to the homomorphisms $f_i \in \mathrm{Hom}_R(D_i, R)$. We shall show that $q_1 = q_2$ iff $f_1 \sim f_2$.

Indeed, let $df_1 = df_2$ under all $d \in D_1 \cap D_2$. Then $dq_1 = dq_2$. Therefore $(D_1 \cap D_2)(q_1 - q_2) = \{0\}$. Since $D_1 \cap D_2$ is a dense right ideal and $(q_1 - q_2)(1) = 1 \cdot h$ for a suitable $h \in H$, we have $(q_1 - q_2)(1) = 0$, i.e. $q_1 = q_2$. The opposite statement is evident. □

Proposition 2. *If D is a left ideal of the ring R, then D is a dense ideal iff for any $r_1, r_2 \in R$, $r_1 \neq 0$ there can be found an $r \in R$ such that $rr_1 \neq 0$ and $rr_2 \in D$.*

Proof. Let us assume that the condition stated holds. Let $Dh = \{0\}$ and $r_1 \in Rh \cap R$. Then $r_1 = r_2 h$, where $r_2 \in R$. If $r_1 \neq 0$ then it is possible to select an $r \in R$ such that $rr_1 \neq 0$ and $rr_2 \in D$. But $rr_1 = rr_2 h \in Dh = \{0\}$, which is a contradiction. Thus D is a dense ideal.

Inversily, let D be a dense left ideal in R. Then, in virtue of lemma 1, Dr_2^{-1} for any $r_2 \in R$ is also a dense right ideal. The mapping $r \mapsto rr_1$ can be extended to a mapping $h : {}_R\mathcal{I} \to {}_R\mathcal{I}$. Consequently, from the equality $(Dr^{-1})r_1 = \{0\}$ follows that $r_1 = 0$. Hence it is shown that our condition is satisfied. □

Corollary 1. *Two dense submodules D and D' of the module ${}_RQ$ are isomorphic iff $Dq = D'$ for some invertible $q \in Q$.* □

Corollary 2. *If D is a dense right ideal of the ring R and $f \in Hom_R(D, R)$ then there exists an element $q \in Q$ such that $df = dq$ for any $d \in D$* □

Corollary 3. *Let us assume that the ring R contains a smallest dense right ideal D_0. Then $Q \cong Hom_R(D_0, D_0)$.*

Proof. We have that $D_0 q^{-1}$ is a dense left ideal for any $q \in Q$. Therefore, $D_0 \subset D_0 q^{-1}$ and $D_0 q \subset D_0$. Thus $Q = (D_0 : D_0) \cong \mathrm{Hom}_R(D_0, D_0)$. □

Note. If R is left artinian then the conditions are satisfied. Besides, from the proof it follows that D_0 is a two–sided ideal. Thus if R is a finite–dimensional simple algebra, then $Q(R) \cong \mathrm{Hom}_R(R, R)$.

5.4. Classical ring of fractions. Let R be a ring satisfying the left Ore condition, i.e. for any $a, b \in R$, where b is a non–zero divisor, there are $a', b' \in R$, where b' is a non–zero divisor, such that $b'a = a'b$. Then R admits a classical ring of fractions.

We shall consider the imbedding $R \subset Q(R)$. Let b be a non–zero divisor in K.

We state that Rb is a dense left ideal. Indeed, let $0 \neq r_1 \in R$ and $r_2 \in R$. In virtue of the Ore condition for $r_2 = a$, there are $a', b' \in R$, where b' is a non–zero divisor, such that $b'r_2 = a'b$ and therefore, $b'r_2 \in Rb$. Since b' is a non–zero divisor, we have $b'r_1 \neq 0$. Thus in virtue of proposition 2 from Subsection 5.3, Rb is a dense left ideal.

Now define $f \subset \mathrm{Hom}_R(Rb, R)$ by putting $rbf = r$. Note that $rb = r'b$ gives $r = r'$. Therefore, in virtue of corollary 2, $bq = bf = 1$ for some $q \in Q$. Thus $Rb(qb - 1) = \{0\}$ and Rb is a dense left ideal. Hence $qb = 1$. (Recall that a non–zero element of the ring Q can not annihilate a dense left ideal of the ring R.) Hence b is an invertible element of the ring Q.

Let the set $Q_{cl} = Q_{cl}(R)$ consist of all elements of the ring $Q = Q(R)$ of the form $b^{-1}a$, where $a, b \in R$, and b is a non–zero divisor. It is easy to see that Q_{cl} is a subring of Q. Indeed,

$$b^{-1}a + d^{-1}c = (d'b)^{-1}(d'a - b'c),$$

where $d'b = d^{-1}c$ and d' is a non–zero divisor, and

$$(b^{-1}a)(d^{-1}c) = (d'b)^{-1}(a'c),$$

where $d'a = a'd$, d' is a non–zero divisor.

It is clear from what was said that Q_{cl} is the classical ring of fractions.

References.

Andrukakievich–Ryabukhin [1], Bokut'–L'vov–Kharchenko [1], Bourbaki [1], Van der Waerden [1], Dixmier [1], Lambek [1], Herstein [1], Chatters [1].

CHAPTER IV

LIE ALGEBRAS

§ 1. Linear Lie algebras

1.1. Definition. Many important problems that are connected with linear operators are solved with a help of transforming the matrix of an operator into some special form. From a course of linear algebra and geometry it is clear that the matrix of any operator over an algebraically closed field can be transformed into triangular form over a suitable basis. This result can not be used for a case of two operators, i.e. in general, there is no basis in which the matrices of two linear operators simultaneously take triangular form. On the other hand, in many cases such a transformation is possible. For instance, this happens in the case of an arbitrary set S of pairwise commuting operators, i.e. for a set such that $\mathcal{AB} - \mathcal{BA} = 0$ for any $\mathcal{A}$, $\mathcal{B} \in S$. The generalization of this result brings a new operation, the commutator of linear operators, and, further, yields the concept of a (linear) Lie algebra.

Let V be a vector space over a field F, $E = \operatorname{End}(V)$ the algebra of linear operators on V. We define the *commutator* of the two operators $\mathcal{A}$, $\mathcal{B} \in E$ by

$$(1) \qquad [\mathcal{A}, \mathcal{B}] = \mathcal{AB} - \mathcal{BA}$$

It is not difficult to check that

$$(2) \qquad [\mathcal{A}, \mathcal{A}] = 0$$

and $[\mathcal{A}, \mathcal{B}] = -[\mathcal{B}, \mathcal{A}]$. The latter property is called *anti–commutativity*. A more complex relation, familiar from elementary analytic geometry, is the so–called *Jacobi identity*

$$(3) \qquad [[\mathcal{A}, \mathcal{B}], \mathcal{C}] + [[\mathcal{B}, \mathcal{C}], \mathcal{A}] + [[\mathcal{C}, \mathcal{A}], \mathcal{B}] = 0.$$

In order to prove this we write

$$[[\mathcal{A},\mathcal{B}],\mathcal{C}] = [\mathcal{AB} - \mathcal{BA},\mathcal{C}] = \mathcal{ABC} - \mathcal{BAC} - \mathcal{CAB} + \mathcal{CBA}$$

and, permuting the letters cyclically,

$$[[\mathcal{B},\mathcal{C}],\mathcal{A}] = \mathcal{BCA} - \mathcal{CBA} - \mathcal{ABC} + \mathcal{ACB},$$
$$[[\mathcal{C},\mathcal{A}],\mathcal{B}] = \mathcal{CAB} - \mathcal{ACB} - \mathcal{BCA} + \mathcal{BAC}.$$

By adding these three equalities term by term we obtain what is required.

It is interesting to note that all identities which are true for commutators of arbitrary operators in arbitrary spaces are the consequences of the two written down above: (2) and (3).

A set L of linear operators from E closed with respect to the operation of taking commutators is called a *(linear) Lie algebra.* In particular the algebra E itself is a linear Lie algebra; in view of its close relation with the complete linear group $GL(V)$ it is usulally denoted by $\mathfrak{gl}(V)$. If $\{e_1,\dots,e_n\}$ is some basis of the finite–dimensional vector space V, then the set T of all operators for which the matrix in this basis is (upper) triangular, is a Lie algebra and is a subalgebra in $\mathfrak{gl}(V)$. Since the commutator $[A, B]$ of two triangular matrices is a matrix with a zero diagonal and the product of degree n (in particular, an n–fold commutator) of triangular matrices with a zero diagonal is equal to zero, we obtain a necessary condition of simultaneous reducibility of a set S of matrices to triangular form: for any $\mathcal{A}_1, \mathcal{A}_2, \dots, \mathcal{A}_{2n} \in S$ the following is satisfied

$$(4) \qquad [\dots[[[\mathcal{A}_1,\mathcal{A}_2],[\mathcal{A}_3,\mathcal{A}_4]],[\mathcal{A}_5,\mathcal{A}_6]],\dots,[\mathcal{A}_{2n-1},\mathcal{A}_{2n}]] = 0$$

This remark shows the importance of the concept of a commutator in the problem of reduction to triangular form. Our main purpose in this paragraph is to prove Lie's theorem which implies, in particular, the sufficiency of this necessary condition in the case of an algebraically closed field.

1.2. Abstract Lie algebras. As often happens, in order to solve the problem it is useful to generalize somewhat the concepts introduced.

We call an arbitrary algebra L a *Lie algebra* over the field F if it has the operations of addition, multiplication by scalars from the base field, and an additional bilinear operation denoted by commutation brackets [,] satisfying the identities

$$(2') \qquad [x, x] = 0,$$

$$(3') \qquad [[x, y], z] + [[y, z], x] + [[z, x], y] = 0$$

Linear Lie algebras are examples of Lie algebras. Another example is the set $\mathbf{R}^3$ of three–dimensional vectors with the usual vector operations of addition, multiplication by scalars, and the vector product. A third example is the Lie algebra W_1 of vector fields on the line. Here the commutator is given by the formula

$$[f\frac{d}{dx}, g\frac{d}{dx}] = (fg' - f'g)\frac{d}{dx}.$$

The concepts of a subalgebra, a quotient algebra are introduced as usual and there are the usual theorems on homomorphisms and isomorphisms. More or less equally standard are the concepts of a *representation* of a Lie algebra and an L–module, where L is a Lie algebra. Thus a representation of a Lie algebra L by linear operators on a vector space V is an arbitrary homomorphism of Lie algebras $\rho : L \to gl(V)$. Thus for any $x, y \in L$ the following equality should be satisfied

$$(5) \qquad \rho([x, y]) = [\rho(x), \rho(y)] = \rho(x)\rho(y) - \rho(y)\rho(x).$$

The space V of the representation ρ is transformed into an *L–module* by writing $xv = \rho(x)(v)$, where $x \in L$, $v \in V$; for any $x, y \in L$, $v \in V$ the equality implied by (5) is satisfied

$$(6) \qquad [x, y]v = x(yv) - y(xv),$$

Inversily, the equality (6) can be used as the definition of an L–module. Then the mapping ρ assigning to an element $x \in L$ the linear operator $\rho(x)$ such that $\rho(x)(v) = xv$, is a representation of the algebra L in the vector space V.

An important example of representation is the adjoint representation ad : $L \to \mathrm{End}(L)$ by linear operators which act on the algebra L itself. If $x \in L$, $v \in L$ then we define

$$(7) \qquad (\text{ad}\, x)(v) = [x, v].$$

Using (2′) and (3′) we verify that equality (5) holds:

$$(\text{ad}[x, y](y) = [[x, y], v] = -[[y, v], x] - [[v, x], y] =$$
$$=[x, [y, v]] - [y, [x, v]] = (\text{ad } x \circ \text{ad } y - \text{ad } y \circ \text{ad } x)(v).$$

The linear operator $\mathcal{A} = \text{ad } x$ has an additional property similar to the rule of differentiation of a product of functions

$$(8) \qquad \mathcal{A}([u, v]) = [\mathcal{A}(u), v] + [u, \mathcal{A}(v)].$$

This is checked in the same way as above by using the basic identities (2′) and (3′):

$$(\text{ad } x)([u, v)] = [x, [u, v]] = -[[u, v], x] =$$
$$= [[v, x], u] + [[x, u], v] = [[x, u], v] + [u, [x, v]] =$$
$$= [(\text{ad } x(u), v] + [u, (\text{ad } x)(v)].$$

A linear operator $\mathcal{A}$ with the property (8) is called a *derivation.*

It should be noted that the concept of a derivation is defined also in the case of an arbitrary (non–associative) algebra over field. The set of all derivations $\text{Der}\,(A)$ of an algebra A is called the algebra of derivations of the algebra A. This set is a Lie algebra with respect to the commutation operation (6).

1.3. Engel's theorem. The first (and main) step of transforming a set of operators into triangular form is the problem of finding a simultaneous eigenvector for all operators of the given set. In this subsection we shall solve the problem presented for the case of nilpotent operators.

Engel's theorem. *Let V be a (non–zero) vector space over a field F, $L \subset \mathfrak{gl}(V)$ a finite dimensional linear Lie algebra consisting of nilpotent operators. Then there exists a non–zero vector $v \in V$ such that $\mathcal{A}(v) = 0$ for any $\mathcal{A} \in L$.*

In order to prove this with a help of induction on the dimension of the algebra it is convenient to paraphrase the theorem in the following way:

Theorem. *Let $\rho : L \to End(V)$ be a non–zero representation of a finite–dimensional algebra L such that for any $x \in L$ the operator $\rho(x)$ is nilpotent. Then there exsists the vector $v \neq 0$ such that $\rho(x)(v) = 0$ for any $x \in L$.*

It should be noted that the algebra L in the statement of the original theorem has a built–in "identity" representation by linear operators on the space V.

We now show that the adjoint representation of the algebra satisfies the conditions of the theorem. Note that the operator ad $\mathcal{A}$ is representable in the form ad $\mathcal{A} = l_{\mathcal{A}} - r_{\mathcal{A}}$, where $l_{\mathcal{A}}(\mathcal{B}) = \mathcal{AB}$ and $r_{\mathcal{A}}(\mathcal{B}) = \mathcal{BA}$ (here the result of action of the operators $l_{\mathcal{A}}$ and $r_{\mathcal{A}}$ can take us out of the algebra L, but this is inessential). The operators $l_{\mathcal{A}}$ and $r_{\mathcal{A}}$, clearly, commute. Therefore in order to determine the degree of the operator $l_{\mathcal{A}} - r_{\mathcal{A}}$, the Newton binomial formula can be applied. If n is such that $\mathcal{A}^n = 0$ then

$$(\text{ad } \mathcal{A})^{2n-1} = \sum (-1)^s \binom{2n-1-s}{s} (l_{\mathcal{A}})^{2n-1-s} (r_{\mathcal{A}})^s =$$

$$= \sum (-1)^s \binom{2n-1-s}{s} l_{\mathcal{A}^{2n-1-s}} r_{\mathcal{A}^s}. \tag{9}$$

When $s < n$ we have $2n - s - 1 \geq n$ and $\mathcal{A}^{2n-s-1} = 0$, when $s \geq n$ then we have $\mathcal{A}^s = 0$. This demonstrates that every member of the sum on the right in (9) is equal to zero. Thus also $(\text{ad } \mathcal{A})^{2n-1} = 0$.

Lemma. *Let H be some subalgebra in L different from L. Then there is a subalgebra K, $\dim K = \dim H + 1$ in which H is contained as an ideal.*

Proof. Within the framework of the general induction assumption with respect to the dimension of the algebra L, we can assume that for any representation of the algebra H by nilpotent operators on some space there exists a non–zero vector annihilated by all operators of the representation . As such a representation for H we consider the homomorphism $\pi : H \to \text{End}(L/H)$ given by the formula ($\mathcal{A} \in H$, $\mathcal{B} \in L$):

$$\pi(\mathcal{A})(\mathcal{B} + H) = (\text{ad } \mathcal{A})(\mathcal{B}) + H = [\mathcal{A}, \mathcal{B}] + H. \tag{10}$$

This is well defined because H is a subalgebra, i.e. $[\mathcal{A},\mathcal{B}] \in H$, if $\mathcal{B} \in H$. Condition (5) follows from the corresponding property proved above for the case of adjoint representation. Now let $\mathcal{C} + H \in L/H$ be a non–zero vector annihilated by all operators $\pi(\mathcal{A})$, where $\mathcal{A} \in H$. Then $\mathcal{C} \notin H$, however $[\mathcal{A},\mathcal{C}] + H = 0$ for any $\mathcal{A} \in H$, i.e. $[\mathcal{A},\mathcal{C}] \in H$. We shall take as K the linear span $K = \langle \mathcal{C}, H \rangle$. Any element from K has the form $\lambda\mathcal{C} + \mathcal{A}$, where $\lambda \in F$, $\mathcal{A} \in H$. Taking two elements of this kind, $\lambda_1\mathcal{C} + \mathcal{A}_1$ and $\lambda_2\mathcal{C} + \mathcal{A}_2$, we obtain

$$[\lambda_1\mathcal{C} + \mathcal{A}_1, \lambda_2\mathcal{C} + \mathcal{A}_2] = \lambda_1\lambda_2[\mathcal{C},\mathcal{C}] + \lambda_1[\mathcal{C},\mathcal{A}_2] - \lambda_2[\mathcal{C},\mathcal{A}_1] + [\mathcal{A}_1,\mathcal{A}_2] \in H.$$

Therefore K is a subalgebra. The same argument shows that H is an ideal in K. The lemma is proved. □

Now we can immediately pass to the proof of Engel's theorem.

Proof. The initial step of the induction on $\dim L$ is provided by $\dim L = 0$. Select in L a proper algebra H of maximal dimension. According to the lemma we can take $\dim H = \dim L - 1$ and H is an ideal in L. Let $\mathcal{C}$ be an element from L that is not in H. Then $L = \langle \mathcal{C} \rangle \oplus H$. According to the induction assumption there is a $v \neq 0$ in V such that $\mathcal{A}(v) = 0$ for any $\mathcal{A} \in H$. We denote by V_0 the subspace in V consisting of all vectors annihilated by the operators from H. Hence $V_0 \neq \{0\}$. The subspace V_0 is invariant under $\mathcal{C}$. Indeed, if $v \in V_0$, $\mathcal{A} \in H$ then (see (1))

$$\mathcal{A}(\mathcal{C}(v)) = [\mathcal{A},\mathcal{C}](v) - \mathcal{C}(\mathcal{A}(v)).$$

Since $[\mathcal{A},\mathcal{C}] \in H$, both members on the right side are equal to zero and $\mathcal{C}(v) \in V_0$.

Consider the restriction of the operator $\mathcal{C}$ to V_0. By assumption there exists a natural m such that $\mathcal{C}^m = 0$ (and $\mathcal{C}^{m-1} \neq 0$). Let $v \in V_0$ be such that $v_0 = \mathcal{C}^{m-1}(v) \neq 0$. Then $\mathcal{C}(v_0) = 0$. Since $L = \langle \mathcal{C} \rangle \oplus H$ for any $\mathcal{A} \in L$ we have $\mathcal{A}(v_0) = 0$, i.e. v_0 is the desired vector. The theorem is proved. □

1.4. Reduction to triangular form

Corollary 1. *Let L be a Lie algebra of linear operators on a finite–dimensional vector space V and such that any operator $\mathcal{A} \in L$ is nilpotent. Then in some* *basis the matrices of all operators from L have a triangular form (with zeros on the main diagonal).*

Proof. We shall carry out an induction on $\dim V$. The starting point is clear ($\dim V = 1$). Now select a $0 \neq e_1 \in V$ for which $\mathcal{A}(e_1) = 0$ for any $\mathcal{A} \in L$ (Engel's theorem). Consider the quotient space $\bar{V} = V/\langle e_1 \rangle$. Its dimension is equal to $n-1$. Any operator $\mathcal{A} \in L$ induces an operator $\bar{\mathcal{A}} : \bar{V} \to \bar{V}$ which acts according to the formula

$$\bar{\mathcal{A}}(\bar{v}) = \bar{\mathcal{A}}(v + \langle e_1 \rangle) = \mathcal{A}(v) + \langle e_1 \rangle$$

(check that this is well defined). It is clear that $\bar{L} = \{\bar{\mathcal{A}} | \mathcal{A} \in L\}$ is a Lie algebra and that it consists of nilpotent operators. Using the induction assumption take a basis $\bar{e}_2 = e_2 + \langle e_1 \rangle, \dots, \bar{e}_n = e_n + \langle e_1 \rangle$, in which all $\bar{\mathcal{A}}$ are triangular. Then $\{e_1, e_2, \dots, e_n\}$ is the desired basis in V. Indeed, if $\mathcal{A} \in L$, $1 < i \leq n$, then

$$\mathcal{A}(e_i) + \langle e_1 \rangle = \bar{\mathcal{A}}(\bar{e}_i) =$$
$$= \lambda_{i-1}\bar{e}_{i-1} + \ldots + \lambda_2 \bar{e}_2 = \lambda_{i-1} e_{i-1} + \ldots + \lambda_2 e_2 + \langle e_1 \rangle.$$

Thus,

$$\mathcal{A}(e_i) = \lambda_{i-1} e_{i-1} + \ldots + \lambda_2 e_2 + \lambda_1 e_1,$$

proving the triangularity of the matrix for $\mathcal{A}$ in the selected basis. □

1.5. Nilpotent Lie algebras. Let L be an arbitrary Lie algebra, M, N ideals in it. We shall define their *commutator*, $[M, N]$, as a linear hull of all commutators of the form $[x, y]$, $x \in M$, $y \in N$.

From the equality (8) for operators of the form $\operatorname{ad} x$, $x \in L$, it follows that $[M, N]$ is an ideal in L. When $M = N = L$ we arrive at the notion of the commutator $L^2 = [L, L]$ of a Lie algebra. A Lie algebra with zero commutator, or in other words, a Lie algebra for which $[x, y] = 0$ for any $x, y \in L$, is called *abelian*. If the sequence of successive commutators, $L \supset L^2 = [L, L] \supset [L^2, L^2] \supset \ldots$, of the algebra L reaches zero in a finite number of steps then L is called a *solvable* (*soluble*) algebra. We also consider the sequence

$$L = L^1 \supset L^2 = [L, L^1] \supset L^3 = [L, L^2] \supset \ldots \tag{11}$$

A Lie algebra such that the sequence (11) reaches zero in a finite number of steps is called *nilpotent*. Simple reasoning shows that $L^n = \{0\}$ iff any n–step commutator of elements from L is equal to zero.

Corollary 2. *Let L be an abstract finite–dimensional Lie algebra with a nilpotent adjoint representation. Then L is a nilpotent Lie algebra.*

Proof. Let $n = \dim L$. Any element from L^{n+1} is a linear combination of *right–normal* commutators, i.e. commutators of the form

$$[x_1, [x_2, \dots, [x_n, x_{n+1}] \dots] = (\text{ad } x_1 \circ \text{ad } x_2 \dots \text{ad } x_n)(x_{n+1}).$$

For suitable basis of the space L the operators ad $x_1, \dots,$ ad x_n are written in the form of triangular matrices of order n with zeros on the main diagonal. The product of these n matrices, evidently, is equal to zero. Thus $L^{n+1} = \{0\}$ and the algebra L is nilpotent. □

1.6. Lie's theorem. Any nilpotent Lie algebra is, in particular, solvable. If we denote by $L^{(s)} = [L^{(s-1)}, L^{(s-1)}]$ the iterated commutator subalgebra of the algebra L of level s ($L = L^{(0)}$), then from formula (11) it is evident that $L^{(s)} \subset L^{s+1}$. Thus if $L^n = \{0\}$ then also $L^{(n-1)} = \{0\}$. Note that the identity (4) from Subsection 1.1 implies the solvability of any Lie algebra in which this identity is fulfilled. Indeed, according to this identity, an n–step commutator of elements from $[L, L]$ is equal to zero, i.e. $[L, L]$ is nilpotent and so it is solvable. It is clear that then L is also a solvable Lie algebra.

Reducibility to the triangular form will be proved for solvable linear Lie algebras.

Lie's theorem. *Let F be an algebraically closed field of characteristic* 0, *L a solvable Lie algebra of linear operators on a finite–dimensional vector space V. Then there is a non–zero vector $v \in V$ which is an eigenvector for all operators from L.*

Proof. Let $L^2 = [L, L]$ be the commutator of the algebra L. Any subspace M in L containing L^2, is an ideal. This remark permits to select an ideal M in L such that $\dim L = \dim M + 1$. Using induction on $\dim L$ with an evident starting point we can assume that V contains a non–zero vector v which is an eigenvector for all operators from M. Let $L = \langle \mathcal{A} \rangle \oplus M$. We consider the cyclic subspace W with respect to the operator $\mathcal{A}$ generated by the vector v. It has the form

$$W = \langle v, \mathcal{A}(v), \mathcal{A}^2(v), \dots, \mathcal{A}^s(v)\rangle$$

for suitable $s = \{1, 2, \dots\}$. This subspace is invariant with respect to the operators from M. Or, to be more precise, if $W_t = \langle v, \mathcal{A}(v), \dots, \mathcal{A}^t(v)\rangle$ and $\mathcal{B} \in M$, then $\mathcal{B}W_t \subset W_t$. This statement can be proved with a help of induction on t. When $t = 0$ we have $\mathcal{B}(v) = \lambda(\mathcal{B})v$, i.e. v is an eigen–vector for the operator $\mathcal{B} \in M$ with eigen–values $\lambda(\mathcal{B})$. We now compute $\mathcal{B}(\mathcal{A}^t(v))$. We have

$$\mathcal{B}(\mathcal{A}^t(v)) = \mathcal{B}\mathcal{A}(\mathcal{A}^{t-1}(v)) = [\mathcal{B}, \mathcal{A}](\mathcal{A}^{t-1}(v)) + \mathcal{A}\mathcal{B}(\mathcal{A}^{t-1}(v)). \tag{12}$$

Since $[\mathcal{B}, \mathcal{A}] \in M$ and $\mathcal{A}W_{t-1} \subset W_t$ we obtain the required. In fact, because of relation (12), it is possible to state even more, viz.

$$\mathcal{B}(\mathcal{A}^t(v)) + W^{t-1} = \lambda(\mathcal{B})\mathcal{A}^t(v) + W^{t-1} \tag{13}$$

If $W \neq V$ then using induction on $\dim V$ we find an eigenvector for all operators from L which is also in W. Thus, assume that $V = W$. If $n = \dim V$ then according to (13), the trace, $\operatorname{tr} \mathcal{B}$, of any operator $\mathcal{B} \in M$ is equal to $n\lambda(\mathcal{B})$. If $\mathcal{B} \in L^2 \subset M$ then $\mathcal{B}$ is a linear combination of commutators of linear operators. It is well–known that the trace of a commutator of two matrices is zero. Then $n\lambda(\mathcal{B}) = 0$, i.e. $\lambda(\mathcal{B}) = 0$. We consider now the subspace V_0 in V consisting of vectors annihilated by operators from L^2. According to what was said, $V_0 \neq \{0\}$. This subspace is L–invariant. Indeed, if $\mathcal{B} \in L^2$, $\mathcal{C} \in L$, $w \in L_0$ then $[\mathcal{B}, \mathcal{C}] \in L^2$ and

$$\mathcal{B}(\mathcal{C}(v)) = [\mathcal{B}, \mathcal{C}](w) + \mathcal{C}(\mathcal{B}(w)) = 0.$$

If $V_0 \neq V$ then we can apply induction on $\dim V$. Otherwise for any $w \in V$ and any $\mathcal{B}, \mathcal{C} \in L$

$$(\mathcal{B}\mathcal{C})(w) = [\mathcal{B}, \mathcal{C}](w) + (\mathcal{C}\mathcal{B})(w) = (\mathcal{C}\mathcal{B})(w),$$

i.e. any two operators from L commute. The fact that an eigenvector exists for a set of commuting operators on a finite dimensional vector space over an algebraically closed field is well known. The theorem is proved. □

1.7. More about reduction to triangular form and nilpotency

Corollary 1. *Under the conditions of the theorem there exists a basis in the space V in which all matrices of operators from L are triangular.*

The proof of this corollary is similar to the one of corollary 1 from Subsection 1.5. □

Corollary 2. *Let F be a field of characteristic zero, L a finite–dimensional abstract solvable Lie algebra over F. Then its commutator is nilpotent.*

Proof. Let $\bar{F}$ be an algebraic closure of the field F.

We consider $\bar{L} = \bar{F} \otimes_F L$ and make $\bar{L}$ into Lie algebra, by putting

$$[\alpha \otimes x, \beta \otimes y] = \alpha\beta \otimes [x, y].$$

Checking that the conditions of the postulates (2′) and (3′) hold is trivial. Further, $\bar{L}^2 = \overline{L^2} = \bar{F} \otimes_F L^2$. Thus, if L is solvable, then $\bar{L}$ is also solvable; similarly, we can verify that L^2 is nilpotent if $\bar{L}^2$ is nilpotent. Therefore it is sufficient to restrict ourselves to the case when F is algebraically closed. In virtue of corollary 2 from Subsection 1.5 it is sufficient to prove that for any $y \in L^2$ the operator ad y is nilpotent. According to corollary 1, in some basis the matrices of all operators of the form ad x, $x \in L$ are triangular.

Since

$$\text{ad}\,[x_1, x_2] = (\text{ad}\,x_1)(\text{ad}\,x_2) - (\text{ad}\,x_2)(\text{ad}\,x_1),$$

the matrix of any operator ad y, $y \in L^2$, has zeros on the main diagonal. Thus, such an ad y is nilpotent (even as an operator on L), and L^2 according to corollary 2 from Subsection 1.5 is nilpotent. □

§ 2. Universal enveloping algebra

2.1. Definition. A Lie algebra can be obtained not only from the algebra of linear operators on a vector space but also from any associative algebra. If A is such an algebra then one can put $[x, y] = xy - yx$ and take any subspace which is closed relatively to taking commutators. A literal repetition of the arguments from Subsection 1.1 shows that the algebra obtained satisfies the

Jacobi identity and anti–commutativity (see, (2′) and (3′) in Subsection 1.2). In this section we are interested in the universality of this way of constructing Lie algebras. For convenience the Lie algebra obtained from A by introducing commutators will be denoted by $[A]$.

An associative algebra U with identity over the field F is called a *universal enveloping algebra* for the (abstract) Lie algebra L over F if there exists the homomorphism of Lie algebras $\epsilon : L \to [U]$ such that for any homomorphism of Lie algebras $\eta : L \to [A]$, where A is some associative algebra over F with 1, there exists a unique homomorphism f of associative algebras with 1, $f : U \to A$, such that $\eta = f\epsilon$. In other words f gives a commutative diagram

(1)

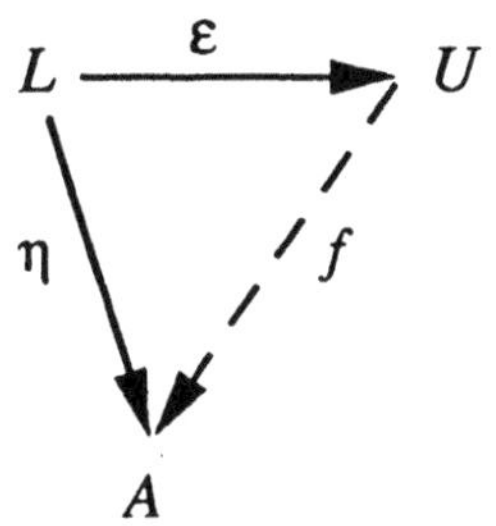

The existence of a universal enveloping algebra is proved with the help of the tensor algebra $T(V)$ of the vector space V (see, § 1.4). Here the vector space of the following form is meant

$$T(V) = T_0(V) \oplus T_1(V) \oplus \ldots \oplus T_n(V) \oplus$$

where $T_0(V) = F \cdot 1$, $T_1(V) = V, \ldots, T_m(V) = T_{m-1}(V) \otimes V$. The operation of multiplication is the usual tensor multiplication, i.e. if $u' = v_1 \otimes \ldots \otimes v_k$, $u'' = v_{k+1} \otimes \ldots \otimes v_{k+m}$ then $u'u'' = v_1 \otimes \ldots \otimes v_{k+m}$. The associativity of this operation is clear. The main property of the algebra $T(V)$ is the following. Let some associative algebra A with 1 be generated by its subspace $V' \subset A$, and let $\phi : V \to V'$ be a linear mapping. Then there exists the unique homomoprhism $\bar{\phi} : T(V) \to A$ of associative algebras with identity (i.e. $\bar{\phi}(1) = 1$) such that $\bar{\phi}|_V = \phi$. If $\{e_1, \ldots, e_n\}$ is a basis in V then any element from $T(V)$ is uniquely represented in the form of linear combination of tensors of the form $e_{i_1} \otimes \ldots \otimes e_{i_m}$. We set $\bar{\phi}(1) = 1$ and

$$\bar{\phi}(e_{i_1} \otimes \ldots \otimes e_{i_m}) = \phi(e_{i_1}) \ldots \phi(e_{i_m}).$$

Trivial checking shows that $\bar{\phi}$ is a homomorphism. The fact that $e_1, \ldots, e_n, \ldots$ generate $T(V)$ as an algebra shows that no other definition is possible.

We consider now in the algebra $T(L)$ the ideal J, which is generated by all elements of the form

$$v_1 \otimes v_2 - v_2 \otimes v_1 - [v_1, v_2], \quad \text{where } v_1, v_2 \in L.$$

We put $U(L) = T(L)/J$.

Proposition. *The algebra $U(L)$ is the universal enveloping algebra of L.*

Proof. For $\epsilon : L \to U(L)$ take the mapping $\epsilon(v) = v + J$ $(v \in L)$. Then

$$\begin{gathered}\epsilon([v_1, v_2]) = [v_1, v_2] + J = \\ = v_1 \otimes v_2 - v_2 \otimes v_1 + J = (v_1 + J)(v_2 + J) - (v_2 + J)(v_1 + J) + J = \\ = \epsilon(v_1)\epsilon(v_2) - \epsilon(v_2)\epsilon(v_1) = [\epsilon(v_1), \epsilon(v_2)].\end{gathered}$$

Further, let $\eta : L \to [A]$ be a homomorphism of Lie algebras, where A is an associative algebra with identity. Define the homomorphism $f' : T(L) \to A$, by setting

$$f'(v) = \eta(v), \quad v \in L.$$

Then

$$\begin{gathered}f'(v_1 \otimes v_2 - v_2 \otimes v_1 - [v_1, v_2]) = \eta(v_1)\eta(v_2) - \eta(v_2)\eta(v_1) - \eta([v_1, v_2]) = \\ = [\eta(v_1), \eta(v_2)] - \eta([v_1, v_2]) = 0 \quad (v_1, v_2 \in L).\end{gathered}$$

Consequently, the kernel of the homomorphism f' contains the ideal J. Thus $f' : T(L) \to A$ factorizes through $U(l)$ to $f : U(L) \to A$

$$f(v + J) = f'(v).$$

It follows that the homomorphism f of associative algebras with identity is the desired one. □

2.2. Basis of the universal enveloping algebra. In order to solve the problem of realizing an arbitrary Lie algebra as a subalgebra of a commutator algebra of some associative algebra the main remaining question is the injectivity of the homomorphism $\epsilon : L \to U(L)$. In virtue of the universal property of the algebra $U(L)$ (see, (1)), the injectivity for ϵ holds given any imbedding $\eta : L \to A$. On the other hand, if $\epsilon : L \to U(L)$ is injective then this is the desired imbedding.

From the practical point of view the problem of injectivity is closely related to the one of finding a basis for the algebra $U(L)$. Indeed, let $E = (e_i)_{i \in \mathcal{I}}$ be some basis for L over the main field F; moreover, let the set of indexes $\mathcal{I}$ be linearly ordered. Since $T(L)$ is generated by the vector space L, the algebra $U(L)$ is generated by the subspace $\epsilon(L)$, thus, by the set $\epsilon(E)$. Therefore $U(L)$ is the linear hull of identity and the set of all monomials of the form

$$\epsilon(e_{i_1})\epsilon(e_{i_2})\ldots\epsilon(e_{i_m}), \quad i_1, i_2, \ldots, i_m \in \mathcal{I}. \tag{2}$$

We show that 1 and a subset of the "ordered" monomials, viz. monomials of the form (2) with the extra condition $i_1 \leq i_2 \leq \ldots \leq i_m$ also generate $U(L)$ as a vector space. Any monomial of the form (2) gives rise to a pair (m, p), where m is the degree of the monomial, p is a number of inversions in the permutation $(i_1, i_2, \ldots, i_m)$.

We assume that the monomials of the form (2) and of degree less than m already have a representation in the form of a linear combination of "ordered" monomials of degree less than m. We can also assume that the monomials of degree m with a number of inversions less than p possess a similar representation. If $p > 0$ then the permutation $(i_1, \ldots, i_m)$ has two neighboring indexes which form an inversion, let us say, $i_k > i_{k+1}$. Let $[e_{i_k}, e_{i_{k+1}}] = \sum_{j \in J} c_j e_j$. Then

$$\begin{aligned}
&\epsilon(e_1)\ldots\epsilon(e_{i_k})\epsilon(e_{i_{k+1}})\ldots\epsilon(e_{i_m}) = \\
&=\epsilon(e_{i_1})\ldots\epsilon([e_{i_k}, e_{i_{k+1}}])\ldots\epsilon(e_{i_m}) + \epsilon(e_{i_1})\ldots\epsilon(e_{i_{k+1}})\epsilon(e_{i_k})\ldots\epsilon(e_{i_m}) = \\
&= \sum_j c_j\epsilon(e_{i_1})\ldots\epsilon(e_j)\ldots(e_{i_m}) + \epsilon(e_{i_1})\ldots\epsilon(e_{i_{k+1}})\epsilon(e_{i_k})\ldots\epsilon(e_{i_m}).
\end{aligned}$$

The terms under the summation sign have lower degree and in the last monomial of degree m the number of inversions is reduced. This allows to apply induction, the starting points being evident ($m = 0$ or $m \neq 0$, $p = 0$). □

The main goal of this paragraph is to prove that the set consisting of 1 and the "ordered" monomials is linearly independent. It then follows that ϵ is injective. Indeed, then the subset of monomials of degree 1, i.e. of elements of the form $\epsilon(e_i)$, $i \in \mathcal{I}$, is linearly independent. Since under the ϵ a basis becomes a linearly independent set this particular mapping is injective.

2.3. Poincaré–Birkhoff–Witt Theorem. *Let L be a Lie algebra over a field F, $E = (e_i)_{i\in\mathcal{I}}$ a basis, $U(L)$ the universal enveloping algebra, $\epsilon : L \to U(L)$ the canonical homomorphism. Then* 1 *and the set of monomials of the form*

$$\epsilon(e_{i_1})\epsilon(e_{i_2})\dots\epsilon(e_{i_m}), \quad i_1 \le i_2 \le \dots \le i_m, \quad m > 0, \tag{3}$$

form a basis of the algebra $U(L)$.

Proof. Any representation $\rho : L \to \operatorname{End} V$ extends to a representation $\bar{\rho} : U(L) \to \operatorname{End} V$. It is sufficient to take in diagram (1) the homomorphism ρ as η. Therefore it is sufficient to construct a representation of the algebra L (or, what is the same, an L–module), for which the set of elements of the form (3) is linearly independent.

In order to achieve this we consider the vector space X with a basis of the form

$$\{x_M | M = (i_1, \dots, i_m), i_1, \dots, i_m \in \mathcal{I}, i_1 \le \dots \le i_m\}.$$

We may also take $M = \emptyset$. For $i \in \mathcal{I}$ and $M = (i_1, \dots, i_m)$, $i_1, \dots, i_m \in \mathcal{I}$, $i_1 \le \dots \le i_m$ we write $|M| = m$, $M' = (i_2, \dots, i_m)$, $i \le M$, if $i \le i_1$. We define an action of the basis elements of L on X, by setting $e_i x_M = x_{iM}$ if $i \le M$ and setting $e_i x_\emptyset = x_i$. Otherwise (guided by the idea that x_M is something similar to $e_{i_1} e_{i_2} \dots e_{i_m}$) we define $e_i x_M$ as

$$e_i x_M = [e_i, e_{i_1}]x_{M'} + e_{i_1}(e_i x_{M'}). \tag{4}$$

In order to give meaning to the right side of the equality (4), it should be noted that $|M'| < m$, and $[e_i, e_{i_1}]$ is a linear combination of the basis elements. This gives meaning to the expression $[e_i, e_{i_1}]x_{M'}$ if it is assumed, that the definition is introduced by induction with respect to $|M|$. Further, we can assume that the defined action is such that $e_i x_M$ is of the form $x_{\bar{M}} + u$, where the index $\bar{M}$ is obtained by ordering the set $\{i\} \cup M$ and $u = \sum_N \lambda_N x_N$, where $|N| \leq |M|$. Then $e_i x_{M'}$ is defined in virtue of induction with respect to $|M|$ and has the form $x_{\widetilde{M}} + \sum_N \mu_N x_N$, where $\widetilde{M}$ is obtained by ordering the set $\{i, i_2, \dots, i_m\}$, and $|N| < |M|$. Now we apply induction with respect to $|M|$,

$$e_{i_1}(e_i x_{M'}) = e_{i_1} x_{\widetilde{M}} + e_{i_1} \sum_N \mu_N x_N = x_{i_1 \widetilde{M}} + \sum_N \mu_N e_{i_1} x_N.$$

Our additional assumption on $e_i x_M$ is now verified by formula (4), since according to this formula, on the right side of that equality the x_P with the largest P is $x_{i_1 \widetilde{M}}$, and $i_1 \widetilde{M}$ is obtainable by ordering the set $\{i, i_1, i_2, \dots, i_m\}$. By the same token the action of the algebra L by linear operators on the space X is defined.

We shall demonstrate that X is an L–module. It is necessary to check the relation (we shall use induction with respect to $|M|$)

$$[e_i, e_j] x_M = e_i(e_j x_M) - e_j(e_i x_M). \tag{5}$$

In virtue of anti–commutativity law we can assume that $i < j$. If $i \leq M$ then we apply (4) replacing M with iM, and M' with M. Otherwise $M = kM'$, $k < i \leq j$. Then (5) turns into

$$[e_i, e_j] e_k x_{M'} = e_i(e_j(e_k x_{M'})) - e_j(e_i(e_k x_{M'})).$$

We apply induction with respect to $|M'|$ and the Jacobi identity to the left side:

$$\begin{aligned}
&[e_i, e_j] e_k x_{M'} = [[e_i, e_j], e_k] x_{M'} + e_k [e_i, e_j] x_{M'} = \\
&= [[e_i, e_j], e_k] x_{M'} + e_k(e_i(e_j x_{M'})) - e_k(e_j(e_i x_{M'})) = \\
&= -[[e_j, e_k], e_i] x_{M'} - [[e_k, e_i], e_j] x_{M'} + e_k(e_i(e_j(x_{M'})) - e_k(e_j(e_i x_{M'})) = \\
&= -[e_j, e_k] e_i x_{M'} + e_i [e_j, e_k] x_{M'} - [e_k, e_i] e_j x_{M'} + e_j [e_k, e_i] x_{M'} + \\
&+ e_k e_i e_j x_{M'} - e_k e_j e_i x_{M'} = [e_k, e_j] e_i x_{M'} + e_i e_j e_k x_{M'} - e_i e_k e_j x_{M'} + \\
&+ [e_i, e_k] e_j x_{M'} + e_j e_k e_i x_{M'} - e_j e_i e_k x_{M'} + e_k e_i e_j x_{M'} - e_k e_j e_i x_{M'} = \\
&= \{e_i e_j e_k x_{M'} - e_j e_i e_k x_{M'}\} + \{[e_k, e_j] e_i x_{M'} + e_j e_k e_i x_{M'} - e_k e_j e_i x_{M'}\} + \\
&+ \{[e_i, e_k] e_j x_{M'} + e_k e_i e_j x_{M'} - e_i e_k e_j x_{M'}\}.
\end{aligned}$$

It remains to show that the second and the third braces are zero.

We shall start with the second brace. As was noted above, $e_i x_{M'} = x_{\widetilde{M}} + \sum_N \mu_N x_N$, where $\widetilde{M}$ is obtained by ordering the sequence (i, M'), and $|N| < |M|$. In this case, using induction and (4), we get

$$e_j e_k e_i x_{M'} = e_j x_{k\widetilde{M}} + \sum_N \mu_N e_j e_k x_N =$$
$$= e_j x_{k\widetilde{M}} + \sum \mu_N [e_j, e_k] x_N + \sum \mu_N e_k e_j x_N =$$
$$= [e_j, e_k] x_{\widetilde{M}} + e_k e_j x_{\widetilde{M}} + \sum \mu_N [e_j, e_k] x_N + \sum \mu_N e_k e_j x_N =$$
$$= [e_j, e_k] e_i x_{M'} + e_k e_j e_i x_{M'},$$

and this proves the required. Analogously, the third brace is equal to zero.

Thus, X is really an L–module.

In the L–module X the following equality holds

$$e_{i_1} e_{i_2} \dots e_{i_m} x_\emptyset = x_{(i_1, i_2, \dots, i_m)},$$

if $i_1 \leq i_2 \leq \dots \leq i_m$. If u_M is an element of the form (3) $M = (i_1, i_2, \dots, i_m)$ then $u_M x_\emptyset = x_M$. Taking a linear combination $u = \sum \mu_M u_M$ which is equal to zero in $U(L)$ and acting by it on $x_\emptyset$ in the module X, we obtain $\sum \mu_M x_M = 0$, i.e. $\mu_M = 0$, for any M. This proves the linear independence of the elements of the form (3) and concludes the proof of the theorem. □

Corollary. *The canonical homomorphism $\epsilon : L \to U(L)$ is an imbedding.* □

In practice this means that L can be seen as a subspace in $U(L)$. In this case the basis of the algebra $U(L)$ is the set consisting of 1 and the elements of the form

$$e_{i_1} e_{i_2} \dots e_{i_m}, \quad \text{where } i_1 \leq i_2 \leq \dots \leq i_m.$$

In a case of the finite–dimensional algebra L with the basis $\{e_1, \dots, e_n\}$ we obtain a basis for $U(L)$ of the form

$$(3') \qquad e_1^{s_1} e_2^{s_2} \dots e_n^{s_n}, \quad s_1, s_2, \dots, s_n \geq 0.$$

§ 3. Magnus theory of free groups

3.1. The Lie ring of a filtered group. Let G be some group. A decreasing series of normal subgroups

$$f : G = G_1 \supset G_2 \supset \ldots \supset G_n \supset G_{n+1} \supset \ldots, \tag{1}$$

such that for any $s, r \geq 1$ the commutator $[G_s, G_r]$ lies in G_{s+r}, is called a *central filtration* in the group G. Here the concept of the *commutator* $[A, B]$ of the two subgroups A, B of the group G is the generalization of the concept of a commutator of a group and is defined as the smallest subgroup in G containing all elements of the form $[a, b] = aba^{-1}b^{-1}$, $a \in A$, $b \in B$. A special case of a central filtration is the *lower central series* of a group, i.e. the set of groups

$$G = \gamma_1(G) \supset \gamma_2(G) \supset \ldots \supset \gamma_n(G) \supset \gamma_{n+1}(G) \supset \ldots, \tag{2}$$

inductively defined by the equality $\gamma_{n+1}(G) = [\gamma_n(G), G]$. A group where the series (2) reaches identity in a finite number of steps, is called *nilpotent*. If n is the smallest integer with the property $\gamma_{n+1}(G) = \{e\}$ then we say that G is nilpotent of class n.

In order to prove properties of groups related to the calculation of commutators it is necessary to have several group–theoretic identities. In these identities the symbol x^y denotes yxy^{-1}.

Proposition (Hall identities). *For any elements a, b from an arbitrary group G the following identities are valid*

(*i*) $[b, a] = [a, b]^{-1}$;

(*ii*) $[a, bc] = [a, b][a, c]^b$, $[ab, c] = [b, c]^a[a, c]$;

(*iii*) $[[a, b], c^b][[b, c], a^c][[c, a], b^a] = e$.

If A, B, C are normal subgroups in some group G then

$$[[A, B], C] \subset [[B, C], A][[C, A], B].$$

Proof. The equality (*i*) is evident. Note that $[a, b] = b^a b^{-1}$. Then $[a, bc] = (bc)^a(bc)^{-1} = b^a c^a c^{-1} b^{-1} = b^a b^{-1} b c^a c^{-1} b^{-1} = [a, b] b [a, c] b^{-1} = [a, b][a, c]^b$.

Here we have used the fact that the mapping $b \mapsto b^a$ is an (interior) isomorphism. The second equality from (*ii*) is proved in a similar way. In order to prove the identity (*iii*) we shall compute the first commutator and perform on it a cyclic permutation of the letters:

$$\begin{aligned}
&[[a,b],c^b] = aba^{-1}b^{-1}bcb^{-1}bab^{-1}a^{-1}bc^{-1}b^{-1} = aba^{-1}cab^{-1}a^{-1}bc^{-1}b^{-1},\\
&[[b,c],a^c] = bcb^{-1}abc^{-1}b^{-1}ca^{-1}c^{-1},\\
&[[c,a],b^a] = cac^{-1}bca^{-1}c^{-1}ab^{-1}a^{-1}.
\end{aligned}$$

The product of all three members is equal to e.

The relation concerning the subgroups evidently follows from (*ii*) and (*iii*). Indeed, any element from $[[A,B],C]$ is a product of commutators of the form

$$[u_1u_2\dots u_m,c], \quad \text{where } u_i = [a_i,b_i]^{\pm 1}, \quad a_i \in A, b_i \in B, c \in C.$$

Using the fact that $a \mapsto a^b$ is an automorphism and that A, B, C are stable with respect to such automorphisms, we obtain from (*ii*) that our element is a product of commutators of the form $[[a,b],c'^b]$, $a \in A$, $b \in B$, $c' \in C$ (for c' the element $c^{b^{-1}}$ can be taken). It remains only to apply (*iii*). □

Now we shall show that (2) really is a central filtration. In order to prove the relation $[\gamma_r(G), \gamma_s(G)] \subset \gamma_{r+s}(G)$ we shall use induction on $t = \min(r,s)$ starting with the evident case when $t = 1$. If $t = r > 1$ then

$$\begin{aligned}
&[\gamma_r(G),\gamma_s(G)] = [[\gamma_{r-1}(G),\gamma_1(G)],\gamma_s(G)] \subset\\
&\subset [[\gamma_1(G),\gamma_s(G)],\gamma_{r-1}(G)] \cdot [[\gamma_s(G),\gamma_{r-1}(G)],\gamma_1(G)] \subset\\
&\subset [\gamma_{s+1}(G),\gamma_{r-1}(G)][\gamma_{s+r-1}(G),\gamma_1(G)] \subset \gamma_{s+r}(G).
\end{aligned}$$

Returning to a central filtration (1) we define the Lie ring $L^f(G)$ (a Lie algebra with coefficients from the ring of integers). Its additive group is the direct sum of the abelian groups (called the homogeneous components) $L^f_n(G) = G_n/G_{n+1}$, $n = 1,2,\dots$, i.e.

$$L^f(G) = \bigoplus_{n=1}^{\infty} L^f_n(G).$$

In order to define a distributive commutator (in the sense of Lie rings) it is sufficient to do it for a pair of elements $\xi \in L_r^f(G)$, $\eta \in L_s^f(G)$. If $\xi = xG_{r+1}$, $\eta = yG_{s+1}$ ($x \in G_r$, $y \in G_s$) then we set

$$[\xi, \eta] = [x, y]G_{r+s+1} \tag{3}$$

Theorem. *The group $L^f(G)$ with the commutator operation* (3) *is a well defined Lie ring.*

Proof. We shall verify that (3) is well defined. If $\xi = xzG_{r+1}$, where $z \in G_{r+1}$ then according to *(ii)*

$$[xz, y]G_{r+s+1} = [z, y]^x[x, y]G_{r+s+1} = [x, y][z, y]^x G_{r+s+1} =$$
$$=[x, y][x, [zy]][z, y]G_{r+s+1} = [x, y][x, [z, y]]G_{r+s+1} = [x, y]G_{r+s+1}.$$

Analogously, we can replace the representative in the class η, and thus in both classes ξ and η simultaneously. With a help of the Hall identities the distributivity of the operation (3) is verified. So if $\xi = xG_{r+1}$, $\eta_1 = y_1G_{s+1}$, $\eta_2 = y_2G_{s+1}$ (if η_1 and η_2 belong to the different homogeneous components then distributivity is fulfilled automatically,

$$[\xi, \eta_1 + \eta_2] = [xG_{r+1}, y_1y_2G_{s+1}] = [x, y_1y_2]G_{r+s+1} =$$
$$=[x, y_1][x, y_2]^{y_1}G_{r+s+1} = [x, y_1][y_1[x, y_2]][x, y_2]G_{r+s+1} =$$
$$=[x, y_1][x, y_2][y_1, [x, y_2]]G_{r+s+1} = [x, y_1][x, y_2]G_{r+s+1} =$$
$$=([x, y_1]G_{r+s+1})([x, y_2]G_{r+s+1}) = [\xi, \eta_1] + [\xi, \eta_2].$$

Anti–commutativity of the operation is evident. Finally, in order to check the Jacobi identity, *(iii)* should be applied. If $\xi = xG_{r+1}$, $\eta = yG_{s+1}$, $\zeta = zG_{t+1}$ then

$$[[\xi, \eta], \zeta] = [[x, y], z]G_{r+s+t+1} = [[x, y], (z^y)^{y^{-1}}]G_{r+s+t+1} =$$
$$=[[x, y], y^{-1}z^y y]G_{r+s+t+1} = [[x, y], z^y[y, z^{-1}]]G_{r+s+t+1} =$$
$$=[[x, y], z^y][[x, y], [y, z^{-1}]]^{z^y}G_{r+s+t+1} =$$
$$=[[x, y], z^y]G_{r+s+t+1} \cdot [[x, y], [y, z^{-1}]]^{z^y}G_{r+s+t+1} =$$
$$=[[x, y], z^y]G_{r+s+t+1}.$$

From the computation presented it is clear that

$$[[\xi,\eta],\zeta]+[[\eta,\zeta],\xi]+[[\zeta,\xi],\eta]=$$
$$=[[x,y],z^y][[y,z],x^z][[z,x],y^x]G_{r+s+t+1}=0.$$

The theorem is proved. □

3.2. Magnus algebra. Let $X=\{x_1,x_2,\dots,x_n,\dots\}$ be some set, R a commutative ring with identity. We shall consider the set $\widehat{A(X)}$ of formal expressions of the form

$$u=\sum_w \lambda_w w,$$

where $\lambda_w\in R$ and w is a word in the alphabet X.

Take $\lambda u=\sum_w \lambda\lambda_w w$. If $v=\sum_w \mu_w w$, then we set

$$u+v=\sum_w(\lambda_w+\mu_w)w \quad\text{and}\quad uv=\sum_w\left(\sum_{w=w'w''}\lambda_{w'}\mu_{w''}\right)w.$$

In particular, the product of the two words w' and w'' is a result of composing them one to another in the indicated order: $w'w''$.

An evident check shows that relative to the operations introduced $\widehat{A(X)}$ is an algebra over the commutative ring R, moreover the identity is the empty word with the coefficient 1. This algebra is called the *Magnus algebra*. The elements of this algebra, naturally, are called power series in the (non–commutative) set of variables X. If $u=\sum_w \lambda_w w$ then we set $u_n=\sum_{|w|=n}\lambda_w w$. Then $u=\sum_{n=0}^{\infty}u_n$ and $(uv)_n=\sum_{m=0}^{\infty}u_m v_{n-m}$. The set $M(X)$ of series without constant term is an ideal in $A(X)$. The set $G(X)=1+M(X)$ is called the *Magnus group*. The fact is that $G(X)$ is evidently closed relative to the operation of multiplication; moreover, for the series $1+u$, $u\in M$, an inverse series is defined as follows

$$(1+u)^{-1}=1-u+u^2-\dots$$

Theorem. *The subgroup $F(Y)$ of the Magnus group $G(X)$ generated by the set Y of the elements of the form $1+x$ ($x\in X$) is a free group with free generating set Y.*

Proof. We shall limit ourselves to the case of a ring R without zero divisors. According to § 2.3 it is sufficient to prove that for any n and any non–zero integers $m_1, \dots, m_n$ the element

$$(1 + x_{i_1})^{m_1}(1 + x_{i_2})^{m_2} \dots (1 + x_{i_n})^{m_n},$$

where $i_s \neq i_{s+1}$ for any $s = 1, 2, \dots, n-1$ is not the identity. According to the Newton binomial formula

$$(1 + x_{i_s})^{m_s} = 1 + \alpha_s x_{i_s} + \beta_s x_{i_s}^2 + \dots$$

and there can be found a q_s such that the coefficient γ_s of $x_{i_s}^{q_s}$ is not equal to zero. Then the coefficient of $x_{i_1}^{q_1} x_{i_2}^{q_2} \dots x_{i_n}^{q_n}$ is equal to $\gamma_1 \gamma_2 \dots \gamma_n \neq 0$. The theorem is proved. □

If $F(X)$ is the free group with free generating set X then the mapping $x \mapsto 1+x$ extends to the injective homomorphism $\imath : F(X) \to G(X)$ the image of which is $F(Y)$. This imbedding allows the proof of important results about the structure of the group $F(X)$. To start with we shall study the problem of the intersection of the terms of the lower central series in the group $F(X)$. It should be noted that for any n the set $1 + M^n$ is a subgroup in $G(X)$. Having taken the intersection of this group with the group $F(X)$ (identified with its image in $G(X)$) we obtain a subgroup F^n; the elements from F^n are series of the form $1 + u_n + u_{n+1} + \dots$ situated in $\imath(F(X))$. Later on we shall show that $F^n = \gamma_n(F(X))$. For now we shall prove a simpler statement.

Lemma. *$(F^n)_{n \in \mathbb{N}}$ is a central filtration in the group $F(X)$.*

Proof. Let $u = 1 + u_r + u_{r+1} + \dots \in F^r$, $v = 1 + v_s + v_{s+1} + \dots \in F^s$.
Then using direct calculation we obtain

$$\begin{aligned} c =&[u, v] = uvu^{-1}v^{-1} = \\ =&(1 + u_r + u_{r+1} + \dots)(1 + v_s + v_{s+1} + \dots)\cdot \\ &\cdot(1 + u_r + u_{r+1} + \dots)^{-1}(1 + v_s + v_{s+1} + \dots)^{-1} = \\ =&1 + (u_r v_s - v_s u_r) + c_{s+r+1} + \dots, \quad \text{where } c_{s+r+1} \in M^{s+r+1}. \end{aligned} \tag{4}$$

□

3.3. Corollary. *Let X be an arbitrary non–empty set, $F(X)$ the free group with free generating set X. Then the lower central series*

$$F(X) = \gamma_1(F(X)) \supset \gamma_2(F(X)) \supset \dots$$

of this group has a trivial intersection.

Proof. We note that for any central filtration of an arbitrary group G:

$$G = G_1 \supset G_2 \supset \dots \supset G_n \supset G_{n+1} \dots$$

the imbedding $G_n \supset \gamma_n(G)$, $n = 1, 2, \dots$ is allways fulfilled. Indeed, $G_1 = \gamma_1(G)$; if it is already proved that $\gamma_n(G) \subset G_n$ then

$$\gamma_{n+1}(G) = [\gamma_n(G), G] \subset [G_n, G] \subset G_{n+1}.$$

It remains only to show that the filtration $(F^n)_{n=1,2,\dots}$ has a trivial intersection. In fact

$$\bigcap_{n=1}^{\infty} F^n \subset \bigcap_{n=1}^{\infty}(1 + M^n) = \{1\}.$$

□

3.4. The Lie ring of free group. We define the free Lie ring $L(X)$ with the free generating set X in a standard way, i.e. as a Lie ring which is generated by the set X and in addition has the property that any mapping of the set X into an arbitrary Lie ring P extends (in unique way) to a homomorphism of Lie rings $\bar{\phi} : L(X) \to P$. We shall not go into the standard proof of existence of a free Lie ring (see, the proof in Subsection 2.1 of existence of the universal enveloping algebra), but indicate its realization in terms of the Magnus algebra over **Z**.

Proposition. *The set X in the Magnus algebra $\widehat{A(X)}$ generates a free Lie ring relative to the commutator operation.*

Proof. It should be noted that the set X does not generate the Magnus algebra because the ring generated by the set X contains only series with a finite number of terms. On the other hand, if $\langle X \rangle$ is the free **Z**–module (free abelian group) with the basis X then from the explicit form of a basis in the tensor algebra $T(\langle X\rangle)$ and in $\widehat{A(X)}$ then we can see that the identity mapping of the set X extends to an injective imbedding $T = T(\langle X\rangle) \to \widehat{A(X)}$ that allows us to identify T with the associative subring in $\widehat{A(X)}$ generated by the set X. We transform T into a Lie ring by introducing the commutator operation. Then we have a homomorphism of Lie rings $\epsilon : L(X) \to T$ under which X is mapped identically. If now $\eta : L(X) \to [B]$ is a homomorphism of Lie rings and B is an associative algebra with 1 then in virtue of the property of the tensor algebra, the mapping $x \mapsto \eta(x)$ extends to the homomorphism f of the associative rings with identity. Besides $f\epsilon = \eta$. Thus, $T = U(L(X))$. According to the Poincaré–Birkhoff–Witt theorem, ϵ is an injective mapping and our proposition is proved. □

Theorem. *Let $F(X)$ be a free group. $L(X)$ the free Lie ring with free generating set X. $L(F)$ is the Lie ring constructed by means of the lower central series of the group $F(X)$. Then the mapping $x \mapsto x\gamma_2(F)$ $(x \in X)$ extends to an isomorphism $L(X) \cong L(F)$.*

Proof. As was already mentioned, the series f of normal subgroups $(F^n)_{n\in\mathbf{N}}$ constructed in Subsection 3.2, is a central filtration. Take, with respect to this series, the Lie ring $L^f(F)$. We form a sequence of homomorphisms of Lie rings

$$L(X) \xrightarrow{\phi} L(F) \xrightarrow{\psi} L^f(F) \xrightarrow{\chi} \widehat{A(X)}$$

in the following way. The homomorphism $\phi : L(X) \to L(F)$ is the (surjective) homomorphism extending the mapping $x \mapsto x\gamma_2(F)$. It remains to prove the injectivity of this homomorphism. Further, $L_n(F) = \gamma_n(F)/\gamma_{n+1}(F)$ and $L^f_n(F) = F^n/F^{n+1}$. Since $\gamma_s(F) \subset F^s$ for any s, and the operation in the rings $L(F)$ and $L^f(F)$ are defined similarly, we see that the mapping $x\gamma_{n+1}(F) \to xF^{n+1}$ is well defined and extends to a homomorphism ψ of Lie rings.

Finally, any homogeneous element ξ from $L^f(F)$ of degree s has the form $\xi = (1+u_s+u_{s+1}+\ldots)F^{s+1}$, and by taking its Lie commutator with the homogeneous element $\eta = (1+v_r+v_{r+1}+\ldots)F^{r+1}$ of degree r we obtain according to formula (4) the homogeneous element $[\xi,\eta] = (1+(u_sv_r - v_ru_s)+\ldots)F^{r+s+1}$. This shows that if we associate to the element $\xi \in L^f_s(F)$ the element u_s then the element $[\xi,\eta]$ is mapped to $u_sv_r - v_ru_s = [u_s, v_r]$. By extending this mapping to sums of homogeneous elements we hence obtain a homomorphism of Lie rings χ. All three homomorphisms are now constructed.

Now let us find the image of the generator $x \in X$ with respect to the composition of three homomorphisms:

$$\chi\psi\phi(x) = \chi\psi(x\gamma_2(F)) = \chi((1+x)F^2) = x.$$

Since a homomorphism of the Lie ring $L(X)$ is determined by its restriction to the generating set X, $\chi\psi\phi$ is the injective homomorphism ϵ of the previous proposition. From what has been said we can conclude to the injectivity of the mapping $\phi : L(X) \to L(F)$. The theorem is proved. □

Now denote by $L_n(X)$ the additive group in $L(X)$ generated by the monomials of degree n with respect to X.

3.5. Factors of the lower central series

Corollary 1. *The abelian groups $L_n(X)$ and $\gamma_n(F(X))/\gamma_{n+1}(F(X))$ are isomorphic for any $n \geq 1$. Besides $F^n = \gamma_n(F)$.*

Proof. Since the set of elements of the form $\{x\gamma_2(F),\ x \in X\}$ generates $L(F)$, the mapping $\phi : L(X) \to L(F)$ is not only injective but also surjective. In addition, $\phi(L_n(X)) \subset L_n(F)$. The latter proves the first part of the statement. Further, bijectivity of the mapping ϕ and injectivity of the mapping $\chi\psi\phi$ gives that ψ is also injective. This permits to prove that $\gamma_n(F)$ coincides with F^n by induction on n starting with the evident case $\gamma_1(F) = F^1$. If $F^n = \gamma_n(F)$ and $x \in F^{n+1} \setminus \gamma_{n+1}(F)$ then $x\gamma_{n+1}(F) \in L_n(F)$, $\psi(x\gamma_{n+1}(F)) = xF^{n+1} = 0$ in $L_n^f(F)$, which is impossible. □

We shall conclude this section with a corollary of which the first statement follows from the theorem on subgroups of a free abelian group, and the second and third statements are presented without proof.

$*$ **Corollary 2.** 1) *For any $n \geq 1$ the group $\gamma_n(F)/\gamma_{n+1}(F)$ is free abelian.*

2) *If $X = \{x_1, \dots, x_d\}$ then the rank $l_d(n)$ of the group $\gamma_n(F)/\gamma_{n+1}(F)$ can be calculated according to the formula*

$$l_d(n) = \frac{1}{n}\sum_{m|n} \mu(m) d^{n/m},$$

where μ is the Möbius function, i.e. $\mu(1) = 1$, $\mu(n) = 0$ if n is divisible by a square of a prime number, $\mu(n) = (-1)^s$ if $n = p_1 \dots p_s$, where $p_1, \dots, p_s$ are the different prime numbers.

3) *As a basis for $L(X)$ there can be chosen any Hall set $\mathcal{H}(X)$ with its homogeneous parts inductively defined in the following way: $\mathcal{H}_1(X) = X$, if $\mathcal{H}_1(X), \dots, \mathcal{H}_n(X)$ are already selected and their union $\mathcal{H}_1(X) \cup \dots \cup \mathcal{H}_n(X)$ is totally ordered, then $\mathcal{H}_{n+1}(X)$ is the set of all commutators of degree $n+1$ of the form $[u, v]$, $u \in \mathcal{H}_s(X)$, $v \in \mathcal{H}_{n+1-s}(X)$, such that $u > v$ and, additionally, if $u = [u_1, u_2]$ then $u_2 \leq v$; the set $\mathcal{H}_1(X) \cup \dots \cup \mathcal{H}_n(X) \cup \mathcal{H}_{n+1}(X)$ is totally ordered by retaining the previous order for $\mathcal{H}_1(X) \cup \dots \cup \mathcal{H}_n(X)$ and in such a way that any element from $\mathcal{H}_1(X) \cup \dots \cup \mathcal{H}_n(X)$ is smaller than any element from $\mathcal{H}_{n+1}(X)$.* □

We shall present an example of such basis set up to degree 5 inclusively:

$$
\begin{aligned}
&y < x < [x,y] < [[x,y],y] < [[x,y],x] < [[[x,y],y],y] < \\
&<[[[x,y],y],x] < [[[x,y],x],x] < [[[[x,y],y],y],y] < \\
&<[[[[x,y],y],y],x] < [[[[x,y],y],x],x] < [[[[x,y],x],x],x] < \\
&<[[[x,y],y],[x,y]] < [[[x,y],x],[x,y]].
\end{aligned}
$$

The obtained results are of interest as they present a convenient unique–valued notation of elements modulo the members of the lower central series. For instance, in the case of the example any element from $F(x, y)$ modulo $\gamma_5(F)$ is equal to precisely one element of the form

$$
\begin{aligned}
&y^{\alpha_1} x^{\alpha_2} [x,y]^{\alpha_3} [[x,y],y]^{\alpha_4} [[x,y],y]^{\alpha_5} \cdot \\
&\cdot [[[x,y],y],y]^{\alpha_6} [[[x,y],y],x]^{\alpha_7} [[[x,y],x],x]^{\alpha_8},
\end{aligned}
\tag{5}
$$

where $\alpha_1, \alpha_2, \dots, \alpha_8$ are integers. In particular, any element in a nilpotent group of class 4 with the two generators x, y can be re presented in the form (5). $*$

§ 4. Lie algebras with triangular decomposition

4.1. Motivation. In Subsection 1.6 we have seen that any solvable Lie algebra L of linear operators on a finite–dimensional vector space V over an algebraically closed field F of characteristic zero is reducible to triangular form. In other words, in a suitable basis of the space V all operators from L have triangular matrices with zeros below the diagonal.

Passing to the unsolvable algebras (with the same conditions on dimension and field) we can find situations where a Lie algebra has the form of the sum of two subalgebras which are reducible to the triangular form. We have this in the case of the Lie algebra $\mathfrak{gl}_n(F)$ of all matrices of order n over the field F or the Lie algebra $\mathfrak{sl}_n(F)$ of matrices with a zero trace, or direct sums of algebras of this type.

In the case of the Lie algebra $\mathfrak{gl}_n(F)$, as in the case of Lie algebra $\mathfrak{t}_n(F)$ of all triangular matrices, it is easy to notice a characteristic subalgebra H (the so–called *torus*) which consists of all diagonal matrices. Its pecularity is that this torus is commutative and all operators $\operatorname{ad} x$, where $x \in H$, act on L in a semisimple way (for the definition of a semisimple operator, see Subsection 6.3.1; there it is also proved that in the case of a perfect field any operator $\mathcal{A}$ can be decomposed into a sum $\mathcal{A} = \mathcal{A}_s + \mathcal{A}_n$ of a semisimple operator $\mathcal{A}_s$ and a nilpotent operator $\mathcal{A}_n$ which commute). We shall check, for instance, that the operator $\operatorname{ad} x$, where $x = \lambda_1 E_{11} + \ldots + \lambda_n E_{nn}$, is semisimple in $\mathfrak{gl}_n(F)$. To that end we shall show that the natural basis of elementary matrices E_{ij} is a basis of eigenvectors for $\operatorname{ad} x$. We have

$$(\operatorname{ad} x)(E_{ij}) = (\lambda_1 E_{11} + \ldots + \lambda_n E_{nn})E_{ij} - E_{ij}(\lambda_1 E_{11} + \ldots + \lambda_n E_{nn}) =$$
$$= \sum_k \lambda_k \delta_{ki} E_{kj} - \sum_k \lambda_k \delta_{jk} E_{ik} = (\lambda_i - \lambda_j)E_{ij}.$$

Thus E_{ij} is an eigenvector for $\operatorname{ad} x$, $x = \sum_{k=1}^n \lambda_k E_{kk}$ with the eigenvalue $\lambda_i - \lambda_j$, i.e. $\operatorname{ad} x$ is indeed a semisimple operator. The mapping $\alpha_{ij} : H \to F$ which assigns to every $x \in H$ the eigenvalue $\lambda_i - \lambda_j$ of the operator $\operatorname{ad} x$ on E_{ij}, is a linear function, i.e. an element from H^*. We have $(\operatorname{ad} x)(E_{ij}) = \alpha_{ij}(x)E_{ij}$. In general, any function $\alpha \in H^*$ such that for some non–zero $y \in L$ we have

$$(\operatorname{ad} x)(y) = \alpha(x)y, \tag{1}$$

is called a *root* of the algebra L with respect to the torus H. The set of all y satisfying condition (1) is then a non–zero subspace L_α in L and is called a *root subspace* (relative to H).

In the case under consideration $L = \mathfrak{gl}_n(F)$, we have the decomposition

$$L = \sum_{\alpha \in H^*} L_\alpha = H \oplus \sum_{\alpha \in H^* \setminus \{0\}} L_\alpha. \tag{2}$$

The set Φ of all non–zero roots of the algebra L relative to the subalgebra H, is called the *root system* of the algebra L. Note that if $\alpha \in \Phi$ then also $-\alpha \in \Phi$, if $\alpha = \alpha_{ij}$ then $-\alpha = \alpha_{ji}$. Moreover, Φ contains a basis Δ of roots of the form

$$\alpha_1 = \alpha_{12}, \alpha_2 = \alpha_{23}, \ldots, \alpha_{n-1} = \alpha_{n-1,n}.$$

All elements from Φ, relative to Δ, split into two components: $\phi = \Phi_- \cup \Phi_+$, where $\Phi_- = -\Phi_+$, and $\alpha \in \Phi_+$, iff $\alpha = m_1\alpha_1 + \ldots + m_{n-1}\alpha_{n-1}$, where $m_1 \geq 0, \ldots, m_{n-1} \geq 0$ (in fact for any i either $m_i = 0$ or $m_i = 1$). If $\alpha \in \Phi_+$ we write $\alpha > 0$, otherwise $\alpha < 0$.

Set $L_+ = \sum_{\alpha>0} L_\alpha$ and $L_- = \sum_{\alpha<0} L_\alpha$ to obtain a *triangular decomposition* of the algebra $L = \mathfrak{gl}_n(F)$:

$$L = L_- \oplus H \oplus L_+ \tag{3}$$

Finally, we should note that the mapping $\sigma : L \to L$ which turns every matrix $x \in L = \mathfrak{gl}_n(F)$ into $\sigma(x) = {}^t x$, the transposed matrix, possesses the properties $\sigma^2 = 1_L$, $\sigma([x, y]) = [\sigma(y), \sigma(x)]$, and it is related to the decomposition (3) by the properties

$$L_- = \sigma(L_+) \quad \text{and} \quad \sigma|_H = 1_H. \tag{4}$$

These heuristic considerations give rise to the following definitions.

4.2. Definition. Let L be a (not necessarily finite–dimesional) Lie algebra over a field F. The ingredients of a triangular decomposition of the algebra are the following:

(*i*) an abelian subalgebra $H \subset L$;
(*ii*) a subalgebra $L_+ \subset L$;
(*iii*) a lattice $Q_+ = \bigoplus_{j\in J} \mathbf{N}\alpha_j \setminus \{0\} \subset H^*$, where the set $\{\alpha_j\}_{j\in J}$ is $\mathbf{Z}$–linearly independent;
(*iv*) an involution $\sigma : L \to L$, i.e. a linear mapping with the properties $\sigma^2 = 1_L$, $\sigma|_H = 1_H$, $\sigma([x, y]) = [\sigma(y), \sigma(x)]$, $x, y \in L$.
We then say that L *possesses a triangular decomposition* if

$$L = \sigma(L_+) \oplus H \oplus L_+$$

and

$$L_+ = \bigoplus_{\alpha\in Q_+} L_+^\alpha \tag{5}$$

where

$$L_+^\alpha = \{x \in L_+ | [h, x] = \alpha(h)x \quad \text{for all} \quad h \in H\}.$$

□

Let L be a Lie algebra with a triangular decomposition. For any funcion $\mu \in H^*$ we define

$$L^\mu = \{x \in L | [h, x] = \mu(h)x \quad \text{for all } h \in H\}.$$

A direct computation shows that $\sigma(L^\mu) = L^{-\mu}$. Therefore

$$L = \bigoplus_{\mu\in H^*} L^\mu,$$

where $L^\alpha = L_+^\alpha$ and $L^{-\alpha} = \sigma(L_+^\alpha)$ if $\alpha \in Q_+$, $L^0 = H$, and $L^\alpha = \{0\}$, otherwise. Write

$$L_- = \sigma(L_+) \quad \text{and} \quad Q = \bigoplus_{j\in J} \mathbf{Z}\alpha_j.$$

The free abelian group Q is called the *root lattice*. From the Jacobi identity follows

$$[L^\mu, L^\lambda] \subset L^{\mu+\lambda} \tag{6}$$

for all $\mu, \lambda \in H^*$. In fact, if $h \in H$, $x \in L^\mu$, $y \in L^\lambda$ then

$$[h,[x,y]] = [[h,x],y] + [x,[h,y]] = [\mu(h)x, y] + [x, \lambda(h)y] = (\mu+\lambda)(h)[x,y].$$

A decomposition of an algebra A

$$A = \bigoplus_{s \in S} A_s$$

into a sum of subspaces A_s labeled by the elements from an additive subgroup S, such that $A_s A_r \subset A_{s+r}$ for any $s, r \in S$, is called a *grading* or, more exactly, an S–grading. In this connection we need not pressupose that A is an associative algebra, or a Lie algebra, etc..

The relations (5), (6) and also the notation introduced show that the algebra L_+ is Q_+–graded, L_- is Q_-–graded and L is Q–graded.

By using the Poincaré–Birkhoff–Witt theorem we are also able to transform the universal enveloping algebra $U(L)$ of a Lie algebra L with a triangular decomposition into a Q–graded algebra. The fact is that the subalgebra H lies in $U(L)$ and acts on the elements of this algebra by the interior derivations:

$$\begin{aligned}[h, x_1x_2\ldots x_k] &= hx_1x_2\ldots x_k - x_1x_2\ldots x_kh = \\ &=[h,x_1]x_2\ldots x_k + x_1[h,x_2]\ldots x_k + \ldots + x_1x_2\ldots[h,x_k].\end{aligned} \tag{7}$$

This action is semisimple, because if we select a basis for L from the root vectors (i.e. from the subspaces L^α) then the standard basis of the algebra $U(L)$ consisting of ordered products of vectors of the basis mentioned also consists of eigenvectors. Therefore, if $x_1 \in L_{\mu_1}, \ldots, x_k \in L_{\mu_k}$ then according to (7), we obtain

$$[h, x_1x_2\ldots x_k] = (\mu_1 + \ldots + \mu_k)(h)x_1\ldots x_k.$$

Thus,

$$U(L) = \bigoplus_{\mu \in H^*} U(L)^\mu.$$

In the above $U(L)^\mu = \{0\}$ if $\mu \notin Q$. The inclusions L_+, $L_- \subset L$ allow us to consider $U(L_+)$ and $U(L_-)$ as subalgebras in $U(L)$. These subalgebras are themselves graded by the elements of the semigroup $Q_+ \cup \{0\}$:

$$U(L_+) = \bigoplus_{\alpha \in Q_+ \cup \{0\}} U(L_+)^\alpha, \quad U(L_-) = \bigoplus_{\alpha \in Q_+ \cup \{0\}} U(L_-)^{-\alpha}.$$

An ideal $\mathcal{I}$ of a graded algebra $A = \oplus_{s \in S} A_s$ is called *graded* if $I = \oplus_{s \in S}(\mathcal{I} \cap A_s)$.

Proposition. *Any ideal R of a Lie algebra L with a triangular decomposition is Q–graded, i.e.*

$$R = \bigoplus_{\alpha \in Q} (R \cap L^\alpha). \tag{8}$$

Proof. If the non–zero element $x \in R$ has the form

$$x = x_1 + \ldots + x_m,$$

where $x_i \in L^{\mu_i}$, and the $\mu_1, \ldots, \mu_m$ are pairwise different, then for any $h \in H$ we have

$$[h, x] = \mu_1(h)x_1 + \ldots + \mu_m(h)x_m \in R. \tag{9}$$

The functions $\mu_1, \ldots, \mu_m$ differ on some finite–dimensional space $H' = \langle e_1, \ldots, e_s \rangle \subset H^*$. Each of these functions restricted to H' is a polynomial in the coordinates $t_1, \ldots, t_s$ of the vector in the given basis. The polynomial

$$q = \mu_1 \ldots \mu_s \prod_{i \neq j} (\mu_i - \mu_j)$$

is non–zero as a polynomial of $t_1, \ldots, t_s$, therefore, there is an $h \in H$ such that all $\mu_1(h), \ldots, \mu_m(h)$ are pairwise different. Repeatedly applying (8), in addition to (9), we obtain an equality

$$\mu_1(h)^k x_1 + \ldots + \mu_m(h)^k x_m \in R, \quad k = 0, 1, \ldots, m-1. \tag{10}$$

Applying the Vandermonde determinant to the system (10) we obtain that

$$x_1, x_2, \dots, x_k \in R.$$

This proves the proposition. □

4.3. Further examples. As was already mentioned, the general linear Lie algebra $\mathfrak{gl}_n(F)$ and the special linear Lie algebra $\mathfrak{sl}_n(F)$ are Lie algebras with a triangular decomposition. Other important examples are the Heisenberg algebra, the Witt algebra, the Virasoro algebra and a most significant class of examples, affine Lie algebras.

Example 1. Heisenberg algebra. A Heisenberg algebra is an algebra M with a one–dimensional center Fc such that $[M, M] = Fc$, and such that moreover for any $x \in M \setminus Fc$ we have $[x, M] = Fc$.

Let M be a Heisenberg algebra. We define a bilinear function $\Psi : M \times M \to F$ by setting

$$[x, y] = \Psi(x, y)c.$$

The element c lies in the kernel of the form Ψ. This induces a skew–symmetric form $\Psi : M/Fc \times M/Fc \to F$. This form is non–degenerate by the definition of the algebra M. Inversily, given a non–degenerate skew–symmetric form $\Psi : \bar{M} \times \bar{M} \to F$ a Heisenberg algebra can be constructed on the space $M = \bar{M} \oplus Fc$ by setting

$$[x + \lambda c, y + \mu c] = \Psi(x, y)c.$$

An Heisenberg algebra naturally appears when linear operators on the ring of polynomials $F[(x_i]_{i \in \mathcal{I}}$ are considered. Let the operator l_i be defined as a multiplication by X_i, ∂_i as the partial derivative with respect to X_i, and c as the identical operator. Then

$$M = \sum_{i \in \mathcal{I}} Fl_i \oplus Fc \oplus \sum_{i \in \mathcal{I}} F\partial_i$$

is a Lie algebra, since $[\partial_i, l_j] = \delta_{ij} c$, and c is a central element in M. (It should be noted that in the case of a field of characteristic zero the ring of polynomials is an irreducible module over this Heisenberg algebra.) We consider now the derivation $d : M \to M$ defined by the equalities $d(l_i) = -l_i$, $d(\partial_i) = \partial_i$, $d(c) = 0$. Let

$$\widetilde{M} = M \oplus Fd.$$

We define

$$[x + \lambda d, y + \mu d] = [x, y] - \mu d(x) + \lambda d(y)$$

for all $x, y \in M$, $\lambda, \mu \in F$. Then $\widetilde{M}$ is a Lie algebra called a *generalized Heisenberg algebra*. We shall construct a triangular decomposition for $\widetilde{M}$. To this end let $H = Fc \oplus Fd$ and define $\alpha \in H^*$ by $\alpha(c) = 0$, $\alpha(d) = 1$. We note that

$$[\lambda c + \mu d, \partial_i] = \mu d(\partial_i) = \mu \partial_i = \alpha(\lambda c + \mu d)\partial_i$$

and

$$[\lambda c + \mu d, l_i] = \mu d(l_i) = -\mu - l_i = (-\alpha)(\lambda c + \mu d)l_i.$$

Thus, $\partial_i \in \widetilde{M}^\alpha$ and $l_i \in \widetilde{M}^{-\alpha}$ for all $i \in \mathcal{I}$.

Finally, we define $\sigma : \widetilde{M} \to \widetilde{M}$, by setting $\sigma(l_i) = \partial_i$, $\sigma(c) = c$, $\sigma(d) = d$. As a result we obtain the triangular decomposition

$$\widetilde{M}_+ = \widetilde{M}^\alpha = \sum_{i \in \mathcal{I}} F\partial_i, \quad H = \widetilde{M}^0, \quad \widetilde{M}^{-\alpha} = \sum_{i \in \mathcal{I}} Fl_i.$$

Example 2. Witt algebra and Virasoro algebra. Consider

$$W = \bigoplus_{i \in \mathbf{Z}} Fx_i.$$

We set $[x_i, x_j] = (j - i)x_{i+j}$. An immediate check shows that the given definition turns W into Lie algebra called the *Witt algebra*. A triangular decomposition is obtained if we set $H = Fx_0$ and define $\alpha \in H^*$ by $\alpha(x_0) = 1$. Then

$$[x_0, x_n] = nx_n = (n\alpha)(x_0)x_n,$$

i.e. $x_n \in L^{n\alpha}$. The anti–involution $\sigma : L \to L$ is defined by $\sigma(x_i) = x_{-i}$. In this case we have a triangular decomposition of the form

$$W = (\bigoplus_{i<0} Fx_i) \oplus Fx_0 \oplus (\bigoplus_{i>0} Fx_i).$$

The Virasoro algebra is the vector space $V = W \oplus Fc$, in addition,

$$[x_i, x_j] = (j-i)x_{i+j} + \tau_{ij}c, \quad [c, V] = 0,$$

where $\tau_{ij} = 0$, if $i \neq -j$ and $\tau_{ij} = (j^3 - j)/12$, if $i = -j$. A triangular decomposition of the algebra is obtained by setting $H = Fx_0 \oplus Fc$ and extending α to H by $\alpha(c) = 0$.

Example 3. Affine Kac–Moody algebras. Let $\mathfrak{g}$ be a finite–dimensional simple Lie algebra over $\mathbf{C}$, for instance, $\mathfrak{g} = \mathfrak{sl}_n(\mathbf{C})$. We define the *current algebra* $G = \mathfrak{g} \otimes \mathbf{Z}[t, t^{-1}]$, where $\mathbf{C}[t, t^{-1}]$ is the algebra of Laurent polynomials in the variable t which is also the group algebra of the infinite cyclic group $\langle t \rangle$ over the field $\mathbf{C}$. The multiplication in G is given by the formula

$$[x \otimes f(t), y \otimes g(t)] = [x, y] \otimes f(t)g(t).$$

We define a one–dimensional central extension $\tilde{\mathfrak{g}}$ of the algebra G as the vector space $\mathfrak{g} = G \oplus \mathbf{C}c$ with multiplication

$$[x \otimes f(t) + \alpha c, y \otimes g(t) + \beta c] =$$

$$= [x, y] \otimes f(t)g(t) + \left(\text{Res } \frac{df(t)}{dt} g(t)\right) K(x, y)c. \tag{11}$$

Here Res $h(t)$ is the coefficient of t^{-1} of the Laurent polynomial $h(t)$ and $K(x, y)$ is the value of the so–called Killing form of the algebra $\mathfrak{g}$ on the elements $x, y \in \mathfrak{g}$. The Killing form is the bilinear symmetric form on the space $\mathfrak{g}$ given by the formula

$$K(x, y) = \text{tr } (\text{ad } x)(\text{ad } y).$$

The Killing form possesses an invariance property

$$K([x, z], y) = K(x, [z, y]). \tag{12}$$

Immediate checking, using property (12), demonstrates that the operation (11) really makes $\tilde{\mathfrak{g}}$ into a Lie algebra. Finally, we consider the derivation $d : \tilde{\mathfrak{g}} \to \tilde{\mathfrak{g}}$ which acts as zero on c and satisfies

$$d(x \otimes t^n) = nx \otimes t^n.$$

The vector space $\hat{\mathfrak{g}} = \tilde{\mathfrak{g}} \oplus \mathbf{C}d$ becomes a so–called affine Kac–Moody algebra if we assume that the commutator of an element d with the element from $\tilde{\mathfrak{g}}$ is equal to the action of d on that element. The product in $\tilde{\mathfrak{g}}$ remains as before. A triangular decomposition for $\hat{\mathfrak{g}}$ is given by

$$\hat{\mathfrak{h}} = \mathfrak{h} \oplus \mathbf{C}c \oplus \mathbf{C}d$$

$$\hat{\mathfrak{g}}_+ = \mathfrak{g}_+ \otimes 1 + \mathfrak{g} \otimes t\mathbf{C}[t], \quad \hat{\mathfrak{g}}_- = \mathfrak{g}_- \otimes 1 + \mathfrak{g} \otimes t^{-1}\mathbf{C}[t^{-1}],$$

where $\mathfrak{g} = \mathfrak{g}_- \oplus \mathfrak{h} \oplus \mathfrak{g}_+$ is a triangular decomposition for $\mathfrak{g}$.

The explicit form of the root system can be illustrated in the case of the example of the algebra $\hat{\mathfrak{g}} = \widehat{\mathfrak{sl}_2(\mathbf{C})}$, i.e. in this case $\mathfrak{g} = \mathfrak{sl}_2(\mathbf{C})$. Identifying $x \otimes f(t)$ with $f(t)x$ we find a basis for $\hat{\mathfrak{g}}$ of the form

$$t^n f, \quad t^n h, \quad t^n e \,(n \in \mathbf{Z}),\ c,\ d.$$

The anti–involution is defined by the formula $\hat{\sigma}(t^n x) = t^{-n}\sigma(x)$. The lattice $\hat{Q}_+$ is generated by the roots $\alpha = h^*$ and $\beta = d^* - h^*$. In fact, we have

$$\hat{\mathfrak{g}}_+ = \sum_{\alpha \in \hat{Q}_+} \hat{\mathfrak{g}}_+^{\lambda}. \tag{13}$$

All subspaces in (13) are one–dimensional; more precisely:

$\mathbf{C}(t^n e)$ corresponds to the root $(n+1)\alpha + n\beta$ $(n \geq 0)$,
$\mathbf{C}(t^n f)$ corresponds to the root $(n-1)\alpha + n\beta$ $(n > 0)$,
$\mathbf{C}(t^n h)$ corresponds to the root $n\alpha + n\beta$ $(n > 0)$.

∗ Kac–Moody algebras have been very useful in applications in different fields of mathematics and physics. In number theory the computation of characters of irreducible representations of such algebras led to a new way of proving the well–known number–theoretic identities of Euler, Jacobi, Gauss and other classical identities and also to the deduction of new identities. We present one of the identities mentioned, the well-known Jacobi triple product identity:

$$\prod_{n\geq 1}(1-u^n v^n)(1-u^{n+1}v^n)(1-u^n v^{n+1}) =$$
$$= \sum_{m\in\mathbf{Z}}(-1)^m u^{m(m+1)/2} v^{m(m-1)/2}.$$

We also note that the Virasoro algebra is related to the theory of the relativistic string. A detailed presentation of the material mentioned can be found in Kac [1], Moody–Pianzola [1]. *

The following result is presented without proof.

Theorem. *A finite-dimensional simple Lie algebra over an algebraically closed field of characteristic zero possesses a triangular decomposition.*

The complete classification of such algebras can be found in many books (see, e.g., Jacobson [1]).

4.4. Highest weight modules. Let L be a Lie algebra with a triangular decomposition. We say that the L-module V is a *highest weight* module with highest weight $\lambda \in H^*$, if there exists an element $v_+ \in V$ such that

(1) v_+ generates V as an L-module, i.e. $U(L)(v_+) = V$;
(2) $L_+ v_+ = \{0\}$;
(3) $hv_+ = \lambda(h)v_+$ for any $h \in H$.

We define V^μ, $\mu \in H^*$ by

$$V^\mu = \{v \in V \mid hv = \mu(h)v \quad \text{for all } h \in H\}.$$

Then

$$V = \bigoplus_{\mu\in H^*} V^\mu.$$

This follows from the root decomposition for $U(L)$ over H.

Further, we set $P(V) = \{\mu \in H^* | V^\mu \neq 0\}$. Then:

(2.1) If $\mu \in P(V)$ then $\mu = \lambda - \sum_{j\in J} c_j\alpha_j$, where $c_j \in \mathbf{N}$ for all $j \in J$;

(2.2) $V^\lambda = Fv_+$;

(2.3) $V^{\lambda-\alpha} = U(L_-)^{-\alpha}v_+$ for all $\alpha \in Q_+ \cup \{0\}$;

(2.4) $U(L)^\alpha V^\mu \subset V^{\mu+\alpha}$.

The elements from the set $P(V)$ are called *weights* of the module V.

The module V is graded with the grading group $\mathbf{Z}\lambda + Q$.

A construction of a module of highest weight is as follows. Let $B = H \oplus L_+$. We consider the one–dimensional space Fv_+. Turn it into a B–module by setting

$$\begin{aligned} xv_+ &= 0 \quad \text{for } x \in L_+, \\ hv_+ &= \lambda(h)v_+ \quad \text{for } h \in H. \end{aligned}$$

We shall consider Fv_+ as a left $U(B)$–module and $U(L)$ as a right $U(B)$–module by identifying $U(B)$ with a subring of $U(L)$. We set

$$M(\lambda) = U(L) \otimes_{U(B)} Fv_+.$$

Then $M(\lambda)$ is a module of the highest weight λ with highest weight vector $1 \otimes v_+$. It is called the *Verma module* with highest weight λ.

$*$ The *universal property of Verma modules*. If M' is an L–module with highest weight λ and $M' = U(L)v'_+$ then there exists a unique epimorphism of L–modules $\psi : M(\lambda) \to M'$ such that $\psi(1 \otimes v_+) = v'_+$. $*$

The uniqueness of Verma modules (with given highest weight) is evident.

$M(\lambda) \cong U(L_-)$ as a vector space. Let N be a submodule in $M(\lambda)$. Then N is graded:

$$N = \bigoplus_{\mu \in H^*} M^\mu, \quad \text{where } N^\mu = N \cap M(\lambda)^\mu.$$

The submodule N is proper iff $N^\lambda = \{0\}$. Thus the sum $N(\lambda)$ of all proper submodules in $M(\lambda)$ is a proper submodule, so that

$$L(\lambda) = M(\lambda)/N(\lambda)$$

is an irreducible module. Moreover, it is a module of highest weight λ.

Theorem. *Let L be an algebra with a triangular decomposition over an algebraically closed field F, V a finite–dimensional irreducible L–module. Then $V \cong L(\lambda)$ for a suitable weight $\lambda \in H^*$.*

Proof. The set of linear operators on the space V which corresponds to the elements of the subalgebra H is commutative. It is well–known (see, Introduction 3.12) that in that case all these operators have common eigenvector $v \neq 0$. If $\mu(h)$ is an eigenvalue of the operator which corresponds to the element h, then $\mu \in H^*$ and $V^\mu \neq \{0\}$. We consider $V' = \sum_{\mu \in H^*} V^\mu$. If $v \in V^\mu$, $x \in L^\alpha$, $h \in H$ then

$$h(xv) = xhv + [h, x]v = \mu(h)xv + \alpha(h)xv = (\mu + \alpha)(h)(xv),$$

i.e. $xv \in V^{\mu+\alpha}$. Thus, V' is an L–submodule, i.e. $V = V' = \sum_{\mu \in H^*} V^\mu$. Further, since V is a finite–dimensional L–module, there is a weight λ such that $\lambda + \alpha_j$ cannot be a weight for any $j \in J$. Then according to Subsection 4.2, $L_+ V^\lambda = \{0\}$. In virtue of irreducibility $V = U(L)v_+$ for any $v_+ \in V^\lambda$, $v_+ \neq 0$. According to the property of universality, V is a quotient module of the Verma module $M(\lambda)$. Since V is simple, we have $V \cong L(\lambda)$. □

∗ In particular, this is how the finite–dimensional simple L–modules are constructed, when L is a finite–dimensional simple algebra over an algebraically closed field of characteristic 0. ∗

§ 5. Lie algebras and Lie groups

5.1. Lie algebra of vector fields. Let M be a n–dimensional analytic manifold. For every point $a \in M$ there is defined (see, Set–theoretic supplement, § 15) the tangent space $T_a(M)$. This is a copy of the space $\mathbf{R}^n$ as a vector space. We consider $\mathbf{R}^n$ as an n–dimensional analytic manifold with its natural analytic structure. Let $T(M)$ be the *tangent bundle* of the manifold M. We denote by p the canonical projection $\mathcal{T}(M) \to M$. Any subset $X \subset \mathcal{T}(M)$ such that $p : X \to M$ is a bijection, is called a *vector field* on M. Thus the vector space X consists of pairs (a, X_a), where $a \in M$, $X_a \in T_a(M)$. If $p^{-1} : M \to X \subset T(M)$ is a morphism of analytic manifolds then the vector field X is called *analytic*. It is clear that X is analytic iff for any function $f \in \mathcal{A}(M)$ we have $Xf \in \mathcal{A}(M)$. Here $(Xf)(a) = X_a f$. If we consider an arbitrary analytical vector space locally, i.e. within the domain of a certain chart, then

$$X = f_1(x)\frac{\partial}{\partial x_1} + \ldots + f_n(x)\frac{\partial}{\partial x_n}, \tag{1}$$

where the $f_1(x), \ldots, f_n(x)$ are analytic functions. The fact is that each of the functions $f_i(x)$ is the result of the action of the field X on the function x_i, $i = 1, \ldots, n$.

Now consider the set $\mathcal{A}(M)$ of analytic functions on M as an associative–commutative algebra over **R** with pointwise operations of multiplication, addition and multiplication by scalars. We denote by Vect (M) the set of analytic vector fields on M. Our problem consists of turning Vect (M) into a Lie algebra over **R**.

Proposition. *Any element from Vect(M) is a derivation of the algebra $\mathcal{A}(M)$. Inversily, any derivation of the algebra $\mathcal{A}(M)$ is an element from Vect(M), i.e. is an analytic vector field.*

Proof. It is known (see, formula (3), Set–theoretic supplement, § 15) that for any point a the tangent vector X_a is a linear mapping from $\mathcal{A}_a(M)$ into **R** satisfying the condition $X_a(fg) = (X_a f)g(a) + f(a)(X_a g)$. Letting a run over all points of the manifold M and taking functions $f, g \in \mathcal{A}(M)$, we obtain

$$X(fg) = (Xf)g + f(Xg). \tag{2}$$

The linearity of the mapping $X : f \mapsto Xf$ is also clear. From the definition of an analytic field it follows that X is a derivation on $\mathcal{A}(M)$. Inversily, any derivation $X : \mathcal{A}(M) \to \mathcal{A}(M)$ for any point $a \in M$ gives a mapping $X_a : f \to (Xf)(a)$. From the condition (2) it is clear that $X_a \in T_a(M)$.

Since $Xf \in \mathcal{A}(M)$, for all $f \in \mathcal{A}(M)$, then according to the definition X is an analytic field, i.e. an element from Vect(M). □

Using the proposition above the vector fields on a manifold are often *defined* as the derivations of the algebra $\mathcal{A}(M)$ of analytical functions on a manifold.

As was noted before, the set Der $\mathcal{A}(M)$ with the operation of commutators of derivations is a Lie algebra. Thus we now have the concept of the Lie algebra Vect(M) of vector fields on an analytic mannifold M. The commutator operation is written locally in the following way. If

$$X = \sum_{i=1}^{n} f_i(x)\frac{\partial}{\partial x_i}, \quad Y = \sum_{i=1}^{n} g_i(x)\frac{\partial}{\partial x_i},$$

then $Z = [X, Y]$ is the vector field

$$(3) \qquad Z = \sum_{i=1}^{n}\left(\sum_{k=1}^{n} f_k(x)\frac{\partial f_i(x)}{\partial x_k} - g_k(x)\frac{\partial f_i(x)}{\partial x_k}\right)\frac{\partial}{\partial x_i}.$$

Formula (3) can also be obtained by immediate verification (apply Z to x_k).

It should be also noted that the algebra Vect(M) is a left module over $\mathcal{A}(M)$, i.e. for any $f \in \mathcal{A}(M)$ and any $X \in$ Vect(M) the field fX, for which $(fX)g = f(Xg)$, is defined. At the same time Vect(M) is not a Lie algebra over $\mathcal{A}(M)$, i.e. the equality $[fX, gY] = fg[X, Y]$ is not always valid.

5.2. Lie algebra of vector fields on a Lie group. The operator $l(a) : x \mapsto ax$ on a Lie group G is a morphism in the sense of analytic manifolds. If $X \in$ Vect(G), then using the tangent mapping in every point one associates to the field X a field $T(l(a))X$ and the action of this field on a function $f \in \mathcal{A}(M)$ in the point $ab \in G$ given by

$$(4) \qquad [T(l(a))X]_{ab} f = X_b(f \circ l(a)).$$

Being a composition of analytic functions the function $f \circ l(a)$ is itself analytic. Therefore $T(l(a))X \in$ Vect(G). We call the field X *left invariant* if for any $a \in G$ we have $T_a(l(a))X = X$. Taking $a = g$, $b = e$ in (4), we obtain for a left–invariant field X the equality

$$X_g f = X_e(f \circ l(g)) \tag{5}$$

This equality shows that a left–invariant vector field on a Lie group is completely determined by its value in the point e (however, instead of e any element of the group G can be taken).

Lemma. *The commutator* $[X,Y]$ *of two left–invariant vector fields* X, Y *on a Lie group* G *is itself left–invariant.*

Proof. It is sufficient to verify the equality

$$[X,Y]_g = [X,Y]_e(f \circ l(g)).$$

We have

$$\begin{aligned} [X,Y]_g f =&(XY - YX)_g f = X_g(Yf) - Y_g(Xf) = \\ =&X_e((Yf) \circ l(g)) - Y_e((Xf) \circ l(g)). \end{aligned} \tag{6}$$

On the other hand,

$$[X,Y]_e(f \circ l(g)) = X_e(Y(f \circ l(g)) - Y_e(X(f \circ l(g)). \tag{7}$$

However, we show that the functions $(Yf) \circ l(g)$ and $Y(f \circ l(g))$ in the vicinity of the point e coincide:

$$((Yf) \circ l(g))(x) = Y_{gx} f = Y_x(f \circ l(g)) = (Y(f \circ l(g))(x).$$

Therefore (6) and (7) coincide and lemma is proved. □

In virtue of the lemma the set $\mathfrak{g}$ of the left–invariant vector fields on the Lie group G is a Lie algebra. It is called the *Lie algebra of the group* G. As was noted before, see (5), every left–invariant vector field is defined by its value in the point e. On the other hand, for any $\xi \in T_e(G)$ formula (5) allows to construct a left–invariant field X such that $X_e = \xi$. Therefore, $\dim \mathfrak{g} = \dim T_e(G) = \dim G = n$.

The commutator of left–invariant fields defines an operation on $T_e(G)$ if for $\zeta = [\xi,\eta]$ we take $\zeta = Z_e$, where $Z = [X,Y]$, and $X_e = \xi$, $Y_e = \eta$. The interpretation of the commutator in geometrical terms with the help of tangent vectors in the point e is presented in the following result.

Proposition. *Let X, Y be left–invariant vector fields on a group G such that $X_e = \xi$, $Y_e = \eta$. We consider curves $x(t)$, $0 \le t \le 1$, $y(t)$, $0 \le t \le 1$, $x(0) = y(0) = e$ for which ξ and η are the tangent vectors in the point e. If $Z = [X,Y]$ and $\zeta = Z_e$ then ζ is the tangent vector at e of the curve*

$$z(t) = x(\sqrt{t})y(\sqrt{t})x(\sqrt{t})^{-1}y(\sqrt{t})^{-1}. \tag{8}$$

Proof. In the vicinity of the point e the multiplication in G is given by analytic functions. If $xy = z$ then the i–th coordinate, $f_i(x,y)$, of the point z is an analytic function of the coordinates $(x_1,\dots,x_n)$, $(y_1,\dots,y_n)$ of the points x, y. From the condition $f_i(x,e) = x_i$, $f_i(e,y) = y_i$ follows that the development of this function into a series at the point zero has the form

$$\begin{aligned} f_i = x_i + y_i + \sum_{j,k} a^i_{jk} x_j y_k + \sum_{j,k,l} g^i_{jkl} x_j x_k y_l + \\ + \sum_{j,k,l} h^i_{jkl} x_j y_k y_l + \epsilon. \end{aligned} \tag{9}$$

The numbers $c^i_{jk} = a^i_{jk} - a^i_{kj}$ are called the *structure constants* of the group G in the given system of coordinates. We shall show that if $X_e = \xi$, $Y_e = \eta$, $Z = [X,Y]$ then $Z_e = \zeta$, with $\zeta_i = \sum_{j,k} c^i_{jk}\xi_j\eta_k$. Indeed, according to the definition $\zeta_i = Z_e x_i = X_e(Yx_i) - Y_e(Xx_i)$. In virtue of the invariance of the field Y we have $(Yx_i)(u) = Y_u x_i = Y_e(x_i \circ l(u))$. The function $x_i \circ l(u)$ in the point v is equal to

$$(x_i \circ l(u)(v) = x_i(uv) = f_i(u,v) = u_i + v_i + \sum_{j,k} a^i_{jk} u_j v_k + \dots$$

Taking the derivative of Y_e with respect to v, subsequently of X_e with respect to u and setting $u = v = 0$ we obtain

$$X_e(Yx_i) = \sum_{j=1}^{n}\sum_{k=1}^{n} a^i_{jk}\xi_j\eta_k.$$

Similarly,

$$Y_e(Xx_i) = \sum_{j=1}^{n}\sum_{k=1}^{i} a^i_{jk}\eta_j\xi_k.$$

Finally, forming the difference and renumbering variables we get

$$\zeta_i = \sum_{j,k} c^i_{jk}\xi_j\eta_k. \tag{10}$$

Now compute the function $h_i(x,y)$, the i–th coordinate of the point $xyx^{-1}y^{-1}$. The condition $f_i(x,x^{-1}) = 0$ implies that the i–th coordinate of the point x^{-1} has the form $-x_i + \epsilon_i$, where ϵ_i consists of the second and higher order terms. Immediate substitution in (9) shows that $h_i(x,y) = f_i(x,y) - f_i(y,x)$ up to terms of the third order. Now consider the curve $x(\sqrt{t})y(\sqrt{t})x(\sqrt{t})^{-1}y(\sqrt{t})^{-1}$. The i–th coordinate of its tangent vector has the form

$$\lim_{t\to 0}\frac{\sum_{j,k} c^i_{jk}x_j(\sqrt{t})y_k(\sqrt{t})}{t} = \lim_{s\to 0}\sum_{j,k} c^i_{jk}\frac{x_j(s)}{s}\frac{y_k(s)}{s} = \sum_{j,k} c^i_{jk}\chi_j\eta_k.$$

Comparing this with (10) we can see that the proposition is proved. □

Evidently, we can also write down our result in the vector form:

$$[\xi,\eta] = \frac{\partial^2}{\partial t\partial s}x(t)y(s)x(t)^{-1}y(s)^{-1}|_{t=s=0} \tag{11}$$

For the i–th coordinate of the commutator we have

$$[\xi,\eta]_i = \sum_{j,k}\frac{\partial^2 h_i}{\partial x_j\partial y_k}\bigg|_{(e,e)}\xi_j\eta_k. \tag{12}$$

If $h_i^*(x,y)$ is the sum of the terms of the second order in $h_i(x,y) = f_i(x,y) - f_i(y,x)$ then formula (10) shows that $\zeta_i = h_i^*(\xi,\eta)$.

5.3. Examples.

1. If G is commutative then $[\xi,\eta] = 0$ for any ξ, $\eta \in \mathfrak{g}$.

2. The tangent space M for $GL_n(\mathbf{R})$ in the point E, as is shown in the Set–theoretic supplement, § 15, is $M_n(\mathbf{R})$. In a small vicinity of E the matrix X is a set of coordinates of the matrix $E+X$. The multiplication law has the form

$$(E+X)(E+Y) = E+X+Y+XY.$$

The function $h(X,Y)$ of multiplication in the coordinates has the form

$$h(X,Y) = X+Y+XY.$$

The function $h^*(X,Y) = h(X,Y) - h(Y,X)$ which defines the multiplication in Lie algebra then has the form $h^*(X,Y) = XY - YX$. Therefore the commutator in Lie algebra $\mathfrak{g} = M_n(\mathbf{R})$ of the Lie group $GL_n(\mathbf{R})$ coincides with the usual commutator

$$[X,Y] = XY - YX. \tag{13}$$

3. *Lie algebra of a Lie subgroup.* From the proposition of Subsection 5.2 (formula (8)), is evident that if H is a Lie subgroup of the group G then the Lie algebra $\mathfrak{h} = T_e(H)$ is a subalgebra of the Lie algebra $\mathfrak{g}$ of the group G.

In particular, if H is a Lie subgroup in $GL_n(\mathbf{R})$ then $\mathfrak{h}$ is a subspace in $M_n(\mathbf{R})$ and the commutator is computed according to formula (13).

We should note that if the subgroup H is defined in a neighborhood of the identity of the group G by the system of equations $F_i(x_1,\dots,x_n) = 0$, $i = 1,\dots,m$, then $T_e(H)$ is defined in $T_e(G)$ by the system of equations

$$\sum_{j=1}^{n} \frac{\partial F_i}{\partial x_j}\bigg|_{x=e} \xi_j = 0,$$

where the $(\xi_1,\dots,\xi_n)$ are the coordinates in $T_e(G)$ corresponding to the system $(x_1,\dots,x_n)$ on the group G (according to the Set–theoretic supplement, § 15, $\xi_i = \frac{\partial}{\partial x_i}$, $i = 1,\dots,n$). For instance, the subgroup O_n is a Lie subgroup in $GL_n(\mathbf{R})$ and is defined by the system of equations $A^tA = E$.

Differentiating at $A = E$, we obtain $X + {}^tX = 0$, i.e. the Lie algebra of the group O_n is the Lie algebra of skew–symmetric matrices under commutator differences.

In a similar way, the Lie algebra of the group $SL_n(\mathbf{R})$ of matrices with determinant 1 is the Lie algebra $\mathfrak{sl}_n(\mathbf{R})$ of matrices with zero trace.

5.4. Adjoint representation. As we have seen in Subsection 2.6.5 a representation of a connected Lie group is determined by its tangent representation. In order to find the connection between a representation of a Lie group and the corresponding representation of its Lie algebra the following is necessary

Proposition. *The tangent mapping $d\phi : T_e(G) \longrightarrow T_e(H)$ of a homomorphism of Lie groups $\phi : G \to H$ is a homomorphism of Lie algebras $\mathfrak{g} \to \mathfrak{h}$.*

Proof. Choose curves $g(t)$, $h(s)$ which define the vectors ξ, $\eta \in \mathfrak{g} = T_e(G)$. Using (11) we can write down the image of the commutator $\zeta = [\xi, \eta]$ under the tangent mapping $d\phi$ in the form

$$(d\phi)\left(\frac{\partial^2}{\partial s \partial t} g(t)h(s)g(t)^{-1}h(s)^{-1}\bigg|_{t=s=0}\right).$$

We shall carry out the differentiation first with respect to s and afterwards with respect to t:

$$(d\phi)\left(\frac{\partial}{\partial s}[g(t), h(s)]\bigg|_{s=0}\right) = \frac{\partial}{\partial s}\phi([g(t), h(s)])\bigg|_{s=0} =$$
$$= \frac{\partial}{\partial s}([\phi(g(t)), \phi(h(s))])\bigg|_{s=0}.$$

Since $d\phi : \mathfrak{g} \to \mathfrak{h}$ is linear and continuous, it commutes with taking a derivative. Consequently,

$$(d\phi)\left(\frac{\partial^2}{\partial t \partial s}[g(t), h(s)]\bigg|_{t=s=0}\right) =$$
$$= \frac{\partial}{\partial t}(d\phi)\left(\frac{\partial}{\partial s}[g(t), h(s)]\bigg|_{s=0}\right)\bigg|_{t=0} =$$
$$= \frac{\partial^2}{\partial t \partial s}[\phi(g(t)), \phi(h(s))]\bigg|_{t=s=0},$$

i.e. $(d\phi)[\xi, \eta] = [(d\phi)\xi, (d\phi)\eta]$, as was required. □

Corollary. *Let $\rho : G \to GL(V)$ be a linear representation of a Lie group G. Then $d\rho$ is a linear representation of the Lie algebra $\mathfrak{g}$.*

If G is connected then ρ is entirely defined by the representation $d\rho$, moreover ρ is irreducible iff this is valid for $d\rho$.

The validity of the latter follows from the results presented in Subsection 2.6.5. □

In Subsection 2.6.5 we considered the adjoint representation of the group $GL_n(V)$ by linear operators on the space $\mathrm{End}(V)$. This representation can be generalized to the case of an arbitrary group $G : \imath_g(x) = gxg^{-1}$. We consider the tangent mapping $d\imath_g : \mathfrak{g} \to \mathfrak{g}$. We then write $\mathrm{Ad}\ g = d\imath_g \in GL(\mathfrak{g})$. Since $\imath_{g_1}\imath_{g_2} = \imath_{g_1 g_2}$, we have $\mathrm{Ad}\ (g_1 g_2) = (\mathrm{Ad}\ g_1)(\mathrm{Ad}\ g_2)$. Moreover, from the definition of Lie group it is clear that $\mathrm{Ad}\ g$ depends smoothly on g. Thus, the mapping $\mathrm{Ad} : G \to GL(\mathfrak{g})$ is a representation of the Lie group G. It is called the *adjoint representation* of G. It is easy to check that the given definition is compatible with the previous one in the case of the group $GL(V)$ or any of its subgroups.

The tangent representation for the representation Ad of the group G is called the *adjoint representation of the Lie algebra* $\mathfrak{g}$ and is denoted by ad . It is possible to demonstrate that like in the case $G \subset GL(V)$, the following equality is valid

$$(\mathrm{ad}\ \xi)(\eta) = [\xi, \eta].$$

References.

Bahturin [1], Bourbaki [2], Vinberg [1], Jacobson [1], Dixmier [1], Magnus–Karrass–Solitar [1], Pontryagin [1], Serre [1], Moody–Pianzola [1], Sagle–Walde [1].

CHAPTER V

HOMOLOGICAL ALGEBRA

§ 1. Complexes of modules

1.1. Homology of a complex. Let A be some associative ring. An *increasing chain complex* $\mathcal{K}$ of modules of A (here all modules are left ones) is a sequence of A–modules and homomorphisms of the form

$$\ldots \leftarrow M_{-1} \stackrel{\partial_0}{\leftarrow} M_0 \stackrel{\partial_1}{\leftarrow} M_1 \stackrel{\partial_2}{\leftarrow} M_2 \leftarrow \ldots \tag{1}$$

such that, moreover, $\operatorname{Ker}\partial_{n-1} \supset \operatorname{Im}\,\partial_n$ for any n. A special case is an *exact* complex, i.e. a sequence (1) such that $\operatorname{Ker}\partial_{n-1} = \operatorname{Im}\,\partial_n$ for any n. *Decreasing* complexes can also be regarded. *A short exact sequence* is an exact sequence of the form

$$0 \to M \xrightarrow{\tau} N \xrightarrow{\epsilon} P \to 0. \tag{2}$$

Here N contains the submodule $\tau(M)$ which is isomorphic to M, and the quotient module with respect to this submodule is isomorphic to P.

Returning to the case of an arbitrary complex we note that the elements from M_n are called *n–dimensional chains* of the complex $\mathcal{K}$ ((1)), the elements from $\operatorname{Ker}\partial_n \subset M_n$ are the *n–dimensional cycles*, the elements from $\operatorname{Im}\,\partial_{n-1}$ the *n–dimensional boundaries*, ∂_n is the *n–dimensional boundary operator*. The group $\operatorname{Ker}\partial_{n-1}/\operatorname{Im}\,\partial_n$ is called the n–th *homology* group of the complex $\mathcal{K}$ and is denoted by $H_n(\mathcal{K})$. If n is left unspecified, then one writes $H_*(\mathcal{K})$, the graded module of all homology groups of the complex.

If M is an A–module then for any $n = 0, \pm 1, \pm 2, \ldots$ we consider the group $P_n = \operatorname{Hom}_A(M_n, M)$ of homomorphisms of A–modules. We define $\delta_n : P_n \to P_{n+1}$ by $\delta_n(f) : M_{n+1} \to M_n$ is the function $(\delta^n f)(x) = f(\partial_{n+1}x)$, $x \in M_{n+1}$. Then $\delta^n f \in \operatorname{Hom}(M_{n+1}, M)$, and

$$((\delta^{n+1}\delta^n)f)(x) = f(\partial_{n+1}\partial_{n+2}x) = 0.$$

Thus the sequence Hom($\mathcal{K}, M$) of the modules P_n with boundary operators δ^n is a chain complex. Its homology $H_*(\mathrm{Hom}(\mathcal{K}, M))$ is called the *cohomology* of the complex $\mathcal{K}$ with values in the module M and is denoted as $H^*(\mathcal{K}, M)$. It is customary to call the n–dimensional chains, cycles, boudaries of the complex Hom($\mathcal{K}, M$) *cochains, cocycles, coboundaries* of the complex $\mathcal{K}$ with values in the module M. The operator δ^n is called the *coboundary operator*.

1.2. Chain homotopy and chain transformation. In the computation of the homology of a complex an important role is played by the concept of a (contracting) homotopy. Namely, a *homotopy* of the complex (1) is a set of A–module homomorphisms $s_n : M_n \to M_{n+1}$ ($n = 0, \pm1, \pm2, \ldots$) such that for any n in the diagram of mappings

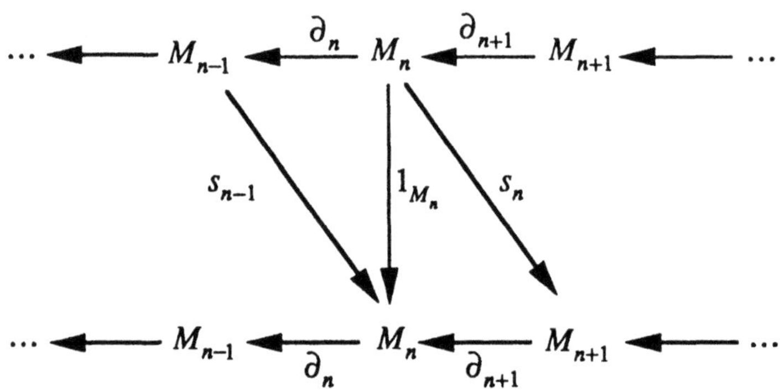

the following relation is valid

(3) $$s_{n-1}\partial_n + \partial_{n+1}s_n = 1_{M_n}.$$

Proposition 1. *The existence of a contracting homotopy in a given complex implies that the homology of this complex is equal to zero (in other words, the complex is exact).*

Proof. Let x be a n–dimensional cycle, i.e. $\partial_n x = 0$. Now (3) implies the relation

$$s_{n-1}\partial_n x + \partial_{n+1} s_n x = x$$

Hence $x = \partial_{n+1}(s_n x)$, i.e. x is a boundary. □

If $\mathcal{K}'$ is another complex of A–modules

$$\dots \overset{\partial'_{-1}}{\leftarrow} M'_{-1} \overset{\partial'_0}{\leftarrow} M'_0 \overset{\partial'_1}{\leftarrow} M'_1 \overset{\partial'_2}{\leftarrow} M'_2 \leftarrow \dots$$

then a set $f = (f_n)_{n\in\mathbf{Z}}$ of homomorphisms of A–modules $f_n : M_n \to M'_n$ is called a *morphism of complexes* if for any n

$$\partial'_n f_n = f_{n-1}\partial_n,$$

i.e. the following diagram of mappings is commutative

$$\begin{array}{ccc} M_{n-1} & \overset{\partial_n}{\longleftarrow} & M_n \\ {\scriptstyle f_{n-1}}\downarrow & & \downarrow{\scriptstyle f_n} \\ M'_{n-1} & \underset{\partial'_n}{\longleftarrow} & M'_n \end{array}$$

Proposition 2. *If $f : \mathcal{K} \to \mathcal{K}'$ is a morphism of complexes, and M is some A–module then f induces homomorphisms $H_*(f)$ and $H^*(f)$ of the homology groups and of the cohomology groups with values in the module M*

$$H_n(f) : H_n(\mathcal{K}) \to H_n(\mathcal{K}'),$$
$$H^n(f) : H^n(\mathcal{K}', M) \to H^n(\mathcal{K}, M).$$

Indeed, if $x + \mathrm{Im}\ \partial_{n+1} \in H_n(\mathcal{K})$ then $f_n(x) + \mathrm{Im}\ \partial'_{n+1} \in H_n(\mathcal{K}')$, because $\partial'_n f_n(x) = f_{n-1}\partial_n(x) = 0$ and

$$f_n(x + \partial_{n+1}y) = f_n(x) + f_n(\partial_{n+1}y) =$$
$$= f_n(x) + \partial'_{n+1} f_{n+1}(y) \in f_n(x) + \mathrm{Im}\ \partial'_{n+1}.$$

So we can set $H_n(f)(\bar{x}) = \overline{f_n(x)}$.

Further, let $\phi + \text{Im }\delta'^{n-1} \in H^n(\mathcal{K}', M)$. Consider $\phi f_n + \text{Im }\delta^{n-1}$. Then $\phi f_n \in \text{Hom}(M_n, M)$. Further, if $x \in M_{n+1}$ then

$$\begin{aligned}\delta^n(\phi f_n)(x) = \phi f_n(\partial_{n+1}x) = \phi\partial'_{n+1}f_{n+1}(x) = \\ = (\delta'^n\phi)(f_{n+1}(x)) = 0.\end{aligned}$$

Hence (ϕf_n) is a cocycle. If ϕ is a coboundary, i.e. $\phi = \delta'^{n-1}\psi$ then

$$\begin{aligned}(\phi f_n)(x) = (\delta'^{n-1}\psi)f_n(x) = \psi\partial'_n f_n(x) = \\ = \psi f_{n-1}\partial_n(x) = \delta^{n-1}(\psi f_{n-1})(x).\end{aligned}$$

So $\phi f_n = \delta^{n-1}(\psi f_{n-1}) \in \text{Im }\delta^{n-1}$. Consequently, the mapping

$$\phi + \text{Im }\delta'^{n-1} \mapsto \phi f_n + \text{Im }\delta^{n-1}$$

is a well defined homomorphism of A–modules. □

Two chain mappings $f, g : \mathcal{K} \to \mathcal{K}'$ are called homotopic if there exists a set of homomorphisms $(s_n)_{n\in\mathbf{Z}}$, $s_n : M_n \to M'_{n+1}$ such that for any $n \in \mathbf{Z}$ we have

$$s_{n-1}\partial_n + \partial_{n+1}s_n = f_n - g_n.$$

It is clear that we get a contracting homotopy of the complex if in the given definition we take $\mathcal{K}' = \mathcal{K}$, $f = 1$, $g = 0$.

Proposition 3. *Suppose that $f, g : \mathcal{K} \to \mathcal{K}'$ are homotopic chain mappings. Then $H_n(f) = H_n(g)$ and $H^n(f) = H^n(g)$ for any $n \in \mathbf{Z}$ and any module of coefficients for the cohomology.*

Proof. Let $x \in \text{Ker }\partial_n$. Then

$$\begin{aligned}H_n(f)(\bar{x}) = f_n(x) + \text{Im }\partial_{n+1} = \\ = g_n(x) + s_{n-1}\partial_n x + \partial_{n+1}s_n(x) + \text{Im }\partial_{n+1} = \\ = g_n(x) + \text{Im }\partial_{n+1} = H_n(g)(\bar{x}),\end{aligned}$$

as $s_{n-1}\partial_n x = 0$, because $x \in \operatorname{Ker}\partial_n$, and $\partial_{n+1}s_n(x) \in \operatorname{Im}\partial_{n+1}$. Further, let $\phi \in \operatorname{Ker}\delta^n$, $x \in M'_n$. Then

$$\begin{aligned}(H^n(f)\phi)(x) &= \phi(f_n(x)) + \operatorname{Im}\ \delta^{n-1} = \\ &= \phi(g_n(x)) + \phi(s_{n-1}\partial_n x) + \phi(\partial_{n+1}s_n(x)) + \operatorname{Im}\ \delta^{n-1} = \\ &= (H^n(g)\phi(x) + \delta^{n-1}(\phi s_{n-1})(x) + (\delta^n\phi)(s_n(x)) + \operatorname{Im}\ \delta^{n-1} = \\ &= (H^n(g))(x) + \psi(x) + \operatorname{Im}\ \delta^{n-1}, \quad \text{where } \psi \in \operatorname{Im}\ \delta^{n-1}.\end{aligned}$$

Thus $\overline{H^n(f)} = \overline{H^n(g)}$. □

1.3. Resolutions. Let A be some ring, M a left A–module. An exact complex of the form

$$0 \leftarrow M = M_0 \overset{\partial_1}{\leftarrow} M_1 \overset{\partial_2}{\leftarrow} M_2 \leftarrow \dots \tag{4}$$

is called a *resolution* of the A–module M. Of particular interest are resolutions where all modules $M_1, M_2, \dots$ belong to some "good" class of modules. For instance, if all modules $M_1, M_2, \dots$ are free, then (4) is called a *free* resolution. Such a resolution always exists. Indeed, there always exists a free A–module M_1 which has a homomorphic image equal to M. We denote this homomorphism by ∂_1. If $\operatorname{Ker}\partial_1$ is the kernel of this homomorphism then there exists a free module M_2 and a surjective homomorphism $\partial_2 : M_2 \to \operatorname{Ker}\partial_1$. Having taken its kernel $\operatorname{Ker}\partial_2$, we construct M_3 and $\partial_3 : M_3 \to \operatorname{Ker}\partial_2$, etc. If $A = \mathbf{Z}$ is the ring of integers then for any M we can select a free resolution where all terms starting with M_3 are zero. Indeed, any submodule of a free $\mathbf{Z}$–module (i.e. actually a subgroup of a free abelian group) is free. Therefore in the construction presented above we can set $M_2 = \operatorname{Ker}\partial_1$ and $\partial_2 = 1_{M_2}$. Then $\operatorname{Ker}\partial_2 = 0$ and the subsequent terms are zero. In this case (4) takes the form of a short exact sequence (2).

The concept of a resolution is important in the definition of such central concepts as cohomology of rings, groups, Lie algebras, etc.. We shall define this for a ring A. In the case of a group G the integer group ring $A = \mathbf{Z}[G]$ of the group G should be considered. In the case of a Lie algebra G over the field F the universal enveloping algebra $U(G)$ should be considered. The details are presented in the following sections and subsections.

So let A be a ring, M a left module. The group $H^n(A, M)$ is defined as the n–th cohomology group of a free resolution of the A–module M with values in M and is called the *n–th cohomology group* of the ring A with values in the module M.

The given definition is noticeably arbitrary in the choice of a free resolution. Besides, in some cases the free resolution is unwieldy and the possibility of using more simple resolution would be desirable. In this connection the "independence of resolution" theorem (see, Subsection 1.5) is important. In order to formulate this theorem it is necessary to introduce the concept of a projective module.

1.4. Projective module. A left A–module P is called *projective* if for any A–modules M' and M'' and any homomorphisms $\phi : M' \to M''$, $\psi : P \to M''$, where ϕ is surjective, there exists a homomorphism $\chi : P \to M'$ which makes the following diagram commutative

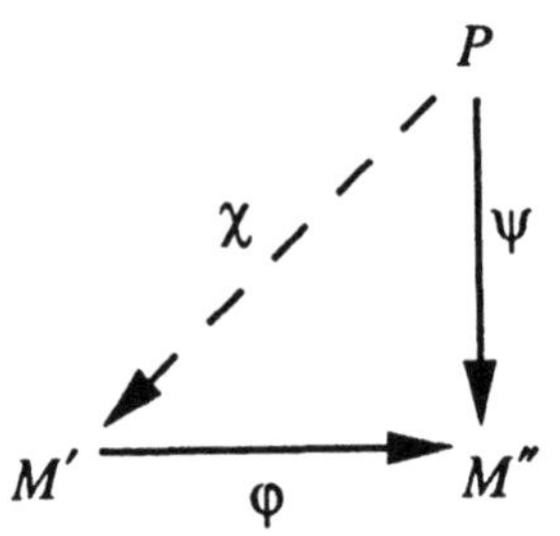

Proposition. *A module P is projective iff it is a direct summand of a free A–module F. In particular, a free module is projective.*

Proof. Supppose that P is projective. We present it as a homomorphic image of a free module F by a homomorphism $\phi : F \to P$; define $\psi : P \to P$ as the identity isomorphism. Then there exists $\chi : P \to F$ such that $\phi\chi = 1_P$, and the intersection of $F' = \operatorname{Ker}\phi$ and $\chi(P)$ is zero. If $x \in F$ then $y = x - \chi\psi^{-1}\phi(x)$ lies in $\operatorname{Ker}\phi$. In fact,

$$\phi(y) = \phi(x) - \phi\chi\psi^{-1}\phi(x) = \phi(x) - \psi\psi^{-1}\phi(x) = 0,$$

i.e.

$$x = y + \chi(\psi^{-1}\phi(x)) \in \operatorname{Ker}\phi \oplus \operatorname{Im}\ \chi \cong F' \oplus P$$

and $F \cong F' \oplus P$, as required.

Inversily, let $F = F' \oplus P$, where F is a free A–module. Let ϕ, ψ, M', M'' be as in the definition above, and π be the projection of the module F on P. We consider $\psi' = \phi\pi : F \to M''$. If X is a basis for F then any mapping of the set X into an arbitrary A–module extends to a homomorphism of A–modules. For any $x \in X$ select an element $y \in M'$ such that $\phi(y) = \psi\pi(x)$. According to what was said before, there exists a homomorphism of A–modules $\chi' : F \to M'$ which takes x into y, as above. If $\imath : P \to F$ is the identity imbedding of the module P into F then the desired homomorphism χ is obtained if we set $\chi = \chi'\imath$. Ideed, we have $\phi\chi' = \psi\pi$. Then for $x \in P$ we get

$$\phi\chi(x) = \phi\chi\imath(x) = \phi\chi'(x) = \psi\pi(x) = \psi(x).$$

Thus, we find a homomorphism $\chi : P \to M'$ such that $\phi\chi = \psi$. □

We call a resolution (4) *projective* if $M_1, M_2, \ldots$ are projective modules.

1.5. Theorem. *Let $\mathcal{K}$, $\mathcal{K}'$ be the two projective resolutions of a module M over the ring A. Then for any $n \geq 0$ the groups $H^n(\mathcal{K}, M)$ and $H^n(\mathcal{K}', M)$ are isomorphic.*

Proof. Let $\mathcal{K}$ be a projective resolution of the form (4) and $\mathcal{K}'$ another projective resolution of the module M

$$0 \leftarrow M = M_0' \overset{\partial_1'^{1}}{\leftarrow} M_1' \overset{\partial_2'}{\leftarrow} M_2' \leftarrow \ldots \qquad (4')$$

First we construct a chain homomorphism $f : \mathcal{K} \to \mathcal{K}'$ such that $f_0 = 1_{M_0}$.

$$\begin{array}{ccccccccccccc}
0 & \longleftarrow & M & = & M_0 & \overset{\partial_1}{\longleftarrow} & M_1 & \overset{\partial_2}{\longleftarrow} & M_2 & \overset{\partial_3}{\longleftarrow} & M_3 & \longleftarrow & \ldots \\
 & & & & \downarrow f_0 = 1_{M_0} & & \downarrow f_1 & & \downarrow f_2 & & \downarrow f_3 & & \\
0 & \longleftarrow & M & = & M_0' & \overset{\partial_1'}{\longleftarrow} & M_1' & \overset{\partial_2'}{\longleftarrow} & M_2' & \overset{\partial_3'}{\longleftarrow} & M_3' & \longleftarrow & \ldots
\end{array}$$

In virtue of the projectivity of the module M_1 and the surjectivity of the boundary operator ∂_1 there exists a homomorphism $f_1 : M_1 \to M_1'$ such that $f_0\partial_1 = \partial_1' f_1$.

Now we examine $f_1\partial_2(M_2)$; because of the previous,

$$\partial_1' f_1\partial_2(M_2) = f_0\partial_1\partial_2(M_2) = 0.$$

By the exactness of the sequence (4), $f_1\partial_2 M_2 \subset \partial_2' M_2'$. Since M_2 is projective, there exists a $f_2 : M_2 \to M_2'$ which makes the following diagram commutative

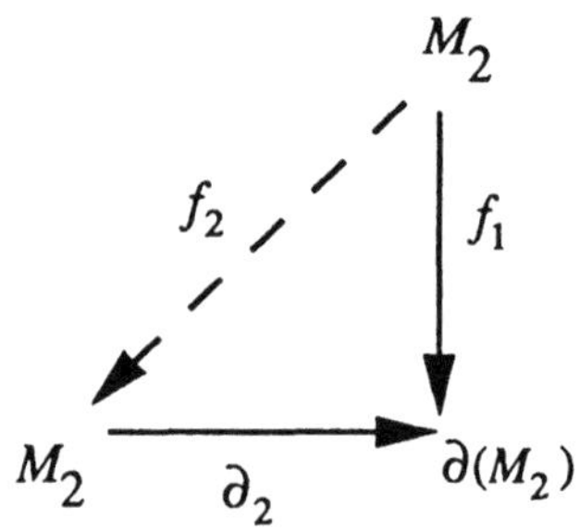

Thus $f_1\partial_2' = \partial_2 f_2$. Continuing in the same way we construct the remaining homomorphisms $f_n : M_n \to M_n'$.

Similarly, we can construct a chain homomorphism $f' : \mathcal{K}' \to \mathcal{K}$ such that $f_0' = 1_M$.

The proof of the theorem will be complete if we show that ff' is homotopic to $1_{\mathcal{K}}$, and that $f'f$ is homotopic to $1_{\mathcal{K}'}$. These two statements amount to the same. Consider $g = ff' - 1_{\mathcal{K}}$. Then $g_0 = 0$. It is sufficient to construct homomorphisms $s_n : M_n \to M_{n+1}$ for any $n > 0$, such that

$$\partial_{n+1}s_n + \partial_n s_{n-1} = g_n.$$

Pictorially,

Set $s_0 = 0$. Because of the commutativity of the squares above $\partial_1 g_1(M) = g_0\partial_1 M = \{0\}$; hence, by exactness, $g_1(M_1) \subset \partial_2 M_2$. In virtue of the projectivity of the module M_1, there exists an s_1 such that $\partial_2 s_1 = g_1$ or

$$s_0\partial_1 + \partial_2 s_1 = g_1.$$

Now consider $g_2 - s_1\partial : M_2 \to M_2$. We have

$$\partial_2(g_2 - s_1\partial_2) = \partial_2 g_2 - \partial_2 s_1 \partial_2 = \partial_2 g_2 - g_1\partial_2 = 0.$$

Therefore $(g_2 - s_1\partial_2)(M_2) \subset \partial_3 M_3$. Because of the projectivity of the module M_2, there exists an s_2 such that $g_2 - s_1\partial_2 = \partial_3 s_2$. In other words,

$$s_1\partial_2 + \partial_3 s_2 = g_2.$$

In terms of diagrams,

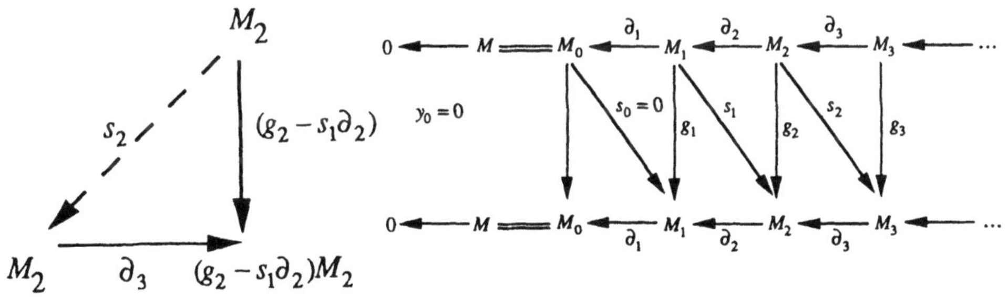

The remaining homomorphisms s_n, $n \geq 3$, are constructed in a similar way.

Thus according to proposition 3,

$$H^*(1) = H^*(ff') = H^*(f)H^*(f') = H^*(f')H^*(f).$$

Therefore $H^*(f) : H^*(\mathcal{K}', M) \to H^*(\mathcal{K}, M)$, and $H^*(f') : H^*(\mathcal{K}, M) \to H^*(\mathcal{K}', M)$ are mutually inverse homomorphisms of the cohomology groups. □

§ 2. Cohomology of groups

2.1 Concept of a G–module. Let G be a group, $A = \mathbf{Z}[G]$ its integer group ring. An abelian group M is called a *G–module* if there is given a homomorphism $\phi : G \to \mathrm{Aut}(M)$ of the group G into the group of automorphisms of the group M. Then we can define a multiplication of elements from M with elements from G, by setting $g \circ x = \phi(g)(x)$, where $g \in G$, $x \in M$ (this is often written as $g \circ x = gx$). In addition, the usual axioms of a module are supposed to hold (see, Introduction § 3). Note that the given definition automatically makes M into an A–module. Indeed,

$$\left(\sum_{g\in G} n_g g\right) x = \sum_{g\in G} n_g(gx), \tag{1}$$

where the n_g are integers, $x \in M$. Inversily, if M is a $\mathbf{Z}[G]$–module then from (1) it is evident that M is also a G–module. Situations when it is appropriate to study G–modules occur in the analysis of intrinsic structure of a group. For instance, if G is a group, T an abelian normal subgroup then for $g \in G$, $x \in T$, we can define:

$$g \circ x = gxg^{-1}.$$

It is evident that T becomes a G–module. Moreover, the elements from T act on T in a trivial way and this allows one to consider T also as a G/T–module

$$(gT)x = gxg^{-1}.$$

The study of such structure is especially useful in the case of solvable groups.

Another situation when it is appropriate to consider G–modules occur in the study of integral representations of groups.

2.2. Free resolutions in the case of groups. Take a group G and a G–module M. Consider the group $\mathbf{Z}$ as a trivial G–module, by setting $g \circ x = x$. Now consider $\mathbf{Z}$ and M as $A = \mathbf{Z}[G]$–modules as described above. Let $\mathcal{K}$ be a projective resolution of the A–module $\mathbf{Z}$:

$$0 \leftarrow \mathbf{Z} \leftarrow P_1 \leftarrow P_2 \leftarrow \ldots \leftarrow P_n \leftarrow P_{n+1} \leftarrow \ldots \tag{2}$$

We define the n–dimensional cohomology group of the group G with coefficients from the module M as the group $H^n(\mathcal{K}, M)$ and denote it by $H^n(G, M)$. In order to obtain an interpretation of the cohomology groups we introduce a standard (free) resolution which is constructed as follows. For $n > 0$ consider the free G–module F_n with as basis all possible sets $(x_1, x_2, \dots, x_n)$, where $x_1, \dots, x_n$ are non–unit elements of the group G. For convenience we shall also consider sets including the unit element. These sets are assumed to be equal to zero in F_n. When $n = 0$ we take F_0 to be the free cyclic A–module generated by the symbol $()$. Thus each element from F_n is a linear combination of elements of the form $x(x_1, x_2, \dots, x_n)$, where $x, x_1, x_2, \dots, x_n \in G$. Define the boundary operator $\partial : F_n \to F_{n-1}$ as the homomorphism of A–modules which is given on the basis of F_n by the formula

$$\begin{aligned} &\partial_n(x_1, x_2, \dots, x_n) = \\ &= x_1(x_2, \dots, x_n) - (x_1x_2, x_3, \dots .x_n) + (x_1, x_2x_3, \dots, x_n) + \dots \\ &\dots + (-1)^{n-1}(x_1, \dots, x_{n-2}, x_{n-1}x)n) + (-1)^n(x_1, x_2, \dots, x_{n-1}). \end{aligned} \tag{3}$$

In particular,

$$\partial_1(x_1) = x_1(\,) - (\,) = x_1 - 1, \partial_2(x_1, x_2) = x_1(x_2) - (x_1x_2) + (x_1).$$

Formula (3) makes sense also when some of x_i's are equal to 1.

The complex $\mathcal{F}$ of the form

$$0 \xleftarrow{\epsilon} F_0 \xleftarrow{\partial_1} F_1 \xleftarrow{\partial_2} F_2 \longleftarrow \dots \longleftarrow F_n \xleftarrow{\partial_{n+1}} F_{n+1} \longleftarrow \dots \tag{4}$$

where

$$\epsilon(x(\,)) = 1,$$

is exact. Indeed, a homotopy $(s_n)_{n \geq -1}$ contracting it can be defined in the following way (here $s_{-1} : \mathbf{Z} \longrightarrow F_0$):

$$s_{-1}(m) = m(\,), \ s_n(x(x_1, \dots, x_n)) = (x, x_1, \dots, x_n).$$

Let us show the validity of the conditions defining homotopies:

$$\epsilon s_{-1} = 1, \ \partial_1 s_0 + s_{-1}\epsilon = 1, \ \partial_{n+1}s_n + s_{n-1}\partial_n = 1 \ (n \geq 1).$$

Indeed,

$$(\partial_1 s_0 + s_{-1}\epsilon)(x(\,)) = \partial_1(x) + s_{-1}(1) = x(\,) - (\,) + (\,) = x(\,),$$
$$(\partial_{n+1}s_n + s_{n-1}\partial_n)(x(x_1,\dots,x_n)) = \partial_{n+1}(x,\ x_1,\dots,\ x_n)+$$
$$+s_{n-1}x(x_1(x_2,\dots,\ x_n) + \sum_{i=1}^{n-1}(-1)^i(x_1,\dots,\ x_ix_{i+1},\dots,\ x_n)+$$
$$+(-1)^n(x_1,\dots,x_{n-1})) = x(x_1,\dots,x_n) - (xx_1,x_2,\dots,x_n)+$$
$$+(-1)^{n+1}(x,x_1,\dots,x_n) + (xx_1,x_2,\dots,x_n)+$$
$$+\sum_{i=1}^{n-1}(-1)^i(x,x_1,\dots,x_ix_{i+1},\dots,x_n)+$$
$$+(-1)^n(x,x_1,\dots,x_n) = x(x_1,\dots,x_n).$$

Note that F_{n+1} is generated as an A–module by the subgroup s_nF_n. Therefore, in order to prove the equalities $\delta_n\delta_{n+1} = 0$, it is sufficient to calculate $\partial_n\partial_{n+1}s_n$. We have

$$\partial_n\partial_{n+1}s_n = \partial_n(1 - s_{n-1}\partial_n) = \partial_n - (\partial_n s_{n-1})\partial_n =$$
$$=\partial_n - (1 - s_{n-2}\partial_{n-1})\partial_n = (\partial_n - \partial_n) + s_{n-2}(\partial_{n-1}\partial_n) = 0.$$

The last equality is verified by induction on n. Let us state what has been proved.

Proposition. *For any group G the complex $\mathcal{F}$ defined by* (4), *is a free resolution of the trivial G–module* $\mathbf{Z}$. □

2.3. Cohomology groups of small dimension. Let G be a group, M a G–module, $A = \mathbf{Z}[G]$, $\mathcal{F}$ the standard resolution of the G–module $\mathbf{Z}$. We examine the complex $\mathrm{Hom}(\mathcal{F}, M)$:

$$0 \to \mathrm{Hom}_A(\mathbf{Z}, M) \xrightarrow{\epsilon^*} \mathrm{Hom}_A(F_0, M) \xrightarrow{\delta^0}$$
$$\to \mathrm{Hom}_A(F_1, M) \xrightarrow{\delta^1} \mathrm{Hom}_A(F_2, M) \xrightarrow{\delta^2} \mathrm{Hom}_A(F_3, M) \to \dots$$

If $\phi \in \mathrm{Hom}_A(F_n, M)$ then ϕ is completely determined by its values on the sets $(x_1, x_2, \dots \dots, x_n)$, $x_i \in G$. Therefore ϕ can be regarded as the function $\phi : G^n \to M$ which vanishes on those sets where at least one of the elements $x_1, \dots, x_n$ is equal to the unit element. The coboundary operator is defined by the formula

$$
\begin{aligned}
&(\delta^n \phi)(x_1, \dots, x_{n+1}) = x_1\phi(x_2, \dots, x_{n+1}) - \\
&-\phi(x_1x_2, x_3, \dots, x_{n+1}) + \phi(x_1, x_2x_3, \dots, x_n) - \\
&- \dots + (-1)^n \phi(x_1, x_2, \dots, x_n x_{n+1}) + \\
&+(-1)^{n+1}\phi(x_1, x_2, \dots, x_n) = x_1\phi(x_2, \dots, x^{n+1}) + \\
&+\sum_{i=1}^{n}(-1)^i \phi(x_1, \dots, x_i x_{i+1}, \dots, x_{n+1}) + \\
&+(-1)^{n+1}\phi(x_1, \dots, x_n).
\end{aligned} \tag{5}
$$

In particular, an element ϕ from $\mathrm{Hom}(F_0, M)$ is a cocycle if for any x we have $(\delta^0\phi)(x) = x\phi() - \phi() = 0$, i.e. $x\phi() = \phi()$. Identifying ϕ with $\phi() \in M$ we obtain that the set of zero–dimensional cocycles is isomorphic to the set M^G of fixed points of the G–module M.

A one–dimensional cochain $\phi \in \mathrm{Hom}(F_1, M)$, which is defined as the mapping $\phi : G \to M$ with the condition $\phi(1) = 0$, is a cocycle iff for any $x, y \in G$ we have

$$x\phi(y) - \phi(xy) + \phi(x) = 0,$$

in other words

$$\phi(xy) = x\phi(y) + \phi(x). \tag{6}$$

Formula (6) reminds one of the derivatives of a product and a function $\phi : G \to M$, satisfying this condition, is called a *derivation* of the group G in the G–module M. The set of all differentiations is defined by $\mathrm{Der}(G, M)$.

Finally a two–dimensional cochain $\psi : G \times G \to M$ is a 2–cocycle if for any x, y, $z \in G$ we have

$$x\psi(y, z) - \psi(xy, z) + \psi(x, yz) - \psi(x, y) = 0. \tag{7}$$

The set of all 2–cocycles is denoted by $Z^2(G, M)$.

Now we consider the explicit form of the coboundaries. When $n = 0$, by definition the coboundaries are equal to zero. Therefore we have $H^0(G, M) \cong M^G$. For $n = 1$ a function $\psi : G \to M$ is a coboundary if there exists the element $m = \phi() \in M$ for which $\psi(x) = xm - m$. It is clear that such a function is a derivation from G into M; it is called an *inner* derivation and is defined by the element $m \in M$. A set of all inner derivations from G into M is denoted $\operatorname{Int}(G, M)$. The quotient group

$$\operatorname{Out}(G, M) = \operatorname{Der}(G, M)/\operatorname{Int}(G, M)$$

is called the group of outer derivations from G into M. Thus

$$H^1(G, M) \cong \operatorname{Out}(G, M).$$

Further, a 2–cochain $\psi : G \times G \to M$ is a 2–coboundary if there exists a 1–cochain $\phi : G \to M$ such that for any $x, y \in G$ we have

$$\psi(x, y) = x\phi(y) - \phi(xy) + \phi(x). \tag{8}$$

The set of all 2–coboundaries is denoted by $B^2(G, M)$. We have

$$H^2(G, M) \cong Z^2(G, M)/B^2(G, M).$$

2.4. Interpretation of the group H^1. One of the interpretations of the group H^1 is related to the problem of module extensions. Let M and N be two G–modules. We consider a short exact sequence

$$0 \to M \xrightarrow{\imath} P \xrightarrow{\epsilon} N \to 0. \tag{9}$$

In the module P there is the submodule $\imath(M)$ which is isomorphic to M and the quotient module with respect to $\imath(M)$ isomorphic to N. In this situation it is said that P is an *extension* of the module M by means of N. Let

$$0 \to M \to P' \to N \to 0 \tag{9'}$$

be another extension of M by means of N. We then say that (8) and (9′) are *equivalent* if there exists an isomorphism $\pi : P \to P'$ which makes the following diagram commutative

$$\begin{array}{ccccccccc} 0 & \longrightarrow & M & \xrightarrow{\imath} & P & \xrightarrow{\epsilon} & N & \longrightarrow & 0 \\ & & \downarrow 1_M & & \downarrow \pi & & \downarrow 1_N & & \\ 0 & \longrightarrow & M & \xrightarrow{\imath'} & P' & \xrightarrow{\epsilon'} & N & \longrightarrow & 0 \end{array} \tag{10}$$

i.e. $\pi\imath = \imath' 1_M$ and $1_N \epsilon = \epsilon' \pi$. The extension

$$0 \to M \xrightarrow{\alpha} M \oplus N \xrightarrow{\beta} N \to 0, \tag{11}$$

where α is the identification with the first summand and β is a projection on the second summand, is called *trivial*.

Finally, the extension is called *split* if it is equivalent to the trivial one.

We shall limit ourselves by studying the extensions where N is a trivial G–module, i.e. $g \circ x = x$ for any $g \in G$ and $x \in N$. Moreover, we shall assume that extensions are trivial as extensions of abelian groups (though not as G–modules). Such extensions are called $\mathcal{A}$–extensions. Thus, if (9) is an $\mathcal{A}$–extension then $P = \imath(M) \oplus N'$, moreover $\epsilon|_{N'} : N' \to N$ is an isomorphism of abelian groups.

Theorem. *The set of equivalence classes of $\mathcal{A}$–extensions of a G–module M by means of a trivial G–module N is in one–to–one correspondence with the set of elements of the group $H^1(G, \mathrm{Hom}_{\mathbf{Z}}(N, M))$.*

Proof. Take a 1–cocycle $\phi : G \to \mathrm{Hom}(N, M)$. According to (6) for any $x, y \in G$, $n \in N$ we have

$$(\phi(xy))(n) = (x\phi(y))(n) + \phi(x)(n).$$

We can turn the abelian group $M \oplus N$ into a G–module by setting

$$x(m, n) = (xm + \phi(x)(n), n), \quad x \in G,\ m \in M,\ n \in N.$$

If $y \in G$ then

$$\begin{aligned}(xy)(m,n) =&((xy)m + \phi(xy)(n), n) = \\ =&(x(ym) + x\phi(y)(n) + \phi(x)(n), n) = \\ =&x(ym + \phi(y)(n), n) = x(y(m,n)).\end{aligned}$$

Thus we obtain a G–module which can be denoted by $P = (M, N, \phi)$. The mapping $\imath : M \to P$ is $m \mapsto (m, 0)$, and $\epsilon : P \to N$ is $(m, n) \mapsto n$. It is clear that $\imath$, ϵ are homomorphisms. Let us assume that the cocycle $\phi' : G \to \mathrm{Hom}(N, M)$ is cohomologous to the cocycle ϕ, i.e. $\phi'(x) - \phi(x) = xf - f$ for a suitable $f \in \mathrm{Hom}(N, M)$, (x is any element from G). Then the mapping $\pi : (M, N, \phi) = P \to P' = (M, N, \phi')$ which is given by the formula $\pi(m, n) = (m + f(n), n)$ makes the extensions (9) and (9′) equivalent. Indeed,

$$\begin{aligned}\pi\imath(m) = \pi(m, 0) = (m, 0) = \imath' 1_M(m), \\ 1_N\epsilon(m, n) = n = \epsilon'(m + f(n), n) = \epsilon'\pi(m, n).\end{aligned}$$

Moreover, π is a homomorphism of G–modules:

$$\begin{aligned}x\pi(m,n) =&x(m + f(n), n) = (xm + xf(n) + \phi(x)(n), n) = \\ =&(xm + \phi'(x)(n) - \phi(x)(n) + f(n) + \phi(x)(n), n) = \\ =&(xm + \phi'(x)(n) + f(n), n) = \pi(xm + \phi'(x)(n), n) = \\ =&\pi x(m, n).\end{aligned}$$

Inversily, let us assume that (9) and (9′) are equivalent extensions of the form $P = (M, N, \phi)$ and $P' = (M, N, \phi')$, and that $\pi : P \to P'$ is an isomorphism which makes the diagram (10) commutative. If $\pi(m, n) = (m', n')$ then $n' = \epsilon'(m', n') = \epsilon'\pi(m, n) = 1_N\epsilon(m, n) = n$. In addition, if we write $\pi(0, n) = (f(n), n)$, then $\pi(m, n) = \pi(m, 0) + \pi(0, n) = \pi\imath(m) + (f(n), n) = \imath' 1_M(m) + (f(n), n) = (m, 0) + (f(n), n) = (m + f(n), n)$.

It is clear that $f \in \mathrm{Hom}_{\mathbb{Z}}(N, M)$. From the fact that $\pi(x(m, n)) = x(\pi(m, n))$ we obtain $(xm+\phi(x)(n)+f(n), n) = (xm+xf(n)+\phi'(x)(n), n)$, i.e. $\phi(x)(n) - \phi'(x)(n) = xf((n) - f(n)$ and hence $\phi - \phi' = \partial f$. Thus the cocycles ϕ and ϕ' are cohomological and (M, N, ϕ) within equivalence, depends only on the class of the cocycle ϕ.

It remains to show that any $\mathcal{A}$–extension (9) of the G–module M by G–module N is equivalent to (M, N, ϕ) for a suitable $\phi \in H^1(G, \mathrm{Hom}(N, M))$. In fact, we present P in the form $P = \imath(M) \oplus N'$. Let $p = \imath(m) + n'$, where $\epsilon(n') = n \in N$ and $x \in G$. Then $x(\imath(m) + n') = x\imath(m) + xn' = \imath(xm) + xn' = \imath(xm) + a(x, n') + n'$, where $a(x, n') \in \imath(M)$ (here the triviality of the action of G on N was used). Now associate to the element $n \in N$ the element $b(x, n') \in M$ such that $a(x, n') = \imath(b(x, n'))$. This turns out to be a homomorphism of the abelian groups $\phi(x) \in \mathrm{Hom}(N, M)$. If $x, y \in G$ then

$$\begin{aligned}
&(xy)(\imath(m) + n') = \imath(xym) + a(xy, n') + n' = \\
&= \imath(xym) + b(xy, n') + n' = \imath(xym + \phi(xy)(n')) + n'; \\
&x(y(\imath(m) + n')) = x(\imath(ym) + a(y, n') + n') = \\
&= \imath(xym + x\phi(y)n' + a(x, n')) + n' = \\
&= \imath(xym + x\phi(y)(n') + \phi(x)(n')) + n'.
\end{aligned}$$

Hence $\phi(xy)(n') = x\phi(y)(n') + \phi(x)(n')$, i.e. $\phi \in H^1(G, \mathrm{Hom}(N, M))$. Therefore, the mapping π which assigns to the element $\imath(m) + n'$ the pair (m, n) is a homomorphism of the G–module P and $P' = (M, N, \phi)$. A simple check shows that this is an equivalence of extensions. The theorem is proved. □

Naturally, there are also other interpretations of the group H^1.

2.5. Interpretation of the group H^2. Let G be a group, M a G–module (see, Subsection 2.1). An extension of the group G with abelian kernel M is a short exact sequence of the form

$$0 \to M \xrightarrow{\imath} P \xrightarrow{\epsilon} G \to 1, \tag{12}$$

where $\imath$, ϵ are homomorphisms of groups, P is a (posssibly, non–commutative) group, $\imath(M)$ a normal subgroup in it which is the kernel of the homomorphism ϵ of the group P to G, and such that the following compatibility condition holds

$$\imath(\epsilon(x) \circ m) = x\imath(m)x^{-1} = \imath(m)^x, \quad x \in P,\, m \in M.$$

This condition means that the structure of the $G = P/\imath(M)$– module induced on $\imath(M)$, by the fact that $\imath(M)$ is an abelian normal subgroup in P (see, Subsection 2.1), is, in fact, the initial G–module structure on M.

As in Subsection 2.4, we say that an extension (12) is equivalent to an extension

$$(12') \qquad 0 \to M \xrightarrow{\imath'} P' \xrightarrow{\epsilon'} G \to 1,$$

if there exists the isomorphism $\psi : P \to P'$ of groups which makes the following diagram commutative

$$(13) \qquad \begin{array}{ccccccccc} 0 & \longrightarrow & M & \xrightarrow{\imath} & P & \xrightarrow{\epsilon} & G & \longrightarrow & 1 \\ & & \downarrow 1_M & & \downarrow \psi & & \downarrow 1_G & & \\ 0 & \longrightarrow & M & \xrightarrow{\imath'} & P' & \xrightarrow{\epsilon'} & G & \longrightarrow & 1 \end{array}$$

For instance, the two non–commutative groups of order 8, namely, the group of quaternions Q_8 and the dihedral group D_4, are examples of inequivalent extensions with an abelian kernel where $G = \mathbf{Z}_2 \oplus \mathbf{Z}_2$, $M = \mathbf{Z}_2$; moreover, the action of the group G on M is trivial (in both cases). The groups $\mathbf{Z}_2 \oplus \mathbf{Z}_2 \oplus \mathbf{Z}_2$ and $\mathbf{Z}_2 \oplus \mathbf{Z}_4$ also belong there. All the extensions mentioned are pairwise inequivalent because this is the case for the group P by itself. However, we can also give an example of two inequivalent extensions where the middle group is the same. Such an example is provided by the extensions

$$0 \to \mathbf{Z}_2 \xrightarrow{\imath} \mathbf{Z}_2 \oplus \mathbf{Z}_4 \xrightarrow{\epsilon} \mathbf{Z}_2 \oplus \mathbf{Z}_2 \to 0,$$

$$0 \to \mathbf{Z}_2 \xrightarrow{\imath'} \mathbf{Z}_2 \oplus \mathbf{Z}_4 \xrightarrow{\epsilon'} \mathbf{Z}_2 \oplus \mathbf{Z}_2 \to 0,$$

where $\imath(\bar{1}) = (\bar{0}, \bar{2})$, $\imath'(\bar{1}) = (\bar{1}, \bar{2})$. Suppose that there is an isomorphism $f : \mathbf{Z}_2 \oplus \mathbf{Z}_4 \to \mathbf{Z}_2 \oplus \mathbf{Z}_4$, which closes the diagram. Then, taking $f(\bar{0}, \bar{1}) = (\bar{\alpha}, \bar{\beta})$, we obtain $f(\bar{0}, \bar{2}) = (\bar{0}, \bar{2}) \neq (\bar{1}, \bar{2})$. Thus the extensions are inequivalent.

Now we construct an extension of groups starting from a group G, a G–module M and a 2–cocycle $f : G \times G \to M$. Set $P = \{(m, g) | \, g \in G, \, m \in M\}$. We define

$$(14) \qquad (m_1, g_1)(m_2, g_2) = (m_1 + g_1 m_2 + f(g_1, g_2), g_1 g_2).$$

A neutral element of this operation is $(0, 1)$. An immediate check gives that an inverse of (m, g) is given by $(-g^{-1}(m + f(g, g^{-1}), g^{-1})$. To check the condition of associativity for the triple (m_1, g_1), (m_2, g_2), (m_3, g_3) we calculate

$$m_1 + g_1 m_2 + f(g_1, g_2) + (g_1 g_2) m_3 + f(g_1 g_2, g_3) =$$
$$= m_1 + g_1(m_2 + g_2 m_3 + f(g_2, g_3)) + f(g_1, g_2 g_3).$$

The condition $(g_1 g_2) m_3 = g_1(g_2 m_3)$ is fulfilled because M is a G–module. Then there remains

$$f(g_1, g_2) + f(g_1 g_2, g_3) = g_1 f(g_2, g_3) + f(g_1, g_2 g_3),$$

or $(\delta^2 f)(g_1, g_2, g_3) = 0$ (see, Subsection 2.3, formula (7)), which holds because f is a cocycle. Therefore P is a group. We fix an imbedding $\imath : M \to P$, by setting $\imath(m) = (m, 1)$, and an epimorphism $\epsilon : P \to G$, by setting $\epsilon(m, g) = g$. The extension obtained is denoted by (M, G, f).

Theorem. *Fix a group G and a G–module M. Then any extension with an abelian kernel for this fixed G and M is equivalent to a suitable extension of the form (G, M, f). The extension (G, M, f) is equivalent to the extension (G, M, f') iff $f - f'$ is a coboundary. In other words the set of (equivalence classes of) extensions of the form described is in a one–to–one corresponence with the elements from the group $H^2(G, M)$.*

Proof. The statement can be checked directly. However, a number of rather long calculations need to be carried out. For instance, let the extension (G, M, f) and (G, M, f') be equivalent and $\phi : P \to P'$ be an isomorphism which closes the commutative diagram (13). If $\phi(m, g) = (m', g')$ then it is evident that $g = g'$. Moreover, $\phi(m, 1) = (m, 1)$. Let $\phi(0, g) = (\psi(g), g)$, $\psi(g) \in M$.

Since $(m, g) = (m, 1)(0, g)$, we have $\phi(m, g) = (m, 1)\psi(g), g) = (m+\psi(g), g)$. Calculate the product of (m_1, g_1) and (m_2, g_2) and, subsequently, of their images. So we obtain

$$m_1 + g_1 m_2 + f(g_1, g_2) + \psi(g_1 g_2) = m_1 + \psi(g_1) + g_1 m_2 + g_1 \psi(g_2) + f'(g_1, g_2).$$

Hence

$$(f - f')(g_1, g_2) = g_1\psi(g_2) - \psi(g_1g_2) + \psi(g_1).$$

Thus $f - f'$ is a coboundary (the condition $\psi(1) = 0$ is evident). Inversily, if $f - f' = \delta\psi$ then the mapping $(m, g) \mapsto (m + \psi(g), g)$ gives the desired equivalence.

Let now (12) be an arbitrary extension of the form considered. We want to construct a $f \in Z^2(G, M)$ such that (1) is equivalent to the extension (G, M, f). To achieve this, select from each coset $\epsilon^{-1}(g) \subset P$ one element $\widetilde{g}$, $g \in G$. In this case any element from P is uniquely represented in the form $m\widetilde{g}$, $m \in M$, $g \in G$. Multiply two elements of that form:

$$(m_1\widetilde{g}_1)(m_2\widetilde{g}_2) = m_1\widetilde{g}_1m_2\widetilde{g}_1^{-1}\widetilde{g}_1\widetilde{g}_2 = (m_1 + g_1 \circ m_2)\widetilde{g}_1\widetilde{g}_2.$$

Further, $\epsilon(\widetilde{g}_1\widetilde{g}_2) = \epsilon(\widetilde{g_1g_2})$. We set $f(g_1, g_2) = \widetilde{g}_1\widetilde{g}_2(\widetilde{g_1g_2})^{-1} \in M$. Then

$$(m_1\widetilde{g}_1)(m_2\widetilde{g}_2) = (m_1 + g_1 \circ m_2 + f(g_1, g_2))\widetilde{g_1g_2}. \tag{15}$$

Formula (15), when compared with (14) shows that if $f \in Z^2(G, M)$ then the mapping $(m, g) \mapsto m\widetilde{g}$ defines an equivalence of extensions.

So it remains to show that f is a 2–cocycle. We use the associativity of multiplication in the group P:

$$\begin{aligned}
&((m_1\widetilde{g}_1)(m_2\widetilde{g}_2))(m_3\widetilde{g}_3) = ((m_1 + g_1m_2 + f(g_1, g_2))\widetilde{g_1g_2})(m_3g_3) = \\
&=(m_1 + g_1m_2 + f(g_1g_2) + g_1g_2m_3 + f(g_1g_2, g_3))\widetilde{g_1g_2g_3}, \\
&(m_1\widetilde{g}_1)((m_2\widetilde{g}_2)(m_3\widetilde{g}_3)) = (m_1\widetilde{g}_1)((m_2 + g_2m_3 + f(g_2, g_3))\widetilde{g_2g_3} = \\
&=(m_1 + g_1g_2m_3 + g_1f(g_2, g_3) + f(g_1, g_2g_3))\widetilde{g_1g_2g_3}.
\end{aligned}$$

Hence

$$f(g_1, g_2) + f(g_1g_2, g_3) = g_1f(g_2, g_3) + f(g_1, g_2g_3).$$

Therefore, f indeed is a 2–cocycle and the theorem is completely proved. □

The group $P = (G, M, 0)$ is denoted by $P = G \leftthreetimes M$ and is called the *semi–direct product* of the groups G and M. The set of elements of the form $\{(m, 1) | \, m \in M\}$ constitute a normal subgroup in P which is isomorphic to

M, and $\{(0, g) | g \in G\}$ is a subgroup isomorphic to G. If we identify these groups, then $P = MG$, and every element x from P is uniquely representable in the form $x = mg$, $m \in M$, $g \in G$.

An extension which is equivalent to a semi-direct product is called *split*. When considering split extension of groups one can omit the commutativity of the group M. Indeed, if the group P has a normal subgroup M then P is an extension of the group M with the help of $G = P/M$.

We then say that P is a *split* extension if it contains a subgroup $\widetilde{G}$ such that $P = \widetilde{G}M$ and, moreover, $\widetilde{G} \cap M = \{1\}$. It is clear that in this case $G \cong \widetilde{G}$. (see, Introduction 1).

2.6. Application to finite groups. At present, homological algebra is a major tool in many areas in mathematics and other fields of science. In this section we consider an application to a problem about the extension of finite groups.

Schur's theorem. *Let G be a finite group with a normal subgroup H of order m and let the quotient group G/H be of order n. If the numbers m and n are mutually prime then G contains a subgroup of order n. In other words, any extension of groups of mutually prime order splits.*

First we shall prove that it is sufficient to limit ourselves to the case when the subgroup H is abelian. For that we shall use induction on the order m of the group H. Thus we assume that for the case of abelian H the theorem is proved. Our aim is to find a subgroup of order n. Let us take a prime number p which divides m and a maximal p-subgroup P in G. The normalizer of the group P in G is denoted by N. By definition, N is the largest subgroup of the group G in which P is normal. According to the second Sylow theorem, the index $(G : N)$ is the number of Sylow p-subgroups of the group G. All these subgroups are conjugate and lie in H. The intersection $H \cap N$ is the normalizer for P in H. The index $(H : H \cap N)$ is again equal to the number of Sylow p-subgroups in H; hence also in G. Thus, $(G : N) = (H : H \cap N)$. Writing down the index $(G : H \cap N)$ in two ways and using the previous equality, we obtain

$$(G : H \cap N) = (G : N)(N : N \cap H) = (G : H)(H : H \cap N),$$

i.e.

$$(N : N \cap H) = (G : H) = n.$$

Going over to the group N, we see that P and $H \cap N$ are normal subgroups of this group and that the quotient group N/P is an extension of the group $H \cap N/P$ with the help of the group $N/N \cap H$ of order n.

According to the induction assumption, there exists a subgroup of order n in N/P which can be represented in the form K/P, where $P \subset K \subset N$ and $(K : P) = n$. The center C of the subgroup P (it is not trivial) is normal in N, all the more in K. Therefore, K/C is an extension of the group P/C with the help of the group K/P of order n. According to the induction assumption K/C has a subgroup of order n which is representable in the form $L/C \subset K/C$. This group is an extension of the abelian p–group C with a help of the group L/C of order n. According to our assumption an extension with abelian kernel splits. This splitting extracts a subgroup M of order n in L, i.e. also in G, as required.

When dealing with the special abelian case it is possible to reduce to a lemma which we present below.

Lemma. *Let G be a group of order k, M a G–module, Then for any $n > 0$ the group $H^n(G, M)$ is periodic; moreover, the periods of all elements divide the number k.*

Proof. For any n–dimensional cochain $f(x_1, \dots, x_n)$, define an $(n-1)$–cochain g by setting

$$g(x_1, \dots, x_{n-1}) = \sum_{x \in G} f(x_1, \dots, x_{n-1}, x).$$

Recall that

$$(\partial^n f)(x_1, \dots, x_{n+1}) = (-1)^{n+1}(x_1 f(x_2, \dots, x_{n+1}) +$$
$$+ \sum_{i=1}^{n} (-1)^i f(x_1, \dots, x_i x_{i+1}, \dots, x_{n+1}) + (-1)^{n+1} f(x_1, \dots, x_n)).$$

In this equality let x_{n+1} run through all elements of the group G, sum over x_{n+1}, to obtain

$$\sum_{x\in G} f(\ldots, x_{n-1}, x_n x) = \sum_{x\in G} f(\ldots, x_{n-1}, x) = g(\ldots, x_{n-1}).$$

Consequently, we have

$$\sum_{x\in G} (\partial f)(x_1, \ldots, x_n, x) = -(\partial g)(x_1, \ldots, x_n) + kf(x_1, \ldots, x_n).$$

If $\partial f = 0$ then this last relationgives that the element $kf = \partial g$ is a coboundary and so the class of the cochain kf is equal to zero. □

In order to conclude the proof of the theorem in the abelian case it remains to apply the theorem from Subsection 2.5, according to which an extension G is equivalent to an extension of the form $(G/H, H, f)$, $\bar{f} \in H^2(G/H, H)$. Since f is a function with values in an abelian group of order m, evidently, $mf = 0$. By the previous lemma we also have $n\bar{f} = 0$. Since $(m, n) = 1$, we have $\bar{f} = 0$. By the same theorem from Subsection 2.5, $(G/H, H, f)$ splits. The theorem is proved. □

§ 3. Splitting of the radical in a finite–dimensional algebra

3.1. Cohomology of associative algebras. From the theory developed in chapter 3 we know that every artinian associative algebra R over a field F has a nilpotent Jacobson radical N and that the quotient $A = R/N$ is semisimple and equal to a direct sum of matrix algebras over skew fields. We say that the radical of the algebra R splits if there exists a subalgebra S in R such that $R = S + N$ and $S \cap N = \{0\}$. It is clear that then $S \cong A$. In this subsection we shall prove a theorem on the splitting of the radical by applying methods of homological algebra, for the case of the finite–dimensional algebras over an algebraically closed field (the formulation presented below of the theorem is a generalized one).

We do not develop the cohomology theory of associative algebras in general but limit ourselves to the definition of the group H^n and the interpretation of the groups H^1 and H^2. Let A be an associative algebra with identity, M a (A, A)–bimodule, i.e. a left and right A–module with the compatibility condition $(am)b = a(mb)$, a, $b \in A$, $m \in M$. We call an n–linear mapping $f : A \times \ldots \times A \to M$ an *n–cochain.* Let $C^n(A, M)$ be the space of such n–cochains. We define the *coboundary operator* ∂_n by setting

$$(\partial_n f)(x_1,\dots,x_{n+1}) = x_1 f(x_2,\dots,x_{n+1})+$$
$$+\sum_{i=1}^{n}(-1)^i f(x_1,\dots,x_i x_{i+1},\dots,x_{n+1})+$$
$$+(-1)^{n+1} f(x_1,\dots,x_n)x_{n+1} \tag{1}$$

It can be verified that $\partial^2 = 0$. The space $Z^n(A,M) = \operatorname{Ker}\partial_n$ is called the space of n–cocycles of the algebra A with values in the (A,A)-bimodule M. The space $B^n(A,M) = \operatorname{Im}\,\partial_{n-1}$ is the space of *n–coboundaries*. Finally, there is the n–dimensional cohomology group

$$H^n(A,M) = Z^n(A,M)/B^n(A,M).$$

An interpretation of the groups H^1 and H^2 is presented in the following theorem.

Theorem. *Let A be an algebra with identity over a field F.*

1) *Let M, N be left A–modules. We turn $T = Hom_F(M,N)$ into a (A,A)–bimodule by setting $(\tau a)m = \tau(am)$, $(a\tau)m = a(\tau m)$, $m \in M$, $\tau \in T$, $a \in A$. Then the group $H^1(A,T)$ can be interpreted as the set of equivalence classes of modules with kernel M and quotient module N.*

2) *Let N be an (A,A)–bimodule. Then the group $H^2(A,T)$ can be interpreted as a set of classes of extensions of algebras with kernel N such that $N^2 = \{0\}$, with quotient algebra A, while the action A on N in such an extension coincides with the (A,A)–bimodule structure on N.*

Proof. We do not go into the details of the verification (it is similar to the case of groups), but only define the bijections involved.

1) If $f \in Z^1(A,T)$ then, according to formula (1), $f(ab) = af(b) + f(a)b$, i.e. f is a so–called *(generalized) derivation*. If $n \in N$ is a zero–dimensional cochain then $(\partial_0 n)(a) = an - na = [a,n]$ is called an *inner derivation*. Now, given a 1–cocycle f, construct the extension

$$0 \to N \xrightarrow{\alpha} U \xrightarrow{\beta} M \to 0.$$

In this extension $U = N \oplus M$ is a direct sum of vector spaces and becomes an A–module, with $a \circ m = am + f(a)m$. The mappings α and β are the natural ones. If the class of the cocycle f is equal to zero, i.e. if there exists a $\tau \in T = \text{Hom}\,(M, N)$ such that $f = \partial\tau$, then the elements of the form $m + \tau(m)$ form a submodule $M' \cong M$ and $U = N \oplus M'$ (a direct sum of A–modules), i.e. the extension is trivial. Inversily, given an extension (2), let ψ be a linear mapping $M \to U$ such that $\beta\psi = 1$. For $a \in A$, we define $f(a)$ as $f(a)m = \psi(am) - a\psi(m)$. Then $\psi \in Z^1(A, T)$. If f is an interior derivation then the splitting can be proved in the same way as above. Inversily, if the extension splits, i.e. if there exists an A–homomorphism $\psi^* : M \to U$ satisfying $\phi\psi^* = 1_M$ then $\tau(m) = \psi^*(m) - \psi(m)$ can be taken in the capacity of $\tau \in T$. All the trivial verifications are left to the reader.

2) If R is an extension with kernel N such that $N^2 = 0$ and with quotient module $A = R/N$, $a = r + N$, then the definitions $an = rn$ and $na = nr$ are unambigous and they turn N into a (A, A)–bimodule. Take an arbitrary linear mapping $\psi : A \to R$ which is inverse from the left to the natural epimorphism. We set

$$f(a, b) = \psi(ab) - \psi(a)\psi(b), \quad a, b \in A.$$

Then, because of associativity of multiplication, we obtain

$$\psi(a(bc)) = f(a, bc) + \psi(a)\psi(bc) = f(a, bc) + \psi(a)f(b, c) + \psi(a)\psi(b)\psi(c),$$
$$\psi((ab)c) = f(ab, c) + \psi(ab)\psi(c) = f(ab, c) + f(a, b)\psi(c) + \psi(a)\psi(b)\psi(c).$$

By subtracting the equalities obtained and taking into account that $\psi(a)n = an$, $n\psi(a) = na$, we get

$$0 = af(b, c) - f(ab, c) + f(a, bc) - f(a, b)c = (\partial_2 f)(a, b, c),$$

i.e. $f \in Z^2(A, N)$.

If $f \in B^2(A, N)$, i.e. $f = \partial_1\phi$, $\phi \in C^1(A, N) = \text{Hom}_F(A, N)$, then the set of elements $\psi(a) + \phi(a)$ forms a subalgebra in R which is isomorphic to A. The main point to test is stability with respect to multiplication: if $a, b \in A$ then

$$(\psi(a) + \phi(a))(\psi(b) + \phi(b)) = \psi(a)\psi(b) + \psi(a)\phi(b) + \phi(a)\psi(b) =$$
$$=\psi(ab) - f(a, b) + a\phi(b) + \phi(a)b = \psi(ab) + \phi(ab) + a\phi(b) - \phi(ab)+$$
$$+\phi(a)b - f(a, b) = \psi(ab) + \phi(ab) + (\partial_1\phi)(a, b) - f(a, b) = \psi(ab) + \phi(ab).$$

If A, N, f are given then the extension R is constructed as the direct sum $R = N \oplus A$ of vector spaces with the natural multiplication

$$(n_1 + a_1)(n_2 + a_2) = n_1a_2 + a_1n_2 + f(a_1, a_2) + a_1a_2.$$

We leave it as an exercise for the reader to complete the proof. □

3.2. Two lemmas. We now turn our attention to the problem about the splitting of the radical in finite–dimensional algebras.

Lemma 1. *Let A be a semisimple algebra over the field F which is equal to the direct sum of several matrix algebras over the field F. Then we can select in A two bases over F, $\{a_1, \dots, a_t\}$ and $\{a'_1, \dots, a'_t\}$, such that $\sum_{i=1}^t a_ia'_i = 1$ and for any $a \in A$ from the equality $a_ia = \sum_{j=1}^n \lambda_{ij}(a)a_j$, $\lambda_{ij}(a) \in F$ there follows $aa'_i = \sum_{j=1}^n a'_j\lambda_{ji}(a)$.*

Proof. Evidently, it is sufficient to select such bases in every direct summand, i.e. in a matrix algebra of the form $M_n(F)$. As the first one of such a basis pair take the system of elementary matrices $\{E_{ij}\}$, and as the second one $\{\frac{1}{n}E_{ij}\}$. Then

$$\sum_{i,j=1}^n E_{ij}\frac{1}{n}E_{ji} = \sum_{i=1}^n E_{ii} = E = 1.$$

Further, if $a = \sum a_{kl}E_{kl}$,

$$E_{ij}a = \sum_{l=1}^n a_{jl}E_{il}, \quad \text{i.e. } \lambda_{ij,kl} = \delta_{ik}a_{jl}.$$

Moreover, $a\frac{1}{n}E_{ij} = \sum_{l=1}^n \frac{1}{n}E_{li}a_{lj}$, ie.

$$\lambda'_{kl,ij} = \delta_{ik}a_{lj}.$$

Then $\lambda_{kl,ij} = \delta_{ki}a_{lj} = \lambda'_{kl,ij}$ what was required to be proved. □

Lemma 2. *Let the algebra A satisfy the condition of lemma 1 and let N be an (A, A)–bimodule. Then the groups $H^1(A, N)$ and $H^2(A, N)$ are zero.*

Proof. Let $f : A \to N$ be a generalized derivation (1–cocycle). Now, consider $u = \sum_{i=1}^{t} a'_i f(a_i)$ and compute $[a, u]$. We obtain

$$\begin{aligned}
[a, u] =& au - ua = \sum_{i=1}^{t} aa'_i f(a_i) - \sum_{i=1}^{t} a'_i f(a_i)a = \\
=& \sum_{i} aa'_i f(a_i) - \sum_{i} a'_i f(a_i a) + \sum_{i} a'_i a_i f(a) = \\
=& \sum_{i,j} a'_j \lambda_{ji}(a) f(a_i) - \sum_{i,j} a'_i \lambda_{ij}(a) f(a_j) + f(a) = f(a).
\end{aligned}$$

Thus f is an inner derivation, i.e. a coboundary. Therefore, its cohomology class is equal to zero.

We now consider a 2–cocycle $h : A \times A \to N$. We shall demonstrate that this cocycle is a coboundary for the 1–cochain f which is given by the formula

$$f(a) = \sum_{i=1}^{t} h(a, a'_i)a_i.$$

We shall compute the coboundary for f (by applying the cocycle property of h):

$$\begin{aligned}
& af(b) - f(ab) + f(a)b = \\
=& \sum_{i} ah(b, a'_i)a_i - \sum_{i} h(ab, a'_i)a_i + \sum_{i} h(a, a'_i)a_i b = \\
=& \sum_{i} h(ab, a'_i)a_i - \sum_{i} h(a, ba'_i)a_i + \sum_{i} h(a, b)a'_i a_i + \\
+& \sum_{i} h(a, a'_i)a_i b - \sum_{i} h(ab, a'_i)a_i = h(a, b).
\end{aligned}$$

So $h = \partial f$, i.e. the class of the cocycle h is equal to zero. □

3.3. The main theorem. We call a field F *perfect* if any of its finite extensions is separable. In other words, F is either a field of characteristic zero or char $F = p > 0$ and $F^p = F$. Among the perfect fields there are the algebraically closed and finite fields (see, § 1.2). The proof of the following theorem will be carried out only in the case of an algebraically closed field.

Theorem. *Let F be a perfect field, R a finite–dimensional algebra over F with radical N. Then there exists a semisimple subalgebra $S \subset R$ such that $R = S \oplus N$. If S_1 and S_2 are two subalgebras in R with the property $R = S_1 \oplus N = S_2 \oplus N$ then there exists an element $n \in N$ such that $S_1 = (1-n)S_2(1-n)^{-1}$.*

Proof. The algebra $A = R/N$ satisfies the conditions of the lemma 2. If $N^2 = \{0\}$ then N is a (A, A)–bimodule as in the theorem from Subsection 3.1. By using lemma 2 above we see that R as an extension of the ideal N splits with a help of A, i.e. the desired subalgebra S exists. Let now $N^2 \neq \{0\}$. Consider the algebra R/N^2. Its radical is equal to N/N^2. The square of the radical is equal to zero. According to what has been said, R contains a subalgebra S_1 such that $R/N^2 = S_1/N^2 \oplus N/N^2$, i.e. $S_1 \cap N = N^2$. The dimension of the algebra S_1 is strictly lower than the dimension of the algebra R, $S_1/N^2 \cong A$, i.e. N^2 is the radical for S_1. Using induction with respect to the dimension of the algebra S_1 we find a subalgebra S satisfying $S_1 = S \oplus N^2$, $S_1 \cong A$. We have $R = S_1 + N = S + N^2 + N = S + N$. Since S is semisimple and N is nilpotent, we have $S \cap N = \{0\}$, $R = S \oplus N$, and the subalgebra S is the desired one.

We shall prove the conjugacy of two semisimple subalgebras S_1 and S_2 satisfying $R = S_1 \oplus N = S_2 \oplus N$. Let ϕ be the natural homomorphism $R \to A$. Let ψ_1, ψ_2 be two homomorphisms $\psi_i : A \to R$ such that $\phi\psi_i = 1_A$ and $\psi_i(A) = S_i$, $i = 1, 2$. We turn N into a (A, A)–bimodule by setting $na = n\psi_2(a)$, $an = \psi_1(a)n$.

Consider the function $f ; A \to R$ which is defined by the formula $f(a) = \psi_1(a) - \psi_2(a)$. Since $\phi\psi_i = 1_A$, $i = 1, 2$, we have $f(a) \in N$ for all $a \in A$. Now

$$\begin{aligned} f(ab) =& \psi_1(ab) - \psi_2(ab) = \\ =& \psi_1(a)(\psi_1(b) - \psi_2(b)) + (\psi_1(a) - \psi_2(a))\psi_2(b) = \\ =& \psi_1(a)f(b) + f(a)\psi_2(b) = af(b) + f(a)b. \end{aligned}$$

Thus f is a generalized derivation. According to lemma 2 there exists an $n \in N$ such that

$$f(a) = \psi_1(a) - \psi_2(a) = an - na = \psi_1(a)n - n\psi_2(a).$$

Rewrite the latter equality in the form $\psi_1(a)(1-n) = (1-n)\psi_2(a)$. Since the element n is nilpotent, the element $(1-n)$ is invertible. Consequently, $\psi_1(a) = (1-n)\psi_2(a)(1-n)^{-1}$ for all $a \in A$, i.e. $S_1 = (1-n)S_2(1-n)^{-1}$. □

§ 4. Brauer group

4.1. Factor system. Let L be an extension of a field K. If A_1 and A_2 are central simple algebras over K and $A_1 \sim A_2$ then $L \otimes_K A_1 \sim L \otimes_K A_2$. Moreover, $L \otimes_K (A \otimes_K B) \cong (L \otimes_K A) \otimes_L (L \otimes_K B)$. From this it follows that $\{A\} \to \{L \otimes_K A\}$ is a homomorphism from Br (K) into Br (L). Its kernel $\mathrm{Br}^L(K)$ consists of those $\{A\}$ such that L is a splitting field for A. We shall study the situation described in the case when L is a Galois extension of the field K.

Let Δ be a division algebra from $\{A\}$, $(\Delta : K) = n^2$ and $(L : K) = m$. Then according to theorem 2 of Subsection 3.4.4, $n | m$, and if $n = mr$ then L is a subfield in $A = M_r(\Delta)$ such that $C_A(L) = L$. It is clear that $A \in \{\Delta\}$. Let $G = \mathrm{Gal}\,(L/K)$. If $s \in G$ then according to the theorem from Subsection 3.4.3, there exists an invertible element $u_s \in A$ such that $u_s l u_s^{-1} = s(l)$ for all $l \in L$. Then $u_s u_t l u_t^{-1} u_s^{-1} = (st)(l) = u_{st} l u_{st}^{-1}$. Thus, $u_{st}^{-1} u_s u_t$ commutes with any $l \in L$, i.e. it is a non–zero element from L. Thus we find a $\rho_{s,t} \in L$ such that

$$u_s u_t = \rho_{s,t} u_{st}. \tag{1}$$

These relations and

$$u_s l = s(l) u_s \tag{2}$$

show that the set A' consisting of elements of the form $\sum_{s \in G} l_s u_s$, where $l_s \in L$, is a subalgebra in A. Since the different automorphisms of a field are linearly independent, we see that from $\sum_s l_s u_s = 0$ follows $l_s = 0$, for all $s \in G$. From that we get that A' is an m–dimensional space over L, where $m = |G|$. Thus $(A' : K) = (A' : L)(L : K) = m^2$. Since $(A : K) = (M_r(\Delta) :$

$K) = r^2n^2 = m^2$, we have $A' = A$. So A coincides with the set of elements of the form $\sum_{s\in G} l_s u_s$.

Using the associativity condition one can write $(u_s u_t)u_v = u_s(u_t u_v)$. Applying (1) we obtain

(3) $$\rho_{s,t}\rho_{st,v} = s(\rho_{t,v})\rho_{s,tv}.$$

We remark also that the elements u_s are non–uniquely defined. They can be replaced by the set $v_s = \mu_s u_s$, where μ_s is any non–zero element from L. Other replacements are impossible because $C_A(L) = L$. Taking $v_s = \mu_s u_s$, $s \in G$, we obtain

(4) $$v_s v_t = \mu_s u_s \mu_t u_t = \mu_s s(\mu_t)\rho_{s,t} u_{st} = \mu_s s(\mu_t)\mu_{st}^{-1}\rho_{s,t} v_{st}.$$

Therefore the elements $\rho_{s,t}$ are replaced by $\rho'_{s,t} = \mu_s s(\mu_t)\mu_{st}^{-1}\rho_{s,t}$.

4.2. Crossed product. We recall the definition of the group $H^2(G, L^*)$, where $L^* = L \setminus \{0\}$ in the multiplicative case. A mapping $(s,t) \mapsto \rho_{s,t}$ from $G \times G$ into L^*, for which (3) is fulfilled, is called a *2–cocycle*. If ρ and σ are two such mappings then their product is defined by the formula $(s,t) \mapsto \rho_{s,t}\sigma_{s,t}$. Moreover, $1 : (s,t) \mapsto 1$ and $\rho^{-1} : (s,t) \mapsto \rho_{st}^{-1}$ are also 2–cocycles.

As a result we obtain the group $Z^2(G, L^*)$. It is easy to note that if $\mu : s \to \mu_s$ is a mapping from G into L^*, then $(s,t) \to \mu_s s(\mu_t)\mu_{st}^{-1}$ is a 2–cocycle which is called a *coboundary* and the set of these is a subgroup $B^2(G, L^*)$. The quotient group $H^2(G, L^*) = Z^2(G, L^*)/B^2(G, L^*)$ is the *second cohomology group* of the group G with the values in L^*.

The main result which we shall prove consists of the fact that there exists a canonical isomorphism between $H^2(G, L^*)$ and the kernel $\mathrm{Br}^L(K)$ of the mapping from Br (K) into Br (L).

Let $\rho \in Z^2(G, L^*)$. We construct an algebra $A = (L, G, \rho)$ which is called the *crossed product* of the field L with the group G and the factor set ρ. The algebra A is the vector space consisting of elements of the form

$$\sum_{s\in G} l_s u_s,$$

where the elements u_s are one–to–one with G ($s \mapsto u_s$ is bijective). The summation in A and the multiplication by scalars from K are defined in the natural way. Further,

(5) $$\left(\sum_s l_s u_s\right)\left(\sum_t m_t u_t\right) = \sum_{s,t} l_s s(m_t)\rho_{s,t} u_{st}.$$

It is easy to check that (5) is an associative operation. From the condition (3) it follows that $\rho_{s,1} = \rho_{1,1} = \rho_{1,s} = s_1(\rho_{1,1})$, so $1 = \rho_{1,1}^{-1} u_1$ is an identity for A. Thus A is an associative algebra with identity.

Theorem. *$A = (L, G, \rho)$ is a central simple algebra over K with $(A : K) = n^2$, $n = (L : K)$. Moreover, L is a splitting field for A.*

Proof. Let B be an ideal in A. We write $\bar{a} = a + B$. Assume that $B \neq A$. Then $l \mapsto \bar{l}$ is a monomorphism from L into $\bar{A}$. Since the elements u_s are invertible in A, the elements $\bar{u}_s$ are invertible in $\bar{A}$. Thus, $\bar{A}$ consists of the sums $\sum \bar{l}_s \bar{u}_s$. From the lemma on the linear independence of automorphisms follows that the elements $\bar{u}_s$ are linearly independent over $\bar{L}$. Therefore $(A : K) = (\bar{A} : \bar{L})(\bar{L} : \bar{K}) = n^2$. Taking $B = \{0\}$ we get $(A : K) = n^2$. Since $(A : K) = (\bar{A} : K)$, we have $B = \{0\}$ and A is simple.

Let now $l \in L$, and let $\sum_s l_s u_s$ commute with l. Then $\sum_s (l l_s - s(l) l_s) u_s = 0$. So $s(l) = l$, for every s satisfying $l_s \neq 0$. It follows that if $\sum_s l_s u_s$ commutes with any l then $\sum_s l_s u_s = l_1 u_1 = l_1 \rho_{1,1} \in L$. Therefore $C_A(L) = L$. If c is an element from the center of the algebra A then $c \in C_A(L) = L$ and $cu_s = u_s c$. Hence $s(c) = c$, for any $s \in G$. Thus $c \in K$. So K is the center and A is a central simple algebra over K. The fact that L is a splitting field follows from theorem 2, Subsection 3.4.4. □

4.3. We say that the cycles ρ and ρ' are *cohomologous* and write $\rho \sim \rho'$, if they differ by a coboundary, i.e. if they define the same element $\{\rho\}$ in $H^2(G, L^*)$. In this case there exists $\mu_s \in L^*$ such that formula (4) holds. From the considerations that led to this formula, is evident that $A = (L, G, \rho) \cong A' = (L, G, \rho')$. This gives a well defined mapping from $H^2(G, L^*)$ into $\mathrm{Br}^L(K)$, which takes $\{\rho\}$ into $\{(L, G, \rho')\}$. The construction which led to the definition of $H^2(G, L^*)$ shows that if Δ is a finite–dimensional central division algebra that is split by the field L, then $\Delta \sim (L, G, \rho)$ for suitable ρ. Therefore the mapping $\{\rho\} \mapsto \{(L, G, \rho)\}$ is surjective to $\mathrm{Br}^L(K)$.

We have to show that it is an isomorphism.

Theorem 1. $(L, G, \rho) \otimes_K (L, G, \sigma) \sim (L, G, \rho\sigma)$.

Proof. The tensor product on the left side has the form $A_1 A_2$, where A_i is a subalgebra and $a_1 a_2 = a_2 a_1$ for $a_i \in A_i$, $i = 1, 2$. Moreover, if $x_1, \ldots, x_r \in A_i$ are linearly independent over K, these elements are linearly independent over A_j, $j \neq i$, i.e. $x_1 y_1 + \ldots + x_r y_r = 0$, $y_k \in A_j$ iff all $y_k = 0$. In addition, we have the isomorphism $l \mapsto l_i$ of the field L onto a subfield L_i in A_i and elements $u_s \in A_1$, $v_s \in A_2$, $s \in G$ such that

$$\begin{aligned} l_1 u_s &= s(l)_1 u_s, \quad l_2(v_s) = s(l)_2 v_s, \quad l \in L, \\ u_s u_t &= (\rho_{s,t})_1 u_{st}, \quad v_s v_t = (\sigma_{s,t})_2 v_{st}. \end{aligned} \tag{6}$$

Since L is separable over K, it has a primitive element θ, i.e. $L = K(\theta)$. Let $f(X)$ be a minimal polynomial for θ over K.

Then $f(X) = \prod_s (X - s(\theta))$ in $L(X)$ and, consequently, $f(X) = \prod_s (X - s(\theta))_1$ and $f(X) = \prod_s (X - s(\theta))_2$ in $L_1[X]$ and $L_2[X]$, respectively. We consider the Lagrange interpolation polynomial

$$f_s(X) = \frac{f(X)}{(X - s(\theta)_1) f'(s(\theta)_1)} = \frac{\prod_{t \neq s} (X - t(\theta)_1)}{\prod_{t \neq s} (s(\theta)_1 - t(\theta)_1)}$$

in $L_1[X]$. Since $1 - \sum_s f_s(X)$ has a degree $\leq n - 1$ and becomes zero for any $X = s(\theta)_1$, this polynomial is equal to zero, i.e.

$$\sum_s f_s(X)) = 1.$$

Moreover, if $s \neq t$ then $f_s(X) f_t(X) \equiv 0 (\bmod f(X))$. Consider now the commutative subalgebra $L_1 L_2$ in $A_1 A_2$. Let

$$e_s = f_s(\theta_2) \in L_1 L_2. \tag{7}$$

Then $e_s \neq 0$, because $1, \theta_2, \ldots, \theta_2^{n-1}$ are linearly independent over L_1. According to what was said before, we have $\sum_s e_s = 1$ and $e_s e_t = 0$ for $s \neq t$. Multiplying by e_t we obtain $e_t^2 = e_t$. So we have

$$e_s^2 = e_s, \quad e_s e_t = 0 \text{ if } s \neq t, \quad \sum_s e_s = 1. \tag{8}$$

From the definition of the element e_s we have $(\theta_2 - s(\theta)_1)e_s = 0$. In this case for $\rho_2 \in L_2$ we obtain

$$\rho_2 e_s = s(\rho)_1 e_s. \tag{9}$$

Thus $L_1L_2 = \bigoplus_s L_1 e_s$ and $L_1 e_s \cong L$. We obtained a (unique) representation of the algebra L_1L_2 in the form of a direct sum of simple algebras.

Now consider the inner automorphisms $x \mapsto u_s x u_s^{-1}$, $x \mapsto v_s x v_s^{-1}$ of the algebra A_1A_2 which satisfy the conditions (6). These automorphisms stabilize L_1L_2 and, hence, permute the simple components of the algebras L_1L_2 just as the elements e_s. Applying these automorphisms to the relations (9) and taking into account that the first ones fix the elements from A_2 and the second ones the elements from A_1, we obtain the following:

$$\rho_2(u_t e_s u_t)^{-1} = (ts)(\rho)_1(u_t e_s u_t^{-1}),$$
$$t(\rho_2)(v_t e_s v_t^{-1}) = s(\rho)_1(v_t e_s v_t^{-1}).$$

Comparing this with (9) we get

$$u_t e_s = e_{ts} u_t, \qquad v_t e_s = e_{st^{-1}} v_t. \tag{10}$$

Now set $e_{s,t} = e_s u_s u_t^{-1}$. Then from (10) follows that the n^2 elements $e_{s,t}$ form a set of elementary matrices:

$$e_{s,t} e_{s',t'} = \delta_{ts'} e_{st'}, \quad \sum_s e_{s,s} = 1.$$

From this follows that $A_1A_2 \cong M_n(B)$, where B is the centralizer in A_1A_2 of the set of elements $e_{s,t}$. In this case $B \cong e_{11}(A_1A_2)e_{11}$. We compute the latter algebra. We already know that $\dim_K A_1A_2 = n^4$, so that $\dim_K B = n^2 = \dim_K e_{11}A_1A_2e_{11}$. Note that $e_{11}A_1A_2e_{11}$ contains $Le_1 = L_1e_{11}$. In addition, $e_{11}A_1A_2e_{11}$ contains the element $w_s = u_s v_s e_{11} = u_s v_s e_1 = e_1 u_s v_s$. Further,

$$\begin{aligned}
w_s(\rho_1 e_{11}) &= e_s v_s \rho_1 e_1 = (s(\rho)_1 e_1) u_s v_s = (s(\rho)_1 e_1) w_s,\\
w_s w_t &= u_s v_s u_t v_t e_1 = u_s u_t v_s v_t e_1 = (\rho_{s,t})_1 u_{st} (\sigma_{s,t})_2 v_{st} e_1 =\\
&= (\rho_{s,t})_1 (\sigma_{s,t})_2 e_1 w_{st} = (\rho_{s,t})_1 (\sigma_{s,t})_2 e_1 w_{st} = (\rho_{s,t}\sigma_{s,t})_1 e_1 w_{st}.
\end{aligned}$$

Thus, $e_{11}A_1A_2e_{11}$ contains a subalgebra which is isomorphic to the cross product $(L, G, \rho\sigma)$. Comparing dimensions gives $e_{11}A_1A_2e_{11} \cong (L, G, \rho\sigma)$. So that finally, $(L, G, \rho) \otimes (L, G, \sigma) \sim (L, G, \rho\sigma)$. □

From the proof it follows that $\{\rho\} \mapsto \{(L, G, \rho)\}$ is a group homomorphism from $H^2(G, L^*)$ onto $\mathrm{Br}^L(K)$. Let us verify that it is in fact an isomorphism.

Theorem 2. *If* $(L, G, \rho) \sim 1$ *then* $\rho \sim 1$, *i.e.* $Br^L(K) = H^2(G, L^*)$.

Proof. To start with note that since $\{\rho\} \to \{(L, G, \rho)\}$ is a group homomorphism, from $\rho \sim 1$ it follows that $(L, G, \rho) \sim 1$. Hence $(L, G, \rho) \sim M_n(K)$. Thus if $(L, G, \rho) \sim 1$ then $(L, G, \rho) \sim (L, G, 1)$.

Consider now an isomorphism $a \mapsto a'$ from $(L, G, 1)$ onto (L, G, ρ). Then the image L' of the field L is a subfield in (L, G, ρ), and for any $s \in G = \mathrm{Gal}(L/K)$ there is an invertible element $v'_s \in (L, G, \rho)$ such that $v'_s l' = (s(l))' v'_s$ for $l' \in L'$ and $v'_s v'_t = v'_{st}$. The isomorphism $l' \mapsto l$ of the field L' on L extends (see, Subsection 3.4.3) to an automorphism η of the algebra (L, G, ρ). Under this automorphism v'_s maps into an element v_s, and we have $v_s l = s(l) v_s$, $v_s v_t = v_{st}$. But, also $u_s l = s(l) u_s$ and $u_s u_t = \rho_{s,t} u_{st}$. Thus, $u_s = \mu_s v_s$. $\mu_s \in L$, and $\rho_{s,t} = \mu_s s(\mu_t) \mu_{st}^{-1}$, i.e. $\rho \sim 1$. □

4.4. Periodicity of the Brauer group. We have seen that every central division algebra has a separable splitting field. Since an extension of a splitting field is also a splitting field, we can see that for Δ there exists a splitting field L which is a finite–dimensional Galois extension of the field K. In this case $\Delta \sim (L, G, \rho)$, for suitable factor system ρ. According to the multiplicative version of the lemma of Subsection 2.6, concerning periodicity of the cohomology groups of a finite group we see that $\rho^n \sim 1$, where n is the order of the group G. According to theorem 2, $\{\Delta\}^n = \{(L, G, \rho^n)\} = \{(L, G, 1)\} = 1$. It should be noted that according to theorem 2 of Subsection 3.4.4, the field L is a subfield in an algebra of the form $A = M_r(\Delta)$ and that $n = mr$, where $\dim_K \Delta = m^2$. Moreover, $C_A(L) = L$ and $A = (L, G, \rho)$. We want to strengthen this result by proving that, in fact, $\{\Delta\}^m = 1$.

Identify A with $\mathrm{End}_\Delta(V)$, where V is an r–dimensional space over Δ. We have

$$\dim_K V = (\dim_\Delta V)(\dim_K \Delta) = rm^2 = (\dim_L V)(\dim_K L) = mr \dim_L V$$

Hence $\dim_L V = m$. Let $\{x_1, \dots, x_m\}$ be a basis for V over L, and $\{u_s\}$ a set of elements of the algebra $A = (L, G, \rho)$ such that $u_s l = s(l)u_s$, $u_s u_t = \rho_{s,t} u_{st}$.
Let $u_s x_i = \sum_{j=1}^n m_{ij}(s) x_j$. Then

$$u_s u_t x_i = u_s \left(\sum_{j=1}^n m_{ij}(t) x_j \right) = \sum_{j,k} s(m_{ij}(t)) m_{jk}(s) x_k,$$

$$\rho_{s,t} u_{st} x_i = \rho_{s,t} \sum_k m_{ik}(st) x_k.$$

This gives the matrix relation

$$\rho_{s,t} M(st) = s(M(t)) M(s).$$

Take the determinants, to obtain

$$\rho_{s,t}^m \mu_{st} = s(\mu_t) \mu_s.$$

Since u_s is a non–zero element from the field L, we see that each matrix $M(s)$ is invertible, so $\mu_s \neq 0$. Therefore from the previous equality it follows that $\rho^m \sim 1$. Consequently, $\{\Delta\}^m = 1$, according to theorem 2, Subsection 4.3.

Theorem. *Let Δ be a central division algebra of dimension m^2 over a field K. Then $\{\Delta\}^m = 1$.* □

∗ To conclude note that the theorems on the structure of finite–dimensional central skew fields have specific deficiencies. We know that each such field is similar to some cross product. However a cross product is not necessary a skew field. Thus arose the problem of the existence of isomorphisms of finite–dimensional central fields with suitable cross products. An example which gives a negative solution of this problem is due to S. Amitsur and can be found in Amitsur [1], Jacobson [2]; see also Bokut'–L'vov–Kharchenko [1]. ∗

§ 5. Hopf algebras

5.1. Definition. Examples. Let R be some associative algebra with unit over a field F; let M, N be vector spaces over F with the structure of a left R–module. The tensor product $M \otimes_F N$ becomes in a natural way a left $R \otimes_F R$–module if we put $(r \otimes s)(m \otimes n) = rm \otimes sn$. In order to make $M \otimes_F N$ into a module over R it is necessary to have a mapping (comultiplication) Δ: $R \to R \otimes_F R$, which satisfies some definite properties. Thus, the property $\Delta(rs) = (\Delta r)(\Delta s)$ should hold in order to get a R–module. This means that Δ is a homomorphism of algebras. Since it is natural that the unit acts in a trivial way we have $\Delta(1) = 1 \otimes 1$, i.e. Δ should be a homomorphism of algebras with unit. If L is another R–module then it is natural to have the property $(M \otimes N) \otimes L \cong M \otimes (N \otimes L)$ which gives $(\Delta \otimes 1_R)\Delta(r) = (1_R \otimes \Delta)\Delta(r)$ for any $r \in R$. The "usual" property $M \otimes N \cong N \otimes M$ leads to the axiom $\Delta r = \tau \Delta r$, where $\tau : a \otimes b \mapsto b \otimes a$ is the switch automorphism of the factors in the tensor square $R \otimes R$. In order to define a natural left R–module structure $M^* = \mathrm{Hom}_F(M, F)$ it is necessary to have a homomorphism η: $R \to R^{op}$ into the opposite ring (see, § 3.4). The fact is that the definition $(rf)(x) = f(rx)$ and its consequence $r(sf) = (sr)f$ show that R^{op} acts indeed on M^* from the left. If we want to stay with R itself then the an anti–endomorphism (antipode) $\eta : R \to R$ is needed. Finally, the possibility of an action on F naturaly reduces to the existence of a homomorphism $\epsilon : R \to F$. The natural isomorphisms $F \otimes M \cong M \cong M \otimes F$ then lead to the axioms $(\epsilon \otimes 1)\Delta r = r = (1 \otimes \epsilon)\Delta r$. Finally, the natural requirement that the action of the ring R on the linear hull $F \cdot 1_M$ of the identity mapping in $\mathrm{Hom}\,(M, M)$ yields the trivial representation ϵ leads to the axioms

$$(\eta \otimes 1)\Delta(r) = \epsilon(r) \cdot 1_R = (1 \otimes \eta)\Delta(r).$$

One of the most natural situations when such mapppings are defined is the case of the group algebra F of a group G over the field F (see, Introduction 2.2). In this case the co–multiplication Δ is given by the rule $\Delta g = g \otimes g$, $g \in G$; the co–identity ϵ is given by the mapping $\epsilon(g) = 1$, for all $g \in G$; the antipode η maps g into g^{-1}, for any $g \in G$. That the various axioms presented do indeed hold follows from the definitions of a tensor product of representations, of a trivial module and of the dual representation.

In Subsection 5.2 we shall consider similar mappings Δ, ϵ, η for the universal envelopping algebra $U(\mathfrak{g})$ of a Lie algebra $\mathfrak{g}$.

Abstracting from the concrete situation of modules we present the definition of a *Hopf algebra* H over a field F as an associative algebra with unit with multiplication law $\mu : H \otimes H \to H$ and a mapping $u : F \to H$ which takes λ into $\lambda \cdot 1$, with two homomorphisms of algebras with unit: the *co–multiplication* $\Delta : H \to H \otimes H$ and the *co–unit* $\epsilon : H \to F$, and with a anti–endomorphism $\eta : H \to H$ which is called an *antipode*, such that the following axioms are satisfied:

(1) $(\Delta \otimes 1_H)\Delta = (1_H \otimes \Delta)\Delta$ – co–associativity;

(2) $(\epsilon \otimes 1_H)\Delta = 1_H = (1_H \otimes \epsilon)\Delta$;

(3) $\mu(\eta \otimes 1_H) = u\epsilon = \mu(1_H \otimes \eta)\Delta$.

A Hopf algebra satisfying

(4) $\Delta = \tau\Delta$, where $\tau(a \otimes b) = b \otimes a$,

is called *co–commutative*.

In the language of diagrams of mappings these axioms are given by the property of commutativity of the diagrams

(1)
$$\begin{array}{ccc} H & \xrightarrow{\Delta} & H \otimes H \\ {\scriptstyle\Delta}\downarrow & & \downarrow{\scriptstyle 1\otimes\Delta} \\ H \otimes H & \xrightarrow{\Delta\otimes 1} & H \otimes H \otimes H \end{array}$$

(2)
$$\begin{array}{ccccc} H \otimes H & \xleftarrow{\Delta} & H & \xrightarrow{\Delta} & H \otimes H \\ \downarrow{\scriptstyle \epsilon\otimes 1} & & \| & & \downarrow{\scriptstyle 1\otimes\epsilon} \\ F \otimes H & \longrightarrow & H & \longleftarrow & H \otimes F \end{array}$$

$$
(3)\qquad
\begin{array}{ccccc}
H \otimes H & \xleftarrow{\Delta} & H & \xrightarrow{\Delta} & H \otimes H \\
 & & \downarrow{\scriptstyle \epsilon} & & \\
\downarrow{\scriptstyle \eta\otimes 1} & & F & & \downarrow{\scriptstyle 1\otimes\eta} \\
 & & \downarrow{\scriptstyle u} & & \\
H \otimes H & \xrightarrow{\mu} & H & \xleftarrow{\mu} & H \otimes H
\end{array}
$$

$$
(4)\qquad
\begin{array}{ccc}
H & \xrightarrow{\Delta} & H \otimes H \\
\| & & {\scriptstyle \tau}\downarrow \\
H & \xrightarrow{\Delta} & H \otimes H
\end{array}
$$

If we do not require that H is an algebra and just provide it with a structure of co–multiplication and co–identity with the axioms (1), (2) then we arrive at the concept of a co–associative coalgebra. If we keep these requirements with the exception of the antipode η and the corresponding axiom (3) we get the concept of a bialgebra (some authors, see, MacLane [1], identify bialgebras and Hopf algebras).

Let H be a Hopf algebra, and K a subalgebra of it. We say that K is a Hopf subalgebra if $\Delta(K) \subset K \otimes K$ and $\eta(K) \subset K$. It is clear that then K is a Hopf algebra with respect to the same operations. For instance, if G_1 is a subgroup in G then $K = F[G_1]$ is a Hopf subalgebra in $H = F[G]$.

A homomorphism $\phi : H \to H'$ of Hopf algebras is a homomorphism of algebras with unit such that $(\phi \otimes \phi)\Delta = \Delta'\phi$, $\eta'\phi = \phi\eta$ and $\epsilon'\phi = \epsilon$. The kernel K of a homomorphism of Hopf algebras, i.e. the set of elements with zero image, is a *Hopf ideal* in the sense that K is an ideal of the algebra H, $\epsilon(K) = \{0\}$, $\eta(K) \subset K$, and

$$\Delta K \subset K \otimes H + H \otimes K.$$

In order to prove this fact we use the usual methods from § 3.5, familiar to the reader. If K is a Hopf ideal of the Hopf algebra H then the quotient algebra

H/K is provideded with co–unity, a co–multiplication, and an antipode if we put

$$\epsilon(x+K)=\epsilon(x), \quad \eta(x+K)=\eta(x)+K,$$
$$\Delta(x+K)=\Delta x+(K\otimes H+H\otimes K).$$

In the last definition the following isomorphism is used

$$(H\otimes H)/(K\otimes H+H\otimes K)\cong(H/K)\otimes(H/K),$$

which, in fact, was already presented above.

5.2. Universal enveloping algebra as a Hopf algebra. Let L be a Lie algebra over a field F, $U(L)$ its universal envelopping algebra. We define a co–multiplication in $U(L)$, by setting $\Delta x = x\otimes 1+1\otimes x$, for $x\in L$; the co–identity $\epsilon : U(L)\to F$ is defined by $\epsilon(x)=0$, for $x\in L$, and the antipode $\eta : U(L)\to U(L)$ by $\eta(x)=-x$, for $x\in L$. We call an element u of the algebra $U(L)$ *primitive* if $\Delta u = u\otimes 1+1\otimes u$. It is clear that the elements from the algebra L are primitive.

Theorem. *The operations introduced turn $U(L)$ into a co–commutative Hopf algebra. In the case of a field F of characteristic zero an element $u\in U(L)$ is primitive iff $u\in L$.*

Proof. Because of the universal property of the algebra $U=U(L)$ to the effect that a homomorphism of the algebra L into an associative algebra with the bracket operation extends uniquely, it is sufficient to verify that Δ, η and ϵ are homomorphisms from L into $U\otimes U$, U^{op} and F. We have

$$\Delta[x,y]=[x,y]\otimes 1+1\otimes[x,y]=(xy-yx)\otimes 1+1\otimes(xy-yx)=$$
$$=(x\otimes 1+1\otimes x)(y\otimes 1+1\otimes y)-(y\otimes 1+1\otimes y)(x\otimes 1+1\otimes x)=$$
$$=(\Delta x)(\Delta y)-(\Delta y)(\Delta x)=[\Delta x,\Delta y],$$
$$\eta([x,y])=-[x,y]=-[-x,-y]=[-y,-x]=[\eta(y),\eta(x)],$$
$$\epsilon([x,y])=0=[\epsilon(x),\epsilon(y)].$$

Since all the mappings Δ, η, ϵ are (anti)homomorphisms, it is sufficient to verify the axioms (1)–(4) on the generating elements of the algebra $U(L)$, i.e. on the elements of the algebra L. We have

$$(\Delta \otimes 1)\Delta x = (\Delta \otimes 1)(x \otimes 1 + 1 \otimes x) =$$
$$= x \otimes 1 \otimes 1 + 1 \otimes x \otimes 1 + 1 \otimes 1 \otimes x = (1 \otimes \Delta)\Delta x,$$
$$\tau\Delta x = \tau(x \otimes 1 + 1 \otimes x) = 1 \otimes x + x \otimes 1 = \Delta x,$$
$$(\epsilon \otimes 1)\Delta x = (\epsilon \otimes 1)(x \otimes 1 + 1 \otimes x) = 1 \otimes x = x = (1 \otimes \epsilon)\Delta x,$$
$$\mu(\eta \otimes 1)\Delta x = \mu(\eta \otimes 1)(x \otimes 1 + 1 \otimes x) = \mu(-x \otimes 1 + 1 \otimes x) =$$
$$-x + x = 0 = \mu(1 \otimes \eta)\Delta x = u\epsilon(x).$$

The first part of the theorem is proved.

Let us assume now that u is a primitive element of the algebra $U(L)$. We introduce on the basis $E = (e_\alpha)_{\alpha\in\mathcal{I}}$ of the algebra L a linear ordering and take the basis for $U(L)$ which consists of the ordered monomials of the variables from the set E. On this set of monomials introduce the lexicographic ordering: the monomial $e_{\alpha_1}e_{\alpha_2}\dots e_{\alpha_n}$ is larger than the monomial $e_{\beta_1}e_{\beta_2}\dots e_{\beta_m}$ if either $n > m$ or $n = m$ and there exists k such that $\alpha_1 = \beta_1, \dots, \alpha_k = \beta_k$, but $\alpha_{k+1} > \beta_{k+1}$. This choice of a basis in $U(L)$ determines a natural choice of basis in $U(L) \otimes U(L)$. Write u in the form $u = \lambda_0 u_0 + \lambda_1 u_1 + \dots$, where u_0 is the lexicographically highest monomial. If the degree of the monomial u_0 is equal to 1 then we are through. Otherwise we consider the tensors of the form $e_{\gamma_1}e_{\gamma_2}\dots e_{\gamma_t} \otimes e_{\gamma_{t+1}}$ in the decomposition of the element Δu, and such that $\gamma_1 \geq \gamma_2 \geq \dots \geq \gamma_t \geq \gamma_{t+1}$. The tensors of this form obtained from u_0 are linearly independent of the analogous tensors which are obtained from other elements. Indeed,

$$\Delta(x_1 \dots x_{t+1}) = (x_1 \otimes 1 + 1 \otimes x_1)\dots(x_t \otimes 1 + 1 \otimes x_t) =$$
$$= x_1 \dots x_t \otimes 1 + \sum_{i=1}^{t-1} x_1 \dots x_{i-1}x_{i+1} \dots x_t \otimes x_t +$$
$$+ \sum_{s<t-1} x_{i_1} \dots x_{i_s} \otimes x_{j_1} \dots x_{j_{t-s}}.$$

Comparing the monomials which occur as summand in the first sum lexicographically proves our statement. Now write u_0 in the form $u_0 = e_{\alpha_1}^{m_1} \dots e_{\alpha_s}^{m_s}$, where $\alpha_1 > \ldots > \alpha_s$. Then the lexicographically highest of the tensors under consideration in Δu_0 will be obtaind from

$$e_{\alpha_1}^{m_1} \ldots e_{\alpha_{s-1}}^{m_{s-1}} \Delta e_{\alpha_s}^{m_s} = e_{\alpha_1}^{m_1} \ldots e_{\alpha_{s-1}}^{m_{s-1}} (e_{\alpha_s} \otimes 1 + 1 \otimes e_{\alpha_s})^{m_s}.$$

Therefore, it is equal to $m_s e_{\alpha_1}^{m_1} \ldots e_{\alpha_{s-1}}^{m_{s-1}} \otimes e_{\alpha_s}^{m_s}$ (we use the Newton binomial formula). Since the characteristic of the base field is equal to zero, this term is non–zero. It is linearly independent from the rest, i.e. in the decomposition of Δu apart from $u \otimes 1 + 1 \otimes u$ there exists other terms, and thus u is a non–primitive element.

5.3. Associating a group with a Lie algebra. Let X be a non–empty set, $\widehat{A(X)}$, the Magnus algebra, an associative algebra with 1 over the field F of characteristic 0, and let $L(X)$ be the Lie subalgebra of it which is generated by the set X with respect to the operation of commutation. We know (see, Subsection 4.3.1) that $L(X)$ is the free Lie algebra with free generating set X.

The elements of the form $u = 1 + \sum_{n=1}^{\infty} a_n = 1 + \bar{u}$ form a subgroup $G(X)$ in $\widehat{A(X)}$. In fact, the inverse for u is the well defined series $u^{-1} = 1 - \bar{u} + \bar{u}^2 - \bar{u}^3 + \ldots$. If u is a series without a constant then the following series are well defined

$$e^u = \sum_{n=0}^{\infty} \frac{u^n}{n!}, \quad \ln(1+u) = \sum_{n=1}^{\infty} (-1)^{n-1} \frac{u^n}{n}.$$

The following relations hold for these series

$$\ln e^u = u, \quad e^{\ln(1+u)} = 1 + u. \tag{5}$$

The validity of these relations follows from the fact that the homogeneous componenets of the series $\ln e^x$ and $e^{\ln(1+x)}$ are polynomials which are zero on an infinite set of rational numbers. The relations (5) show that the functions e and $\ln$ realize inverse bijections between set $M(X)$ of the series without constant term and the set $G(X)$ of the series with constant term 1. In particular, for $u, v \in M(X)$ there exists an $w \in M(X)$ such that $e^u e^v = e^w$. This fact shows that $e^{M(X)}$ is a subgroup in $G(X)$. In particular, if u, v commute then $w = u + v$. We are interested in the element w for given $u, v \in L(X)$. To get an answer to this problem we introduce the Lie subalgebra $\widehat{L(X)}$ in $\widehat{A(X)}$ which consists of the series $\sum_{n=1}^{\infty} a_n$ where $a_n \in L_n(X)$, $n = 1, 2, \ldots$

Theorem. $e^{\widehat{L(X)}}$ *is a subgroup in the group of invertible elements of the algebra* $\widehat{A(X)}$.

Proof. We extend the co–multiplication $\Delta : A(X) \to A(X) \otimes A(X)$ to $\widehat{A(X)}$ by setting $\Delta(\sum_n a_n) = \sum_n \Delta a_n$. This definition is correct, since Δa_n is a homogeneous element of degree n. It is evident that Δ is a (continuous) homomorphism and that $\Delta u = u \otimes 1 + 1 \otimes u$ iff $u \in \widehat{L(X)}$. The second statement follows from the theorem from Subsection 5.2 above and from the fact that two elements in $\widehat{A(X)}$ are equal iff their homogeneous components are equal. Now, let $a, b \in \widehat{L(X)}$. Consider w satisfying $e^w = e^a e^b$. We can write down $w = \ln e^a e^b$. Apply the co–multiplication to w, to obtain

$$
\begin{aligned}
&\Delta \ln(e^a e^b) = \ln \Delta(e^a e^b) = \ln((\Delta e^a))\Delta e^b)) = \\
&= \ln(e^{\Delta a} e^{\Delta b}) = \ln(e^{a\otimes 1 + 1\otimes a} e^{b\otimes 1 + 1\otimes b}) = \\
&= \ln(e^{a\otimes 1} e^{1\otimes a} e^{b\otimes 1} e^{1\otimes b}) = \\
&= \ln(e^{a\otimes 1} e^{b\otimes 1} e^{1\otimes a} e^{1\otimes b}) = \\
&= \ln((e^a \otimes 1)(e^b \otimes 1)(1 \otimes e^a)(1 \otimes e^b)) = \\
&= \ln((e^a e^b \otimes 1)(1 \otimes e^a e^b)) = \ln(e^a e^b \otimes 1) + \ln(1 \otimes e^a e^b) = \\
&= \ln(e^a e^b) \otimes 1 + 1 \otimes \ln(e^a e^b) = w \otimes 1 + 1 \otimes w.
\end{aligned}
$$

According to what was said above we can conclude that $w \in \widehat{L(X)}$. □

If $X = \{x, y\}$ then $e^x e^y \in e^{H(x,y)}$, where $H(x,y) = \widehat{L(x,y)}$. The series $H(x,y)$ is called the *Hausdorff series*. Its homogeneous components are linear combinations of commutators of x, y of fixed degree. In order to compute the series there exists various formulas among which the best known is the so–called Campbell–Baker–Hausdorff formula (see, for example, Serre [1]). It is useful to know a few terms of the series

$$H(x,y) = x + y + \frac{1}{2}[x,y] + \frac{1}{12}[x,[x,y]] + \frac{1}{12}[y,[y,x]] + \dots$$

The importance of the Hausdorff series lies in the fact that any Lie algebra where the series is defined for all pairs of elements, becomes a group if one defines $x \circ y = H(x,y)$ (all axioms are verified automatically starting from the fact that $e^{\widehat{L(x,y)}}$ is a group). Thus, if the Lie algebra L is nilpotent then the commutators of high enough degree are equal to zero and we obtain a nilpotent group L°. If the series $H(x, y)$ converges only in certain neighborhood of the zero then not all products are defined and we obtain a *local* group. These are important in investigating the correspondence between Lie algebras and Lie groups.

References.

Bourbaki [5], Curtis–Reiner [1], MacLane [1], Herstein [1], Bergman [1].

CHAPTER VI

ALGEBRAIC GROUPS

§ 1. Hopf algebras and algebraic groups

1.1. Algebra of representing functions. Let S be a non–empty set, F a field. We denote by F^S the algebra of functions on S with values in F. The (pointwise) operations of addition, multiplication of functions, and multiplication by scalars from F are:

$$\begin{aligned}(f+g)(s) &= f(s)+g(s),\\ (fg)(s) &= f(s)g(s),\\ (\lambda f)(s) &= \lambda f(s).\end{aligned}$$

Here $f, g \in F^S$, $s \in S$, $\lambda \in F$. If $S = G$ is a semigroup then F^G becomes a left and right G–module if for $x \in G$, $f \in F^G$ and $y \in G$ we set

$$(xf)(y) = f(yx), \quad \text{and } (fx)(y) = f(xy).$$

A function $f \in F^G$ is called *representing* if the linear hull of the set $Gf = \{xf \mid x \in G\}$ is finite–dimensional. The set of representative functions $\mathcal{R}_F(G)$ (or simply $\mathcal{R}(G)$, if F is fixed) is an F–subalgebra in F^G and is called the *algebra of representing functions* of the subgroup G with values in the field F. We shall verify, for instance, that if $f, g \in \mathcal{R}(G)$ then also $fg \in \mathcal{R}(G)$. Indeed, if $\{e_1, \ldots, e_k\}$ is the basis of a linear hull for Gf and $\{e'_1, \ldots, e'_l\}$ for Gg then $x(fg) = (xf)(xg)$, i.e. $G(fg)$ lies in the linear hull of the set $\{e_i e'_j \mid i = 1, \ldots, k; j = 1, \ldots, l\}$. It is also clear that $\mathcal{R}(G)$ is a two–sided G–submodule in F^G. Our problem in this section is to introduce operations for $\mathcal{R}(G)$ which transform it in a Hopf algebra under the assumption that G is a group.

We denote by the letter m the law of composition on G, $m : G \times G \to G$, and by π the mapping $\pi : F^G \otimes F^G \to F^{G\times G}$ which is given by the formula

$$\pi\left(\sum f \otimes g\right)(x, y) = \sum f(x)g(y). \tag{1}$$

Here $f, g \in F^G$, $x, y \in G$. It is clear that this is well defined. In order to introduce a coproduct in $\mathcal{R}(G)$ it is necessary to have the mapping $m^* : F^G \to F^{G\times G}$ which is induced by the law m of composition in G. It is naturally given by the formula

$$(m^* f)(x, y) = f(m(x, y)) = f(xy). \tag{2}$$

Now set

$$\delta(f) = (\pi^{-1} \circ m^*)(f), \quad f \in \mathcal{R}(G). \tag{3}$$

Proposition. *The mapping δ is well defined and its image is in $\mathcal{R}(G) \otimes \mathcal{R}(G)$.*

Proof. First we shall show that π is injective and that Im $m^* \subset$ Im π. Indeed, let $u = \sum_{j=1}^n f_j \otimes g_j \in \operatorname{Ker} \pi$. We denote by the letter V the subspace in F^G which is spanned by $\{f_1, \dots, f_n\}$. Let $\dim V = m$. Using induction on n take a basis $\{v_1, \dots, v_m\}$ in V and points $x_1, \dots, x_m$ in G such that $v_i(x_j) = 0$ for $i \neq j$, and $v_i(x_i) = 1$, $i = 1, 2, \dots, m$. For v_1' we choose an arbitrary non–zero function from V and as x_1 any point such that $v_1'(x_1) = 1$. Further, the subspace

$$V_1 = \{f \in V \mid f(x_1) = 0\}$$

has codimension 1 in V which allows us to apply induction and select $\{v_2, \dots, v_m\}$ and $\{x_2, \dots, x_m\}$ with the required properties. The systems $\{x_1, x_2, \dots, x_m\}$ and $\{v_1, v_2, \dots, v_m\}$, where $v_1 = v_1' - v_1'(x_2)v_2 - \dots - v_1'(x_m)v_m$, are the desired ones. Returning to the element u from the kernel, we can write $u = \sum_{i=1}^m v_i \otimes \widetilde{g}_i$. If $y \in G$ then computing u in the point (x_j, y), we obtain $0 = \pi(u)(x_j, y) = v_j(x_j)\widetilde{g}_j(y) = \widetilde{g}_i(y)$. Thus all the functions $\widetilde{g}_1, \dots, \widetilde{g}_m$ are zero, so that also $u = 0$.

Let now $f \in \mathcal{R}(G)$. Then Gf lies in a finite–dimensional subspace with the basis $\{u_1, \dots, u_m\}$. Let the functions $g_1, \dots, g_m \in F^G$ be such that for any $y \in G$ we have

$$yf = g_(y)u_1 + \dots + g_n(y)u_n.$$

In this case

$$\pi\left(\sum_{i=1}^{n} u_i \otimes g_i\right)(x,y) = \sum_{i=1}^{n} u_i(x)g_i(y) =$$
$$= \left(\sum_{i=1}^{n} g_i(y)u_i\right)(x) = (yf)(x) = f(xy) = (m^*f)(x,y).$$

So Im $m^* \subset$ Im π, and the mapping δ is well defined.

Let now $f \in \mathcal{R}(G)$. Reasoning as above we select $v_1, \dots, v_n \in F^G$ and $x_1, \dots, x_n \in G$ such that

$$\delta(f) = \pi^{-1}(m^*(f)) = \sum_{i=1}^{n} v_i \otimes h_i, \tag{4}$$

where $v_i(x_j) = \delta_{ij}$, $i, j = 1, \dots, n$. If $y \in G$ then computing m^*f on (x_j, y) we get

$$f(x_j y) = h_j(y), \quad \text{i.e. } h_j = f x_j.$$

Thus all coefficients h_j in the decomposition (4) lie in $\mathcal{R}(G)$. Hence $\delta(f) \in F^G \otimes \mathcal{R}(G)$. In a completely similar way we have $\delta(f) \in \mathcal{R}(G) \otimes F^G$. Therefore, of course, $\delta(f) \in \mathcal{R}(G) \otimes \mathcal{R}(G)$, proving the proposition. □

If $f \in \mathcal{R}(G)$ then $\pi(\delta(f)(x \otimes y)) = f(xy)$. Thus if $\delta(f) = \sum_i f_i' \otimes f_i''$ then

$$f(xy) = \sum_i f_i'(x) f_i''(y).$$

The latter formulas show that if the functions from $\mathcal{R}(G)$ separate the points in G then the coproduct in $\mathcal{R}(G)$ defines the product in G.

Now assume that G is a group. Define the mapping $\eta : \mathcal{R}(G) \to \mathcal{R}(G)$ by setting $(\eta(f))(x) = f(x^{-1})$. This mapping plays the part of the antipode in $\mathcal{R}(G)$. Finally, we define the co–unit ϵ in $\mathcal{R}(G)$ by setting $\epsilon(f) = f(e)$, where e is the identity element of the group G.

In order to formulate the main result of this section we note by $\mu : \mathcal{R}(G) \otimes \mathcal{R}(G) \to \mathcal{R}(G)$ the pointwise multiplication in $\mathcal{R}(G)$, by $u : F \to \mathcal{R}(G)$ the unit in $\mathcal{R}(G)$, i.e. the mapping $u(\alpha) = \alpha \cdot 1_G$, where $\alpha \in F$. We shall identify F with the subalgebra in $\mathcal{R}(G)$ of the form $F \cdot 1_G$.

Theorem. *The mappings introduced above μ, u, δ, ϵ, η make the algebra $\mathcal{R}(G)$ of representing functions on the group G into a Hopf algebra.*

Proof. We shall verify the axioms (1)–(3) of the definition of a Hopf algebra. To do this we shall use an identification π of the algebra $\mathcal{R}(G) \otimes \mathcal{R}(G)$ with a subalgebra of the algebra of functions $F^{G\times G}$.

Let $f \in \mathcal{R}(G)$, $x, y, z \in G$. If $\delta(f) = \sum_i f_i' \otimes f_i''$ then

$$(\delta \otimes 1) \circ (f)(x \otimes y \otimes z) = \left(\sum_i \delta(f_i') \otimes f_i''\right)(x \otimes y \otimes z) =$$
$$= \sum_i f_i'(xy) f_i''(z) = f((xy)z).$$
$$(1 \otimes \delta)(f)(x \otimes y \otimes z) = \left(\sum_i f_i' \otimes \delta(f_i'')\right)(x \otimes y \otimes z) =$$
$$= \sum_i f_i'(x) f_i''(yz) = f(x(yz)).$$

Because of the associativity of the multiplication in a group we have $((\delta \otimes 1) \circ (f) = (1 \otimes \delta) \circ (f)$ proving that co–associativity (1) holds. Moreover,

$$((\epsilon \otimes 1) \circ \delta) f(x) = \left(\sum_i \epsilon(f_i') f_i''\right)(x) = \sum_i f_i'(e) f_i''(x) = f(x),$$
$$((1 \otimes \epsilon) \circ \delta) f(x) = \left(\sum_i f_i' \epsilon(f_i'')\right)(x) = \sum_i f_i'(x) f_i''(e) = f(x),$$

i.e. the condition (2) is fulfilled. Finally,

$$(\mu \circ (\eta \otimes 1) \circ \delta)(f)(x) = \sum_i f_i'(x^{-1}) f_i''(x) = f(x^{-1}x) = f(e) = \epsilon(f),$$

$$(\mu \circ (1 \otimes \eta) \circ \delta)(f)(x) = \sum_i f_i'(x) f_i''(x^{-1}) = f(xx^{-1}) = f(e) = \epsilon(f).$$

Thus

$$(\mu \circ (\eta \otimes 1) \circ \delta)(f) = \epsilon(f) 1_G = (u \circ \epsilon)(f) = (\mu \circ (1 \otimes \eta) \circ \delta)(f).$$

The theorem is proved. □

1.2. Group associated with Hopf algebra. Let (C, δ, ϵ) be a co–algebra, (A, μ, u) an algebra, both of them over the field F. Then the set $H = \mathrm{Hom}_F(C, A)$ of all F–linear mappings from C into A becomes an algebra over F if we define for $h, k \in H$ their product as equal to

$$hk = \mu \circ (h \otimes k) \circ \delta.$$

Using the co–associativity $(1 \otimes \delta) \circ \delta = (\delta \otimes 1) \circ \delta$ in C and the associativity $\mu \circ (1 \otimes \mu) = \mu \circ (\mu \otimes 1)$, we obtain that H is associative:

$$\begin{aligned}(hk)l &= \mu \circ (\mu \circ (h \otimes k) \circ \delta \otimes l) \circ \delta = \\ &= \mu \circ (\mu \otimes 1) \circ ((h \otimes k) \otimes l) \circ (\delta \otimes 1) \circ \delta = \\ &\mu \circ (1 \otimes \mu) \circ (h \otimes (k \otimes l)) \circ (1 \otimes \delta) \circ \delta = \\ &= \mu \circ (h \otimes (\mu \otimes (k \otimes l) \circ \delta) \circ \delta = h(kl).\end{aligned}$$

Since $(\epsilon \otimes 1) \circ \delta = \mathrm{id}_C = (1 \otimes \epsilon) \circ \delta$ and $\mu \circ (u \otimes 1) = \mathrm{id}_A = \mu \circ (1 \otimes u)$, we see that $u \circ \epsilon$ is the identity in H:

$$\begin{aligned}h \cdot (u \circ \epsilon) &= \mu \circ (h \otimes u \circ \epsilon) \circ \delta = \mu \circ (1 \otimes u) \circ (h \otimes 1) \circ (1 \otimes \epsilon) \circ \delta = \\ &= h \otimes 1 = h = 1 \otimes h = \mu \circ (u \otimes 1) \circ (1 \otimes h) \circ (\epsilon \otimes 1) \circ \delta = \\ &= \mu \circ (u \circ \epsilon \otimes h) \circ \delta = (u \circ \epsilon) \cdot h.\end{aligned}$$

Here we have used the natural identification of F with $F \otimes F$, and of C with $F \otimes C$ and $C \otimes F$, of A with $A \otimes F$ and $F \otimes A$. If $A = F$ we then define $C^{\circ} = \mathrm{Hom}\,(C, F)$.

Now let $A = (A, \mu, u, \delta, \epsilon, \eta)$ be a Hopf algebra. We define by $\mathcal{G}(A)$ the set of homomorphisms from A into F as a F–algebra. An element h is in $\mathcal{G}(A)$ if $\mu'(h \otimes h) = h\mu$, where μ' is the multiplication in F. Considering F as a F–algebra, A as an F–co–algebra we see, as above, that the operation

$$hk = (h \otimes k) \circ \delta$$

(we omitted μ') is associative and has a co–identity, as above. Further, $\mathcal{G}(A)$ is closed with respect to this operation. In fact, if $h, k \in \mathcal{G}(A)$ then $h \otimes h = h\mu$, $k \otimes k = k\mu$ and

$$\begin{aligned} hk \otimes hk =&(h \otimes k) \circ \delta \otimes (h \otimes k) \circ \delta = (h \otimes k \otimes h \otimes k) \circ (\delta \otimes \delta) = \\ =&(h \otimes h \otimes k \otimes k) \circ (\delta \otimes \delta) = (h \circ \mu \otimes k \circ \mu) \circ (\delta \otimes \delta) = \\ =&(h \otimes k) \circ (\mu \otimes \mu) \circ (\delta \otimes \delta) = (h \otimes k) \circ \delta \circ \mu = (hk) \circ \mu. \end{aligned}$$

We have used the commutativity of the multiplication in F and the fact that $\delta : A \to A \otimes A$ is a homomorphism of algebras.

We now show that $\mathcal{G}(A)$ is a group. It remains to check that if $h \in \mathcal{G}(A)$ and η is the antipode in A then $h \circ \eta$ is inverse to h. We have

$$\begin{aligned} h(h \circ \eta) =&(h \otimes h \circ \eta) \circ \delta) = (h \otimes h) \circ (1 \otimes \eta) \circ \delta = \\ =&h\mu \circ (1 \otimes \eta) \circ \delta = h \circ u \circ \epsilon = u \circ \epsilon = \epsilon. \end{aligned}$$

Similarly, one checks that $(h \circ \eta)h = \epsilon$.

Thus we have proved

Proposition 1. *Let A be a Hopf algebra over the field F. The set $\mathcal{G}(A)$ of homomorphisms from A into F with the multiplication operation $(h, k \in \mathcal{G}(A))$*

$$hk = (h \otimes k) \circ \delta$$

is a group. □

The construction of the Hopf algebra $\mathcal{R}(G)$ of representative functions of the group G and of the group $\mathcal{G}(A)$ of the Hopf algebra A are mutually inverse in certain sense. In order to determine the exact sense of this statement it is necessary to prove an important auxiliary statement.

We shall make A into a two–sided A°–module (A° is considered as an algebra relatively to the operation introduced above). For that with any $\tau \in A^{\circ}$ associate two operators $\tau_{[}$ and $\tau_{]}$ which act on A, by setting

$$\tau_{[} = (1_A \otimes \tau) \circ \delta, \quad \tau_{]} = (\tau \otimes 1_A) \circ \delta \tag{5}$$

Proposition 2. *Let $A = (A, \mu, u, \delta, \epsilon, \eta)$ be a Hopf algebra over the field F. The mappings $\tau \mapsto \tau_{[}$, $\tau \mapsto \tau_{]}$ are an injective homomorphism and antihomomorphism from A° in $End(A)$. Each $\tau_{[}$ commutes with each $\sigma_{]}$, $\tau, \sigma \in A^{\circ}$, i.e. A is a bimodule over A°. The bimodule A is locally finite, i.e. it coincides with the union of the family of its finite–dimensional A°–subbimodules.*

Proof. Note that for any $\tau, \sigma \in A^{\circ}$ we have

$$\tau\sigma = (\tau \otimes \sigma) \circ \delta = \tau \circ (1_A \otimes \sigma) \circ \delta = \tau \circ \sigma_{[}.$$

Further,

$$\delta \circ \sigma_{[} = \delta \circ (1_A \otimes \sigma) \circ \delta = (\delta \otimes \sigma) \circ \delta = (1_A \otimes 1_A \otimes \sigma) \circ (\delta \otimes 1_A) \circ \delta =$$
$$= (1_A \otimes 1_A \circ \sigma) \circ (1_A \otimes \delta) \circ \delta = (1_A \otimes \sigma_{[}) \circ \delta$$

Multiplying by $1_A \otimes \tau$ we obtain

$$\tau_{[} \circ \sigma_{[} = (1_A \otimes (\tau \circ \sigma_{[})) \circ \delta = (1_A \otimes \tau\sigma) \circ \delta = (\tau\sigma)_{[}.$$

Similarly, $\tau_{]} \circ \sigma_{]} = (\sigma\tau)_{]}$. Further, $\epsilon_{[} = 1_A$ ($[1_A \otimes \epsilon) \circ \delta] = 1_A$). In addition, $\epsilon \circ \tau_{[} = \tau$, i.e. $\tau_{[}$ is an injective mapping. Further,

$$\sigma_{]} \circ \tau_{[} = (\sigma \otimes 1_A \otimes \tau) \circ (\delta \otimes 1_A) \circ \delta = (\sigma \otimes 1_A \otimes \tau) \circ (1_A \otimes \delta) \circ \delta = \tau_{[} \circ \sigma_{]}.$$

This demonstrates that A is indeed an A°–bimodule.

It remains to prove that A is locally finite. Let $a \in A$, $\delta(a) = \sum_i a'_i \otimes a''_i$.

From the definition it is clear that the left A°–orbit $A^\circ_{[}(a)$ lies in the F–space which is spanned by the elements a'_i, and the right A°–orbit $A^\circ_{]}(a)$ is in the F–hull of the elements a''_i. Hence each of these orbits is finite–dimensional. Hence also $\dim A^\circ_{[}(A^\circ_{]}(a)) < \infty$. □

Let now A be a Hopf algebra, $\mathcal{G}(A)$ the corresponding group, $B = \mathcal{R}(\mathcal{G}(A)$. We define a mapping $\Theta : A \to F^{\mathcal{G}(A)}$ by setting

$$\Theta(a)(f) = f(a),$$

where $a \in A$, $f \in F^{\mathcal{G}(A)}$.

Corollary. *The image of the mapping Θ lies in B and this mapping is a morphism of Hopf algebras.*

Proof. If $a \in A$ then it is necessary to prove that $\Theta(a)\mathcal{G}(A)$ lies in a finite–dimensional F–subspace. Indeed, if $g \in \mathcal{G}(A)$, $f \in \mathcal{G}(A)$ then

$$\begin{aligned}(\Theta(a)g)(f) &= \Theta(a)(gf) = \Theta(a)((g \otimes f) \circ \delta) = ((g \otimes f) \circ \delta)(a) = \\ &= f((g \otimes 1_A) \circ \delta(a)) = \Theta(g_{]}(a))(f).\end{aligned}$$

The set $g_{]}(a)$, $g \in \mathcal{G}(A)$, according to the proposition 2, lies in a finite–dimensional F–subspace, This proves the first part of the corollary. The second part is proved by direct computation. We shall check, for instance, that $\Theta : A \to B$ preserves the co–product, i.e.

$$(\Theta \otimes \Theta) \circ \delta = \delta \circ \Theta.$$

In fact, let $a \in A$ and $\delta(a) = \sum_i a'_i \otimes a''_i$. If $f, g \in F^{\mathcal{G}(A)}$ then

$$[((\Theta \otimes \Theta) \circ \delta)(a)](f \otimes g) = \sum_i f(a'_i)g(a''_i) = (f \otimes g) \circ \delta(a).$$

On the other hand,

$$[\delta \circ \Theta(a)](f \otimes g) = (fg)(a) = (f \otimes g) \circ \delta(a),$$

as required to prove. □

Proposition. *The set $X \subset F^G$ of homomorphisms of the group G into the multiplicative group F^* of the field F is linearly independent.*

Proof. Let $\chi_1, \dots, \chi_n$ be a pairwise different elements from χ. Applying induction with respect to n we assume that

$$\alpha_1\chi_1 + \dots + \alpha_n\chi_n = 0,$$

where $\alpha_1, \dots, \alpha_n$ are non–zero elements from the field F. For any g, $g_1 \in G$ we write down

$$\alpha_1\chi_1(g_1 g) + \dots + \alpha_n\chi_n(g_1 g) = 0,$$

and

$$[\alpha_1\chi_1(g_1)\chi_1 + \dots + \alpha_n\chi_n(g_1)\chi_n](g) = 0,$$

i.e.

$$\alpha_1\chi_1(g_1)\chi_1 + \dots + \alpha_n\chi_n(g_1)\chi_n = 0.$$

By assumption there exists a g_1 such that $\chi_1(g_1) \neq \chi_2(g_2)$. Multiplying the first of these equalities by $\chi_1(g_1)$ and subtracting the result term by term from the last equality we obtain

$$\beta_2\chi_2 + \dots + \beta_n\chi_n = 0,$$

where $\beta_i = \alpha_i(\chi_i(g_1) - \chi_1(g_1))$, $i = 2, \dots, n$, moreover $\beta_2 \neq 0$. By applying induction we obtain a contradiction. □

1.3. Affine algebraic sets. A non–empty set S together with a finitely generated subalgebra $\mathcal{P}(S) \subset F^S$ is called an *affine algebraic* F–set, where F is a field, if $\mathcal{P}(S)$ separates points in S and any homomorphism from $\mathcal{P}(S)$ into F coincides with taking a value of a function in a suitable point $s \in S$. Because of the Hilbert theorem any algebraic set $S \subset F^n$ (in the sense of § 1.5) is affine algebraic with $\mathcal{P}(S) = F[X_1, \dots, X_n]/\mathcal{I}(S)$.

Let s^*, for $s \in S$, denote the evaluation mapping $s^* : F^S \to F$. The second condition now takes the form: for any homomrphism $\phi : \mathcal{P}(S) \to F$ there exists $s_\phi \in S$ such that $\phi(f) = f(s_\phi)$ for any function $f \in F^S$, i.e. $\phi = s_\phi^*$.

For instance, if $S = \{s_1, \dots, s_n\}$ is a finite set then the algebra F^S is finite–dimensional. Take $\mathcal{P}(S) = F^S$. Then $F^S = \langle \chi_1, \dots, \chi_n \rangle$, where $\chi_i(s_j) = \delta_{ij}$, $i, j = 1, \dots, n$. Further, $\chi_i \chi_j = \delta_{ij}\chi_i$. If $f : F^S \to F$, f is non–zero, then from the last condition it follows that there is an i in $\{1, \dots, n\}$ such that $f(\chi_i) = 1$, $f(\chi_j) = 0$ when $j \neq i$. Let f_i be equal to such function, then $f_i = s_i^*$.

Another example. Let S be the set of points of the line with the point 0 deleted. In F^S we select the subset $\mathcal{P}(S) = F[X, X^{-1}]$ of Laurent polynomials. The polynomials $X - a$ and $X - b$ separate the points a, b. Any homomorphism from $F[X, X^{-1}]$ is defined by the mapping X to a number a, which differs from zero, i.e. defined by the computation of the polynomial in the point $a \in S$.

It should be remarked that the set S in the second example is not algebraic in the sense of the definition which is given in the first chapter.

The affine algebraic set S is provided with the Zariski toplogy: the closed subsets are the annihilators (the sets of zeros) of subsets from $\mathcal{P}(S)$. If T is a non–empty closed subset from S then its annihilator J_T is a proper ideal in $\mathcal{P}(S)$; T becomes an algebraic set if we take as $\mathcal{P}(T)$ the set of restrictions to T of functions from $\mathcal{P}(S)$. It is clear that $\mathcal{P}(T) \cong \mathcal{P}(S)/J_T$.

If A and B are affine algebraic sets then we call any mapping $\alpha : A \to B$, for which $\mathcal{P}(B) \circ \alpha \subset \mathcal{P}(A)$, a *morphism* from A into B. A morphism is a continuous mapping in the sense that the preimage $\alpha^{-1}(T)$ of a closed set (with the annihilator J_T) is a closed set (with the annihilator $J_T \circ \alpha$).

The elements from $\mathcal{P}(S)$ are called the *polynomial functions* on S. A morphism of affine algebraic sets is called a *polynomial mapping*.

Let S and T be two algebraic sets with algebras of polynomial functions $\mathcal{P}(S)$ and $\mathcal{P}(T)$. We shall make $S \times T$ into an algebraic set. To do that we should note that according to Subsection 1.1 the mapping $\pi : F^S \otimes F^T \to F^{S \times T}$ which is given by the formula

$$\pi\left(\sum f \otimes g\right)(s,t) = \sum f(s)g(t),$$

is injective. We define $\mathcal{P}(S \times T)$ as $\pi(\mathcal{P}(S) \otimes \mathcal{P}(T))$. If $(s_1, t_1) \neq (s_2, t_2)$ then there are $f \in \mathcal{P}(S)$ and $g \in \mathcal{P}(T)$ such that if $s_1 \neq s_2$ then $f(s_1) \neq f(s_2)$ and if $t_1 \neq t_2$ then $g(t_1) \neq g(t_2)$. It is clear that in the first case $\pi(f \otimes 1)$ and in the second case $\pi(1 \otimes g)$ separate the given points. If now ψ is a homomorphism $\psi : \mathcal{P}(S \times T) \to F$ then its restriction to $\mathcal{P}(S) \otimes 1$ lifts to a homomorphism $s^* : \mathcal{P}(S) \to \mathcal{P}(S) \otimes 1 \to F$ for suitable $s \in S$. Analogously, there is also a

$t \in T$ such that $t^* : \mathcal{P}(T) \to 1 \otimes \mathcal{P}(T) \to T$ coincides with the restriction to $1 \otimes \mathcal{P}(T)$.

Note that $\pi(\mathcal{P}(S) \otimes \mathcal{P}(T))$ is generated by the subalgebras $\mathcal{P}(S) \otimes 1$ and $1 \otimes \mathcal{P}(T)$.

In this case $\phi(f \otimes g) = \phi(f \otimes 1)\phi(1 \otimes g) = f(s)g(t) = \pi((f \otimes g))(s,t)$.

Thus, $S \times T$ is an algebraic set with the algebra of polynomial functions $\pi(\mathcal{P}(S) \otimes \mathcal{P}(T))$. The projections π_1, π_2 from $S \times T$ into S and T are morphisms: $\mathcal{P}(S) \circ \pi_1 \subset \pi(\mathcal{P}(S) \otimes \mathcal{P}(T))$ and $\mathcal{P}(T) \circ \pi_2 \subset \pi(\mathcal{P}(S) \otimes \mathcal{P}(T))$. Indeed, for instance, if $(s,t) \in S \times T$, $f \in \mathcal{P}(S)$, then $(f \circ \pi_1)(s,t) = f(s) = \pi(f \otimes 1)(s,t)$, i.e. $f \otimes \pi_1 = \pi(f \otimes 1)$. Besides, if $\sigma : A \to S$, $\tau : A \to T$ are morphisms then $\sigma \times \tau : A \to S \times T$ is a morphism. Really, if $f \in \mathcal{P}(S)$, $g \in \mathcal{P}(T)$, $a \in A$, then

$$\pi(f \otimes g) \circ (\sigma\tau)(a) = f(\sigma(a))g(\tau(a)) = [f \circ \sigma)(g \circ \tau)](a).$$

Since $f \circ \sigma \in \mathcal{P}(A)$, $g \circ \tau \in \mathcal{P}(A)$ then also $(f \circ \sigma)(g \circ \tau) \in \mathcal{P}(A)$, i.e. $\pi(f \otimes g) \circ (\sigma \times \tau) \in \mathcal{P}(A)$.

Hence the structure of an algebraic set on $S \times T$ coincides with our intuitive ideas.

1.4. Affine algebraic group. Let G be a group and, simultaneously, an algebraic set over the field F with algebra of polynomial functions $\mathcal{P}(G)$. We then say that G is an *algebraic* F–group if the multiplication $G \times G \to G$ and the inverse mapping $G \to G$ are morphisms (polynomial mappings). It should be remarked that in this case $\mathcal{P}(G)$ is a subalgebra in the Hopf algebra $\mathcal{R}_F(G)$. Indeed, $\mathcal{P}(G \times G) = \pi(\mathcal{P}(G) \otimes \mathcal{P}(G)) \subset \pi(F^G \otimes F^G)$. Let $m : G \times G \to G$, $m^* : F^G \to F^{G \times G}$; then since m is a morphism, we have for any $f \in \mathcal{P}(G)$, $m^*(f) \subset \mathcal{P}(G \times G) \subset \pi(F^G \times F^G)$. In an analogous way to the proofs in Subsection 1.1, it follows that $f \in \mathcal{R}_F(G)$. The fact that $\mathcal{P}(G)$ is closed with respect to the operations of Hopf algebra follows from the definition of the operations. That it is also closed with respect to antipode follows from the polynomiality of the inverse element mapping in G.

Now assume that G is an algebraic F–group, $A = \mathcal{P}(G)$ its Hopf algebra of polynomial functions. We consider the group $\mathcal{G}(A)$ of the algebra A. Its elements are the homomorphisms from $\mathcal{P}(G)$ into F. According to the definition of an algebraic group, each such homomorphism has the form g^*, $g \in G$. It is evident, that $\lambda : g \to g^*$ is an isomorphism of the groups G and $\mathcal{G}(A)$. As was noted in Subsection 1.2, there exists a canonical morphism from $\mathcal{P}(G)$ into $\mathcal{R}(\mathcal{G}(A))$. Identifying its image with $\mathcal{P}(G)$ we see that

$\mathcal{G}(A)$ is isomorphic to G; moreover the algebras of polynomial functions are isomorphic to $\mathcal{P}(G)$. This allows to identify the elements from G with the homomorphisms of the algebra $\mathcal{P}(G)$ into F. Besides, if $a \in \mathcal{P}(G) = \mathcal{P}(\mathcal{G}(A))$ and $g \in G$ then $(a \circ \lambda)(g) = a(g^*) = g^*(a) = a(g)$, i.e. $\mathcal{P}(G) \circ \lambda \subset \mathcal{P}(G)$. Thus $\lambda : G \to \mathcal{G}(\mathcal{P}(G))$ is an isomorphism of algebraic groups.

Some main examples of algebraic groups are the following. Let F be an additive group of the field F. We set $\mathcal{P}(F) = F[x]$, where x is the identity mapping of the group G. (It should be remarked that $x \neq 1$, because $1(a) = 1$ and $x(a) = a$.) Now $\delta(x) = \pi^{-1} \circ m^* \circ x$. By virtue of the continuity of operation, $\delta(x) \subset \mathcal{P}(F) \otimes \mathcal{P}(F)$. Because of the injectivity of $\pi : \mathcal{P}(F) \otimes \mathcal{P}(F) \to \mathcal{P}(F \times F)$ and the fact that $[\pi(\delta(x))](a, b) = x(a + b) = a + b$, we see that for $\delta(x)$ we should take $\delta(x) = x \otimes 1 + 1 \otimes x$. Analogously, $\eta(x) = -x$. Besides, from the identity $(\epsilon \otimes 1_{\mathcal{P}(F)}) \circ \delta = 1_{\mathcal{P}(F)}$ we derive

$$x = (\epsilon \otimes 1_{\mathcal{P}(F)})(x \otimes 1 + 1 \otimes x) = \epsilon(x) \cdot 1 + 1 \cdot x = \epsilon(x) + x,$$

i.e. $\epsilon(x) = 0$. Further, it is clear that x separates the points from F. Then any homomorphism from $\mathcal{P}(F)$ into F is defined by the image of the element x. If $\phi(x) = a \in F$ then $\phi = a^*$. According to the definition of multiplication in the group $\mathcal{G}(\mathcal{P}(F))$ we have

$$\begin{aligned}(a^* b^*)(x) &= (a^* \otimes b^*) \circ \delta(x) = (a^* \otimes b^*)(x \otimes 1 + 1 \otimes x) = \\ &= a^*(x) + b^*(x) = a + b,\end{aligned}$$

i.e. $a^* b^* = a + b$, thus, $\mathcal{G}(\mathcal{P}(F)) \cong F$. If $f \in \mathcal{P}(F)$, $a^* \in \mathcal{G}(\mathcal{P}(F))$, then putting $f^*(a^*) = a^*(f) = f(a)$, we obtain an isomorphism of the algebras $\mathcal{P}(F)$ and $\mathcal{P}(\mathcal{G}(\mathcal{P}(F)))$. Hence the algebraic group F is identified with the algebraic group of homomorphisms from $F[x]$ into F.

Further, the group $F^* = F \setminus \{0\}$ with respect to the multiplication of numbers is algebraic with the algebra of polynomial functions $\mathcal{P}(F^*) = F[u, u^{-1}]$, where $u : F^* \to F^*$ is an identical mapping. Reasoning as above, we obtain $\delta(u) = u \otimes u$, $\epsilon(u) = 1$.

We present a generalization of the previous example. Let E be a finite–dimensional F–algebra, E^* the group of invertible elements of the algebra E.

We define $\mathcal{P}(E^*)$ as the Hopf subalgebra in $\mathcal{R}_F(E^*)$ which is generated by the restrictions to E^* of the elements from $E^\circ = \mathrm{Hom}(E, F)$. It is clear that the points of the set E^* are separated by the functions from $\mathcal{P}(E^*)$ (even from $E^\circ|_{E^*} \subset \mathcal{P}(E^*)$).

The fact that any homomorphism from $\mathcal{P}(E^*)$ is given by the value it takes in a suitable point from E^* is clear from the explicit construction for $\mathcal{P}(E^*)$. In order to describe this construction we consider a faithful representation ρ of the algebra E by linear operators on a finite dimensional space V. Let σ be the restriction of ρ to E^*. We denote by $S(\sigma)$ the space of all representing functions which are associated with σ, i.e. of the functions of the form $\gamma \circ \sigma$, where $\gamma \in (\mathrm{End}_F(V))^\circ$ (if we identify the operators on V with matrices in some basis then we are concerned with the matrix elements of the matrices which are associated with the operators from $\sigma(E^*)$). We denote by d_σ the function on E^* which maps the element x into the determinant $d_\sigma(x)$ of the operator $\sigma(x)$. The function d_σ is a polynomial in the elements from $S(\sigma)$. Since the elements from $S(\sigma)$ lie in the image of the space E° in $\mathcal{P}(E^*)$, we see that $d_\sigma \in \mathcal{P}(E^*)$. As $d_\sigma(x^{-1}) = d_\sigma(x)^{-1}$, $\eta(d_\sigma)(x) = d_\sigma(x^{-1})$ and $\eta d_\sigma \in \mathcal{P}(E^*)$, we obtain $d_\sigma^{-1} \in \mathcal{P}(E^*)$. The explicit formula for the inverse matrix shows that $\eta(f) = d_\sigma^{-1} g$, where g is a polynomial in elements from $S(\sigma)$. Now let A be the subalgebra in $\mathcal{P}(E^*)$ which is generated by the mapping d_σ^{-1} and the elements from $S(\sigma)$; then A is stable with respect to antipode; since σ is a homomorphism, A is stable also with respect to the right– and left–shifts by elements from E^* (the shifts even stabilize $S(\sigma)$). Thus A coincides with $\mathcal{P}(E^*)$. Now let $\{v_1, \dots, v_n\}$ be a basis in V.

For $x \in V$ we write

$$\rho(x)(v_j) = \sum_{i=1}^{n} f_{ij}(x) v_i.$$

This defines the elements $f_{ij} \in E^\circ$. Let $g_{ij} = f_{ij}|_{E^*}$. Then $\mathcal{P}(E^*)$ as a F–algebra is generated by the functions g_{ij} $(1 \le i, j \le n)$ and $\det_\sigma^{-1} = d_\sigma^{-1}$. We have

$$\delta(g_{ij}) = \sum_{k=1}^{n} g_{ik} \otimes g_{kj}, \quad \epsilon(g_{ij}) = \delta_{ij} \quad \text{(Kronecker symbol)}$$

Now let ϕ be a homomorphism from $\mathcal{P}(E^*)$ into F. Choose a maximal linear independent subsystem $\{f_{i_k j_k}\}$, $k = 1, \dots, l$, and a system of the elements $x_1, \dots, x_l \in E$ such that $f_{i_k j_k}(x_m) = \delta_{km}$. If $\phi(f_{ij}) = f_{ij}(y)$, $\phi(d_\sigma^{-1})\phi(d_\sigma) = 1$, i.e. $\phi(d_\sigma^{-1}) d_\sigma(y) = 1$, i.e. $d_\sigma(y) \neq 0$ and $y \in E^*$. Hence $\phi = y^*$ and that means that $\mathcal{P}(E^*)$ is the algebra of polynomial functions on E^*.

Let G be an algebraic group with the algebra of functions $\mathcal{P}(G)$. A subgroup K in G is called *algebraic* if it is closed in G or, which is the same, is the annihilator in G of some subset S from $\mathcal{P}(G)$. It should be noted that the annihilator J_K of the subgroup $K \subset G$ is a Hopf ideal, i.e. J_K is an ideal in $\mathcal{P}(G)$ and

$$(6)\quad \epsilon(J_K) = 0, \quad \delta(J_K) \subset J_K \otimes \mathcal{P}(G) + \mathcal{P}(G) \otimes J_K, \quad \eta(J_K) \subset J_K.$$

Indeed, let $f \in J_K$. Since $u\epsilon = \mu(1 \otimes \eta)\delta$ it follows from $\delta(f) = \sum f_i' \otimes f_i''$ that

$$\epsilon(f) \cdot 1 = \sum f_i'(\eta f_i'').$$

Evaluating this on $e \in K$ we obtain $\epsilon(f) = f(ee^{-1}) = f(e) = 0$.

Thus $\epsilon(J_K) = \{0\}$. Evidently one also has the inclusion $\eta(J_K) \subset J_K$.

Now let $f \in J_K$, and let V be a finite–dimensional subspace in $\mathcal{P}(G)$ such that $\delta(f) \subset V \otimes V$. We select a basis in $J_K \cap V$ and supplement it with functions $u_1, \ldots, u_m$ to a basis in V. Proceeding in the same way as in the proposition 1 we select points $x_1, \ldots, x_m \in K$ such that $u_i(x_j) = \delta_{ij}$. Then $\pi(u_k \otimes u_l)(x_i, x_j) = \delta_{ik}\delta_{jl}$. If now $\delta(f) = \sum \alpha_{kl} u_k \otimes u_l + v$, where $v \in J_K \otimes \mathcal{P}(G) + \mathcal{P}(G) \otimes J_K$, then for any i, j

$$(\pi\delta f)(x_i, x_j) = f(x_i, x_j) = 0,$$

and on the other hand, $(\pi\delta f)(x_i, x_j) = \alpha_{ij}$. This proves that

$$\delta f = v \in J_K \otimes \mathcal{P}(G) + \mathcal{P}(G) \otimes J_K.$$

We should remark that if conditions (6) are satisfied the quotient algebra $\mathcal{P}(G)/J_K$ is a Hopf algebra (with respect to the "quotient operations"). This algebra is isomorphic to the algebra of polynomial functions on the annihilator $\bar{K}$ of the ideal J_K in G. From property (6) we derive that $\bar{K}$ is a subgroup in G; so if $x, y \in \bar{K}$, $f \in J_K$, then $f(xy) = \mu[(\delta(f))(x \otimes y)] \subset \mu(J_K \otimes \mathcal{P}(G) + \mathcal{P}(G) \otimes J_K)(x \otimes y) = \{0\}$. Therefore $xy \in \bar{K}$. Analogously, $x^{-1} \in \bar{K}$, if $x \in \bar{K}$. The subgroup $\bar{K}$ is algebraic with an algebra of polynomial functions which is isomorphic to $\mathcal{P}(G)/J_K$.

The following important statement states more exactly what is implied by the previous argument.

Proposition. *Let G be an algebraic group, K a submonoid in G (i.e. a set containing* 1 *and closed with respect to multiplication). Then the closure, $\bar{K}$, of K in G is an algebraic subgroup in G.*

Proof. According to the previous arguments the annihilator J_K satisfies the first two of the conditions (6) and the closure $\bar{K}$ is a submonoid in G. Now let $x \in \bar{K}$. Then $x^* \in \mathcal{P}(G)^\circ$, and the shift operator $x^*_{[}$ on $\mathcal{P}(G)$ can be investigated. If $f \in J_K$ then

$$\begin{aligned} x^*_{[}(f) &=((1 \otimes x^*) \circ \delta)(f) = \\ &=(1 \otimes x^*) \left(\sum_i p_i \otimes q_i + \sum_j q'_j \otimes p'_j \right) = \sum p_i q_i(x) \subset J_K. \end{aligned}$$

Here p_i, p'_j are elements from J_K. Thus $x^*_{[}$ induces on J_K a linear operator. Because of local finiteness of the $\mathcal{P}(G)^\circ$–module $\mathcal{P}(G)$ (it is also a locally finite G–module) we see that the operator $(x^{-1})^*_{[}$ which is inverse to $x^*_{[}$, also leaves J_K invariant. In particular, the function

$$[(1 \otimes (x^{-1})^*) \circ f](y) = f(yx^{-1})$$

belongs to J_K, if $f \in J_K$. Hence $f(x^{-1}) = 0$ for any $f \in J_K$, i.e. $x^{-1} \in \bar{K}$. The proposition is proved. □

If G, H are two algebraic groups then the structure of the direct product of algebraic sets on $G \times H$ together with the structure of the direct product of groups make $G \times H$ into an algebraic group with algebra of polynomial functions $\mathcal{P}(G) \otimes \mathcal{P}(H)$. So $F^n = F \oplus \ldots \oplus F$ is an algebraic group with an algebra of functions for which the ring $F[X_1, \ldots, X_n]$ of polynomials in the variables $X_1, \ldots, X_n$ is used.

If V is a finite–dimensional F–space, $E = \mathrm{End}_F(V)$, then $E^* = GL(V)$ is an algebraic group. A polynomial representation of an algebraic group G in V is a homomorphism of groups $\rho : G \to GL(V)$, which is a polynomial mapping. The concept of polynomial G–module is defined in a natural way.

1.5. Irreducible algebraic groups. Let S be an algebraic set, T a non–empty subset. We say that T is *irreducible* if T is not representable in the form of union of two proper closed subsets (see, Chapter 1, § 5). Here is an equivalent condition: any pair of non–empty open subsets in T has a non–empty intersection. Here U is closed (open) in T iff $U = V \cap T$, where V is closed (open) in S. In Chapter 1, § 5, is proved that T is irreducible iff its annihilator J_T in $\mathcal{P}(S)$ is a prime ideal (i.e. $\mathcal{P}(S)/J_T$ has no zero divisors).

The ring $\mathcal{P}(S)$ is noetherian, i.e. increasing chains of ideals stabilize. This means that chains of closed subsets terminate. Finally, the latter means that increasing chains of open subsets from S stabilize in S. In other words, S is a noetherian topological space. In § 1.5 there was proved

Proposition. *Let S be an algebraic set. Then S contains a finite number of maximal irreducible subsets, moreover S is its union.* □

These maximal irreducible subsets, $T_1, \dots, T_m$ are called *irreducible components* of the algebraic set S. In the case of an arbitrary algebraic set its components can have a non–empty intersection. However, if $S = G$ is an affine algebraic group, the following is correct.

Theorem. *Let G be an affine algebraic group. The irreducible components of the group G are mutually disjoint. The component G_1 containing the neutral element of the group G is a closed normal group, and the irreducible components are the co–sets of the group G with respect to the subgroup G_1. In addition, G_1 is the unique irreducible closed subgroup of finite index in G.*

Lemma. (1) *If $\alpha : S \to T$ is a polynomial mapping and A is irreducible in S then $\alpha(A)$ is irreducible in T.*

(2) *If S and T are irreducible algebraic sets then $S \times T$ is an irreducible set.*

Proof. If $f, g \notin J_{\alpha(A)}$ and $fg \in J_{\alpha(A)}$ then $f \circ \alpha,\ g \circ \alpha \in \mathcal{P}(S) \setminus J_A$. Indeed, if $(f \circ \alpha)(a) = 0$, for all $a \in A$, then $f(\alpha(a)) = 0$, for all $\alpha(a) \in \alpha(A)$.

Thus, in fact $f \circ \alpha \in J_A$. Then $(f \circ \alpha)(g \circ \alpha) = (fg) \circ \alpha \in J_A$. This shows that if $J_{\alpha(A)}$ is not a prime ideal then J_A is not prime also. So the reducibility of the set $\alpha(A)$ implies the reducibility of the set A and (1) is proved.

In order to prove (2) it is sufficient to show that $\mathcal{P}(S) \otimes \mathcal{P}(T)$ the algebra of polynomial functions on $S \times T$, has no zero divisors. Indeed, for $\mathcal{P}(T)$ one can take any integral domain B. So let $u,\ v \in \mathcal{P}(S) \otimes B$ be such that $uv = 0$.

Write $u = \sum_{i=1}^{n} u_i \otimes b_i$, $v = \sum_{i=1}^{n} v_i \otimes b_i$, where $b_1, \dots, b_n$ is suitable linear independent set in B, $u_i, v_i \in \mathcal{P}(S)$. Taking the values of u_i, v_i in a point $s \in S$ determines a homomorphism $s^B : \mathcal{P}(S) \otimes B \to B$. If for some i the element $v_i(s)$ differs from zero then $s^B(v) \neq 0$, i.e. $s^B(u) = 0$ and $u_j(s) = 0$, for every $j = 1, \dots, n$. Thus for any $s \in S$ and any pair of indexes i, j we have $u_i(s)v_j(s) = 0$, i.e. $u_i v_j = 0$. If for some j we have $v_j \neq 0$ then $u_i = 0$ for all i, so $u = 0$. Therefore, $\mathcal{P}(S) \otimes B$ have no zero divisors and (2) is proved. □

Proof of the theorem. We shall demonstrate that the neutral element e of the algebraic group G is contained in only one of its irreducible components. In fact, if $e \in U, V$, where U, V are the irreducible components then UV is the image of the Cartesian product $U \times V$ under the multiplication morphism from $G \times G$ in G. According to statement (1) of the lemma, UV is irreducible. Since $U \ni e$, we have $UV \supset U$; analogously $UV \supset V$; by maximality then $U = UV = V$. Denote by G_1 the irreducible component which contains e. As above, $G_1 G_1 \subset G_1$. Since $x \mapsto x^{-1}$ is a morphism, we see that G_1^{-1} is a maximal irreducible set containing e. Therefore $G_1^{-1} = G_1$, i.e. G_1 is a subgroup.

Further, it should noted that in any Hopf algebra A the mappings $\tau_{[} = (1_A \otimes \tau) \circ \delta$ and $\tau_{]} = (\tau \otimes 1_A) \circ \delta$, where $\tau \in A^{\circ}$, are linear endomorphisms of this algebra. If $A = \mathcal{P}(G)$, $\tau = g^*$, where $g \in G$, then $g^* f$ is the function of the form $(g_{[}^* f)(x) = f(gx)$; analogously $(f g_{]}^*)(x) = f(xg)$. This shows that the shift mappings $x \mapsto gx$ and $x \mapsto xg$ are morphisms of the group G. The images of the subgroup G_1 under these shifts are the left and right co–sets of the group G with respect to the subgroup G_1. So these classes are irreducible components of the group G, and G_1 is a normal subgroup.

Now let K be any closed irreducible subgroup in G of finite index. Then (the left, right) co–sets with respect to K are irreducible components of the group G. Because of the uniqueness of decomposition into a union of irreducible components by using $e \in K$ we have $K = G_1$.

The theorem is proved. □

§ 2. Action of an algebraic group on a set

2.1. Semi–invariants. Let G be an algebraic F–group, H a subgroup, $g : H \to F^*$ a homomorphism of groups. We say that the function $d \in \mathcal{P}(G)$

is a *semi–invariant* for H if $xd = g(x)d$, for any $x \in H$. The homomorphism g is called the *weight* of the semi–invariant d.

Theorem. *Let G be an algebraic F–group, H an algebraic subgroup. There exists a finite subset D in $\mathcal{P}(G)$ and an element $f \in \mathcal{P}(G)$ of which the restriction to H is a homomorphism of groups from H into F^* such that*
(i) $xd = f(x)d$ for any $x \in H$ and $d \in D$;
(ii) if $x \in G$ and $xd \in Fd$ for all $d \in D$ then $x \in H$.

We shall make few preliminary remarks. If V is an F–vector space which is a G–module then the exterior algebra (Grassmann algebra) $\Lambda(V)$ also is a G–module if we put

$$g(v_1 \wedge \ldots \wedge v_s) = (gv_1) \wedge \ldots \wedge (gv_s),$$

where $g \in G$, $v_1, \ldots, v_s \in V$. Precisely such G–modules are important in the proof of our theorem.

Further, let V be a locally finite G–module and $V^\circ = \mathrm{Hom}(V, F)$. For $\gamma \in V^\circ$ and $v \in V$ we define the mapping $\gamma/v : G \to F$ by setting

$$(\gamma/v)(x) = \gamma(xv).$$

Starting from the G module structure on F^G we write $g(\gamma/v) = \gamma/gv$. Since V is a locally finite G–module, we have $\gamma/v \in \mathcal{R}(G)$. Remark that if $\rho : G \to GL(V)$ is the representation corresponding to the given module then the functions γ/v generate a space $S(\rho)$.

We shall pass directly to the proof of the theorem.

Proof. Let $\mathcal{I} = J_H$ be the annihilator of H in $\mathcal{P}(G)$. By virtue of the local finitiness of the G–module $\mathcal{P}(G)$ and of the Hilbert basis theorem, there exists a finite–dimensional G–invariant susbspace V in $\mathcal{P}(G)$ such that $V \cap \mathcal{I}$ generates $\mathcal{I}$ as an ideal. Let $d = \dim(V \cap \mathcal{I})$. Consider the action of the group G on $\Lambda_d(V)$. Let S be the canonical image of $\Lambda_d(V \cap \mathcal{I})$ in $\Lambda_d(V)$. Then $\dim S = 1$ and $S = Fs$, where s is an arbitrary non–zero element from S.

Since $H \cdot \mathcal{I} \subset \mathcal{I}$, it is clear that S is a H–invariant subspace in $\Lambda_d(V)$. Thus, for any $x \in H$, we have $x \cdot s = g(x)s$, where g is a homomorphism from H into F^*. Choose a function $\sigma \in (\Lambda_d(V))^\circ$ for which $\sigma(s) = 1$. We denote by f the representing function $\sigma/s : G \to F$. The restriction of the function f to H coincides with g. It is important to note that $f \in \mathcal{P}(G)$. Indeed, if

$x \in G$ and $\delta s = \sum s_i' \otimes s_i''$ then $xs = \sum_i s_i''(x)s_i'$, $\sigma(xs) = \sum_i \sigma(s_i')s_i''(x)$ and $\sigma/s = \sum_i \sigma(s_i')s_i'' \in \mathcal{P}(G)$. Further, if $s = {}_1f \wedge \ldots \wedge {}_df$, $\delta({}_if) = \sum_j {}_if_j' \otimes {}_if_j''$ then $\sigma(xs) = \sum \sigma({}_1f_{j_1}' \wedge \ldots \wedge {}_df_{j_d}')({}_1f_{j_1}'' \cdots {}_df_{j_d}'')(x)$.

Let now $\{\sigma_1, \ldots, \sigma_p\}$ be a F–basis of the annihilator of S in $(\Lambda_d(V))^\circ$. We shall show that σ_i/s is a semi–invariant for H. Indeed, if $x \in H$, then

$$x(\sigma_i/s) = \sigma_i/(xs) = g(x)(\sigma_i/s).$$

We shall show that the set of semi–invariants $D = \{\sigma_1/s, \ldots, \sigma_p/s\}$ is the desired one. Suppose, that $x(\sigma_i/s) = \lambda_i(\sigma_i/s)$, for $x \in G$, $\lambda_i \in F$. Computing the value of the function $x(\sigma_i/s)$ on the neutral element we then obtain $\sigma_i(xs) = 0$, for every $i = 1, \ldots, p$. Then $xs \in S$, i.e. $xS = S$. Consider any $t \in V \cap \mathcal{I}$. Then in $\Lambda_{d+1}(V \cap \mathcal{I}) = \{0\}$ we have $t \wedge S = \{0\}$ and hence $x(t \wedge S) = \{0\}$. At the same time

$$x(t \wedge S) = (xt) \wedge xS = (xt \wedge S).$$

Hence $(xt) \wedge S = \{0\}$, i.e. $xt \in V \cap \mathcal{I}$. Since $V \cap \mathcal{I}$ generates $\mathcal{I}$ as an ideal, we have $x\mathcal{I} \subset \mathcal{I}$, i.e. $x \in H$. Thus, D is the desired set. □

2.2. Invariants of normal groups. An important application of the theorem from the previous section is the following

Theorem. *Let H be a normal algebraic subgroup in the algebraic group G. There exists a finite set $Q \subset \mathcal{P}(G)$ consisting of invariants for $H$and its left stabilizer in G is equal to H.*

Proof. Let D be the finite set of semi–invariants for H which is determined by the theorem of Subsection 2.1. We shall denote by g the common weight for the action of the the group H on D. Let J be the smallest G invariant subspace in $\mathcal{P}(G)$ containing D. Because of the local finitiness of the G–module $\mathcal{P}(G)$ we have $\dim_F J < \infty$. Further, we denote by J_1 the set of H–semi–invariants of weight g in J; it is clear that J_1 is an H–submodule. Let now $x \in G$. We consider the subset xJ_1; if $h \in H$, $s \in J_1$ and $g_x(h) = g(x^{-1}hx)$ then

$$h(x \circ s) = x(x^{-1}hx) \circ s = xg(x^{-1}hx) \circ s = g_x(h)(xs),$$

i.e. the elements from xJ_1 are the H–semi–invariants of weight g_x. Since, $J = \sum_{x \in G} xJ_1$, because of the finite–dimensionality of the subspace J, and from the usual properties of weight decomposition, we obtain $J = x_1 J_1 \oplus \ldots \oplus x_k J_1$, for suitable $x_1, \ldots, x_k$ such that $\{g_{x_1}, \ldots, g_{x_k}\}$ is the set of all pairwise different weights of the H–module J.

Now denote by U the subalgebra in $\operatorname{End}(J)$ consisting of endomorphisms which stabilize every $J_i = x_i J_i$. Let $\rho : G \to \operatorname{End}(J)$ be the polynomial representation of the group G which is induced by the action of the group G on $\mathcal{P}(G)$ from left. If $u \in U$ and $x \in G$ then it is clear that $\rho(x) u \rho(x)^{-1} \in U$. So we have defined the representation $\sigma : G \to \operatorname{End}(U)$ which is given by the formula

$$\sigma(x)(u) = \rho(x) u \rho(x)^{-1}.$$

We shall show that the set $S(\sigma)$ of representative functions which are associated with σ, lies in $\mathcal{P}(G)$. In fact, U is the linear hull of tensors of the form $u = l \otimes d$, where $l \in J^*$, $d \in J$ (with $u(f) = l(f)d$). If $\gamma \in J^\circ$, $x \in G$, $f \in \mathcal{P}(G)$,

$$\delta(f) = \sum_i f_i' \otimes f_i'', \quad \delta(d) = \sum_i d_i' \otimes d_i'',$$

then

$$\begin{aligned}
&\gamma \circ \rho(x) u \rho(x)^{-1} \circ f = \gamma \circ (1 \otimes x^*) \circ \delta \circ u \circ (1 \otimes (x^{-1})^*) \circ \delta \circ f = \\
&= \gamma \circ (1 \otimes x^*) \circ \delta \circ (l \otimes d) \circ \sum_i (\eta f_i'')(x) f_i' = \\
&= \gamma \circ (1 \otimes x^*) \circ \delta(d) \left(\sum_i (\eta f_i'')(x) l(f_i') \right) = \\
&= \gamma \left(\sum_j d_j' d_j''(x) \right) \left(\sum_i (\eta f_i'')(x) l(f_i') \right) = \\
&= \sum_{i,j} \gamma(d_j') l(f_i') [d_j''(\eta f_i'')](x).
\end{aligned}$$

This demonstrates that the space $S(\sigma)$ of representing functions of σ lies in $S(\rho)(\eta S(\rho))$, where $S(\rho)$ is a set of representing functions of ρ and both sets are in $\mathcal{P}(G)$. Since $\epsilon(xf) = (xf)(e) = f(x)$, we see that $J \subset S(\rho)$; moreover, it is the minimal left and right G–invariant subspace containing J. Since the elements from H act as scalar operators on every J_i, it is clear that H lies in the kernel of the representation σ. So $S(\sigma)$ lies in $\mathcal{P}(G)^H$, the set of functions which are stable with respect to the left (and right) action of the subgroup H.

Now let Q be a finite subset in $\mathcal{P}(G)^H$ which generates $S(\sigma)$. We assume that x is an element from G such that $xq = q$, for any $q \in Q$. Then $x \in \operatorname{Ker}\sigma$, i.e. $\rho(x)$ commutes with any element from U. It follows that any J_i is invariant with respect to $\rho(x)$ and the restriction of $\rho(x)$ to J_i is a scalar multiplication. Since $D \subset J_1$, we obtain that D consists of semi–invariants for x. We can apply the theorem from Subsection 2.1 and then $x \in H$. So Q is the sought for finite set of invariants for H. The theorem is proved. □

From theorem just proved it is easy to deduce the following

Corollary. *Let G be an affine algebraic F–group, $\mathcal{P}(G)$ the set of polynomial functions on G. Then the F– algebra $\mathcal{P}(G)$ is isomorphic to the direct sum of the algebras of polynomial functions on the irreducible components of the group G.*

Proof. Let $G = \cup g_i G_1$. According to what has been proved, the set $\mathcal{P}(H)^{G_1}$ separates elements from G/G_1. Since the value on the elements from G_1 of any function from $\mathcal{P}(G)^{G_1}$ is constant, this subset is imbedded naturally in F^{G/G_1}. So $\mathcal{P}(G)^{G_1}$ is a subalgebra in F^{G/G_1} which separates the points of this finite set. It is not difficult to see that then $\mathcal{P}(G)^{G_1} = F^{G/G_1}$. In particular, $\mathcal{P}(G)^{G_1}$ contains the characteristic functions $\chi_1, \chi_2, \ldots, \chi_n$ of the co–sets of the group G with respect to the subgroup G_1. For any $i = 1, 2, \ldots, n$, the set $\chi_i\mathcal{P}(G)$ is an ideal in $\mathcal{P}(G)$, moreover $\mathcal{P}(G) = \bigoplus_{i=1}^{n} \chi_i\mathcal{P}(G)$ and $\chi_i\mathcal{P}(G)$ is isomorphic to the algebra of polynomial functions on the irreducible component g_iG_1 of the group G (the annihilator of this component is $\sum_{j\neq i} \chi_j\mathcal{P}(G)$). The corollary is proved. □

2.3. Polynomial mappings and homomorphisms

Theorem. *Let F be an algebraically closed field, G, H affine algebraic groups over F, G an irreducible group, and $\rho : G \to H$ be a polynomial mapping taking the neutral element from G into the neutral element from H. Then the products of finite sequences of elements of $\rho(G)$ form an irreducible algebraic subgroup P in H and there exists the natural number n such that* *any element from P is the product of n elements from $\rho(G)$.*

Proof. Define mappings $\rho_m : \underbrace{G \times \ldots \times G}_{m} \to H$ by setting

$$\rho_m(x_1, \ldots, x_m) = \rho(x_1) \ldots \rho(x_m).$$

It is clear that ρ_m is a polynomial mapping. We denote by J_m the annihilator of $\rho_m(G_m)$ in $\mathcal{P}(H)$, where $G_m = \underbrace{G \times \ldots \times G}_{m}$. By virtue of irreducibility of the group G the group G_m and thus its image $\rho_m(G)$ under the polynomial mapping are irreducible (see, Subsection 2.1). It follows that J_m is a prime ideal. Since $\rho_m(G_m) \subset \rho_{m+1}(G_{m+1})$, we see that $J_{m+1} \subset J_m$. From the proposition of Subsection 1.5.6 concerning the finiteness of a chain of prime ideals we obtain that there is a q such that $J_m = J_q$, when $m \geq q$. Let $P = \bigcup_{m=1}^{\infty} \rho_m(G_m)$. Then J_q is the annihilator for P in $\mathcal{P}(H)$. If Q is the closure of P in H then, according to Subsection 1.4, Q is an algebraic subgroup in H. Since its annihilator is a prime ideal, we obtain, finally, that Q is an irreducible algebraic subgroup in H.

We now prove that each element from Q can be represented in the form of product of no more that $n = 2q$ elements from $\rho(G)$. Consider the subring $\mathcal{P}(H) \circ \rho_q$ in $\mathcal{P}(G_q)$. Using the theorem of Subsection 1.5.5, we find a function $f \in \mathcal{P}(G_q)$ such that each homomorphism of algebras from $\mathcal{P}(H) \circ \rho_q$ into F that does not annihilate f, extends to a homomorphism from $\mathcal{P}(G_q)$ into F. According to the definition, any homomorphism from $\mathcal{P}(G_q)$ in F is a taking of values of functions in a suitable point from G_q.

Let the function $g \in \mathcal{P}(H)$ be such that $f = g \circ \rho_q$. Since f is a non–zero function, $g \notin J_q$. Let $x \in Q$. From the stability of the ideal J_q with respect to the action of the subgroup Q on $\mathcal{P}(H)$ and relatively to the antipode η it is easy to derive that $x\eta(g) \notin J_q$. This means that there is an element $u \in G_q$ such that

$$0 \neq (x\eta(g))(\rho_q(u)) = g(x^{-1}\rho_q(u)^{-1}).$$

At the same time $x^{-1}\rho_q(u)^{-1} \in Q$. This means that the element $x^{-1}\rho_q(u)^{-1}$ defines a homomorphism of F–algebras $\sigma : \mathcal{P}(H) \circ \rho_q \to F$ which is given by the formula

$$\sigma(h \circ \rho_q) = h(x^{-1}\rho_q(u)^{-1}).$$

Under this homomorphism we have $\sigma(f) = g(x^{-1}\rho_q(u)^{-1}) \neq 0$. According to the definition of the element f the homomorphism σ is the restriction to $\mathcal{P}(H) \circ \rho_q$ of a homomorphism $y^* : \mathcal{P}(G_q) \to F$, where y is suitable element from G_q. So for any $h \in \mathcal{P}(H)$, we get $h(x^{-1}\rho_q(u)^{-1}) = (h\rho_q)(y) = h(\rho_q(y))$. Hence $x^{-1}\rho_q(u)^{-1} = \rho_q(y)$, i.e. $x = \rho_q(u)^{-1}\rho_q(y)^{-1}$. Thus every element from Q is the product of $2q$ elements from $\rho(G)$. Finally, this means also that $Q = P$. The theorem is proved. □

2.4. Extension of the coefficient field. Let G be an affine algebraic F–group, K an extension of the field F. We shall construct the Hopf algebra $\mathcal{P}(G) \otimes K$ and investigate the algebraic K–group $\mathcal{G}(\mathcal{P}(G) \otimes K)$ which is associated with the previous one. We denote this group by G^K. If $\phi \in G$ is a homomorphism of F–algebras from $\mathcal{P}(G)$ into F then this homomorphism determines a homomorphism $\phi \otimes 1 : \mathcal{P}(G) \otimes K \to K$. This gives an injective imbedding from G into G^K. The image of the group G into G^K, evidently, distinguishes the functions of the algebra $\mathcal{P}(G) \otimes K$. So $\mathcal{P}(G^K)$ is identified with $\mathcal{P}(G) \otimes K$, and G is a dense subgroup in G^K. If G is irreducible then $\mathcal{P}(G)$ and also $\mathcal{P}(G) \otimes K$ are integral domains, i.e. G^K also is an irreducible algebraic group.

Extension of the field of scalars is useful in proving the following statement. In its formulation $\mathrm{Fr}(B)$ is the field of fractions of the integral domain B.

Lemma. *Let F be a field, G an irreducible algebraic F–group, B a Hopf subalgebra in $\mathcal{P}(G)$. Then $Fr(B) \cap \mathcal{P}(G) = B$.*

Proof. Without restriction of generality it can be assumed that B is a Hopf subalgebra which is finitely generated as an F–algebra. If L is an algebraically closed field containing F and G^L is the irreducible affine algebraic L–group which is obtained from G with a help of extension of the coefficient field, then, clearly, from the equality $\mathrm{Fr}(B \otimes L) \cap (\mathcal{P}(G) \otimes L) = B \otimes L$ there follows the equality which is to be proved. This allows us to assume that the base field F is algebraically closed.

Consider $f \in \mathrm{Fr}(B) \cap \mathcal{P}(G)$. Let J be the non–zero ideal in B which consists of all b and such that $(xf)b$ lies in B, for any $x \in G$. Since $x(uv^{-1}) = (xu)(xv)^{-1}$ and also because of the local finitiness of the G–module $\mathcal{P}(G)$, we obtain that $V = \langle xf \,|\, x \in G\rangle$ is a finite–dimensional subspace in $\mathrm{Fr}(B)$. If $V = \langle u_1 v_1^{-1}, \ldots, u_s v_s^{-1}\rangle$ then we can take $v_1 \ldots v_s$ as a b. It is evident that J is stable with respect to the group G. Let j be a non–zero element from J and $y \in \mathcal{G}(B)$. Then $jy \neq 0$ (this is the result of the action of an element from the group $\mathcal{G}(B)$ on the non–zero function j on this group). Since G distinguishes the functions from $\mathcal{P}(G)$, there is an $x \in G$ such that $(j \cdot y)(x) \neq 0$. However, $(j \cdot y)(x) = (xj)(y)$. Therefore $(x \cdot j)(y) \neq 0$. Since $x \cdot j \in J$, we get that J has no zeros in $\mathcal{G}(B)$. Denote now by J' the radical of J in B. If $J' \neq B$, and F is algebraically closed, then, according to the theorem of Subsection 1.5.5, the homomorphisms from B/J' into F distinguish the elements of this algebra. In particular, there exists a non–zero mapping $\phi : B \to F$ for which $\phi(J) \subset \phi(J') = \{0\}$, i.e. $\phi \in \mathcal{G}(B)$ is a zero for J, but this is impossible. Therefore $J' = B$, i.e. $J = B$ (as $B \ni 1$). Hence it is clear that $f \in B$. The lemma is proved. □

2.5. Homomorphism theorem

Theorem. *Let G be an affine algebraic group, B a Hopf subalgebra in $\mathcal{P}(G)$. Then B is a finitely generated F–algebra. If F is algebraically closed then restriction to B gives a natural surjective mapping $G \to \mathcal{G}(B)$.*

Proof. The algebra $\mathcal{P}(G)$ is finitely generated. First let G be irreducible. Then $\mathcal{P}(G)$ has no zero divisors. We consider the tower of fields

$$F \subset \mathrm{Fr}(B) \subset \mathrm{Fr}(\mathcal{P}(G)).$$

As $\mathrm{Fr}(\mathcal{P}(G))$ is finitely generated over F then, according to the proposition, the same can be said of $\mathrm{Fr}(B)$. Let $\mathrm{Fr}(B) = F(u_1 v_1^{-1}, \dots, u_s v_s^{-1})$. We denote by B_1 the Hopf subalgebra in $\mathcal{P}(G)$ which is generated by the elements $u_1, \dots, u_s$, $v_1, \dots, v_s$. Thus B_1 is finitely generated as an F–algebra and $\mathrm{Fr}(B_1) = \mathrm{Fr}(B)$. Using the previous lemma we obtain

$$B = \mathrm{Fr}(B) \cap \mathcal{P}(G) = \mathrm{Fr}(B_1) \cap \mathcal{P}(G) = B_1.$$

So $B = B_1$ and B is indeed a finitely generated F–algebra.

If G is not irreducible then according to the corollary of the theorem of Subsection 2.2, the algebra $\mathcal{P}(G)$ is representable in the form

$$\mathcal{P}(G) = \mathcal{P}(G)f_1 + \dots + \mathcal{P}(G)f_m,$$

where f_1 is the characteristic function of the maximal irreducible subgroup G_1 and $f_2, \dots, f_m$ are the characteristic functions of the remaining irreducible components $G_2, \ldots, G_m$. The ideal $\mathcal{P}(G)f_1$ is identified with $\mathcal{P}(G_1)$ and Bf_1 is a Hopf subalgebra in $\mathcal{P}(G_1)$. According what has been proved, Bf_1 is a finitely generated F–algebra. Note that if $x_iG_i = G_1$ then $f_i = x_if_1$ and $B_i = x_i(Bf_1)$. So the subalgebra $Bf_1 \oplus \ldots \oplus Bf_m$ of the algebra $\mathcal{P}(G)$ which contains B, is finitely generated over F. Applying corollary 3, Subsection 1.3.1, we obtain that B is a finitely generated F–algebra.

Now assume that F is an algebraically closed field. We consider the morphism of restriction $\rho : G \to \mathcal{G}(B)$. Then according to the theorem of Subsection 2.3, the set $\rho(G_1)$ is an irreducible closed subgroup in $\mathcal{G}(B)$. The set $\rho(G)$ is the union of a finite number of subsets which are obtained by the shifts from $\rho(G_1)$, therefore this set is closed. Finally, G distinguishes the functions from the algebra B, i.e. $\rho(G)$ is dense in $\mathcal{G}(B)$. Therefore $\rho(G) = \mathcal{G}(B)$ and the theorem is proved. □

Combining the results just obtained we get the main theorem.

Theorem. *Let F be an algebraically closed field, G an affine algebraic group, H a normal algebraic subgroup in G. Then G/H has the structure of an affine algebraic group such that its polynomial functions composed with the canonical morphism $\pi : G \to G/H$ yield an isomorphism of the Hopf algebra $\mathcal{P}(G/H)$ with $\mathcal{P}(G)^H$. If $\gamma : G \to K$ is a morphism of affine algebraic F–groups with kernel containing H then the induced homomorphism of groups $\gamma^H : G/H \to K$ which satisfies the condition $\gamma^H \circ \pi = \gamma$, is a morphism of affine algebraic groups.*

Proof. It is clear that $\mathcal{P}(G)^H$ is closed with respect to the left and right action of the group G on $\mathcal{P}(G)$ and also relative to the antipode. It follows that $\mathcal{P}(G)^H$ is a Hopf subalgebra in $\mathcal{P}(G)$. According to the theorem of Subsection 2.2, the kernel of the homomorphism of restriction from G into $\mathcal{G}(\mathcal{P}(G)^H)$ coincides with H. According to the theorem of Subsection 2.5, this homomorphism is surjective and $\mathcal{P}(G)^H$ is a finitely generated F–algebra. In this case $\mathcal{G}(\mathcal{P}(G)^H)$ is canonically isomorphic to G/H. If $f \in \mathcal{P}(G)^H$ then $f(gH) = f(g)$. This provides the group G/H with the structure of an algebraic group with algebra of functions $\mathcal{P}(G)^H$.

Further, let $\gamma : G \to K$ be a morphism of affine algebraic F–groups, i.e. $\mathcal{P}(K) \circ \gamma \subset \mathcal{P}(G)^H$. Writting $\gamma = \gamma^H \circ \pi$ we obtain

$$\mathcal{P}(K) \circ \gamma^H \circ \pi \subset \mathcal{P}(G)^H = \mathcal{P}(G/H) \circ \pi.$$

Hence $\mathcal{P}(K) \circ \gamma^H \subset \mathcal{P}(G/H)$, i.e. γ^H is a morphism of affine algebraic groups. The theorem is proved. □

§ 3. Action of an algebraic group by linear operators

3.1. Jordan decomposition of a linear operator. A linear operator $\phi : V \to V$ of a finite–dimensional vector space V over the field F is called *semisimple* if its minimal polynomial has no multiple roots in the algebraic closure $\bar{F}$ of the field F. In other words $\bar{\phi} \otimes 1 : V \otimes \bar{F} \to V \otimes \bar{F}$ is (similar to) a diagonal operator. Recall that an operator $\psi : V \to V$ is called *nilpotent* if for some suitable natural n we have $\psi^n = 0$. The main theorem of this subsection is the following.

Theorem. *Let F be a perfect field, V a finite–dimensional F–space, $\mathcal{A}$ a linear operator on V. Then there exist linear operators $\mathcal{A}^{(n)}$ and $\mathcal{A}^{(s)}$ of the space V such that $\mathcal{A} = \mathcal{A}^{(s)} + \mathcal{A}^{(n)}$, where $\mathcal{A}^{(s)}$ is semisimple, $\mathcal{A}^{(n)}$ is nilpotent; moreover, each of the operators $\mathcal{A}^{(s)}$, $\mathcal{A}^{(n)}$ is a linear combination of positive powers of the operator $\mathcal{A}$ (so that, in particular, $\mathcal{A}^{(s)}\mathcal{A}^{(n)} = \mathcal{A}^{(n)}\mathcal{A}^{(s)}$). If $\mathcal{A} = \mathcal{S} + \mathcal{N}$, where $\mathcal{N}$ is nilpotent, $\mathcal{S}$ is semisimple and $\mathcal{S}\mathcal{N} = \mathcal{N}\mathcal{S}$ then $\mathcal{S} = \mathcal{A}^{(s)}$, $\mathcal{N} = \mathcal{A}^{(n)}$.*

We preceed the proof by lemma.

Lemma. *Let F be a perfect field, X a variable. We consider a polynomial $f \in F[X] \setminus F$ and a polynomial g, which is a product of pairwise mutually prime and prime polynomials from $F[X]$ containing all prime factors of the polynomial f. Then there is an endomorphism $\pi : F[X] \to F[X]$ such that $\pi(g) \in F[X]f$ and $\pi(X) - X \in F[X]g$.*

Proof. Since F is perfect, the polynomial g has no multiple roots over any extension of the field F. Hence g and its derivative g' are mutually prime. Let $u, v \in F[X]$ be such that $ug' + vg = 1$. Let σ be the endomorphism of the algebra $F[X]$ which takes X into $X - ug$. Then for any non–negative l we obtain $\sigma(X^l) = X^l - lX^{l-1}ug + w_l g^2$, where $w_l \in F[X]$. Hence $\sigma(g) =$

$g-g'ug+wg^2$, $w \in F[X]$. By the choice of u and v we have $\sigma(g) = vg^2+wg^2$, i.e. $\sigma(g) \in F[X]g^2$. Since a suitable power of the polynomial g lies in $F[X]f$, we obtain that for some m we have $\sigma^m(g) \in F[X]f$. Now take $\pi = \sigma^m$. So we get $\pi(g) \in F[X]f$. Further,

$$\pi(X) - X = \sum_{i=0}^{m-1} \sigma^{i+1}(X) - \sigma^i(X) = \sum_{i=0}^{m-1} \sigma^i(\sigma(X) - X) =$$
$$= \sum_{i=0}^{m-1} \sigma^i(-ug) \in F[X]g.$$

The lemma is proved. □

Proof of the theorem. Let $\rho : F[X] \to \mathrm{End}_F(V)$ be the homomorphism of algebras which takes X into $\mathcal{A}$. Its kernel is a principal ideal $F[X]f$. If f is divisible by X then g is defined as the product of prime factors of the polynomial f; otherwise g is defined as the product just mentioned multiplied by X. Let π be an endomorphism of the algebra $F[X]$ as in the lemma. Let $q = \rho(\pi(X))$. Now $\pi(X) \in XF[X]$. According to the lemma $\rho \circ \pi$ also annihilates $F[X]g$. The image of $F[X]$ under this homomorphism is an image of the algebra $F[X]/F[X]g$. Since g is not divisible by a square, the algebra $F[X]/F[X]g$ is a direct sum of extensions of the field F. So q is a semisimple operator on the space V.

Further, $\mathcal{A} - q = \rho(X - \pi(X)) \in \rho(F[X]g)$, so that $\mathcal{A} - q$ is a nilpotent operator. Thus $\mathcal{A}^{(s)} = q$, $\mathcal{A}^{(n)} = \mathcal{A} - q$ satisfy the conditions of the theorem.

Now let $\mathcal{S}$ and $\mathcal{N}$ be as in the theorem. Then $\mathcal{S}$ and $\mathcal{N}$ commute with $\mathcal{A}$, i.e. with $\mathcal{A}^{(s)}$ and $\mathcal{A}^{(n)}$. Hence, $\mathcal{N} - \mathcal{A}^{(n)}$ is nilpotent and $\mathcal{S} - \mathcal{A}^{(s)}$ is semisimple. From the equality of these endomorphisms it follows that $\mathcal{N} - \mathcal{A}^{(n)} = 0 = \mathcal{S} - \mathcal{A}^{(s)}$, i.e. $\mathcal{N} = \mathcal{A}^{(n)}$, $\mathcal{S} = \mathcal{A}^{(s)}$. The theorem is proved. □

The decomposition $\mathcal{A} = \mathcal{A}^{(s)} + \mathcal{A}^{(n)}$ is called a *Jordan decomposition* of the operator $\mathcal{A}$, and the operators $\mathcal{A}^{(s)}$ and $\mathcal{A}^{(n)}$ are called *semisimple* and *nilpotent components* of this operator, respectively.

Now let $\mathcal{A}$ be an automorphism. Then $\mathcal{A}^{(s)}$ also is an automorphism (exercise for the reader). Let $\mathcal{A}^{(u)} = 1_V + (\mathcal{A}^{(s)})^{-1}\mathcal{A}^{(n)}$. Then $\mathcal{A}^{(u)} - 1_V$ is a nilpotent operator, i.e. $\mathcal{A}^{(u)}$ is an *unipotent* operator. We have

$$\mathcal{A}^{(s)}\mathcal{A}^{(u)} = \mathcal{A}^{(s)}(1_V + (\mathcal{A}^{(s)})^{-1}\mathcal{A}^{(n)}) = \mathcal{A}^{(s)} + \mathcal{A}^{(n)} = \mathcal{A},$$

and $\mathcal{A}^{(u)}$, $\mathcal{A}^{(s)}$ are commuting operators. Since $\mathcal{A}$ is an automorphism, we have $1_V \in \mathcal{A}F[\mathcal{A}]$, hence $\mathcal{A}^{(u)} \in \mathcal{A}F[\mathcal{A}]$. The automorphism $\mathcal{A}^{(u)}$ is called the *unipotent component* of the operator $\mathcal{A}$. The decomposition $\mathcal{A} = \mathcal{A}^{(s)}\mathcal{A}^{(u)}$ possesses a similar property of uniqueness as above.

3.2. Unipotent radical of a group. A set S of linear operators on the vector space V is called *nilpotent* if there exists a natural n such that a product of any n operators from S is equal to zero. A linear representation $\rho : G \to GL(V)$ is called *unipotent* if the set of operators of the form $\{1_V - \rho(g) | \, g \in G\}$ is nilpotent. We also say that V is a *unipotent* G–module. A subgroup T of an algebraic group G is called *unipotent* if $\mathcal{P}(G)$ is a locally unipotent T–module, i.e. the action by shifts from T on any finite–dimensional T–invariant subspace V from $\mathcal{P}(G)$ makes V into a unipotent T–module. The largest unipotent normal subgroup G_u of the group G (if it exists) is called a *unipotent radical* of the group G. Our aim is to prove that a unipotent radical always exists in an algebraic group.

The importance of the definition introduced lies in the following.

Proposition. 1) *A subgroup H of an algebraic group G is unipotent iff the image of this subgroup under any polynomial representation is unipotent.*

2) *A unipotent subgroup is nilpotent.*

3) *The closure of a unipotent subgroup is unipotent.*

Proof. 1) If $\rho : G \to GL(V)$ is a polynomial representation then the space $S(\rho)$ of coefficients of this representation (see Subsection 1.) lies in $\mathcal{P}(G)$, and is G–invariant and, consequently, the action by shifts of the subgroup H on $S(\rho)$ is unipotent. So there is a natural number n such that for any $h_1, \dots, h_n \in H$, $\gamma \in (\mathrm{End}\,(V))^\circ$, we have $(1 - h_1)\dots(1 - h_n)(\gamma \circ \rho) = 0$. Using the definition of action and the fact that ρ is a homomoprhism of groups, we get that the value of the left hand side in the point $e \in G$ is equal to zero and has the form $\gamma_0(1_V - \rho(h_1))\dots(1_V - \rho(h_n)) = 0$, for any $\gamma \in ((\mathrm{End}\,(V))^\circ$, i.e.

$$(1_V - \rho(h_1))\dots(1_V - \rho(h_n)) = 0,$$

and the subgroup $\rho(H)$ is unipotent.

2) Let V be a finite–dimensional G–invariant subspace in $\mathcal{P}(G)$ which generates $\mathcal{P}(G)$. Because of the unipotency of the action of the subgroup H on V, this subspace contains a series of H–invariant subspaces

$$\{0\} = V_n \subset V_{n-1} \subset \ldots \subset V_0 = V, \tag{1}$$

on the factors of which H acts trivially. The representation ρ for H in V is faithful and the image is given by triangular matrices. Thus H is nilpotent as an abstract group.

3) It should be remarked that the unipotency of the subgroup H is equivalent to the fact that for any composition series (1) of an arbitrary H–invariant subspace the action of the subgroup H on V_{i-1}/V_i, $i = 1, \ldots, n$, is trivial. If $\rho_i : G \to GL(V_{i-1}/V_i)$ are the induced representations, then this is equivalent to the fact that the set $S(\rho_1) \oplus \ldots \oplus S(\rho_n) \subset \mathcal{P}(G)$ consists of fixed points of action of the group H. Since $S(\rho_1) \oplus \ldots \oplus S(\rho_n)$ is finite–dimensional, its stabilizer is an algebraic subgroup N. Thus the closure $\bar{H}$ for H lies in N and also is unipotent. The proposition is proved. □

Theorem. *Any algebraic group G has a unipotent radical.*

Proof. We define G_u as the subgroup in G which is generated by the set of all normal unipotent subgroups of this group. It is clear that G_u is normal in G. We shall show that G_u is unipotent. So consider a finite–dimensional G_u–invariant subspace V in $\mathcal{P}(G)$; let $\dim V = d$. Consider the product $(e-x_1)\ldots(e-x_d) \in \mathbf{Z}[G]$, $x_i \in G_u$, and its action on V. The set $\{x_1, \ldots, x_d\}$ lies in the product of a finite number of unipotent normal subgroups of the group G.

The triviality of the product written down can be proved if we assume that all elements lie in the product of two unipotent normal subgroups S and T of the group G. Note that if $s \in S$ and $t \in T$ then

$$\begin{gathered} e - st = s(e-t) + e - s, \quad (e-t)s = s(e - s^{-1}ts), \\ (e-t)(e-s) = (e-t) - s(e - s^{-1}ts). \end{gathered} \tag{2}$$

This allows to rewrite any product $(e - s_1t_1)\ldots(e - s_nt_n)$ in the form of a linear combination of products of the form

$$x(e - x_1)\dots(e - x_p)(e - y_1)\dots(e - y_q), \tag{3}$$

where $x_i \in S$ and $y_j \in T$. If $n \geq d^2$ then by dividing the product $(e - s_1t_1)\dots(e - s_nt_n)$ into d groups with d co–factors we see that each of the groups is a linear combination of products of the form (3). If $q = 0$ in this type of product then the product is equal to zero because of the nilpotency of the subgroup S. In the remaining products there is a factor $e - y, y \in T$. Multiplying out such products the number of such co–factors, according to (2), does not decrease and so we are able to apply the property of unipotency for T. Thus, ST is unipotent in V; of course, then the product of d (not d^2) elements of the form $e - x$, $x \in ST$, is equal to zero. Hence, G_u is unipotent. It should be mentioned that the closure of G_u is unipotent and normal, and it coincides with G_u. Therefore, G_u itself is a unipotent normal algebraic subgroup. The theorem is proved. □

3.3. The splitting property for algebraic groups. Let G be an algebraic group with Hopf algebra of polynomial functions $\mathcal{P}(G)$. Then according to Subsection 1.4 the group G is identified with the group G^* of homomorphisms of the algebra $\mathcal{P}(G)$ into the field F if we set $g^*(f) = f(g)$. If $x \in \mathcal{P}(G)^\circ$ then $x_[$ is defined as the operator $x_[: \mathcal{P}(G) \to \mathcal{P}(G)$ such that $x_[= (1_{\mathcal{P}(G)} \otimes x) \circ \delta$. We know that $\mathcal{P}(G)$ is a locally finite $\mathcal{P}(G)^\circ$–module with respect to the action by the operators $x_[$, $x \in (\mathcal{P}(G))^\circ$. If V is a finite–dimensional $(\mathcal{P}(G))^\circ$–invariant subspace in $\mathcal{P}(G)$ then $x_[$ has an additive Jordan decomposition. From the condition of uniqueness it follows that $x_[= (x_[)^{(s)} + (x_[)^{(n)}$, where $(x_[)^{(n)}$ is locally nilpotent and $(x_[)^{(s)}$ is a semisimple operator on $\mathcal{P}(G)$. It can be seen that if $f \in \mathcal{P}(G)$ then $(x_[)^{(s)}(f)$ is defined as $(\bar{x}_[)^{(s)}(f)$, where $\bar{x}_[$ is a restriction of the operator $x_[$ to some finite–dimensional $\mathcal{P}(G)$–invariant subspace $\bar{V}$ which contains f. That this is well defined is evident from the fact that if $\bar{\bar{V}}$ is another subspace and $\bar{\bar{x}}_[$ is the restriction of $x_[$ to $\bar{\bar{V}}$ then $f \in \widetilde{V} = \bar{V} + \bar{\bar{V}}$ and $(\bar{x}_[)^{(s)}$, $(\bar{\bar{x}}_[)^{(s)}$ are the restrictions of the operator $(\widetilde{x}_[)^{(s)}$ to $\bar{V}$ and $\bar{\bar{V}}$, respectively. Their actions on $\bar{V} \cap \bar{\bar{V}}$ coincide, i.e. $(\bar{x}_[)^{(s)}(f) = (\bar{\bar{x}}_[)^{(s)}(f)$, as required. □

Lemma. *Let ϕ be a linear operator on $\mathcal{P}(G)$. The operator ϕ has the form $x_[$ for some $x \in \mathcal{P}(G)^\circ$ iff*

$$\delta \circ \phi = (1_{\mathcal{P}(G)} \otimes \phi) \circ \delta. \tag{4}$$

In this case $\phi = (\epsilon \circ \phi)_{[}$, *where* ϵ *is the co–unit of the algebra* $\mathcal{P}(G)$.

Proof. If equality (4) holds then $\phi = (\epsilon \circ \phi)_{[}$, where ϵ is the co–unit of the Hopf algebra $\mathcal{P}(G)$. Indeed, multiply (4) from the left by $1 \otimes \epsilon$, to obtain

$$\phi = (1 \otimes \epsilon) \circ \delta \circ \phi = (1 \otimes \epsilon)(1 \otimes \phi) \circ \delta = (1 \otimes (\epsilon \circ \phi)) \circ \delta = (\epsilon \circ \phi)_{[}.$$

Inversily, let $\phi = x_{[} = (1 \otimes x) \circ \delta$. If $\tau \in (\mathcal{P}(G))^{\circ}$ then according to the lemma of Subsection 1.2, the operator ϕ commutes with $\eta = (\tau \otimes 1) \circ \delta$. Write

$$\phi \circ (\tau \otimes 1_{\mathcal{P}(G)}) \circ \delta = (\tau \otimes 1_{\mathcal{P}(G)}) \circ \delta \circ \phi.$$

Then

$$(\tau \otimes 1)((1 \otimes \phi) \circ \delta - \delta \circ \phi) = 0.$$

Since this equality holds for all $\tau \in (\mathcal{P}(G))^{\circ}$, we obtain

$$\delta \circ \phi = (1 \otimes \phi) \circ \delta.$$

Multiply by $1 \otimes \epsilon$, to obtain $\phi = (\epsilon \circ \phi)_{[}$. The lemma is proved. □

Using this lemma for the action of the operator $x_{[}$ on any finite–dimensional subspace V, we find that $(x_{[})^{(s)}$ and $(x_{[})^{(n)}$, the polynomials in $x_{[}$, which are in $(\mathcal{P}(G))^{\circ}|_V$, satisfy (4). Because of the local finiteness of the $(\mathcal{P}(G))^{\circ}$–module $\mathcal{P}(G)$ this equality holds for $(x_{[})^{(s)}$ and $(x_{[})^{(n)}$ on all of $\mathcal{P}(G)$. Applying the lemma once more we find $x^{(s)}$ and $x^{(n)}$ in $\mathcal{P}(G)$ such that $(x^{(s)})_{[} = (x_{[})^{(s)}$ and $(x^{(n)})_{[} = (x_{[})^{(n)}$. It is clear that the properties $x_{[} = x_{[}^{(s)} + x_{[}^{(n)}$ and $x_{[}^{(s)} x_{[}^{(n)} = x_{[}^{(n)} x_{[}^{(s)}$ are retained. If, additionally, $x_{[}$ is an automorphism then there exists a decomposition $x_{[} = x_{[}^{(s)} x_{[}^{(u)}$. By virtue of injectivity of the mapping $x \mapsto x_{[}$ the elements $x^{(s)}$, $x^{(n)}$, $x^{(u)}$ are completely determined and are called *the semisimple, nilpotent* and *unipotent components* of the element x.

We have already remarked that $G \subset \mathcal{P}(G)^{\circ}$ ($g \mapsto g^*$ and $g^*(f) = f(g)$.

The considerations provided above allow to prove the main result of this paragraph.

Theorem. *Let* F *be a perfect field,* G *an affine algebraic group. For any* $x \in G$ *the multiplicative components* $x^{(s)}$, $x^{(u)}$ *belong to the group* G. *If* ρ *is a morphism of affine algebraic groups then* $\rho(x^{(s)}) = (\rho(x))^{(s)}$, $\rho(x^{(u)}) = (\rho(x))^{(u)}$.

Proof. Let $x \in G$ be considered as a homomorphism from $\mathcal{P}(G)$ into F. Consider the components $x^{(s)}$ and $x^{(u)} \in \mathcal{P}(G)^{\circ}$. It is necessary to prove that these linear functions also are homomorphisms from $\mathcal{P}(G)$ into F, i.e., let us say, $x^{(s)} \otimes x^{(s)} = x^{(s)} \circ m$, where $m : \mathcal{P}(G) \otimes \mathcal{P}(G) \to \mathcal{P}(G)$ is the multiplication in $\mathcal{P}(G)$. First of all, we shall prove that $x_{[}^{(s)}$ and $x_{[}^{(u)}$ are automorphisms of the algebra $\mathcal{P}(G)$ (it is a necessary and sufficient condition). It is more convenient at first to investigate any linear automorphism ϕ of the space $V + V^2$ which stabilizes both V and V^2, where V is a finite–dimensional G–invariant subspace in $\mathcal{P}(G)$. Define a linear automorphism $[\phi]$ of the space $\mathrm{Hom}_F(V \otimes V, V^2)$ by

$$[\phi](h) = \phi \circ h \circ (\phi^{-1} \otimes \phi^{-1}).$$

The operator ϕ is an automorphism of the algebra $\mathcal{P}(G)$ iff $[\phi](m) = m$ in any finite–dimensional G–invariant subspace. We shall show that if ϕ is semisimple then so is $[\phi]$. Indeed, without restriction of generality assume that F is algebraically closed. Then V and V^2 are direct sums of one–dimensional ϕ–invariant subspaces

$$V = \sum_i F u_i, \quad V^2 = \sum_k F v_k.$$

If $h_{ijk}(u_r \otimes u_s) = \delta_{ri}\delta_{sj} v_k$ then $\mathrm{Hom}\,(V \otimes V, V^2) = \sum F h_{ijk}$, moreover

$$\begin{aligned}\phi \circ h_{ijk} \circ (\phi^{-1} \otimes \phi^{-1})(u_r \otimes u_s) &= \lambda_r \lambda_s \phi \circ h_{ijk}(u_r \otimes u_s) = \\ = \lambda_r \lambda_s \delta_{ir} \delta_{js} \phi v_k = \lambda_s \lambda_r \mu_k \delta_{ir} \delta_{js} v_k &= \lambda_s \lambda_r \mu_k h_{ijk})(u_r \otimes u_s).\end{aligned}$$

If now ϕ is a unipotent operator then the selection of the triangular basis for ϕ in V and in V^2 brings the proof of the unipotency of the operator $[\phi]$ to the case when ϕ is the identity operator. Moreover, if $\phi = \phi^{(s)}\phi^{(u)} = \phi^{(u)}\phi^{(s)}$ then

$$[\phi] = [\phi^{(s)}][\phi^{(u)}] = [\phi^{(u)}][\phi^{(s)}].$$

From this it is clear that

$$[\phi]^{(s)} = [\phi^{(s)}] \quad \text{and } [\phi]^{(u)} = [\phi^{(u)}].$$

Now let $x \in G$ and ϕ be the restriction of the operator $x_{[}$ to $V + V^2$. Let $m : V \otimes V \to V^2$ be the multiplication operator. Since $x_{[}$ is an automorphism of the F–algebra $\mathcal{P}(G)$ we have $[\phi] \circ m = m$. In this case both $[\phi]^{(s)}$ and $[\phi]^{(u)}$ stabilize Fm, and hence leave it invariant. Then also $[\phi^{(s)}]$ and $[\phi^{(u)}]$ leave m invariant. This happens for any finite–dimensional subspace, i.e. $x_{[}^{(s)}$ and $x_{[}^{(u)}$ are automorphisms of the F–algebra $\mathcal{P}(G)$. It remains to note that in this case $x^{(s)} = \epsilon \circ x_{[}^{(s)}$ (see the arguments before the formulation of the theorem). As $\epsilon : \mathcal{P}(G) \to F$ is a homomorphism of algebras, we see that $x^{(s)}$ and $x^{(u)}$ are homomorphisms of algebras from $\mathcal{P}(G)$ into F, i.e. $x^{(s)}$, $x^{(u)} \in G$, as required.

If $\rho : G \to H$ is a morphism of affine algebraic groups then $\mathcal{P}(H)$ naturally becomes a polynomial G–module. We already know that any polynomial G–module is semisimple with respect to $x^{(s)}$ and is locally unipotent with respect to $x^{(u)}$. In particular, $\mathcal{P}(H)$ is semisimple with respect to $\rho(x^{(s)})$ and is locally unipotent with respect to $\rho(x^{(u)})$. These automorphisms commute and their product is equal to $\rho(x)$.Therefore $\rho(x^{(s)}) = \rho(x)^{(s)}$ and $\rho(x^{(u)}) = \rho(x)^{(u)}$. The theorem is proved. □

3.4. Semi–direct product. A subgroup R of the group G is called *linearly reductive* if $\mathcal{P}(G)$ is a semisimple R–module. This property is equivalent to the fact that any polynomial G–module is R–semisimple.

According to Subsection 2.1.1 and Subsection 1.4, if $N \triangleleft R$ then N is also a reductive subgroup. According to the results from Subsection 3.2 a polynomial representation of a unipotent subgroup is triangular and, thus, we obtain that the unipotent radical G_u of a reductive group G is always trivial. Finally, it should be noticed that a direct product of two linearly reductive groups is itself linearly reductive.

The simplest example of a linearly reductive group is the multiplicative group F^* of the field F: its algebra of functions $F[X, X^{-1}]$ as an F^*–module looks like

$$F[X, X^{-1}] = \bigoplus_{n \in \mathbf{Z}} FX^n.$$

According to what was said before $\underbrace{F^* \times \ldots \times F^*}_{n}$ is a linearly reductive group. It is called the *n–dimensional algebraic torus*.

One of the results of this section is a representation of an arbitrary nilpotent algebraic group in the form of the direct product of a torus and its unipotent radical. In order to describe the structure of solvable groups we need a more subtle construction – a semi–direct product of affine algebraic groups.

Let F be a field, G an affine algebraic group, N a normal algebraic subgroup in it. According to the theorem of Subsection 2.5, $\mathcal{P}(G)^N$ is a Hopf subalgebra which is finitely generated as an F–algebra. Let K be the group $\mathcal{G}(\mathcal{P}(G)^N)$; it is an affine algebraic group with $\mathcal{P}(K) = \mathcal{P}(G)^N$. The mapping by restriction of homomorphisms from G into $K = \mathcal{G}(\mathcal{P}(G)^N)$ in the case of an algebraically closed field is surjective according to Subsection 2.5. If the field is not algebraically closed then the statement will not necessary be valid. In order to define semi–direct products we need to consider normal subgroups with precisely this property; these subgroups are called properly normal.

Definition. Let N be a proper normal subgroup of the affine algebraic group G and R an algebraic subgroup in G with the properties: (1) $G = NR$; (2) $N \cap R = \{e\}$; (3) the restriction to R of the canonical homomorphism $G \to G/N$ is an isomorphism of algebraic groups. In this case G is a semi–direct product of the groups N and R; this is written as $G = N \leftthreetimes R$ or $G = R \rightthreetimes N$.

The main result follows

Theorem. *Let G be an algebraic group, R a linearly reductive subgroup in G such that $G = G_u R$. Then R is an algebraic subgroup in G and $G = G_u \leftthreetimes R$.*

Before starting the proof of the theorem we shall make some remarks about semi–direct products. If $G = N \leftthreetimes R$, $f \in \mathcal{P}(G)$, then we denote by f_R, f_N the restrictions of the function f to R and N, respectively, i.e. $f_R \in \mathcal{P}(R)$, $f_N \in \mathcal{P}(N)$. Let also $\pi : G \to G/N$ be the natural homomorphism, $\alpha : G/N \to R$ the isomorphism of algebraic groups such that $\alpha\pi|_R = 1_R$ and $\pi\alpha = 1_{G/N}$. A mapping which is conjugate with α is an isomorphism of Hopf algebras $\sigma : \mathcal{P}(R) \to \mathcal{P}(G)^N$. Define the polynomial projection $\beta : G \to N$ by $\beta(x) = x(\alpha(\pi(x))^{-1}$. If $y \in R$, then, clearly, $\beta(xy) = \beta(x)$. A mapping conjugate to β is an isomorphism of F–algebras $\sigma : \mathcal{P}(N) \to \mathcal{P}(G)^R$; the restriction mapping is inverse to it.

Note that the affine set G is the direct product of the affine sets N and R. Using $\sigma \otimes \rho$ we see that the isomorphism $\mathcal{P}(N) \otimes \mathcal{P}(R) \to \mathcal{P}(G)$ will be proved if we can show that the natural mapping $\mathcal{P}(G)^R \otimes \mathcal{P}(G)^N \to \mathcal{P}(G)$ is an isomorphism of F–algebras. First of all, injectivity of the latter mapping follows from the fact that the set $N \times R$ distinguishes elements from the algebra

$\mathcal{P}(G)^R \otimes \mathcal{P}(G)^N$. In order to prove surjectivity we take $f \in \mathcal{P}(G)$ and write $\delta(f) = \sum_i u_i \otimes v_i$. Then for any $x, y \in G$, we obtain $f(x \cdot y) = \sum_i u_i(x)v_i(y)$. If x runs over N and y over R then

$$v_i(y) = (\rho(v_i|_R))(xy), \quad u_i(x) = (\sigma(u_i|_N))(xy).$$

Hence,

$$f = \sum_i \sigma((u_i|_N))\rho((v_i|_R))$$

lies in the image of the algebra $\mathcal{P}(G)^R \otimes \mathcal{P}(G)^N$ in $\mathcal{P}(G)$.

Now let's pass to the proof of the theorem.

Proof. Note that the closure R^+ in G of the linearly reductive subgroup R is itself linearly reductive. The intersection $R^+ \cap G_u$ is a normal subgroup in R^+, i.e. it is a linearly reductive group. By virtue of unipotency, $R^+ \cap G_u = \{e\}$, and so $R = R^+$.

Select a finite–dimensional G–submodule V in $\mathcal{P}(G)$ which generates $\mathcal{P}(G)$ as an algebra, and consider a composition series of this module

$$\{0\} = V_n \subset V_{n-1} \subset \ldots \subset V_0 = V.$$

Construct the exterior direct sum V' with the summands V_i/V_{i+1}, $i = 0, 1, \ldots$ $\ldots, n-1$. Since V is semisimple as an R–module, there is a homomorphism of R–modules $\gamma : V' \to V$. Let τ and τ' denote the representations of G in V and V', respectively.

Let α be the unique homomorphism of the groups $\alpha : G \to R$ such that $\alpha(x)G_u = xG_u$, for any $x \in G$. We shall show that α is a morphism of algebraic groups.

As was noted before, the unipotent radical lies in the kernel of any irreducible representation of the group, therefore the action of the subgroup G_u on V_i/V_{i+1} is trivial, i.e. $\tau' = \tau' \circ \alpha$. So for any $x \in G$ the following holds

$$\gamma\tau'(x)\gamma^{-1} = \gamma\tau'(\alpha(x))\gamma^{-1} = \tau(\alpha(x)).$$

Consequently, for any $v \in V$ we get

$$v(\alpha(x)) = (\alpha(x)v)(e) = \epsilon(\tau(\alpha(x))(v)) =$$
$$= \epsilon((\gamma\tau'(x)\gamma^{-1})(v)) = (\epsilon \otimes v) \circ (\gamma\tau'(x)\gamma^{-1}).$$

Thus, $v \circ \alpha$ is a representing function of the polynomial representation τ', i.e. $v \circ \alpha \in \mathcal{P}(G)$. Since V generates $\mathcal{P}(G)$ as an algebra, we conclude that $\mathcal{P}(G) \circ \alpha \subset \mathcal{P}(G)$. However $\mathcal{P}(G) \circ \alpha = \mathcal{P}(R) \circ \alpha$, i.e. $\mathcal{P}(R) \circ \alpha \subset \mathcal{P}(G)$ and α is a morphism of algebraic groups.

It is clear that $\mathcal{P}(G) \circ \alpha$ is elementwise invariant under the action by translates of G_u. Moreover, if $f \in \mathcal{P}(G)^{G_u}$ then $f = f \circ \alpha$. It follows that a mapping which is conjugate to α gives an isomorphism of Hopf algebras $\rho : \mathcal{P}(R) \to \mathcal{P}(G)^{G_u}$. Now define a polynomial mapping $\beta : G \to G_u$, by $\beta(x) = x\alpha(x)^{-1}$. Then the isomorphism of algebras

$$\sigma(\mathcal{P}(G_u)) \to \mathcal{P}(G)^R.$$

is conjugate to β.

Applying arguments from the previous theorem we see that the mapping of the product from $\mathcal{P}(G)^R \otimes \mathcal{P}(G)^{G_u}$ into $\mathcal{P}(G)$ is an isomorphism of algebras. It instantaneously follows that the mapping from G into $\mathcal{G}(\mathcal{P}(G)^{G_u})$ is surjective, i.e. G_u is a proper normal subgroup in G.

The subgroup G_u is the kernel of the earlier constructed morphism α, so it induces in a natural way a morphism of algebraic groups $G/G_u \to R$. It is clear that this moprhism is inverse to the restriction to R of a canonical morphism from G onto G/G_u. Now it is clear that $G = G_u \rightthreetimes R$. □

3.5. Structure of abelian groups

Theorem 1. *Let F be a perfect field, G an abelian algebraic affine F–group. There exists a unique linearly reductive algebraic subgroup G_s in G such that G is the direct product $G_u \times G_s$. The elements from G_s are the semisimple elements from G and any linearly reductive subgroup of the group G is contained in G_s.*

Proof. Let R be a linearly reductive subgroup in G. Since F is perfect and $\mathcal{P}(G)$ is locally finite as an R–module, we can apply proposition 2.1.5 and deduce that for any $K \supset F$ the tensor product $\mathcal{P}(G) \otimes K$ is a semisimple R–module as before. Take K to be algebraically closed and then consider a finite–dimensional G–invariant K–subspace V in $\mathcal{P}(G) \otimes K$. As R is abelian, there exists a finite family of group homomorphisms ρ from R into K^* and a decomposition $V = \sum V_\rho$ such that for any x an element from R acts on V_ρ by scalar multiplication with $\rho(x)$. If R' is another linearly reductive subgroup in

G then every V_ρ is stable with respect to R' and allows a decomposition into a sum of proper subspaces with respect to R'. Combining this with the first decomposition we obtain a decomposition for V into a direct sum of subspaces and both R and R' act as scalars on everyone of these subspaces.

Let now $(R_\alpha)_{\alpha \in \mathcal{I}}$ be a set of linearly reductive subgroups in G, completely ordered with respect to inclusion. Since V is finite–dimensional the process of decomposition which is described above, stabilizes in a finite number of steps, i.e. V is a sum of one–dimensional subspaces which are stable with respect to the union of the family $(R_\alpha)_{\alpha \in \mathcal{I}}$. Since this is satisfied for any V, we see that $\mathcal{P}(G)$ is a semisimple module for the union of the family $(R_\alpha)_{\alpha \in \mathcal{I}}$.

From the lemma of Subsection 3.4, it follows that this union is a linearly reductive subgroup in G. At the same time, from what was said above, it is clear that a subgroup generated by a pair of linearly reductive subgroups is itself linearly reductive.

From the Zorn lemma and the latter remark follows the existence of a largest linearly reductive subgroup G_s in G. In particular, any semisimple element from the group G will be in G_s. It is also clear from the study of the action in $\mathcal{P}(G) \otimes K$ that any element from G_s is semisimple.

Let now $x \in G$. Consider the multiplicative Jordan decomposition $x = x^{(u)}x^{(s)}$. We have $x^{(s)} \in G_s$. Since G is abelian, we have $x^{(u)} \in G_u$. So $G = G_u G_s$. According to the theorem of Subsection 3.4, the group G is a semi–direct product and by virtue of the abelian property, it is also the direct product of the groups G_u and G_s. □

Theorem 2. *If F is an algebraically closed field then any abelian linearly reductive affine algebraic F–group G is a torus (i.e. $G \cong F^* \times \ldots \times F^*$).*

Proof. Let G be such a group, V a finite–dimensional G–invariant subspace in $\mathcal{P}(G)$ which generates $\mathcal{P}(G)$ as a F–algebra. The conditions for G mean that $V = F_{v_1} \oplus \ldots \oplus F_{v_q}$, where all the one–dimensional subspaces F_{v_i} are G–invariant. For any element $x \in G$ one has $xv_i = \gamma_i(x)v_i$, where $\gamma_i \in \mathcal{P}(G)$, i.e. it is a group homomorpism from G into F^*. From now on we call the elements γ_i *polynomial characters.*

We have $v_i = \epsilon(v_i)\gamma_i$. Since V generates $\mathcal{P}(G)$ as a F–algebra, we see that $\mathcal{P}(G)$ is the linear hull of a set Σ which is the multiplicative subgroup generated by the elements γ_i. Recall the simple fact that a set of group homomorphisms from an arbitrary group into the multiplicative group of a field are linearly independent as functions with values in this field (see, § 1.2).

Hence any element from $\mathcal{P}(G)$ which is a homomorphism of groups from G into F^* belongs to Σ. In other words, Σ is the group of all polynomial characters of the group G.

Let σ be an arbitrary polynomial character of the group G. It is evident that $\sigma : G \to F^*$ is a morphism of algebraic groups. According to the theorem of Subsection 2.3, $\sigma(G)$ is an irreducible algebraic subgroup in F^*. From the evident dimension reason either $\sigma(G) = F^*$ or $\sigma(G) = \{1\}$. This shows, in particular, that Σ is a torsion–free group. Since Σ is finitely generated, it is free abelian with the basis $\{\sigma_1, \dots, \sigma_n\}$. From the linear independence of the elements of the group Σ as functions follows the algebraic independence of the elements $\sigma_1, \dots, \sigma_n$. Hence

$$\mathcal{P}(G) = F[\sigma_1, \dots, \sigma_n, \sigma_1^{-1}, \dots, \sigma_n^{-1}].$$

This algebra is the tensor product of the Hopf subalgebras $F[\sigma_i, \sigma_i^{-1}]$ each of which can be identified with $\mathcal{P}(F^*)$. So G is the direct product of n copies of the group F^*. □

3.6. Nilpotent groups

Theorem. *Let F be an algebraically closed field, G an irreducible nilpotent algebraic affine F–group. Then the semisimple elements make up a central algebraic subgroup G_s such that $G = G_u \times G_s$. The group G_s is a F–torus and contains any linearly reductive subgroup in G.*

Proof. Let C be a center of the group G. Since G is nilpotent, the subgroup C_1 is trivial only in a case when G is trivial, because the lower central series of the group ends at unity; according to the theorem of Subsection 2.3, all of its terms are irreducible algebraic subgroups in G. We shall start by proving that every semisimple element from the group G is central. We can use induction on the dimension of the group G (see Subsection 1.5.5) with an evident starting point.

Since C_1 is non–trivial, the dimension of the group G/C is lower than the dimension of the group G. Let x be a semisimple element in G and $y \in G$. The canonical image of the element x in G/C is semisimple and according to the induction assumption $yxy^{-1} = cx$, where $c \in C$. It is clear that the components of the multiplicative decomposition of c lie in C. Therefore the unipotent component of the element c coincides with a unipotent component of the element yxy^{-1}. Since x is semisimple, this is also true of yxy^{-1}, i.e. c is a semisimple element. On the other hand, c lies in the commutator subgroup of the group G, so that according to the corollary of Subsection 4.1, $c \in G_u$. Thus c is the identity and $x \in C$.

From the fact that the semisimple elements of the group G lie in its center and from theorem 1 of Subsection 3.5, we easily obtain that these elements form a unique maximal linearly reductive subgroup C_s in C. So $G_s = C_s$. From the corollary of Subsection 4.1, we know that the unipotent elements of the group G lie in G_u. From the Jordan decomposition we get $G = G_u G_s$. According to the theorem of Subsection 3.4, we obtain that $G = G_u \leftthreetimes G_s$. Since G_s is central in G, it follows that $G = G_u \times G_s$. Because of the irreducibility of the group G, G_s and G_u are irreducible. In particular, from the theorem 2 of Subsection 3.5, follows that the group G_s is torus.

Finally, let R be any linearly reductive subgroup in G. The image of this subgroup by projection on G_u is a linearly reductive subgroup in G_u, i.e. is trivial. So $R \subset G_s$. □

§ 4. Solvable groups

4.1. Irreducible representations of solvable groups

Theorem. *Let F be an algebraically closed field, G an irreducible solvable affine algebraic F–group. Then any simple polynomial G–module is one–dimensional.*

Proof. By applying the theorem of Subsection 2.3, to the polynomial mapping from $G \times G$ into G, which takes the pair (x, y) into $xyx^{-1}y^{-1}$, we see that the commutator subgroup K of the group G is an irreducible algebraic subgroup in G. Since G is solvable, we have $K \neq G$, if $G \neq \{1\}$. Carrying out induction with respect to dimension of the group G we assume that the theorem is valid for K.

Now consider a simple polynomial G–module V. From the condition of normality for K in G it easily follows that V is a semisimple K–module (see, Subsection 3.4). According to the induction assumption, V is a direct sum of a family of one–dimensional K–invariant subspaces. By grouping direct summands, if it is necessary, we can write $V = V_1 \oplus \ldots \oplus V_q$, where $V_1, \ldots, V_q$ are proper subspaces which correspond to pairwisely different polynomial characters $\gamma_1, \ldots, \gamma_q$ of the group K and such that each element x from K acts on every V_i by scalar multiplication with $\gamma_i(x)$. As K is normal in G, the elements from G permute the subspaces V_i among themselves. Therefore the stabilizer of V_1 in G has finite index in G. Since G is irreducible, it follows that the stabilizer of V_1 in G coincides with G. As V is a simple G–module, it follows that $V_1 = V$. This means that G is scalar on V.

The determinants of operators of the space V corresponding to the elements from K are equal to 1, as these elements are products of commutators. So if $n = \dim V$ then $\gamma_1(x)^n = 1$, for any $x \in K$. This shows that $\operatorname{Ker} \gamma_1$ has finite index in K. From the irreducibility of the subgroup K follows that γ_1 is the trivial character. So V can be seen as a G/K–module. Since G/K is abelian, V is a simple G–module, and F is an algebraically closed field, we have $\dim V = 1$. □

Corollary. *Let the group G satisfy the conditions of the theorem. Then G/G_u is abelian and every unipotent element from G lies in G_u.*

Proof. According to the theorem the commutator subgroup K of the group G acts trivially on any simple polynomial G–module. Therefore $K \subset G_u$, i.e. G/G_u is abelian.

Let x be a unipotent element from G and H the subgroup of the group G which is generated by the element x and by the subgroup G_u. From the theorem of Subsection 3.2 follows that H is unipotent. Since G/G_u is abelian, H is normal in G. Therefore $H = G_u$, i.e. $x \in G_u$. □

4.2. Splitting of the unipotent radical

Lemma. *Let F be a field, G a unipotent affine algebraic F–group, q a natural number not divisible by the characteristic of the field F. Then the mapping $x \mapsto x^q$ is bijective as a mapping $G \to G$.*

Proof. To start with, let the characteristic p of the field F be non–zero. Take a finite–dimensional G–submodule V in $\mathcal{P}(G)$ which generates $\mathcal{P}(G)$ as an F–algebra. Consider an element $x \in G$ and denote by x' the restriction to V of the endomorphism $x_{[}$. Since G is unipotent, there is a natural n such that $(x' - 1_V)^{p^n} = 0$ or $(x')^{p^n} = 1_V$. Since V generates $\mathcal{P}(G)$ as a F–algebra, we see that x^{p^n} is the neutral element in G. By the condition on q there exist r,

s such that $rq + sp^n = 1$. If $y = x^r$ then $y^q = x$. Moreover, from $z^q = y^q$ follows $z = y$.

It remains to investigate the case when F has zero characteristic. We write $\mathbf{1} = 1_{\mathcal{P}(G)}$. From the fact that $\mathcal{P}(G)$ is a sum of submodules, on each of which any element $\mathbf{1} - x_{[}$ ($x \in G$) is nilpotent, it follows that the formal sum

$$\log x_{[} = -\sum_{n>0} \frac{(\mathbf{1} - x_{[})^n}{n}$$

is a well defined linear operator on $\mathcal{P}(G)$. Evidently, this operator is locally nilpotent on $\mathcal{P}(G)$. Similarly, if any locally nilpotent linear operator γ on $\mathcal{P}(G)$ is given then $\exp\gamma$ can be defined as a linear operator on $\mathcal{P}(G)$. A direct computation of formal power series gives, for any m, $\exp(m \log x_{[}) = x_{[}^m$. We want to show that $\log x_{[}$ is a derivation of the algebra $\mathcal{P}(G)$.

Let t be an additional variable. Extend the F–linear endomorphisms of the algebra $\mathcal{P}(G)$ to an $F[t]$–linear endomorphisms of the ring $\mathcal{P}(G)[t]$ in the evident way. It is clear that the operator $\exp(t \log x_{[})$ is well defined. Indeed, let $f, g \in \mathcal{P}(G)$. We consider the expression

$$\exp(t \log x_{[})(fg) - \exp(t \log x_{[})(f) \cdot \exp(t \log x_{[})(g).$$

This is a polynomial $p(t)$ with coefficients from $\mathcal{P}(G)$. According to the previous, $p(m) = 0$, for any integer m. It follows that every coefficient of the polynomial $p(t)$ is equal to zero. Setting the coefficient of t equal to zero we obtain

$$(\log x_{[})(fg) = (\log x_{[})(f)g + f(\log x_{[})(g),$$

as required. Given this, with the help of the usual formal argument it is proved that $u = \exp(q^{-1} \log x_{[})$ is an automorphism of the F–algebra $\mathcal{P}(G)$. We have $u^q = x_{[}$. Consequently, if $y = \epsilon u$, then $y \in G$ and $y^q = x$. Let now $z^q = x$. Then $z_{[}^q = y_{[}^q$ and from there by taking a logarithm, dividing by q, and applying $\epsilon \circ \exp$ we obtain $z = y$. □

Let S be a group M an S–module, f a mapping from S into M. Recall that f is a cocycle for S in M if $f(xy) = xf(y) + f(x)$, for any $x, y \in S$. Such a cocycle is a coboundary if there is an $m \in M$ for which $f(x) = xm - m$, for any $x \in S$.

Proposition. *Let F be an algebraically closed field, G an abelian linearly reductive affine F–group, M a unipotent abelian affine algebraic F–group with the structure of a G–module such that the defining structure mapping from $G \times M$ into M is polynomial. Let f be a polynomial cocycle from G into M. Then f is a coboundary.*

Proof. According to theorem 2 of Subsection 3.5, the irreducible component G_1 of the neutral element from the group G is a torus. Since F is algebraically closed, G_1 is a divisible group. As G is abelian, it is also the direct product of the subgroup G_1 and a finite group. In this case there exists an increasing sequence $\{G(n)\}$ of finite subgroups in G, the union of which is dense in G. In the case that the field F has non–zero characteristic p, the order of every element from every subgroup $G(n)$ is not divisible by p, as each element of the abelian linearly reductive group $G(n)$ is semisimple. Therefore, if q_n is the order of the group $G(n)$ then q_n is not divisible by the characteristic of the field F. From the lemma it follows that if we use additive notation of the operation in M then multiplication by q_n is a bijection $M \to M$.

Consider now the restriction of the cocycle f to $G(n)$. Summing the cocycle thus defined over all elements y of the group $G(n)$ we obtain $q_n f(x) = s - xs$, where s is the sum of all $f(y)$. Since division by q_n is possible, it follows that the restriction of the cocycle f to $G(n)$ is a coboundary.

Let now $M(n)$ be a set of all elements m from M such that $f(x) = xm - m$, here x is arbitrary in $G(n)$. We have just seen that $M(n)$ is non–empty. It is evident that $M(n+1) \subset M(n)$ for any n.

The assumption about the G–module structure for M means that every $M(n)$ is closed in M. As M is a noetherian topological space, the sequence $\{M(n)\}$ stabilizes. This means that there exists $m \in M$ such that $f(x) = xm - m$, for any $x \in \cup G(n)$. Since this union is dense in G, and f and the G–module structure are polynomial, it follows that $f(x) = xm - m$, for any $x \in G$. □

Denote by $[A, B]$ the mutual commutator of the subgroups A, $B \subset G$. Let F be an algebraically closed field, G an affine algebraic F–group. Define a sequence of subgroups in G_u by the equalities $G_u^{(1)} = G_u$, $G_u^{(n+1)} = [G, G_u^{(n)}]$. We denote by G_u^{∞} the intersection of all terms from the series of normal subgroups $G_u^{(0)} \supset G_u^{(1)} \supset G_u^{(2)} \supset \dots$.

Now assume that G is irreducible and solvable. As in 4.3.1, we define the lower central series by setting $\gamma_1(G) = G$, $\gamma_{n+1}(G) = [G, \gamma_n(G)]$. According to the theorem of Subsection 2.3, each subgroup $\gamma_n(G)$ is irreducible algebraic in G and from the corollary in Subsection 4.1 it follows that $\gamma_2(G) \subset G_u$. Therefore for any $n > 1$ we have $G_u^{(n)} \subset \gamma_n(G) \subset G_u^{(n-1)}$. Since G is a noetherian topological space, the chain of closed subsets $\gamma_n(G)$ should stabilize. Hence $G_u^{\infty} = \gamma_n(G)$ for all sufficiently large n. In particular, G_u^{∞} is an irreducible normal algebraic subgroup in G if G is irreducible and solvable.

Theorem. *Let F be an algebraically closed field, G a solvable irreducible affine algebraic F–group, T a maximal algebraic torus in G. Then $G = G_u \leftthreetimes T$, and for any linearly reductive subgroup K in G there exists an element $r \in G_u^{\infty}$ such that $rKr^{-1} \subset T$.*

Proof. First we shall show that G contains a torus T such that $G = G_u \leftthreetimes T$. According to the theorem of Subsection 3.4, it is sufficient for the proof to show that $G = G_uT$. In the case that G is nilpotent, the theorem of Subsection 3.6, contains the required result. Assume now that G is not nilpotent. Then G necessarily contains a non–central semisimple element. In fact, otherwise any commutator of elements of the group whould be a commutator consisting of unipotent elements, i.e. of elements from G_u. It would follow that G is nilpotent.

Let G^s denote the centralizer of s in G. Since G is irreducible and $G^s \neq G$, the dimension of the group G^s is strictly lower than the dimension of the group G. Using induction with respect to dimension we can find a torus T in $(G^s)_1$ such that $(G^s)_1 = ((G^s)_1)_uT$. Now assume that $G = G_uG^s$. Then also $G = G_u(G^s)_1$. According to the corollary of Subsection 4.1, $((G^s)_1)_u \subset G_u$. Consequently, according to the previous, $G = G_uT$. So the decomposition result follows from the equality $G = G_uG^s$.

Let S be the smallest algebraic subgroup in G containing s. Since s is a semisimple element, the group S is a linearly reductive abelian subgroup in G. We denote by $\delta_n(G)$ the n–th term of the commutator series of the group G. Each group $\delta_n(G)$ is an irreducible normal algebraic subgroup in G and $\delta_n(G)$ is trivial for suffciently large n. We also know that $\delta_1(G) \subset G_u$ and $\delta_n(G)$ is unipotent, for every $n > 0$. Now $\delta_n(G)/\delta_{n+1}(G)$ is a unipotent abelian affine algebraic F–group. Consider this group as a S–module by using conjugation. It is clear that then assumptions of the proposition above hold.

Fix $n > 0$ and consider $x \in G$ such that $yxy^{-1}x^{-1} \in \delta_n(G)$ for all $y \in S$. Note that any element satisfies this condition if $n = 1$. Let $f(y)$ denote the canonical image of the element $yxy^{-1}x^{-1}$ in $\delta_n(G)/\delta_{n+1}(G)$. Then f is a polynomial cocycle for S in $\delta_n(G)/\delta_{n+1}(G)$ in the sense of the proposition above. According to that proposition f is a coboundary. Thus, there is an element $z \in \delta_n(G)$ such that $y(zx)y(zx)^{-1} \in \delta_{n+1}(G)$, for any $y \in S$. Since $\delta_{n+1}(G)$ is trivial for sufficiently large n, we can finally find $b \in \delta_1(G) \subset G_u$ such that bx commutes with any element from S and, thus, belongs to G^s. Then $G = G_u G^s$, as required.

Now consider the product $\gamma_n(G)T$, $n > 1$. It is clear that it is an algebraic subgroup and it has the form $\gamma_n(G) \rtimes T$. Consider any linearly reductive subgroup R in G and assume that $R \subset \gamma_n(G)T$, for some $n > 1$. Evidently, $\gamma_{n+1}(G)T$ is a normal subroup in $\gamma_n(G)T$ and the quotient group $\gamma_n(G)T/\gamma_{n+1}(G)T$ is isomorphic to the unipotent affine algebraic group $\gamma_n(G)/\gamma_{n+1}(G)$. Since R is linearly reductive, its canonical image in $\gamma_n(G)/\gamma_{n+1}(G)$ should be trivial. Therefore $R \subset \gamma_{n+1}(G)T$. It follows that any linearly reductive subgroup of the group G lies in $G_u^\infty T$.

Let us simplify our notations. Let us write G instead of $G_u^\infty \rtimes T$ and write $G(0) = G_u^\infty$. Also for any $n > 0$, let $\delta_n(G)$ define the n–th commutator of the new group G. Let K be any linearly reductive subgroup in G. It remains to prove that $G(0)$ has an element r such that $rKr^{-1} \subset T$. Replacing K by its closure in G we can limit ourselves to the case that K is an algebraic subgroup in G. From the fact that G/G_u is abelian it follows that K is abelian. Assume that for some $n > 0$ there is an $r_n \in G(0)$ such that $r_n K r_n^{-1} \subset \delta_n(G)T$. (When $n = 0$, take $r_0 = 1$.) Let $S = r_n K r_n^{-1}$ and note that S is a linearly reductive subgroup in G. For any $x \in S$ we denote by $f(x)$ the canonical image of the projection $\gamma(x)$ of the element x on $\delta_n(G)$ in the semi–direct product $\delta_n(G) \rtimes T$. As above, $\delta_n(G)/\delta_{n+1}(G)$ is a S–module with respect to conjungation and f is a polynomial cocycle. According to the proposition f is a coboundary. Thus there is a $t \in G(0)$ such that $tSt^{-1} \subset \delta_{n+1}(G) \rtimes T$. The element $r_{n+1} = tr_n$ now has the necessary property. □

4.3. Reduction of a solvable linear group to a triangular form. The group $T_n(F)$ of upper triangular matrices, evidently, is solvable. Our aim is to prove a converse to this statement for irreducible algebraic groups. It should be noted that the proof of the following Kolchin theorem contains some overlap with the theorem of Subsection 4.1.

Theorem. *Let F be an algebraically closed field, $G \subset GL_n(F)$ an irreducible solvable algebraic group. Then G is conjugate in $GL_n(F)$ with a subgroup of $T_n(F)$.*

Proof. Note that it is sufficient to prove the theorem only for the case when F^n is irreducible with respect to G. Indeed, assuming otherwise, we can see that G is conjugate to a subgroup of matrices of the form

$$\left\{ g = \begin{pmatrix} A(g) & * \\ 0 & B(g) \end{pmatrix} \middle| g \in G \right\}.$$

The mappings $g \mapsto A(g)$, $g \mapsto B(g)$ are polynomial, i.e. $\{A(g)\}$, $\{B(g)\}$ are closed irreducible algebraic groups.

Using an induction assumption with respect to n we can reduce these groups to triangular form. Then the desired result also holds for G.

According to what was proved in Subsection 2.3, the commutator subgroup G' of the group G is closed and irreducible. Using induction with respect to the solvability length of the group G we can assume that G' is reducible to triangular form. Denote by U the subspace in F^n which is generated by all eigenvectors for G'. This subspace is non–zero and G–admissable. Indeed, let u be a common eigenvector for G', i.e. $g'(u) = \lambda(g')u$, $\lambda(g') \in F$, $g' \in G'$. Then for any $g \in G$:

$$g'(g(u)) = g(g^{-1}g'g)(u) = \lambda(g^{-1}g'g)g(u). \tag{1}$$

Since F^n is irreducible, we have $U = F^n$. This fact means that G' can be diagonalized in F^n; we shall assume that it is diagonal.

If $g' = \operatorname{diag}(\lambda_1, \dots, \lambda_n)$ and $g \in G$ then $g^{-1}g'g \in G'$, and, consequently, is diagonal. Comparing characteristic polynomials we see that

$$g^{-1}g'g = \operatorname{diag}(\lambda_{\sigma_g(1)}, \dots, \lambda_{\sigma_g(n)}),$$

where σ_g is a permutation of indexes. So the number of elements which are conjugate with g' is finite. Thus the closed subgroup $C_G(g')$, the centralizer of the element g' in G, has finite index in G. By virtue of irreducibility of the group G we have $C_G(g') = G$, i.e. G' lies in the center of the group G. We should note that formula (1) now becomes $g'(gu) = \lambda(g')gu$, so that any proper subspace U for G' is G–invariant, i.e. each g' is scalar with a fixed eigen–value $\lambda(g')$. Thus $U = F^n$ and G' acts scalarly on F^n. Finally, $G' \subset SL_n(F)$, i.e. $\lambda(g')^n = 1$ for any n. This shows that $|G'| \leq n$. Because of irreducibility $G' = \{E\}$. So the group G is commutative and, hence, is reducible to triangular form (see, Subsection 3.12 and 4.1.5).

Let us apply the theorem to abstract linear groups.

Corollary. *A solvable group of matrices contains a subgroup of finite index which is reducible to triangular form.*

Proof. Whether the group theoretic identity $X_{i_1}^{m_1} X_{i_2}^{m_2} \ldots X_{i_s}^{m_s} = 1$ holds, evidently, is a polynomial condition. So if a group G is solvable then the same can be said about its closure $\bar{G}$ in the Zariski topology on $GL_n(F)$. Taking the irreducible component of the unit $\bar{G}_1$ we obtain an irreducible solvable group in $\bar{G}$, and, consequently, a reducible to the triangular form subgroup, of finite index. Then $\bar{G}_1 \cap G$ is a triangulizable subgroup of finite index in G. □

∗ An improvement of this result has been given by A.I. Mal'cev who has shown that the index of a triangulizable subgroup in a solvable group of matrices of order n need not exceed a constant which is dependent only on n. This, in particular, restricts the solvability length of solvable subgroups in $GL_n(F)$ to a function of n. ∗

References.

Humphrey [1], Hochschild [1].

CHAPTER VII

VARIETIES OF ALGEBRAS

§ 1. Universal algebras and varieties

1.1. Definition of universal algebra. A non–empty set A is called a *universal algebra* with a set of operations Ω if for any $\omega \in \Omega$ there exists a natural number $n = n(\omega)$ (it is possible that $n = 0$) such that ω is an n–ary (or n–local) operation on A, i.e. ω is a mapping from $A^n = \underbrace{A \times \ldots \times A}_{n}$ into A. For $n = 0$ the mapping ω simply fixes an element from A (it can be assumed that $\omega \in A$). For instance, a multiplicative group G with unit e and inverse element a^{-1} for $a \in G$ is a universal algebra with set of operations $\Omega = \{\omega_0, \omega_1, \omega_2\}$ such that $\omega_0 = e$, $\omega_1(a) = a^{-1}$, $\omega_2(a, b) = ab$ $(a, b \in G)$. It is clear that not each universal algebra with such a set of operations is a group: as a matter of fact, nothing is said about the properties of the operations.

If A_1, A_2 are two algebras with identical set of operations Ω then a mapping $\phi : A_1 \to A_2$ is called a *homomorphism* if for any $\omega \in \Omega$ and any $a_1, \ldots, a_n \in A$, $n = n(\omega)$, we have

$$\phi(\omega(a_1, \ldots, a_n)) = \omega(\phi(a_1), \ldots, \phi(a_n)).$$

A non–empty subset B in a universal algebra is called a *subalgebra* if for any $\omega \in \Omega$ and any $b_1, \ldots, b_n \in B$ $(n = n(\omega))$ we have $\omega(b_1, \ldots, b_n) \in B$. In particular, if $\omega \in A$ is a 0–ary operation then $\omega \in B$.

An equivalence relation $\sim$ on a universal algebra A with set of operations Ω is called a *congruence* if for any $\omega \in A$ and any $a_1, \ldots, a_n$ and $a'_1, \ldots, a'_n$ $(n = n(\omega))$ such that $a_1 \sim a', \ldots, a_n \sim a'_n$, we have

$$\omega(a_1, \ldots, a_n) \sim \omega(a'_1, \ldots, a'_n).$$

The set $A/\sim$ of equivalence classes of elements of the algebra A with respect to a congruence $\sim$ becomes a universal algebra with the same set of operations if for any $\omega \in \Omega$ and any classes $\bar{a}_1, \dots, \bar{a}_n$ with representatives $a_1, \dots, a_n \in A$ we set

$$\omega(\bar{a}_1, \dots, \bar{a}_n) = \overline{\omega(a_1, \dots, a_n)};$$

$A/\sim$ is called the *quotient algebra* of the algebra A with respect to the congruence $\sim$.

If $\phi : A_1 \to A_2$ is a homomorphism of algebras then its kernel $\operatorname{Ker}\phi$ is the congruence $\sim$ such that $a \sim a'$ iff $\phi(a) = \phi(a')$. The following result is valid.

1.2. Homomorphism theorem. 1) *Let $\phi : A_1 \to A_2$ be a homomorphism of universal algebras with fixed set of operations Ω. Then the image $Im\ \phi$ is a subalgebra in A_2, the kernel $Ker\,\phi$ is a congruence in A_1 and the mapping $\bar{a} \mapsto \phi(a)$ ($a \in A$) is an isomorphism of the quotient algebra $A/Ker\,\phi$ and of the algebra $Im\ \phi$:*

$$A/\operatorname{Ker}\phi \cong \operatorname{Im}\ \phi.$$

2) *Let ρ_1 and ρ_2 be two congruences of the algebra A such that $\rho_1 \supset \rho_2$. Consider the algebra A/ρ_2 and define on it a congruence $\bar{\rho}_1 = \rho_1/\rho_2$ by putting $\bar{a}\bar{\rho}_1\bar{b}$ iff $a\rho_1 b$. Then $(A/\rho_2)/(\rho_1/\rho_2) \cong A/\rho_1$.* □

1.3. Algebra of words. In order to define certain types of universal algebras it is necessary to have the concept of identities. The most accurate method for defining identities is to introduce them first in the so-called *algebra of words* $F(X, \Omega)$ on an alphabet X with set of operations Ω. The set $F(X, \Omega)$ is defined as the smallest subset in the set of (non-associative) words W on the alphabet $X \cup \Omega$ containing the set $X \cup \Omega_0$, where Ω_0 is the set of symbols of 0-ary operations that is closed under taking the results of the operations: if $a_1, \dots, a_n \in F$ then $\omega(a_1, \dots, a_n) \in F$, where $\omega \in \Omega\backslash\Omega_0$. It is clear that F is an algebra with a set of operations Ω. The main property of the algebra $F(X, \Omega)$ is the following (the universality property).

Proposition. *Let A be an arbitrary algebra with set of operations Ω, and let $\phi : X \to A$ be an arbitrary mapping. Then there exists a unique homomorphism $\bar{\phi} : F(X, \Omega) \to A$ that is generated by ϕ, i.e. $\bar{\phi}|_X = \phi$.*

Proof. Any element $a \in F$ is either equal to an element from $X \cup \Omega_0$, and then one should put $\bar{\phi}(a) = \phi(a)$, for $a \in X$, and $\bar{\phi}(a) = a$, for $a \in \Omega_0$, or it is represented uniquely in the form $\omega(a_1, \dots, a_n)$, for some $a_1, \dots, a_n \in F$ of shorter length (as words in the alphabet $X \cup \Omega_0$). Applying induction we set

$$\bar{\phi}(\omega(a_1, \dots, a_n)) = \omega(\phi(a_1), \dots, \phi(a_n)).$$

So the mapping $\bar{\phi}$ is determined for all elements from $F(X, \Omega)$. The way it is defined shows that $\bar{\phi}$ is the desired unique homomorphism. □

Example. Let $X = \{x\}$, and Ω consist of one binary operation, viz. juxtaposition. Then

$$F(x, \Omega) = \{x, xx, (xx)x, x(xx), (x(xx))x, \dots\}.$$

If $A = \mathbf{N}$ is the set of natural numbers with the addition operation then the mapping $\phi : x \to 1$ extends to the homomorphism $\bar{\phi}$: $F(x, \Omega) \to \mathbf{N}$ such that $\bar{\phi}(a) = |a|$, where $|a|$ is a length of the word a (the brackets are not counted).

1.4. Identities. An expression

$$u \equiv v, \tag{1}$$

where u, $v \in F(X, \Omega)$ is called an *identity* for the class of algebras with the set of operations Ω depending on the set of variables X. If A is an algebra with the set of operations Ω then the identity (1) holds in A if for any homomorphism $\phi : F(X, \Omega) \to A$ we have $\phi(u) = \phi(v)$. Let $u = u(x_1, \dots, x_n)$, $v = v(x_1, \dots, x_n)$ (i.e. these words are written by means of a finite subset $\{x_1, \dots, x_n\}$ of the set X. If $a_1, \dots, a_n \in A$ then $\phi(u)$ is denoted by $u(a_1, \dots, a_n)$ and is called a result of substituting the elements $a_1, \dots, a_n$ from A into the word $u \in F(X, \Omega)$. Thus

$$u(x_1, \dots, x_n) \equiv v(x_1, \dots, x_n) \tag{1'}$$

is an identity in A if for any $a_1, \dots, a_n \in A$ we have

$$u(a_1, \dots, a_n) = v(a_1, \dots, a_n).$$

Example. If $\Omega_\Gamma = \{\omega_0, \omega_1, \omega_2\}$ is the set of operations (or, as it is also called, the *signature*) of a group then the identities

$$\begin{aligned} &\omega_2(x,\omega_0) \equiv \omega_2(\omega_0,x) \equiv x, \\ (2) \qquad &\omega_2(\omega_1(x),x) \equiv \omega_2(x,\omega_1(x)) \equiv \omega_0 \\ &\omega_2(x,\omega_2(y,z)) \equiv \omega_2(\omega_2(x,y),z) \end{aligned}$$

hold in any group.

1.5. Varieties. Let X be a non–empty set, $F = F(X,\Omega)$ the algebra of words of signature Ω on the alphabet X, $W \subset F \times F$. The class of algebras A, such that $u \equiv v$ is an identity for any pair $(u,v) \in W$, is called the *variety* of algebras $\mathfrak{W}$ defined by the set W or determined by the set of indentities. For instance, the class of all groups is the variety of algebras of signature Ω_Γ which is defined by the set of identities (2). Returning to the ususal description of operations in groups we see that (2) can be rewritten in the form

$$xe \equiv ex \equiv x, \quad x^{-1}x \equiv xx^{-1} \equiv e, \quad x(yz) \equiv (xy)z.$$

Adding the identity $x_1x_2 \equiv x_2x_1$ gives the variety of abelian groups. Among the varieties of suitable signature one also finds associative rings, Lie rings, etc.. The class of fields is not a variety. We can convince ourselves of the latter fact with the help of the following result.

1.6. Birkhoff theorem. *A non–empty class $\mathcal{K}$ of algebras of fixed signature Ω is a variety iff it is closed with respect to transition to subalgebras, homomorphic images and Cartesian products.*

Before starting on the proof of this result we need some additional information. First of all, we define the Cartesian product $C = \prod_{\alpha\in\mathcal{I}} A_\alpha$ of the family of algebras $(A_\alpha)_{\alpha\in\mathcal{I}}$ of fixed signature Ω as the set of functions $f : \mathcal{I} \to \bigcup_{\alpha\in\mathcal{I}} A_\alpha$ such that $f(\alpha) \in A_\alpha$, for any $\alpha \in \mathcal{I}$. If $\omega \in \Omega_n$, $f_1,\dots,f_n \in C$, we set $f = \omega(f_1,\dots,f_n)$ iff

$$f(\alpha) = \omega(f_1(\alpha),\dots,f_n(\alpha)).$$

Further, if $W \subset F(X,\Omega) \times F(X,\Omega)$, where $F(X,\Omega)$ is the algebra of words on signature Ω in the alphabet X then the smallest congruence $\sim$ on an algebra A such that $u(a_1,\dots,a_n) \sim v(a_1,\dots,a_n)$, for any pair $(u,v) \in W$ and any $a_1,\dots,a_n \in A$, is called the *verbal* congruence which is defined by the set

W. That congruence is equal to the intersection of all possible congruences with a described property.

∘ The main property (*the universality property*) of a verbal congruence is:

$$F(Y, \mathfrak{W}) = (F(Y, \Omega)/\sim) \in \mathfrak{W},$$

where $\mathfrak{W}$ is the variety defined by the set W, moreover, any *mapping* $\phi : Y \to A \in \mathfrak{W}$, where Y is some set, uniquely extends to a homomorphism $\bar{\phi} : F(Y, \mathfrak{W}) \to A$. ∘

We call the algebra $F(Y, \mathfrak{W})$ the *free algebra* of the variety $\mathfrak{W}$ with free generating set Y.

1.7. The proof of Birkhoff's theorem. It is evident, that if an identity is satisfied in some algebra then it is satisfied in any of its subalgebras and in any homomorphic image. Further, if the identity is valid in each algebra A_α from certain family $(A_\alpha)_{\alpha \in \mathcal{I}}$ of algebras of the same signature then it is valid in the Cartesian product $C = \prod_{\alpha \in \mathcal{I}} A_\alpha$. This shows that any variety is a class of algebras which is closed with respect to the taking of subalgebras, homomorphic images and direct products.

The proof of the inverse part of the theorem is more complicated. Let $\mathcal{K}$ be a non–empty class of algebras of signature Ω and which is closed with respect to the transitions indicated in the formulation. Let X be a countable alphabet $X = \{x_1, x_2, \ldots\}$. We denote by W the set of pairs $(u, v) \in F \times F$, such that $u \equiv v$ is an identity in each algebra from $\mathcal{K}$. Let $\mathfrak{W}$ be the variety of algebras of signature Ω defined by the set W. It is clear that $\mathcal{K} \subset \mathfrak{W}$. It is necessary to prove the inclusion $\mathcal{K} \supset \mathfrak{W}$. Let Y be an arbitrary non–empty set. The main implication of the universality property (see the end of Subsection 1.6) is that any algebra from $\mathfrak{W}$ is a homomorphic image of suitable free algebra $F(Y, \mathfrak{W})$. Therefore it is sufficient to show that $F(Y, \mathfrak{W}) \in \mathcal{K}$, for any Y. For that take the algebra of words $F(Y, \Omega)$ on the alphabet Y. The totality (of isomorphism classes) $\mathcal{L}$ of algebras from $\mathcal{K}$ for which the cardinality does not exceed $|Y|$, is a set. Correspondingly, the totality of mappings from Y into algebras from $\mathcal{L}$ is also a set. We denote this set by Φ. Let $F_\phi = \phi(F(Y, \Omega))$. Form the Cartesian product

$$C = \prod_{\phi \in \Phi} F_\phi.$$

For any $y \in Y$, let $\bar{y}$ denote the element from C such that $\bar{y}(\phi) = \phi(y) \in F_\phi$. Our aim is to demonstrate that the subalgebra D in C which is generated by the set $\bar{Y} = \{\bar{y} | y \in Y\}$, is isomorphic to the algebra $F(Y, \mathfrak{M})$. We shall prove that D possesses the universality property. The mapping $y \mapsto \bar{y}$ from Y into D extends to a homomoprhism $\psi : F(Y, \Omega) \to D$. Denote by ρ_0 the kernel of this homomorphism. Now let A be an arbitrary algebra from $\mathfrak{M}$ and $\chi : \bar{Y} \to A$ an arbitrary mapping. Consider the homomorphism $\tilde{\chi} : F(Y, \Omega) \to A$ which extends the mapping $y \mapsto \chi(\bar{y})$. Let $\rho = \operatorname{Ker} \tilde{\chi}$. We shall prove that $\rho_0 \subset \rho$. Indeed, let $(u, v) \in \rho_0$, $u = u(y_1, \dots, y_n)$, $v = v(y_1, \dots, y_n)$. Then, for any $\phi \in \Phi$, we have $\phi(u) = \phi(v)$. This means that $u(y_1, \dots, y_n) \equiv v(y_1, \dots, y_n)$ is an identity in $\mathcal{L}$, consequently, in $\mathcal{K}$. Replacing $y_1, \dots, y_n$ by $x_1, \dots, x_n$, we see that $(u(x_1, \dots, x_n), v(x_1, \dots, x_n)) \in W$. Since $A \in \mathfrak{M}$, we obtain that $u(a_1, \dots, a_n) = v(a_1, \dots, a_n)$ in A; in particular

$$\begin{aligned}\tilde{\chi}(u(y_1, \dots, y_n)) &= u(\chi(y_1), \dots, \chi(y_n)) = \\ =&v(\chi(y_1), \dots, \chi(y_n)) = \tilde{\chi}(v(y_1, \dots, y_n)),\end{aligned}$$

i.e. $(u, v) \in \rho$. According to the theorem of Subsection 1.2, $D = F(Y, \Omega)/\rho_0$ is homomorphically mapped on $F(Y, \Omega)/\rho$; moreover, $\bar{y}$ goes to $\chi(\bar{y})$ under the homomorphism mentioned. So D possesses the universality property with respect to $\bar{Y}$. In particular, the identity mapping $Y \to \bar{Y}$ extends to an (invertible) homomorphism from D into $F(Y, \mathfrak{M})$. Thus $D \cong F(\bar{Y}, \mathfrak{M})$. Since D is a subalgebra of a Cartesian product of the family of algebras F_ϕ from $\mathcal{K}$, we see that

$$F(\bar{Y}, \mathfrak{M}) \cong F(Y, \mathfrak{M}) \in \mathcal{K},$$

as required. □

1.8. Free algebras in varieties. We have already met free algebras: free groups (Ch. 2), free associative algebras and free Lie algebras over a field (Ch. 4). We shall pay attention to the signature Ω of *linear algebras* over a field K. This signature contains the signature of rings: 0, – (taking negatives), +, $\cdot$, also for each $\lambda \in K$ there is a unary operation which associates to each x from an algebra its product λx by the scalar λ.

It should be noted, as it was done in the proof of the previous theore, that a variety can be defined not only by a system of identities but also by a system of generating algebras. In fact, if $\mathcal{M}$ is some class of algebras then by var $\mathcal{M}$ we denote the smallest variety which contains all algebras from $\mathcal{M}$. In particular, if $\mathcal{M}$ consists of one algebra A then var A is the smallest variety containing A. According to the Birkhoff theorem (to be more accurate, to its proof) var A consists of homomorphic images of subalgebras of direct powers (Cartesian powers) of the algebra A. Here is one important statement.

Corollary. *Let A be a finite algebra, $\mathfrak{V}$ = var A. Then any finitely generated algebra from $\mathfrak{V}$ is finite.*

Proof. Analyzing the proof of the Birkhoff theorem we see that in the construction of the algebra $D = F(Y, \mathfrak{W})$ we can take as the class $\mathcal{L}$ the one–element class $\mathcal{L} = \{A\}$. Then if $Y = |Y| < \infty$, the set of maps Φ is finite, and all algebras F_ϕ have finite order as subalgebras in A. Therefore, $|F(Y, \mathfrak{W})| < \infty$. It is clear that from this there follows the finiteness of any finitely generated algebra of the variety $\mathfrak{W} = \mathfrak{V}$. □

A variety such that any finitely generated algebra is finite, is called *locally finite*. The problem of describing identities of finite groups naturally leads to the investigation of locally finite varieties.

1.9. Free Cantor algebras. In many important cases the cardinality of a free generating set X of a free algebra $F(X, \mathfrak{V})$ of the variety $\mathfrak{V}$ is an invariant and is called its *rank*. This occurs in the case of varieties of groups, linear algebras, etc.. In this subsection we shall introduce an example of a variety for which the property of invariance does not hold.

Let n be some natural number, $n \geq 2$, Ω_n the signature consisting of unary operations $\alpha_1, \dots, \alpha_n$ and an n–ary operation λ. We denote by $\mathfrak{V}_n$ the variety of algebras of the signature Ω_n which is defined by a system of identities

$$\lambda(\alpha_1(x), \dots, \alpha_n(x)) \equiv x,$$
$$\alpha_i(\lambda(x_1, \dots, x_n)) \equiv x_i, \quad i = 1, \dots, n.$$

An example of an algebra from $\mathfrak{V}_2$ is given by the Cantor process (Cantor diagonalization). With the help of the following numbering of the pairs of natural numbers

1	2	4	7	...
3	5	8	...	
6	9	...		
10	...			
...				

we define on the set **N** a binary operation λ by assigning to a pair (i, j) the number which lies in the intersection of the i–th row and j–th column of the table above. Further, we define $\alpha_1(x)$ as the number of the row where the number x is written. Analogously, $\alpha_2(x)$ is the number of the column, where x is written.

Now let X be a non–empty set and $F_n(X) = F(X, \mathfrak{V}_n)$ the free algebra of the variety $\mathfrak{V}_n$ with the free generating set X. The algebra $F_n(X)$ is called the *n–th free Cantor* algebra.

Theorem. *There is an isomorphism $F_n(X) \cong F_n(Y)$ iff $|X| \equiv |Y| (\mathrm{mod}\ (n-1))$.*

First of all we give an explicit description of the elements from the algebra $F_n(X)$. Let $A = \{\alpha_1, \dots, \alpha_n\}$. We call a word in the alphabet $X \cup A \cup \{\lambda\}$ which satisfies the following conditions a *standard form* over X:

(a) $\alpha_{i_1} \dots \alpha_{i_k} x$ is a standard from if $x \in X$, $k \geq 0$, $1 \leq i_j \leq n$, for $j = 1, 2, \dots, k$.

(b) if $w_1, \dots, w_n$ are standard forms then $\lambda w_1 \dots w_n$ is a standard form with the exception of the case when there exists a standard form u such that $w_i = \alpha_i u$, for $i = 1, 2, \dots, n$.

(c) Other standard forms do not exist.

We shall transform the set S of standard forms into $\mathfrak{V}_n$–algebra by setting

$$\begin{aligned} \alpha_i(\alpha_{i_1} \dots \alpha_{i_k} x) &= \alpha_i \alpha_{i_1} \dots \alpha_{i_k} x, \\ \alpha_i(\lambda w_1 \dots w_n) &= w_i, \\ \lambda(w_1, \dots, w_n) &= \lambda w_1 \dots w_n, \end{aligned}$$

if there does not exist a standard form u such that $w_i = \alpha_i u$, $i = 1, 2, \dots, n$, and if there does exist such a u then

$$\lambda(w_1, \dots, w_n) = \lambda(\alpha_1 u, \dots, \alpha_n u) = u.$$

It is evident that the algebra S of standard forms belongs to the variety $\mathfrak{V}_n$. If $P \in \mathfrak{V}_n$, and $\theta : X \to P$ is a mapping then the homomorphism $\bar{\theta} : S \to P$ is defined as in a case of algebra of words (see Subsection 1.4). Indeed, set

$$\begin{aligned}\bar{\theta}(\alpha_{i_1} \dots \alpha_{i_k} x) &= \alpha_{i_1}(\dots(\alpha_{i_k}(\theta x))\dots),\\ \bar{\theta}(\lambda w_1 \dots w_n) &= \lambda(\theta(w_1), \dots, \theta(w_n)).\end{aligned}$$

Checking that

$$\bar{\theta}(\alpha_i(w)) = \alpha_i(\bar{\theta}(w)) \quad \bar{\theta}(\lambda(w_1, \dots, w_n)) = \lambda(\bar{\theta}(w_1), \dots, \bar{\theta}(w_n)),$$

i.e. that $\bar{\theta} : S \to P$ is a homomorphism, follows directly from the definition and is left to the reader. So $F_n(\mathfrak{V}) \cong S$.

For $X \subset P \in \mathfrak{V}_n$ we denote by $\langle A\rangle(X)$ the subalgebra in P which is generated by the set X with a help of operations of the set $A = \{\alpha_1, \dots, \alpha_n\}$; $\langle\lambda\rangle(X)$ is similar but allows only the operation λ.

Lemma 1. *(a) If $x \in X$ then $F_n(X)$ is generated freely by the set*

$$(X \setminus \{x\}) \cup \{\alpha_i x \,|\, 1 \le i \le n\}.$$

(b) If $x_1, \dots, x_n$ are different elements from the set X then $F_n(X)$ is generated freely by the set

$$(X \setminus \{x_1, \dots, x_n\} \cup \{\lambda x_1 \dots x_n\}.$$

Proof. (a) Let $Y = (X \setminus \{x\} \cup \{\alpha_1 x, \dots, \alpha_n x\}$, and let $\Theta : Y \to P \in \mathfrak{V}_n$ be some mapping. Define $\Theta^* : X \to P$ by setting

$$\begin{aligned}&\Theta^*(y) = \Theta(y), \quad \text{if } y \in X \setminus \{x\},\\ &\Theta^*(x) = \lambda(\Theta(\alpha_1(x)), \dots, \Theta(\alpha_n(x)).\end{aligned}$$

Since X is a free generating set, there exists a unique homomorphism $\widehat{\Theta} : F_n(X) \to P$ which extends Θ^*. However, $\widehat{\Theta}$ also extends Θ since

$$\widehat{\Theta}(y) = \Theta^*(y) = \Theta(y) \quad \text{for } y \in X \setminus \{x\},$$
$$\widehat{\Theta}(\alpha_i(x)) = \alpha_i(\widehat{\Theta}(x)) = \alpha_i(\Theta^*(x)) =$$
$$= \alpha_i(\lambda(\Theta(\alpha_1(x)), \ldots, \Theta(\alpha_n(x))) = \Theta(\alpha_i(x)), \quad i = 1, \ldots, n.$$

Thus the mapping $\Theta : Y \to P$ extends to a homomorphism $\widehat{\Theta} : F_n(X) \to P$. Since $x = \lambda(\alpha_1(x), \ldots, \alpha_n(x))$, the set Y generates $F_n(X)$. So Y is a free generating set for $F_n(X)$.

Statement (b) is proved in a similar way. □

This lemma already implies that the idea of rank for the free Cantor algebra $F_n(X)$ is ambigous. In order to prove the theorem completely more accurate information is needed.

We call the procedure of transition to a new generating set, which is described in statement (a) of the lemma, a simple expansion of the set X, and the procedure of statement (b) a simple contraction. Using a simple expansion (contraction) d times is called a d–multiple expansion (contraction). If Y is the result of a d–multiple expansion of a finite free system of generators X then

$$|Y| = |X| + (n-1)d.$$

Lemma 2. *If a set X generates the $\mathfrak{V}_n$–algebra P, then $P = \langle\lambda\rangle(\langle A\rangle(X))$, and for each $y \in P$ the set $\langle A\rangle(y) \setminus \langle A\rangle(X)$ is finite.*

Proof. It is sufficient to consider the case when the algebra $P = S$ is free on X. Then S is isomorphic to the algebra of standard forms over X and, consequently, the first statement is correct according to the definition of standard form.

In order to prove the second part of the lemma we should remark that if $y \in S$ and the number of occurrences of λ in y (in a standard form) is equal to m then $\alpha_{i_1} \ldots \alpha_{i_r}(y) \in \langle A\rangle(X)$, when $r \geq m$. Consequently, the elements from $\langle A\rangle(y) \setminus \langle A\rangle(X)$ have the form $\alpha_{i_1} \ldots \alpha_{i_r} y$, where $r < m$ and therefore their number is finite. □

Lemma 3. *Let X be a finite set and $U \subset \langle A\rangle(X) \subset F_n(X)$. Then the following conditions are equivalent:*

(*a*) $U = \langle A\rangle(X) \cap \langle A\rangle(Y)$ *for some set Y which generates the algebra $F_n(X)$;*

(*b*) *the set U is A–closed, i.e. $U = \langle A\rangle(U)$, and $\langle A\rangle(X) \setminus U$ is finite;*

(*c*) $U = \langle A\rangle(Z)$ *for some basis Z of the algebra $F_n(Y)$.*

Proof. Assume that $U = \langle A\rangle(X) \cap \langle A\rangle(Y)$, and, moreover, that Y generates the algebra $F_n(X)$. Then U is A–closed as an intersection of A–closed sets. For each $x \in X$, the set $\langle A\rangle(x) \setminus \langle A\rangle(Y)$ is finite, according to lemma 2. Since X is finite, $\langle A\rangle(X) \setminus \langle A\rangle(Y) = \langle A\rangle(X) \setminus U$ is also finite. Thus, from (a) follows (b).

Now let U be U–closed and $\langle A\rangle(X) \setminus U$ be finite. We shall prove that the property (c) holds. Apply induction on $|\langle A\rangle(X) \setminus U|$ with the evident starting point, $|\langle A\rangle(X) \setminus U| = 0$, when one sets $Z = X$. Otherwise take the longest element $w \in \langle A\rangle(X) \setminus U$ (i.e. with the largest number of occurrences of the α_i). Then the set $U^* = U \cup \{w\}$ is A–closed and $|\langle A\rangle(X) \setminus U^*| = |\langle A\rangle(X) \setminus U| - 1$. According to the induction assumption there exists an expansion Z^* of the basis X such that $U^* = \langle A\rangle(Z^*)$. The element w lies in Z^*, as otherwise w would have the form $w = \alpha_{i_1} \ldots \alpha_{i_r}(z)$, where $z \in Z^*$ and $r > 0$, and, consequently, $z \in U^* \setminus \{w\} = U$; however the set U is A–closed, therefore $w \in U$ which is impossible. We set

$$Z = Z^* \setminus \{w\} \cup \{\alpha_i(w) | 1 \le i \le n\}.$$

This set is an expansion of the basis X and according to the choice of the element w we have $\alpha_i(w) \in U$, for every i. Consequently, $U = \langle A\rangle(Z)$, as is required to prove.

The fact that (a) follows from (c) is trivial. □

Proof of the theorem. Let X, Y be two free generating sets for $F_n(X)$, $U = \langle A\rangle(X) \cap \langle A\rangle(Y)$. According to lemma 3 there exists an expansion Z' of X such that $U = \langle A\rangle(Z')$, and an expansion Z'' of Y such that $U = \langle A\rangle(Z'')$. However $\langle A\rangle(Z)$ determines Z uniquely as the set of all elements from $\langle A\rangle(Z)$ which are not representable in the form $\alpha_i(z)$, $1 \le i \le n$. So $Z = Z' = Z''$ is a common expansion for X and Y. For suitable natural r, s we have

$$|Z| = |X| + (n-1)r = |Y| + (n-1)s.$$

Consequently, $|X| \equiv |Y| \pmod{(n-1)}$.

The sufficiency of the condition of the theorem is evident from lemma 1. □

∗ An important application of free Cantor algebras occurs when considering the group of automorphisms $G_{n,r}$ of a free algebra of rank r of the variety $\mathfrak{V}_n$, i.e. of the algebra $F_n(X)$ for $|X| = r$. It turns out that every group $G_{n,r}$ is finitely defined. For even n these groups are simple, and for odd n they contain a simple subgroup of index 2. For $m \neq n$, the groups $G_{m,r}$ and $G_{n,s}$ are not isomorphic. So there exists an infinite family of finitely presented simple infinite groups. ∗

§ 2. Finite basis problem for identities in groups

2.1. Identities in groups. When investigating varieties of certain types of universal algebras: groups, rings, Lie algebras, etc. it is customary to start not with all algebras of given signature but with the variety of all algebras of the given form, such as the variety of all groups, all rings, say, associative ones, all Lie algebras over a fixed field, etc.. Therefore, if we are concerned with the identities of one or another variety of groups (rings and etc.) then, frequently, the identities which define the variety of all groups (rings and etc.) are silently added or removed. In fact, this situation occurs in the statement and solution of the problem concerning finite bases. For convenience of formulation we shall investigate the variety of groups. Let us say, that the identity $u \equiv v$ is a *consequence* of the identity $u' \equiv v'$ if in any group G, where $u' \equiv v'$ is fulfilled, $u \equiv v$ is also satisfied. It should be noted that any identity of the form $u \equiv v$ in a group is reducible with help of multiplication by v^{-1} to an identity of the form $w \equiv e$, where e is the unit of group. If $F(X)$ is a free group with the countable free generating set X, $W \subset F(X) \times F(X)$, then according to what was said, W is equivalent to its subset of pairs of the form $\{(w, e)\}$ as a set of identities. It is clear that if in some group G the identity $w_1 \equiv e$, $w_2 \equiv e$ are fulfilled then also $w_1 w_2^{-1} \equiv e$. Moreover, let $w = w(x_1, \ldots, x_n)$ and $v_1(x_1, \ldots, x_m), \ldots, v_n(x_1, \ldots, x_m) \in F(X)$. If $w(x_1, \ldots, x_n) \equiv e$ is an identity in the group G then also $w(v_1(x_1, \ldots, x_m), \ldots, v_n(x_1, \ldots, x_m)) \equiv e$ is an identity in G. So the set V of all left parts of identities $w \equiv e$, which are fulfilled in a fixed group G (or in any fixed class of groups), is a subgroup of the group $F(X)$. Since

$$u^{-1}w(x_1,\dots,x_n)u = w(ux_1u^{-1},\dots,ux_nu^{-1}),$$

this subgroup V is normal. The reader can easily show that the verbal congruence mentioned in Subsection 1.6 is nothing else but a partition of the group $F(X)$ into the left (or right) cosets of the subgroup V. The group $F(X)/V$ is a free group for the corresponding variety. Thus, if $\mathfrak{V}$ is some variety of groups then it corresponds to a subgroup V in the free group $F(X)$ with countable generating set X. Such a subgroup $V = V(F(X))$ is called *verbal*. It should be noted that an analogous subgroup can be defined in any group G: if V is some set of elements (words) from $F(X)$ then $V(G)$ is the subgroup which is generated by the set of values of words from V, i.e. by the elements of the form $v(g_1,\dots,g_n)$, where $v \in V$, $g_1,\dots,g_n \in G$.

According to its definition $V(G)$ is the smallest normal subgroup in G such that $G/V(G) \in \mathfrak{V}$.

If $\mathfrak{V}_1$ and $\mathfrak{V}_2$ are two varieties of groups, and V_1 and V_2 are the verbal subgroups in the free group $F(X)$ of countable rank corresponding to them, then $\mathfrak{V}_1 \subset \mathfrak{V}_2$ evidently implies $V_2 \subset V_1$. Moreover, if $V_1 = V_2$ then, since any identity involves a countable set of variables, we have $\mathfrak{V}_1 = \mathfrak{V}_2$. Therefore identities in groups and verbal subgroups in $F(X)$, where X is countable, are in a one–to–one inclusion reversing correspondence. If $\mathfrak{V}$ is some identity, $F(X,\mathfrak{V})$ its countable rank free group, then the subvarieties of the variety $\mathfrak{V}$ are in a one–to–one inclusion reversing correspondence with the verbal subgroups of the group $F(X,\mathfrak{V})$.

2.2. Finite basis property. We say that a variety of groups $\mathfrak{V}$ if *finitely based* if the system of all its identities is a consequence of a finite subsystem. For instance, any variety $\mathfrak{V}$ consisting only of abelian groups is finitely based. In fact, it should be noted that the identity $xy \equiv yx$ or $[x,y] = xyx^{-1}y^{-1} \equiv e$ is fulfilled in $\mathfrak{V}$. Let also m be a natural number such that the identity $x^m \equiv e$ is valid in $\mathfrak{V}$ (it is possible that such an identity does not exist; in that case we assume $m = 0$). We shall prove that $x^m \equiv e$ and $[x,y] \equiv e$ imply all other identities of the variety $\mathfrak{V}$. Let $w(x_1,\dots,x_n) \equiv e$ in $\mathfrak{V}$. Since F/F' is abelian $wF' = x_1^{s_1}\dots x_n^{s_n}F'$, where F' is the commutator subgroup of the group F. Then $w = x_1^{s_1}\dots x_n^{s_n}w'$, where $w' \in F'$. By substituting $x_2 = \dots = x_n = e$ we obtain that w' becomes e, and the identity $x_1^{s_1} \equiv e$ is fulfilled in $\mathfrak{V}$. From the minimality of the number m follows that s_1 is divisible by m. Analogously, $s_2,\dots,s_n$ are divisible by m. Further, since w' lies in F', this element is a product of values of commutators $[x,y]$. So w is a product of words $v_1, \ldots, v_r$ such that any identity $v_i \equiv e$ is a consequence of the determined pair of identities. This fact proves the existence of a finite basis of the variety $\mathfrak{V}$.

Similar, but more complex arguments prove the finite bases property of any variety which completely consists of nilpotent groups.

The question whether there exists a variety of groups which has no finite basis for its identities was not solved for 30 years after it was stated by B. Neumann in 1937.

In 1969 examples by A.Yu. Ol'shanski, S.I. Ad'yan and M. R. Vaughan–Lee were published. An example by J.H. Kleiman is one of the most effective examples of varieties of groups without the finite basis property.

2.3. Product varieties. Let $\mathfrak{U}$, $\mathfrak{V}$ be the varieties of groups. The class $\mathfrak{U}\mathfrak{V}$ of extensions of groups from $\mathfrak{U}$ with a help of groups from $\mathfrak{V}$ is called the *product variety* $\mathfrak{U}\mathfrak{V}$. Thus any group $G \in \mathfrak{U}\mathfrak{V}$ contains a normal subgroup $H \in \mathfrak{U}$ such that $G/H \in \mathfrak{V}$. An application of the Birkhoff theorem, Subsection 1.6, shows that $\mathfrak{U}\mathfrak{V}$ is indeed a variety. Let $F(X)$ be a free group of countable rank, U, V verbal subgroups in $F(X)$ which correspond to the varieties $\mathfrak{U}$, $\mathfrak{V}$, respectively. Then $W = U(V(F(X)))$ is a normal subgroup in $F(X)$ which is stable with respect to the replacement of $x_1, \ldots, x_n, \ldots$ by elements from $F(X)$. The latter property means that W can be taken as the verbal subgroup which corresponds to the identities of some variety $\mathfrak{W}$. If $G \in \mathfrak{W}$ then $W(G) = \{e\}$, i.e. $U(V(G)) = \{e\}$; we have $V(G) \in \mathfrak{U}$, $G/V(G) \in \mathfrak{V}$, i.e. $G \in \mathfrak{U}\mathfrak{V}$. Inversily, if $G \in \mathfrak{U}\mathfrak{V}$ (see above) then $V(G) \subset M$, $U(V(G)) \subset U(H) = \{e\}$, i.e. $G \in \mathfrak{W}$. Thus $\mathfrak{W} = \mathfrak{U}\mathfrak{V}$ and $W = U(V(F))$ is the verbal subgroup corresponding to the product of varieties $\mathfrak{U}\mathfrak{V}$.

∗ The description obtained of the identities of product varieties implies easily the proof of associativity of products of varieties. The main theorem on products of varieties states: *each variety which is different from the variety of all groups and from the variety consisting only of the unit group, decomposes uniquely into a product of varieties which admit no further factorization.* ∗

We shall denote by $\mathfrak{B}_n$ $(n \geq 2)$ the variety of groups which is given by the identity $x^n \equiv 1$. Our aim is to prove that the product variety $\mathfrak{B}_4\mathfrak{B}_2$ has no finite basis for its identities. If we take into consideration the structure of the identities for a product of varieties then we obtain that the system of identities

(1) $$(x_1^2)^4 \equiv e, \quad (x_1^2 x_2^2)^4 \equiv e, \ldots, (x_1^2 x_2^2 \ldots x_n^2)^4 \equiv e, \ldots$$

is not reducible to any of its finite subsystems.

Let $u_n = (x_1^2 \ldots x_n^2)^4$. It is clear that every identity $u_n \equiv e$ is a consequence of every identity $u_m \equiv e$, where $m \geq n$. Therefore it is sufficient to prove the following statement: for every $n \geq 1$, system (1) is not a consequence of the identity $u_n \equiv e$. In order to show this it is sufficient to construct a sequence of groups C_m which do not lie in $\mathfrak{B}_4\mathfrak{B}_2$, such that for any n there is an m for which $u_n(C_m) = \{e\}$.

2.4. Construction of a class of groups. Let us give the groups in terms of generators and defining relations

(2) $$A_m = \langle a_1, a_2, \ldots, a_{2m} \mid a_i^2 = e, \ [[a_i, a_j], a_k] = e; \ i, j, k = 1, \ldots, 2m \rangle.$$

Since the group $\mathbf{Z}_2^{2m} = \underbrace{\mathbf{Z}_2 \oplus \ldots \oplus \mathbf{Z}_2}_{2m}$ is a quotient–group of the group A_m, we see that $A_m/A_m' \cong \mathbf{Z}_2^{2m}$. Further, the commutator subgroup A_m' of the group A_m coincides with its center and is a group which is isomorphic to the group $\mathbf{Z}_2^{m(2m-1)}$. Indeed, in a nilpotent group of step 2 we have $[u, v][u, w] = [u, vw]$ (see, § 4.3); in particular, any commutator of generators is either trivial or is an element of order 2.

A realization of the group A_m can be obtained in the following way.

Consider over the field $\mathbf{Z}_2$ the ring R of non–commuting polynomials in the variables $x_1, \ldots \ldots, x_{2m}$. In this ring consider the ideal $\mathcal{I}$ of polynomials of the form $f = \sum_i \mu_i x_i^2 + f'$, where f' does not contain monomials of degree ≤ 2.

Each co–set of the ring R with respect to the ideal $\mathcal{I}$ contains precisely one polynomial of the form

$$\alpha_0 + \sum_i \beta_i x_i + \sum_{i \neq j} \gamma_{ij} x_i x_j.$$

We shall investigate the subgroup G_m in $R/\mathcal{I}$ which is generated by the elements $a_i = 1 + \bar{x}_i$, $i = 1, 2, \ldots, 2m$. Let us show that the defining relations (2) hold. In fact,

$$\begin{aligned} a_i^2 &=(1+\bar{x}_i)^2 = 1, \\ [\bar{a}_i, \bar{a}_j, \bar{a}_k] &= [(1+\bar{x}_i)(1+\bar{x}_j)(1+\bar{x}_i)(1+\bar{x}_j), 1+\bar{x}_k] = \\ &= [1+\bar{x}_i\bar{x}_j + \bar{x}_j\bar{x}_i, 1+\bar{x}_k] = \\ &= (1+\bar{x}_i\bar{x}_j+\bar{x}_j\bar{x}_i)(1+\bar{x}_k)(1+\bar{x}_i\bar{x}_j+\bar{x}_j\bar{x}_i)(1+\bar{x}_k) = \\ &= (1+\bar{x}_k+\bar{x}_i\bar{x}_j+\bar{x}_j\bar{x}_i)^2 = e \end{aligned}$$

At the same time the set of elements $[a_i, a_j] = 1 + \bar{x}_i\bar{x}_j + \bar{x}_j\bar{x}_i$ for $i < j$, (there are $\binom{2m}{2} = m(2m-1)$ of them) is linearly independent over $\mathbf{Z}_2$; this proves the statement about the structure of the group A'_m.

Now we fix an element $a \in A'_m$, $a \neq e$, and define a function $\rho : A'_m \to \{0,1\}$, by setting $\rho(g) = 1$, if $g = a$, and $\rho(g) = 0$, if $g \neq a$. Denote the coset of the element $a \in A_m$ in A_m/A'_m by $\bar{a}$. Given a fixed $a \in A'_m$ we construct a group B_m with as generating set all possible symbols of the form b^g, c^k, where $g \in A_m$, $k \in A_m/A'_m$, and with as set of defining relations

$$\begin{aligned} &(b^g)^2 = (c^k)^2 = e, \quad [b^g, b^h] = e \quad \text{if } \bar{g} \neq \bar{h}, \\ (3) \qquad &[b^g, b^h] = (c^{\bar{g}})^{\rho(gh^{-1})} \quad \text{if } \bar{g} = \bar{h}, \\ &[c^k, b^g] = e \quad \text{for any } k, g. \end{aligned}$$

It is clear that B_m is generated only by just the set of elements b^g, $g \in A_m$, and B'_m is generated by the elements c^k, $k \in A_m/A'_m$. A realization of the group B_m in a ring of polynomials is entirely analogous to the realization of the group A_m (with some extension of the ideal $\mathcal{I}$). The group B'_m is a primary abelian 2–group with the set $\{c^k \mid k \in A_m/A_m^2\}$ as independent generators.

We define an action of the group A_m on B_m by

$$(b^g)^h = b^{gh}, \quad (c^k)^h = c^{k\bar{h}} \quad (g, h \in A_m,\ k \in A_m/A'_m).$$

For any $g \in A_m$ this action preserves the relations (3), therefore it extends to an automorphism α_g of the group B_m. It is clear that $g \mapsto \alpha_g$ is an isomorphism of the group A_m onto a subgroup in Aut (B_m).

We form the semi–direct product $C_m = A_m \leftthreetimes B_m$. According to the definition (see, § 5.2) this is the set of pairs (g, s), $g \in A_m$, $s \in B_m$ with the operation $(g_1, s_1)(g_2, s_2) = (g_1g_2, \alpha_{g_2}(s_1)s_2)$.

Immediate verification shows that C_m is a group in which $A_m \cong \{(g, e) | g \in A_m\}$ is a subgroup and $B_m \cong \{(e, s) | s \in B_m\}$ is a normal subgroup. We shall write $(g, s) = gs$. Evidently, $\alpha_g(s) = gsg^{-1} = s^g$. Consider now the subgroup $D_m = A'_m B_m$ in C_m. Any element $d \in D_m$ can be presented in the form $d = gb^{g_1}b^{g_2} \dots b^{g_n}$, where $g \in A'_m$, $g_i \in A_m$.

Lemma 1. *The relation $d^4 = e$ holds iff either $\rho(g) = 0$, or for any $k \in A_m/A'_m$, the number of elements g_i with the condition $\bar{g}_i = k$ in the expression for d is even.*

Proof. Since $D^4_m \subset B'_m$, then it is clear a priori that d^4 has the form of a product of elements of the form c^k, for different g, k. The elements c^k for various k are independent and each such c^k can be obtained by composing commutators of the form $[b^g, b^{g'}]$, where $\bar{g} = \bar{g}' = k$. Now note that for $\bar{g} \neq \bar{g}'$ the elements b^g, $b^{g'}$ commute, then it is clear that it is sufficient to consider the case of an element $d = gb^{g_1} \dots b^{g_n}$ such that $\bar{g}_1 = \dots = \bar{g}_n = k$.

We have

$$d^2 = b^{g_1 g} \dots b^{g_n g} b^{g_1} \dots b^{g_n}.$$

Switching factors leads to commutators which are central elements from B'_m, of square equal to 1.

Therefore, before taking the square of the element d^2 the factors can be arbitrarily permuted

$$d^4 = (b^{g_1} b^{g_1 g} \dots b^{g_n} b^{g_n g})(b^{g_1} b^{g_1 g} \dots b^{g_n} b^{g_n g}).$$

We shall show that the right side is equal to $(c^k)^{n\rho(g)}$. Indeed, the element $b^{g_1} b^{g_1 g}$ in the second bracket commutes with all $b^{g_i} b^{g_i g}$, $i \neq 1$.

In order to see this calculate the (distributive!) commutator $[b^{g_i} b^{g_i g}, b^{g_1} b^{g_1 g}]$. We have

$$[b^{g_i} b^{g_i g}, b^{g_1} b^{g_1 g}] = [b^{g_i} b^{g_1}][b^{g_i g}, b^{g_1 g}][b^{g_i}, b^{g_1 g}][b^{g_i g}, b^{g_1}] =$$
$$= (c^k)^{\rho(g_i g_1^{-1})}(c^k)^{\rho(g_i g_1^{-1})}(c^k)^{\rho(g_i g^{-1} g_1^{-1})}(c^k)^{\rho(g_i g_1^{-1})} = e,$$

as $g = g^{-1}$ (this element is taken from A'_m). This shows that

$$d^4 = (b^{g_1} b^{g_1 g})^2 \dots (b^{g_n} b^{g_n g})^2 = (c^k)^{n\rho(g)},$$

since

$$\mathfrak{b}^{g_1}\mathfrak{b}^{g_1 g}\mathfrak{b}^{g_1}\mathfrak{b}^{g_1 g} = [\mathfrak{b}^{g_1}, \mathfrak{b}^{g_1 g}] = (c^k)^{\rho(g_1 g^{-1} g_1^{-1})} = (c^k)^{\rho(g)}.$$

□

Thus $d^4 = 1$ is equivalent to the number n being even or $\rho(g) = 0$.

Now we shall find the conditions under which $u_n \equiv e$ holds in C_m.

Lemma 2. *$u_n \equiv e$ is an identity in C_m iff the fixed element $a \in A'_m$ is not representable in the form $a = g_1^2 \dots g_n^2$, for any $g_1, \dots, g_n \in A_m$.*

Proof. Calculating u_n in C_m on the elements $g_1\mathfrak{b}_1, \dots, g_n\mathfrak{b}_n$, we obtain

$$((g_1\mathfrak{b}_1)^2 \dots (g_n\mathfrak{b}_n)^2)^4 = (g_1^2 \dots g_n^2 \widetilde{\mathfrak{b}})^4, \quad \text{where } g_i \in A_m,\ \mathfrak{b}_i, \widetilde{\mathfrak{b}} \in B_m.$$

If a is not representable in the form $g_1^2 \dots g_n^2$ then it is clear that $u_n \equiv e$ is an identity in C_m. Now assume that for some $g_1, \dots, g_n \in A_m$ we have $a = g_1^2 \dots g_n^2$. Then at least one of g_i, say, $g_n \notin A_m$ (otherwise $a = e$). Calculate $u_n(g_1, g_2, \dots, g_{n-1}, g_n\mathfrak{b}^1)$. We then obtain

$$(g_1^2 \dots g_{n-1}^2 (g_n\mathfrak{b}^1)^2)^4 = (a\mathfrak{b}^{g_n}\mathfrak{b}^1)^4 \not\equiv e,$$

according to the lemma 1. Thus $u_n \equiv e$ is not an identity in C_m. □

2.5. Theorem. *The variety $\mathfrak{B}_4\mathfrak{B}_2$ does not possess the finite basis property.*

Proof. As was noted before (see Subsection 2.2, 2.3) it is sufficient to find a group C_m, for any n, that does not lie in $\mathfrak{B}_4\mathfrak{B}_2$ and such that the identity $u_n \equiv e$ holds. Let us establish that for any $m = 1, 2, \dots$ and any choice of $a \in A_m$, $C_m \notin \mathfrak{B}_4\mathfrak{B}_2$. Indeed, since $a \in A'_m = A_m^2$, there is an n such that $a = g_1^2 \dots g_n^2$. Then by lemma 2 of Subsection 2.3, $u_n \equiv e$ is not an identity in C_m. Inversily, we shall show that for $m = n + 1$, $u_n \equiv e$ is an identity in C_m, for suitable $a \in A'_m$. For this it is sufficient to select an element a in A'_m which is not representable in the form $g_1^2 \dots g_n^2$, $g_i \in A_m$. It follows from Subsection 2.3 that the number of elements in A'_m is equal to $2^{\binom{2m}{2}} = 2^{2m^2 - m} = 2^{2n^2+3n+1}$. The number of all possible squares does not exceed the number 2^{2m} of the subsets of $\{a_1, \dots, a_{2m}\}$ (before raising the element $g = a_{i_1} \dots a_{i_p}$ to the power 2 the factors can be permuted, i.e. A'_m is a central subgroup of index 2). The number of products of the form $g_1^2 \dots g_n^2$ does not exceed $(2^{2m})^n = 2^{2n^2+2n}$. This number is lower than the number of all elements in A'_m and the choice of the necessary element a in A'_m is possible. In the corresponding group C_m the identity $u_n \equiv e$ is fulfilled (but $C_m \notin \mathfrak{B}_4\mathfrak{B}_2$). □

∗ **2.6.** We note without proof that the identities of any finite group are finitely based (Oates–Powell theorem). An analogous fact occurs in the case of associative rings, Lie rings and a range of other rings. At the same time the W. Specht's problem remains unsolved: is it true that any variety of associative rings is finitely based? However, a recent achievement of A. R. Kemer says: any associative algebra over a zero charcteristic field possesses the finite basis property. It is also not known if the algebras of all square matrices of size $n \geq 3$ possess the finite basis property (for $n = 2$ the answer is positive). A separate section is devoted to associative algebras with identities which are more important in applications.

To conclude the present section we should note that the simplest examples of not finitely based algebras appear in the class of all groupoids (algebras with one binary operation), there we can find an example of a groupoid of 3 elements without the property mentioned. In a class of semi–groups the minimal order of a semi–group which does not possess the finite basis property, is equal to 6. ∗

§ 3. PI–algebras

3.1. Types of identities. The study of identities of linear algebras over a field has its own particularities. If F is a field, $\{x_1, x_2, \dots\}$ is a countable set of variables, $f(x_1, \dots, x_n)$ is a non–commutative and even non–associative polynomial with coefficients from F, then $f(x_1, \dots \dots, x_n) \equiv 0$ is an identity in the algebra A over the field F, if $f(a_1, \dots, a_n) = 0$, for any $a_1, \dots, a_n \in A$. Moreover, if A is an associative algebra then $f(x_1, \dots, x_n) \equiv 0$ is a non–trivial identity in the associative algebra A, if it holds in A but is not satisfied at least in one associative algebra B over F. Similar terminology is used in the case of Lie algebras etc..

If $\mathfrak{V}$ is some variety of algebras over the field F, X is a non–empty set, $F\langle X\rangle$ is the algebra of all polynomials (non–associative and non–commutative) in X with coefficients from F then the set $\mathfrak{V}(X)$ of polynomials f such that $f \equiv 0$

is an identity in $\mathfrak{V}$, is an ideal in $F\langle X\rangle$. A special property of this ideal is that if $f(x_1,\ldots,x_n) \in \mathfrak{V}(X)$, $g_1(x_1,\ldots,x_m),\ldots,g_n(x_1,\ldots,x_m) \in F\langle X\rangle$ then $f(g_1(x_1,\ldots,x_m),\ldots,g_n(x_1,\ldots,x_m)) \in \mathfrak{V}(X)$. The quotient–algebra $F(X,\mathfrak{V}) = F(X)/\mathfrak{V}(X)$, as can be easily verified, is the free algebra of the variety $\mathfrak{V}$ with the free generating set X. If $\mathfrak{V}$ is the variety of all associative algebras then $F(X,\mathfrak{V}) = A(X)$ is a free associative algebra; if $\mathfrak{V}$ is the variety of Lie algebras then $F(X,\mathfrak{V}) = L(X)$ is a free Lie algebra. If $\mathfrak{V}$ is some variety consisting only of associative algebras then $F(X,\mathfrak{V})$ is a quotient–algebra of the algebra $A(X)$. We shall denote by $T_{\mathfrak{V}}(X)$ the kernel of natural homomorphism $A(X) \to F(X,\mathfrak{V})$. The non–zero elements of the ideal $T_{\mathfrak{V}}(X)$ are the non–trivial associative identities of the variety $\mathfrak{V}$ which depend on the set of variables X. The situation for varieties of Lie algebras is analogous.

We call a polynomial $f(x_1,\ldots,x_n) \in F\langle X\rangle$ *homogeneous* (with respect to the variable x_1) if it is representable in the form of a linear combination of monomials of the same degree (relatively to x_1). If the polynomial is homogeneous with respect to all its variables then it is called *multihomogeneous*. If the degree of homogeneity with respect to every variable is equal to 1 then the polynomial is called *multilinear*. An identity $f(x_1,\ldots,x_n) \equiv 0$ of which the left part is homogeneous, multihomogeneous, multilinear is called, respectively, homogeneous, multihomogeneous, multilinear.

We say that an identity $f(x_1,\ldots,x_n) \equiv 0$ is a *consequence* of the identity $g(x_1,\ldots,x_n) \equiv 0$ if in any algebra, where $g(x_1,\ldots,x_n) \equiv 0$ is fulfilled, $f(x_1,\ldots,x_n) \equiv 0$ also holds. In addition, an identity can be a consequence of another idenitity in the class of associative algebras but not a consequence in a class of Lie algebras and inversely, and this requires one to be cautious in formulations. Two identities are *equivalent* if each is a consequence of the another one. Similar definitions pertain to systems of identities.

Proposition 1. *Let a variety $\mathfrak{V}$ be defined by the system of identities $\{f_\alpha \equiv \mid \alpha \in \mathcal{I}\}$. Then the ideal of identities of the variety $\mathfrak{V}$ depending on variables from the set X coincides with the ideal of the algebra $F\langle X\rangle$ which is generated by the elements of the form $f_\alpha(g_1,\ldots,g_n)$, where the $g_1,\ldots,g_n$ range over all possible elements from $F\langle X\rangle$, $\alpha \in \mathcal{I}$.*

Ideals of identities of varieties of associative algebras, Lie algebras, etc. are constructed in a similar way.

Proof. Let $\mathfrak{V}'(X)$ be the ideal defined in the formulation of the theorem. Then, clearly, $\mathfrak{V}'(X) \subset \mathfrak{V}(X)$. The algebra $F(X)/\mathfrak{V}'(X)$ lies in the variety $\mathfrak{V}$, thus, it is a quotient–algebra of the free algebra $F(X)/\mathfrak{V}(X)$ under the natural mapping on the set X. Hence $\mathfrak{V}(X) \subset \mathfrak{V}'(X)$, i.e. $\mathfrak{V}(X) = \mathfrak{V}'(X)$. □

Proposition 2. *Any system of identities over an infinite field F is equivalent to a system of multihomogeneous ones. If char $F = 0$ then any system of identities is equivalent to a system of multilinear ones.*

Proof. Let $f = f(x_1, \dots, x_n) \equiv 0$ be some identity. If f is not homogeneous with respect to x_1 then we write it in the form $f = f_0 + f_1 + \dots + f_t$, where f_i is homogeneous of degree i with respect to x_1. Choose pairwise different scalars $\lambda_0, \lambda_1, \dots, \lambda_t$. The identities $f(\lambda_i x_1, x_2, \dots, x_n) \equiv 0$ are consequences from $f = 0$, $i = 0, 1, \dots, t$. However $f(\lambda_i x_1, x_2, \dots, x_n) = f_0 + \lambda_i f_1 + \lambda_i^2 f_2 + \dots + \lambda_i^t f_t$. Thus the following $t + 1$ consequences are obtained

$$f_0 + \lambda_0 f_1 + \lambda_0^2 f_2 + \dots + \lambda_0^t f_t \equiv 0,$$

$$f_0 + \lambda_1 f_1 + \lambda_1^2 f_2 + \dots + \lambda_1^t f_t \equiv 0,$$

$$\dots\dots\dots\dots\dots\dots\dots\dots\dots\dots$$

$$f_0 + \lambda_t f_1 + \lambda_t^2 f_2 + \dots + \lambda_t^t f_t \equiv 0.$$

The determinant of this system of equalities is the Vandermonde determinant $W(\lambda_0, \lambda_1, \dots, \lambda_t) \not\equiv 0$, so there is an equivalent system of identities of the form $f_0 \equiv 0$, $f_1 \equiv 0, \dots, f_t \equiv 0$. Since $f \equiv 0$ is a consequence of this system, we arrive at an equivalent system of identities which are homogeneous with respect to x_1. This allows us to carry out induction with respect to the number of variables in the identity.

Now assume that F is a field of characteristic 0. The method used to obtain an equivalent multilinear system of identities recalles the polarization procedure for a quadratic form. According to the first part of the proof the given identity $f(x_1, x_2, \dots, x_n) \equiv 0$ can be assumed multihomogeneous. If the degree of homogeneity with respect to every variable is equal to 1 then the statement is evident. Therefore we assume that some degree of homogeneity, say, t, with respect to the variable x_1, is not equal to 1. Now consider the following consequence of the identity $f(x_1, \dots, x_n) \equiv 0$

$$g(y,z,x_2,\dots,x_n) = f(y+z,x_2,\dots,x_n)-$$
$$-f(y,x_2,\dots,x_n) - f(z,x_2,\dots x_n) \equiv 0.$$

It is clear that the degree of homogeneity of the identity $g(y,z,x_2,\dots,x_n) \equiv 0$ with respect to the variables y, z is strictly less than the degree of homogeneity of the identity $f \equiv 0$ with respect to x_1. Applying induction with respect to the maximum t of degrees of homogeneity and to the number of variables degree of homogeneity > 1 we can assume that $g(y,z,x_2,\dots,x_n) \equiv 0$ is equivalent to a system of multi–linear identities. Now replace y, z by the variable x_1 in $g \equiv 0$ to obtain

$$g(x_1,x_1,x_2,\dots,x_n) = 2^t f(x_1,x_2,\dots,x_n)-$$
$$-2f(x_1,x_2,\dots,x_n) = (2^t-2)f(x_1,x_2,\dots,x_n) \equiv 0.$$

Since $t > 0$, the coefficient $2^t - 2$ is non–zero and $f(x_1,x_2,\dots,x_n) \equiv 0$ is a consequence of the identity $g(y,z,\dots,x_n) \equiv 0$. □

Associative algebras over a field D which satisfy a non–trivial identity are called PI–algebras (this acronym comes from the phrase: polynomial identity).

3.2. Structure of the free algebras of a variety. We denote by $F_n\langle X\rangle$ the subspace in the free algebra $F\langle X\rangle$ which consists of 0 and of the homogeneous polynomials of degree n. Then

$$F\langle X\rangle = \bigoplus_{n=1}^{\infty} F_n\langle X\rangle.$$

If F is an infinite field, $\mathfrak{V}$ is some variety of algebras over F, and $\mathfrak{V}(X)$ is an ideal of identities of the variety $\mathfrak{V}$ which depend on X, then according to proposition 2 of Subsection 3.1,

$$\mathfrak{V}(X) = \bigoplus_{n=1}^{\infty} [\mathfrak{V}(X) \cap F_n(X)].$$

Hence

$$F(X,\mathfrak{V}) = \bigoplus_{n=1}^{\infty} F_n(X,\mathfrak{V}),$$

where $F_n(X,\mathfrak{V})$ is as $F_n(X)$; however, with respect to the image of the set X in $F(X,\mathfrak{V})$ (this image is also denoted by X). The decomposition of the algebra $F(X,\mathfrak{V})$ into a direct sum of multihomogeneous components is analogous. In the case of a field with zero characteristic, where each identity is equivalent to a system of multilinear ones, the most significant are multilinear elements. The collection of these elements in the algebra $F(X,\mathfrak{V})$ (from the set of variables $\{x_1,\dots,x_n\}$) is denoted by $P_{\mathfrak{V}}(x_1,\dots,x_n)$ ($P(x_1,\dots,x_n)$ for the class of all associative algebras). The space $P_{\mathfrak{V}}(x_1,\dots,x_n)$ is finite–dimensional, its dimension is denoted by $c_n(\mathfrak{V})$. The growth of the sequence $c_n(\mathfrak{V})$ is called the growth of the variety $\mathfrak{V}$. For instance, if $\mathfrak{V}$ is the variety of all associative algebras then $P_{\mathfrak{V}}(x_1,\dots,x_n)$ is the set of all multilinear words in the alphabet $\{x_1,\dots,x_n\}$. It is clear that $c_n(\mathfrak{V}) = n!$. If $\mathfrak{V}$ is the variety of all associative commutative algebras then $c_n(\mathfrak{V}) = 1$, for all n.

We call the growth of a sequence $\{c_n\}$ *polynomial* if there is a real number α such that $c_n \le n^\alpha$, for all n; and it is *exponential* if there is a real number β such that $c_n \le \beta^n$, for all n. If such a β does not exists then the growth is called *super–exponential.* Such is the growth of the variety of all associative algebras. Super–exponential growth is not uncommon for varieties of Lie algebras. In the case of associative algebras the following result is valid.

Theorem. *Let $\mathfrak{V}$ be an arbitrary proper variety of associative algebras over some field F (i.e. $\mathfrak{V}$ differs from the variety of all associative algebras). Then its growth is exponential.*

Lemma 1. *Some non–trivial multilinear identity holds in any proper variety of associative algebras.*

Proof. This is an immediate consequence of a polarization procedure from proposition 2 of Subsection 3.2, if we additionally note that the polarization of a non–zero associative polynomial is also non–zero. □

Lemma 2. *The number of permutations $s = (i_1, i_2, \dots, i_n)$ of the first n natural numbers, such that for some fixed d, $2 \le d \le n$, any set of d numbers* *contains at least one ordered pair, does not exceed the number* $(d-1)^{2n}$.

Proof. Let $N_n = \{1, 2, \dots, n\}$. We fix a permutation s of the same form as in the formulation of lemma and define a partial ordering on N_n, by setting that $i \leq j$, if either $i = j$ or (i, j) is ordered pair in s, i.e. $i = i_k$, $j = j_l$, and $k < l$, $i < j$. Then any set of d elements in N_n contain at least one pair of comparible elements. In this case N_n splits into a non–intersecting union of not more than $d-1$ non–intersecting chains by Dilworth's theorem which states the following (in the finite case): let P be a finite partially ordered set. Then the minimal number of non–intersecting chains which together contain all elements from P is equal to the maximal number of elements in an antichain, that is a set whose all elements are pairwise incomparible. (Cf Section 7.2 in Hall [2].) Therefore any permutation of the considered form is defined by the splitting of the set N_n into a union of $d-1$ non–intersecting chains and by the successive composition of permutations preserving the order inside each chain. The number of splittings into $d-1$ chains is not larger than $(d-1)^n$ (the number of mappings from N_n into N_{d-1}). Further, in each of the n places of the permutation s there occurs one of the elements of one of the $d-1$ chains. Therefore the number of ways of composition does not exceed $(d-1)^n$. As a result we obtain the required number. □

Proof of the theorem. We assume that some non–trivial identity of degree d is fulfilled in the variety $\mathfrak{V}$. As it was noted above the polarization procedure allows to find in $\mathfrak{V}$ a non–trivial multilinear identity of degree d. We write it in the form

$$(1) \qquad y_1 y_2 \dots y_d \equiv \sum_{\sigma \neq 1} \lambda_\sigma y_{\sigma(1)} y_{\sigma(d)} \dots y_{\sigma(n)},$$

where σ is a permutation in $\{1, 2, \dots, d\}$, $\lambda_\sigma \in F$. In order to prove the theorem we need to show that the space $P_{\mathfrak{V}}(x_1, x_2, \dots, x_n)$ is generated by monomials $x_{\tau(1)} x_{\tau(2)} \dots x_{\tau(n)}$ for which the permutation $\sigma = (\tau(1), \tau(2), \dots, \tau(n))$ satisfies the conditions of lemma 2. Let u be the lexicographically smallest multilinear associative monomial which does not satisfy the conditions mentioned. Then u contains a subword of the form

$$u' = (x_{\alpha_1} \dots x_{\beta_1})(x_{\alpha_2} \dots x_{\beta_2}) \dots (x_{\alpha_{d-1}} \dots x_{\beta_{d-1}}) x_{\alpha_d},$$

for which $\alpha_1 > \alpha_2 > \ldots > \alpha_{d-1} > \alpha_d$. Substitute in (1) $y_i = x_{\alpha_i} \ldots x_{\beta_i}$, $i = 1, 2, \ldots, d-1$, $y_d = x_{\alpha_d}$; then we obtain that u

$$u = \sum_{\sigma \neq 1} \lambda_\sigma v_\sigma, \quad \lambda_\sigma \in F,$$

in the space $P_{\mathfrak{V}}(x_1, \ldots, x_n)$ has a representation, where every monomial v_σ is lexicographically less than u. A contradiction. □

Corollary (A. Regev). *Let A and B be PI–algebras over the field F. Then $T = A \otimes_F B$ is also a PI–algebra.*

Proof. Let $\mathfrak{V}$ be a variety of algebras over the field F containing both A and B. We have $\mathfrak{V}(A) = \mathfrak{V}(B) = 0$. Let V be the ideal of identities of the variety $\mathfrak{V}$. Suppose that V contains a non–trivial polynomial of some degree $d \geq 2$. Let $m_1(x), \ldots, m_q(x)$ be the monomials which constitute a basis in $P(x_1, \ldots, x_n)$ modulo the ideal V, $q \leq (d-1)^{2n}$. For any permutation $\sigma \in S_n$ we obtain that for suitable $\lambda_1(\sigma), \ldots, \lambda_q(\sigma) \in F$

$$x_{\sigma(1)} \ldots x_{\sigma(n)} \equiv \sum_{i=1}^{q} \lambda_i(\sigma) m_i(x)$$

is an identity in A and B. Let us find a polynomial

$$f(x_1, \ldots, x_n) = \sum_{\sigma \in S_n} \gamma_\sigma x_{\sigma(1)} \ldots x_{\sigma(n)},$$

where $\gamma_\sigma \in F$ such that $f \equiv 0$ is an identity in $A \otimes_F B$. We have

$$f(a_1 \otimes b_1, \ldots, a_n \otimes b_n) = \sum_{\sigma \in S_n} \gamma_\sigma a_{\sigma(1)} \ldots a_{\sigma(n)} \otimes b_{\sigma(1)} \ldots b_{\sigma(n)} =$$

$$= \sum_{\sigma \in S_n} \sum_{i=1}^{q} \sum_{j=1}^{q} \lambda_i(\sigma) \lambda_j(\sigma) \gamma_\sigma m_i(a) \otimes m_j(b).$$

For suffciently large n, we have $q^2 \leq (d-1)^{4n} < n!$, and, therefore, the system of linear equations with respect to the $n!$ unknowns γ_σ

$$\sum_{\sigma \in S_n} \lambda_i(\sigma) \lambda_j(\sigma) \gamma_\sigma = 0 \quad (1 \leq i, j \leq q)$$

has a non–trivial solution $\gamma_\sigma = \gamma_\sigma^0$, and the expression of the form

$$f(x_1, \dots, x_n) = \sum_{\sigma \in S_n} \gamma_\sigma^0 x_{\sigma(1)} \dots x_{\sigma(n)} \equiv 0.$$

is then a non–trivial identity in $A \otimes_F B$. □

3.3 Primitive PI–algebras. One of the basic results on associative PI–algebras is the Kaplansky theorem on the structure of primitive PI–algebras. The key ingredient of this theorem is

Lemma. *No identity of degree less than $2n$ holds in the matrix algebra $M_n(F)$.*

Proof. According to Subsection 3.1, if in some algebra a non–trivial identity of degree d holds then there also is a multilinear identity of degree not larger than d. We consider the multilinear polynomial

$$f = x_1 \dots x_d + \sum_{\substack{\sigma \in S_d \\ \sigma \neq 1}} \alpha_\sigma x_{\sigma(1)} \dots x_{\sigma(d)},$$

where $\alpha_\sigma \in F$ and $d < 2n$.

We shall assign the following values to the variables $x_1, \dots, x_d$: $x_1 = e_{11}$, $x_2 = e_{12}$, $x_3 = e_{22}$, $x_4 = e_{23}, \dots$, where e_{ij} is a usual elementary matrix. As $d \leq 2n - 1$, such a selection of values for $x_1, \dots, x_d$ is possible; moreover, for any permutation which differs from the identity we have $x_{\sigma(1)} \dots x_{\sigma(d)} = 0$. Hence

$$f(e_{11}, e_{12}, \dots) = e_{11} e_{12} e_{22} \dots \neq 0,$$

and, consequently, $f \equiv 0$ is not an identity in $M_n(F)$. Because any non–trivial multilinear identity of degree d is reducible to the form $f \equiv 0$, the statement of lemma is proved. □

Theorem. *Let A be a primitive algebra which satisfies a polynomial identity of degree d. Then A is a finite–dimensional simple algebra over its centre Z, and its dimension over Z does not exceed $[d/2]^2$, where $[d/2]$ is the integer part of the number $d/2$.*

Proof. Since A is primitive, according to the density theorem (see, § 2.1), either A is isomorphic to $M_n(\Delta)$, where Δ is some skew field, or A, for any $m > 0$, contains a subalgebra which homomorphically maps onto $M_m(\Delta)$. The latter is impossible because of the lemma, as in that case the identity of degree d, for any m, would be satisfied by the algebra $M_m(F)$ as a homomorphic image of a subalgebra of the algebra A. Thus, it remains only that $A = M_m(\Delta)$.

Let K be a maximal subfield in Δ. As in § 3.2 the ring $\Delta \otimes_Z K$ is a dense ring of linear operators in the vector space Δ over K. The argument used just now can be transferred to $\Delta \otimes_Z K$, as this ring satisfies all multilinear identities of the ring Δ and, hence, of the ring A. Therefore $\Delta \otimes_Z K \cong M_m(K)$, for some natural m. Hence $A \otimes_Z K \cong M_n(\Delta) \otimes_Z K \cong M_{mn}(K)$. Then $d \geq 2mn$ and $mn \leq [d/2]$. But

$$\dim_K(A \otimes_Z K) = \dim_Z A = (mn)^2.$$

Consequently, $\dim_Z A \leq [d/2]^2$. Since $A = M_n(\Delta)$, A is a simple algebra, and the theorem is proved. □

Corollary 1. *If A is a PI–algebra, and $J(A)$ is its Jacobson radical then the quotient–algebra $A/J(A)$ is isomorphic to a subalgebra of a matrix algebra of suitable order over a commutaive algebra. If A has no non–zero nil ideals then the same conclusion is also valid for A.*

Proof. We consider the case of an algebra B which is semi–simple in the sense of Jacobson (i.e. with $J(B) = \{0\}$). Since $\{0\} = J(B)$ is the intersection of the primitive ideals $\{0\} = \bigcap_\alpha P_\alpha$ and $B_\alpha = B/P_\alpha$ is a primitive PI–algebra, the mapping $\phi : B \to \prod_\alpha B_\alpha$ given by $(\phi(b))(\alpha) = b + P_\alpha$, is an isomorphic imbedding in a direct product of primitive algebras with an identity of fixed degree d. Each of the algebras B_α is an algebra of dimension not higher than some k, which depends only on d, over its centre Z_α. Using the regular representation, i.e. $x \mapsto l(x) : a \mapsto xa$, we can imbed every ring B_α in $M_k(Z_\alpha)$, and $\prod_\alpha B_\alpha$ in $M_k(\prod_\alpha Z_\alpha)$. Because $Z = \prod_\alpha Z_\alpha$ is a commutative ring, we obtain an imbedding for B in $M_k(Z)$.

The possibility of the required imbedding for $B = A/J(A)$ in $M_k(Z)$ follows directly from what was proved. If A has no nil ideals then $B = A[t]$, according to § 3.1, is a semi–simple algebra, and is also a PI–algebra, thus it allows an imbedding in $M_k(Z)$, as it was above. Obviously then, an imbedding of A in $M_k(Z)$ can also be found. □

Denote by $\mathcal{S}_n(x_1, \ldots, x_n)$ the non–commutative polynomial

$$\mathcal{S}_n(x_1, \ldots, x_n) = \sum_{\sigma \in S_n} \operatorname{sign}(\sigma) x_{\sigma(1)} \ldots x_{\sigma(n)}.$$

An identity of the form $\mathcal{S}_n(x_1, \ldots, x_n) \equiv 0$ is called a *standard identity of degree n*. Clearly, in any algebra A generated by m elements over a commutative ring Z all standard identities of degree higher than m are fulfilled. Therefore the following is valid

Corollary 2. *If A is a PI–algebra which has no non–zero nil ideals, then A satisfies a standard identity.* □

∗ The Grassman algebra $\Lambda(V)$ of an infinite–dimensional space V satisfies the non–trivial identity $[x, [y, z]] \equiv 0$, but it does not satisfy any standard identity. ∗

3.4. Radical of a PI–algebra. In § 3.1 various radicals of associative rings were presented and among them there were the upper nil radical (the maximal two–sided nil ideal), the locally nilpotent radical and the lower nil radical. Here we shall show that in the case of PI–algebras all these radicals coincide.

Theorem 1. *Any nil PI–algebra is locally nilpotent.*

Proof. Let A be such an algebra and N its maximal locally nilpotent ideal. Our aim is to show that $A = N$. Assume that $B = A/N \neq \{0\}$. Then B has no two–sided locally nilpotent ideals. We show that a one–sided such ideal also does not exist. Indeed, if $\mathcal{I}$ is a locally nilpotent right ideal in B then we can form $J = B\mathcal{I} + \mathcal{I}$. Let $S = \{x_i\}$ be a finite subset in J. Then $x_i = \sum a_{ij} u_{ij} + u_i$, where u_{ij}, $u_i \in \mathcal{I}$. Let $u_{ijk} = u_i a_{jk}$, $u_{ijkl} = u_{ij} a_{kl}$. The set $T = \{u_i, u_{ij}, u_{ijk}, u_{ijkl}\}$ is a finite subset of elements from $\mathcal{I}$. Consequently, there exists an n such that a product of any n elements from T is equal to zero. Now consider a product of any n elements from S. Such product is a sum of terms each of which contains a product of n elements from T. Such product is equal to zero and, consequently, J is a two–sided locally nilpotent ideal.

Now let a be a non–zero element from B and such that $a^2 = 0$. We consider the right ideal aB. If $aB = \{0\}$ then Fa, F the ground field, is a locally nilpotent ideal which contradicts the assumption. So $aB \neq \{0\}$. Consider a multilinear identity $f(x_1, x_2, \ldots, x_m) \equiv 0$ in the algebra B. It can be written in the form $f_1(x_2, \ldots, x_m)x_1 + f_2(x_1, \ldots, x_m) \equiv 0$, where the variable x_1 does not occur in the last place in the monomials of f_2. Reordering the variables it can be assumed that $f_1(x_2, \ldots, x_m) \not\equiv 0$. Replace $x_1, x_2, \ldots, x_m$ by $a_1 = a$, $a_2, \ldots, a_m \in aB$. Then from $a^2 = 0$ it follows that $f_2(a_1, a_2, \ldots, a_m) = 0$. Hence $f_1(a_2, \ldots, a_n)a = 0$. Let D be the left annihilator of the subalgebra aB in this subalgebra. We see that aB/D satisfies the identity of degree $m-1$ of the form $f_1(x_2, \ldots x_m) \equiv 0$. Applying induction with respect to m the conclusion can be drawn that aB/D is a locally nilpotent algebra. We have arrived at a contradiction. □

Theorem 2. *The upper locally nilpotent and the lower nilpotent radicals of an arbitrary PI–algebra coincide.*

Proof. The lower nil radical N of the algebra A has the form $N = \bigcup_\alpha N_\alpha$, where α runs over some initial interval of ordinals; moreover, $N_0 = \{0\}$, $N_{\alpha+1}/N_\alpha$ is a sum of all nilpotent ideals in A/N_α for a non–limit ordinal α and $N_\beta = \bigcup_{\alpha<\beta} N_\alpha$ for a limit ordinal β. We show that all the radicals mentioned are equal to N_2 (it is clear that $N_2 = N_1$ is possible). It is sufficient to show that if in A an identity of degree d is fulfilled then $B^{[d/2]} \subset N_1$, where B is an arbitrary locally nilpotent subalgebra.

To start with assume that B is nilpotent. Let n be an arbitrary positive integer; we define

$$U_{2i-1} = B^{n-1}AB^{i-1}, \quad U_{2i} = B^{n-i}AB^i, \quad i = 1, 2, \ldots, n.$$

Then if $h \leq 2n$, $U_1U_2 \ldots U_h = (B^{n-1}A)^h B^{[h/2]}$. Moreover, if $j > k$ then $U_jU_k \subset AB^nA$. Let now n be the smallest integer such that the ideal AB^nA is nilpotent (such an n exists because the subalgebra B is nilpotent). Suppose that $n > [d/2]$. Then $2n > d$ and we can take $h = d$ in the formulas indicated above. In this case it should be noted that if $(i_1, i_2, \ldots, i_d)$ is a permutation of the numbers $(1, 2, \ldots, d)$ which is not the identity then $U_{i_1} \ldots U_{i_d} \subset AB^nA$. On the other hand, we have in A some multilinear identity of degree d, from which the following relation is obtained

$$u_1u_2 \ldots u_d = \sum \beta_{i_1 \ldots i_d} u_{i_1} \ldots u_{i_d},$$

where $u_i \in U_i$. From this we get that

$$(B^{n-1}A)^d \subset AB^nA,$$

and therefore $(AB^{n-1}A)^{d+1} \subset AB^nA$. This implies nilpotency for $AB^{n-1}A$ contradicting the minimality of the number n. So it is proved that $AB^{[d/2]}A$ is nilpotent and

$$AB^{[d/2]}A + AB^{[d/2]} + B^{[d/2]}A + B^{[d/2]}$$

is also nilpotent. Since the latter subalgebra is an ideal, we see that $B^{[d/2]} \subset N_1$.

Now let B be a arbitrary nil subalgebra in A. We know that B is locally nilpotent. Consequently, if $\{b_1, \dots, b_r\}$ is any finite subset consisting of $r \geq [d/2]$ elements of the algebra B, and B_0 is the subalgebra generated by the set $\{b_1, \dots, b_r\}$ then B_0 is nilpotent. From what has been proved just now follows that $B_0^{[d/2]} \subset N_1$. In particular, $b_1 \dots b_{[d/2]} \in N_1$. It is clear that then $B^{[d/2]} \subset N_1$. The theorem is proved. □

∗ For a long time the question of when the Jacobson radical of a finitely generated PI–algebra is nilpotent was not solved. Due to the efforts of several authors this problem was recently solved for the case of algebras over an arbitrary noetherian commutative ring (see, Bokut'–L'vov–Kharchenko [1]). The ring $F[[t]]$ of formal power series of one variable t over the field F is an example of a PI–algebra whose Jacobson radical, the set of series without constant term, is not nilpotent. ∗

§ 4. Central polynomials for matrix algebras

4.1. Definition. A non–commutative polynomial $f(x_1, \dots, x_n) \in A(X)$ with coefficients from the field F, the values of which are scalar on the matrix algebra $M_n(F)$ but are not all zero, is called a *central polynomial* of the algebra $M_n(F)$. The existence of such polynomials in the case of a finite field F was proved by V. N. Latyshev and A. L. Shmel'kin. In the case of an infinite field F such polynomials were indicated by E. Formanek and Yu. P. Razmyslov. We shall present the arguments of Formanek which permit the construction of central polynomials.

We shall construct a central polynomial for the algebra $M_n(F)$ with integer coeficients. Consider the ring $\mathbf{Z}[X_1, \dots, X_n]$ and denote by $\sigma_1, \dots, \sigma_n$

the elementary symmetric polynomials in $X_1, \dots, X_n$, i.e. $\sigma_1 = X_1 + \dots + X_n, \dots, \sigma_n = X_1 \dots X_n$. Let $g \in \mathbf{Z}[X_1, \dots, X_n]$ be an arbitrary symmetric polynomial. Then $g(X_1, \dots, X_n) = h(\sigma_1, \dots, \sigma_n) \in \mathbf{Z}[\sigma_1, \dots, \sigma_n]$. The polynomial h in such a representation is unique and using it the polynomial g defines a mapping $G : M_n(F) \to F$ which associates to the matrix $a \in M_n(F)$ the number $h(\operatorname{tr} a, \dots, \det a)$, where $\operatorname{tr} a, \dots, \det a$ is the sequence of coefficients of the characteristic polynomial

$$\chi_a(\lambda) = \det(\lambda e - a) = \lambda^n - (\operatorname{tr} a)\lambda^{n-1} + \dots + (-1)^n \det a$$

of the matrix a.

The mapping G which corresponds to the symmetric polynomial

$$g = d = \prod_{i<j} (X_i - X_j)^2,$$

is denoted by D and is called the *discriminant mapping*. If F is an infinite field then the mappings $a \mapsto \operatorname{tr} a, \dots a \mapsto \det a$ are polynomial functions on the space $M_n(F)$. Then G, which is defined by the formula $a \mapsto h(\operatorname{tr} a, \dots, \det a)$, is also a polynomial function. If F is an algebraically closed field and $\rho_1, \dots, \rho_n$ are the characteristic roots of a then $\operatorname{tr} a = \rho_1 + \dots + \rho_n$, $\det a = \rho_1 \dots \rho_n$ so that $G(a) = h(\sum \rho_i, \dots, \prod \rho_i) = g(\rho_1, \dots, \rho_n)$. Therefore G coincides with the mapping

$$a \mapsto g(\rho_1, \dots, \rho_n),$$

which is defined by the symmetric polynomial g. In particular, if $g = d$ then the corresponding mapping D has the form

$$a \to \prod_{i<j} (\rho_i - \rho_j)^2.$$

It is clear that $D(a)$ differs from zero only in the case when the characterisitc roots of the matrix a are pairwise different. This occurs only when a is similar to a diagonal matrix with pairwise different elements on the diagonal. The set of such matrices is open in the Zariski topology (see, § 1.5).

Now we consider the ring of integer polynomials $\mathbf{Z}[X_1, \dots, X_n, X_{n+1}]$ and the free associative algebra over F of the form $A(x, y_1, \dots, y_n)$. If the polynomial $f \in \mathbf{Z}[X_1, \dots, X_{n+1}]$ is given by

$$f = \sum_{(\nu)} \alpha_{(\nu)} X_1^{\nu_1} X_2^{\nu_2} \dots X_{n+1}^{\nu_{n+1}}, \quad (\nu) = (\nu_1, \nu_2, \dots, \nu_{n+1}),$$

then we set

$$p_f = \sum_{(\nu)} \alpha_{(\nu)} x^{\nu_1} y_1 x^{\nu_2} y_2 \dots x^{\nu_n} y_n x^{\nu_{n+1}} \in A(x, y_1, \dots, y_n).$$

It should be noted that the mapping $f \mapsto p_f$ is additive and that p_f is linear with respect to any y_i, $i = 1, \dots, n$. Perform the substitution

$$x \mapsto \sum_{i=1}^{n} \rho_i e_{ii}, \quad y_1 \mapsto e_{i_1 j_1}, \dots, y_n \mapsto e_{i_n j_n}.$$

Then

$$\begin{aligned} & x^{\nu_1} y_1 x^{\nu_2} y_2 \dots x^{\nu_n} y x^{\nu_{n+1}} \mapsto \\ \mapsto & \rho_{i_1}^{\nu_1} \rho_{i_2}^{\nu_2} \dots \rho_{i_n}^{\nu_n} \rho_{j_n}^{\nu_{n+1}} e_{i_1 j_1} e_{i_2 j_2} \dots e_{i_n j_n} = \\ = & \delta_{j_1 i_2} \delta_{j_2 i_3} \dots \delta_{j_{n-1} i_n} \rho_{i_1}^{\nu_1} \dots \rho_{i_n}^{\nu_n} \rho_{j_n}^{\nu_{n+1}} e_{i_1 j_n}, \end{aligned}$$

from which follows

$$\begin{aligned} & p_f \left(\sum_{i=1}^{n} \rho_i e_{ii}, e_{i_1 j_1}, \dots, e_{i_n j_n} \right) = \\ = & \delta_{j_1 i_2} \delta_{j_2 i_3} \dots \delta_{j_{n-1} i_n} f(\rho_{i_1}, \dots, \rho_{i_n}, \rho_{j_n}) e_{i_1 j_n}. \end{aligned}$$

The obtained value is equal to zero for all sequences $(e_{i_1 j_1}, \dots, e_{i_n j_n})$ with the exception of

$$(e_{i_1 i_2}, e_{i_2 i_3}, \dots, e_{i_{n-1} i_n}, e_{i_n j_n}),$$

and in this case

$$p_f \left(\sum_{i=1}^{n} \rho_i e_{ii}, e_{i_1 i_2}, \dots, e_{i_{n-1} i_n}, e_{i_n j_n} \right) = f(\rho_{i_1}, \dots, \rho_{i_n}, \rho_{j_n}) e_{i_1 j_n}.$$

We should note that as there are only n possibilities of selecting indices two numbers in the sequence $(i_1, i_2, \dots, i_n, j_n)$ must be equal.

Now assume that f satisfies the conditions:

(i) $f(X_1, \dots, X_{n+1})$ is divisible by any $X_i - X_j$, $i \neq j$ with the exception of $X_1 - X_{n+1}$;

(ii) $g(X_1, \dots, X_n) = f(X_1, \dots, X_n, X_1)$ is symmetric with respect to $X_1, \dots, X_n$.

Then $p_f(\sum_i \rho_i e_{ii}, e_{i_1j_1}, \dots, e_{i_nj_n}) = 0$ for all sets $(e_{i_1j_1}, \dots, e_{i_nj_n})$ with the exception of $(e_{i_1i_2}, e_{i_2i_3}, \dots, e_{i_ni_1})$ for different i_j and then

$$p_f(\sum_{i=1}^{n} \rho_i e_{ii}, e_{i_1i_2}, \dots, e_{i_ni_1} = g(\rho_1, \dots, \rho_n) e_{i_1i_1}.$$

Now consider the polynomial

$$q_f(x, y_1, \dots, y_n) = \sum_{i=0}^{n-1} p_f(x, y_{i+1}, \dots, y_{i+n}),$$

where the indices are taken modulo n. Then $q_f(\sum_i \rho_i e_{ii}, e_{i_1j_1}, \dots, e_{i_nj_n}) = 0$ for all choices $(e_{i_1j_1}, \dots, e_{i_nj_n})$ with the exception of $(e_{i_1i_2}, e_{i_2i_3}, \dots, e_{i_ni_1})$ for different i_j, which in this case

(1) $$q_f\left(\sum_{i=1}^{n} \rho_i e_{ii}, e_{i_1i_2}, e_{i_2i_3}, \dots, e_{i_ni_1}\right) = g(\rho_1, \dots, \rho_n) e,$$

where e is the identity matrix in $M_n(F)$.

The simplest choice of a polynomial f with the properties (i) and (ii) has the form

$$c = \prod_{i=2}^{n} (X_1 - X_i)(X_{n+1} - X_i) \prod_{\substack{i,j=2 \\ i<j}}^{n} (X_i - X_j)^2.$$

In this case the polynomial g takes the form

$$d(X_1, \dots, X_n) = c(X_1, \dots, X_n, X_1) = \prod_{1 \le i < j \le n} (X_i - X_j)^2 \neq 0;$$

for such $g = c$ there is an additional property:

(*iii*) for any field F there is an $a \in M_n(F)$ such that $G(a) \neq 0$.

If F is infinite then we can take as a a diagonal matrix with different eigenvalues. If F is finite then we use the fact that in $F[\lambda]$ there exists an irreducible polynomial $\phi(\lambda)$ of degree n with the leading coefficient 1 (see, § 1.2). There exists a matrix a for which $\phi(\lambda)$ is a characteristic polynomial. Since $\phi(\lambda)$ is separable, a has n different characteristic roots in the algebraic closure $\bar{F}$ of the field F. Because the mapping G does not change under extension of the field F it is clear that $G(a) \neq 0$.

4.2. Theorem. *For an $f \in \mathbf{Z}[X_1, \dots, X_{n+1}]$ which satisfies (i), (ii), (iii), the polynomial $q_f(x, y_1, \dots, y_n)$ defined above is a central polynomial for $M_n(F)$. There exists $a \in M_n(F)$ such that $G(a)$ is not equal to zero and there are $b_1, \dots, b_n \in M_n(F)$ such that $q_f(a, b_1, \dots, b_n) \neq 0$. Further, if $l(X_1, \dots, X_n) \in \mathbf{Z}[X_1, \dots, X_n]$ and l is a symmetric polynomial then*

$$q_{lf}(a, b_1, \dots, b_n) = L(a) q_f(a, b_1, \dots, b_n) \tag{2}$$

for any $a, b_1, \dots, b_n \in M_n(F)$, where L is defined as before from the polynomial $l(X_1, \dots, X_n)$.

Proof. To start with assume that the field F is algebraically closed.

We shall show that $[q_f(x, y_1, \dots, y_n), z] \equiv 0$ is an identity in $M_n(F)$ and that (2) is satisfied. This is equivalent to the proof of the fact that the mappings

$$(a, b_1, \dots, b_n, c) \mapsto [q_f(a, b_1, \dots, b_n), c]$$

and

$$(a, b_1, \dots, b_n) \mapsto q_{lf}(a, b_1, \dots, b_n) - L(a) q_f(a, b_1, \dots, b_n)$$

are zero. Since $q_f(x, y_1, \dots, y_n)$ and $[q_f(x, y_1, \dots, y_n), z]$ are linear with respect to $y_1, \dots, y_n, z$, it is sufficient to check our statement for the values $b_1, \dots, b_n, z$ which are equal to the elements of some basis of the algebra $M_n(F)$ over F. For fixed b_i and c the mappings considered are polynomial with respect to a. In other words, there exist the polynomials P_{ij}, Q_{ij} of n^2 variables ξ_{ij} such that the mappings considered have the form

$$\left(a = \sum a_{ij}e_{ij}; b_1, \dots, b_n, c\right) \mapsto \sum_{i,j} P_{ij}(a_{11}, a_{12}, \dots)e_{ij},$$

$$(a, b_1, \dots, b_n, c) \mapsto \sum_{i,j} Q_{ij}(a_{11}, a_{12}, \dots)e_{ij}.$$

So we should show that the polynomial functions $(a, b_1, \dots, b_n, c) \mapsto P_{ij}(a)$, $(a, b_1, \dots, b_n) \mapsto Q_{ij}(a)$ are zero for the fixed selection of elements $b_1, \dots, b_n$, c in the basis considered. It is sufficient to justify this statement for all a in some set of the algebra $M_n(F)$ which is Zariski open. Let this set consist of those a for which $G(a) \neq 0$. Since $g(X_1, \dots, X_n) = f(X_1, \dots, X_n, X_1)$ is divisible by any difference $X_i - X_j$, $i \neq j$, it follows from the condition $G(a) \neq 0$ that all characterisitc roots of the matrix a are pairwise different. Consequently, a is similar to a diagonal matrix, so by using automorphisms of the algebra $M_n(F)$ we can assume that a is diagonal: $a = \sum \rho_i e_{ii}$ and the basis considered is the set of elementary matrices $\{e_{ij}\}$. If we observe that $G(a) = g(\rho_1, \dots, \rho_n)$ then we see that our result is based on (1) and on the previous arguments. So the mappings considered are zero if F is algebraically closed.

If F is an arbitrary field and $\bar{F}$ is its algebraic closure then $F \subset \bar{F}$ and there exists an imbedding $M_n(F)$ in $M_n(\bar{F})$. The proof is valid for $M_n(\bar{F})$ and hence, for $M_n(F)$. Besides, according to (*iii*) there exists the matrix a with the property mentioned. According to (1) there are $b_1, \dots, b_n \in M_n(\bar{F})$ such that $q_f(a, b_1, \dots, b_n) \neq 0$. According to the linearity property of the polynomial q_f with respect to $y_1, \dots, y_n$ and the fact that any basis of the algebra $M_n(F)$ is also a basis for $M_n(\bar{F})$, we can select $b_1, \dots, b_n$ also in $M_n(F)$ and in such a way that $q_f(a, b_1, \dots, b_n) \neq 0$. The theorem is proved. □

It should be noted that a central polynomial for $M_n(F)$ is an identity in $M_{n-1}(F)$.

Indeed, $M_{n-1}(F) \subset M_n(F)$ is the subalgebra with the n–th row and the n–th column consisting of zero. If $q(x_1, \dots, x_m)$ is a central polynomial then $q(a_1, \dots, a_m)$ is a scalar matrix and for $a_1, \dots, a_m \in M_{n-1}(F)$ we have $q(a_1, \dots, a_m) \in M_{n-1}(F)$, i.e. this value is equal to zero.

In Chapter 3 the Goldie theorem on localization of noetherian semi–primary rings was proved. According to this theorem any such ring is a (left, right) order in a classical semi–prime ring; in the case of a prime noetherian ring, it is an order in a classical simple ring, i.e. in a ring of matrices of some order over field, and a noetherian ring without zero divisors has a classical field of fractions.

In this theorem the noetherian condition is necessary to check that the Ore condition holds. In a similar way the existence of an identity can be used. Originally, it was noticed by Amitsur that a PI–ring without zero divisors satisfies the Ore condition. Later, following the scheme of the proof of the Goldie theorem, E. Posner and other authors have proved that a primary PI–ring R possesses a classical ring of fractions $\mathrm{Fr}(R)$, moreover $\mathrm{Fr}(R)$ is a ring of matrices over a field which is finite–dimensional over its own centre, and the identities of the rings R and $\mathrm{Fr}(R)$ coincide.

The use of central polynomials allows to simplify the proof of the Posner theorem and to make the structure of the ring of fractions for a primary PI–ring more precise.

4.3. Central ring of fractions of a prime ring. We call the ring R_S the *central ring of fractions* of the ring R, where S is the set of regular elements of the centre C of the ring R (see, §§ 1.3, 3.3). We should note that if R is prime then S coincides with the centre C of the ring R. Indeed, if $c \neq 0$, $c \in C$, then cR is a two–sided ideal which is annihilated by the two–sided ideal RaR, where a is an element from R such that $ac = 0$. From the primality of the ring follows that $RaR = \{0\}$, i.e. the element c has no zero divisors, i.e. is regular.

The second remark is that from the primality of the ring R with centre C follows that the central ring of fractions R_C is also prime. In fact, it is sufficient to show that if $aR_Cb = \{0\}$ then either $a = 0$, or $b = 0$. Indeed, let $(s^{-1}x)(t^{-1}y)(u^{-1}z) = 0$, for all $t^{-1}y \in R_C$. It follows from this that $xyz = 0$, for all $y, z \in R$, so either $x = 0$, or $z = 0$, because of the primality of the ring R. So either $s^{-1}x = 0$, or $u^{-1}z = 0$, as required.

The third remark is that the centre of the ring R_C is $\mathrm{Fr}(C)$, i.e. it is simply the ring of fractions of C. Indeed, it is evident that the elements of the form $a^{-1}b$, where $0 \neq a$, $b \in C$, are central in R_C. Let $c^{-1}a$ be a central element in R_C. Then $(d^{-1}b)(c^{-1}a) = (c^{-1}a)(d^{-1}b)$, i.e. $(cd)^{-1}ba = (dc)^{-1}ab$. Multiplying by $cd = dc$ we obtain $ab = ba$, for any $b \in R$. Therefore, a is a central element, and our statement is valid.

Now consider the case when R is a prime PI–ring.

Proposition 1. *Let R be a prime F–algebra, where F is some field, C its centre. The identity $f(x_1,\dots,x_n)\equiv 0$ holds in R iff it holds in R_C.*

Proof. Since R is canonically imbedded in R_C, it is sufficient to prove that the identity of the algebra R holds in R_C. We write

$$f(x_1,\dots,x_m)=\sum_{j=0}^{d} f_j(x_1,\dots,x_m),$$

where f_j is homogeneous of degree j. We should note that if m elements of the ring R_C are given then they can be reduced to a common denominator c. Assume we want to verify the following: $f(c^{-1}a_1,\dots,c^{-1}a_m)=0$. Then $f(ca_1,\dots,ca_m)=0$ and $\sum_{j=0}^{d} c^j f_j(a_1,\dots,a_m)=0$. In this equality we replace c by $1,c,\dots,c^d$. Then we obtain the system of equations

$$\begin{gathered}
f_0+f_1+\dots+f_d=0,\\
f_0+cf_1+\dots+c^d f_d=0,\\
\dots\dots\dots\dots\dots\dots\dots\dots\\
f_0+c^d f_1+\dots+c^{d^2} f_d=0,
\end{gathered}$$

where $f_j=f_j(a_1,\dots,a_m)$. The determinant of the given system is the Vandermonde determinant $W(1,c,\dots,c^d)=c^e(1-ch(c))$, where $e=(d^3-d)/6$, and h is a polynomial with integer coefficients. Solving this system of equations with respect to $f_0,\dots,f_d$ we obtain

$$c^e(1-ch(c))f_j=0,\quad 0\le j\le d.$$

Hence $c^{-1}f_j=h(c)f_j$. By applying this equality several times we obtain

$$c^{-j}f_j=h(c)^j f_j.$$

Then

$$f(c^{-1}a_1, \dots, c^{-1}a_m) = \sum_{j=0}^{d} c^{-j} f_j(a_1, \dots, a_m) =$$

$$= \sum_{j=0}^{d} h(c)^j f = \sum_{j=0}^{d} f_j(h(c)a_1, \dots, h(c)a_m) =$$

$$= f(h(c)a_1, \dots, a_m) = 0.$$

The proposition is proved. □

Proposition 2. *Let R be a prime ring, C its centre, $\mathcal{I}$ some non-zero ideal in R. If R is a PI-ring then $\mathcal{I}$ has a non-zero intersection with C.*

Proof. The ring R has no nilpotent ideals. Since R is a PI–ring, it has no nil ideals (see Subsection 3.4). We consider the ring of polynomials $R[X]$ with coefficients from R. According to § 3.1 this ring is Jacobson semisimple. As before $R[X]$ is a PI–algebra. Since $R[X]$ is a subdirect product of primitive PI–algebras of bounded degree, according to Kaplanski theorem (Subsection 3.3) $R[X]$ is a subdirect product of matrix rings of bounded degree over their centre. For any primitive ideal P of the ring $R[X]$ either $P+\mathcal{I}[X] = R[X]$, or $\mathcal{I}[X] \subset P$. Since the intersection of all primitive ideals is zero, there are ideals, for which $\mathcal{I}[X] + P = R[X]$. We select among these ideals such P_0 for which the degree of the matrix algebra $R[X]/P_0$ is the least. Let q be a central polynomial for the algebra $M_{n_0}(F)$ with a constant term equal to zero. Then for any $a_1, \dots, a_m \in R$ and any primitive ideal P of degree not higher than n_0 the element $q(a_1, \dots, a_m) + P$ lies in the centre of the algebra $R[X]/P_0$ and is not equal to zero for suitable set $a_1^0, \dots, a_m^0$. Because $\mathcal{I}[X] + P_0 = R[X]$, such a choice can be taken from $\mathcal{I}[X]$.

In this case for any P either the degree of the ring $R[X]/P$ does not exceed n_0, and then $q(a_1^0, \dots, a_m^0)+P/P$ lies in the centre of this algebra, or the degree of $R[X]/P$ is higher than n_0, then $P \subset \mathcal{I}[X]$ and again $q(a_1^0, \dots, a_m^0)+P/P$ lies in the centre of this algebra. As a result $q(a_1^0, \dots, a_m^0)$ is a central non–zero element from $\mathcal{I}[X]$.

Returning to R we should note that if C is the centre in R then $C[X]$ is the centre in $R[X]$. Further, $\mathcal{I}[X] \cap C[X] = (\mathcal{I} \cap C)[X]$. Since $\mathcal{I}[X] \cap C[X] \neq \{0\}$, we have also $\mathcal{I} \cap C \neq \{0\}$. □

Corollary. *Let R be a prime PI–ring with a field as a centre. Then R is a simple algebra.* □

Now it is not difficult to prove the main result.

4.4. Theorem. *Let R be a prime PI–algebra with centre C, $R_0 = R_C$ its central ring of fractions. Then:*

(1) R_0 is a finite–dimensional simple algebra over its centre F, moreover F is the field of fractions of C;

(2) R_0 is a left and right ring of fractions of R;

(3) The identities of the algebras R and R_0 coincide.

Proof. It was noted above that R_0 is a prime PI–algebra, moreover its centre is the field of fractions F for C. According to the corollary R_0 is a simple algebra. In this case R_0 is primitive and according to the Kaplansky theorem R_0 is finite–dimensional over F. From the finiteness of dimension of the algebra R_0 follows that this algebra is artinian, so it is a matrix ring over field. It is clear then that any regular element in R_0 is invertible. Any element from R_0 has the form $c^{-1}a = ac^{-1}$, $c \in C$, $a \in R$. Thus, any regular element from R is also regular in R_0, consequently, it is also invertible in R_0. It is clear that (2) is fulfilled. The statement (3) is proved in the proposition 1 of Subsection 4.3. □

∗ It should be noted that the given theorem cannot be extended to the case of semi–prime PI–rings (there are counterexamples of Bergman and Rowen). ∗

References.

Bahturin–Ol'shanski [1], Bokut'–L'vov–Kharchenko [1], Cohn [1], Kleiman [1], Kurosh [1,2], Mal'cev [1], Neumann [1], Razmyslov [1], Solvable and simple infinite groups, Jacobson [1].

SET-THEORETIC SUPPLEMENT

1. Sets and classes. According to the definition by the founder of set theory, G. Cantor, a set is "any collection into a whole M of definite and separate objects m of our intuition or our thought".[1]

In order to denote the fact that m is an element of the set M we write $m \in M$; the notation $m \notin M$ means that m does not belong to M (one can also write $M \ni m$ and $M \not\ni m$). If M_1 and M_2 are two sets and moreover the elements from the set M_1 also belong to the set M_2, then we say that M_1 is a subset in M_2 and write $M_1 \subset M_2$. The same can be denoted in the form $M_2 \supset M_1$. A set which contains no elements is called empty and is denoted by $\emptyset$. There exist various methods for representing sets. The most explicit is the method of enumerating elements from a set. For instance, $0 = \emptyset$, $1 = \{\emptyset\}$, $2 = \{0, 1\}$, $3 = \{0, 1, 2\}, \ldots$ is one of the ways for representing the natural sequence $\mathbf{N} = \{0, 1, 2, 3, \ldots\}$. A more general but sometimes dangerous way is a method of representing a set by indicating a common property of its elements. For instance, $\{x \in \mathbf{N} | x = 2m$ for suitable $m \in \mathbf{N}\}$ is a way of representing the set of even numbers. However the notation $V = \{M$ is a set $| M \notin M\}$ does not represent any set because of logical contradictions (Russell paradox).

One of the ways to avoid such contradictions is to introduce classes as collections for some of which the property of "being an element" has no sense. The classes which are the elements of "larger classes" still are called sets. By using this agreement we eliminate Russell paradox, having noted that the collection introduced above is the class of all sets. In order to eliminate some other contradictions it is necessary that the property with a help of which a class is introduced, should be presented by logical formulas and that the variables which correspond to classes of general form would not enter these formulas with the existence and universal quantifiers.

2. Operations over sets and classes. The axiomatics by J. von Neumann–Hilbert–Bernays (NBG, see, Mendelson [1, ch. 4]) postulates the existence of an empty set and rules of constructing any other class from it. In particular, the set of natural numbers can be constructed by using the method which is indicated above (subsect. 1.1). We shall skip the details of these constructions, but note that for any two classes $\mathcal{A}$, $\mathcal{B}$ there are defined the *union* $\mathcal{A} \cup \mathcal{B}$, as a collection of elements from $\mathcal{A}$ or from $\mathcal{B}$, the *intersection* $\mathcal{A} \cap \mathcal{B}$ as a collection of elements from both $\mathcal{A}$ and $\mathcal{B}$, the *complement* $\mathcal{A} \backslash \mathcal{B}$ as the collection of elements from $\mathcal{A}$ but not from $\mathcal{B}$.

[1]G. Cantor, Contributions to the founding of the theory of transfinite numbers, Dover reprint; translation by Ph.E.B. Jourdain in 1915, from G. Cantor, Beiträge zur Begründung der transfiniten Mengenlehre, I, II, Math. Ann. 46 (1895), 481–512; ibid 49 (1897), 207–246 (translator's note).

The *direct (Cartesian) product* $\mathcal{A} \times \mathcal{B}$ is defined as the class of (ordered) pairs (a, b), where $a \in \mathcal{A}$, $b \in \mathcal{B}$. If M is a set then 2^M is a set of all subsets in M. A *correspondence* between the classes $\mathcal{A}$ and $\mathcal{B}$ is a subclass $\Theta \subset \mathcal{A} \times \mathcal{B}$. If $\mathcal{A} = \mathcal{B}$ then a correspondence $\Theta \subset \mathcal{A} \times \mathcal{A}$ is called a *relation*. For instance, the relation Θ for inclusion of sets is a relation on the class $\mathcal{V}$ of all sets: $\Theta \subset \mathcal{V} \times \mathcal{V}$; we have $(M_1, M_2) \in \Theta$ iff $M_1 \subset M_2$.

A correspondence $f \subset \mathcal{A} \times \mathcal{B}$ is called *functional*, or simply a *function*, if for any $x \in \mathcal{A}$ there exists only one $y \in \mathcal{B}$ such that $(x, y) \in f$. The collection of elements x for which such a y exists, is called the domain of definition of the function f. For an element x from the domain of definition we can write $y = f(x)$, if $(x, y) \in f$. The collection $y \in \mathcal{B}$ such that $y = f(x)$, for some $x \in \mathcal{A}$, is called the range of values of the function f and is denoted as $f(\mathcal{A})$. From the axioms of the theory of sets it follows that the domain of definition and the range of values of a function are classes. For instance, the association to any field of its order (number of elements) is a function $f \subset \mathcal{F} \times \mathbf{N}$, where $\mathcal{F}$ is the class of all fields, $\mathbf{N}$ is the set of natural numbers, for which the domain of definition is the class of all finite fields, and the range of values is the set of all numbers of the form p^n, where p is a prime number and n is a natural number.

3. Mappings. If the domain of definition of the function $f \subset \mathcal{A} \times \mathcal{B}$ coincides with $\mathcal{A}$ then we say that f is a *mapping* from $\mathcal{A}$ into $\mathcal{B}$ and write $f : \mathcal{A} \to \mathcal{B}$ or $\mathcal{A} \xrightarrow{f} \mathcal{B}$. If $y = f(x)$ then we say that y is the *image* of the element x, and x is a *preimage* of the element y. The subclass $f(\mathcal{A}) \subset \mathcal{B}$ is called the image of the mapping f. If $\mathcal{A}$ and $\mathcal{B}$ are two classes then the mapping $f : \mathcal{A} \to \mathcal{B}$ is called *injective* if from $f(x_1) = f(x_2)$, where $x_1, x_2 \in \mathcal{A}$, follows that $x_1 = x_2$. We say that f is *surjective* if for any $y \in \mathcal{B}$ there can be found an $x \in M$ such that $y = f(x)$.

A mapping $f : \mathcal{A} \to \mathcal{B}$ which possesses both the properties of injectivity and surjectivity is called *bijective*. An example of a bijective mapping of the class $\mathcal{A}$ onto itself is the identity mapping $1_{\mathcal{A}}$, i.e. the mapping such that $1_{\mathcal{A}}(x) = x$, for any $x \in \mathcal{A}$.

If two mappings are given $f : \mathcal{A} \to \mathcal{B}$, $g : \mathcal{B} \to \mathcal{C}$ then their *product* (or *composition*) will be the mapping $h : \mathcal{A} \to \mathcal{C}$ such that $h(x) = g(f(x))$, for any $x \in \mathcal{A}$. We write $h = gf$. A significant property of the composition of mappings is its associativity: if $f : \mathcal{A} \to \mathcal{B}$, $g : \mathcal{B} \to \mathcal{C}$, $h : \mathcal{C} \to \mathcal{D}$ then

$$(hg)f = h(gf) : \mathcal{A} \to \mathcal{D}.$$

If $f : \mathcal{A} \to \mathcal{B}$ is some mapping then the mapping $h : \mathcal{B} \to \mathcal{A}$ (which may very well not exist) is called *inverse* to f if $fg = 1_{\mathcal{B}}$ and $gf = 1_{\mathcal{A}}$. Then we write $g = f^{-1}$.

Proposition. 1) *A mapping $f : \mathcal{A} \to \mathcal{B}$ is injective iff for any $g_1, g_2 : \mathcal{C} \to \mathcal{A}$ from the equality $fg_1 = fg_2$ it follows that $g_1 = g_2$ (in other words, an injective mapping can be cancelled from the left).*

2) *A mapping $g : \mathcal{C} \to \mathcal{A}$ is surjective iff for any f_1, $f_2 : \mathcal{A} \to \mathcal{B}$ it follows from $f_1 g = f_2 g$ that $f_1 = f_2$. (Cancellation from the right is then possible.)*

4. Operations over classes of sets. In passing to infinite collections of classes one should be cautious in connection with the restrictions mentioned in subsection 2. The expression a "collection of classes" itself is already ambigous. Therefore, below we shall be concerned only with operations over a class of sets. If $(M_\alpha)_{\alpha\in\mathcal{I}}$ is such class then $\bigcup_{\alpha\in\mathcal{I}} M_\alpha$ is the class of elements which belong at least to one of the sets M_α, $\alpha \in \mathcal{I}$, and $\bigcap_{\alpha\in\mathcal{I}} M_\alpha$ is a set of elements which belong to all M_α, $\alpha \in \mathcal{I}$. It is more complicated with the direct product $\prod_{\alpha\in\mathcal{I}} M_\alpha$. In the case when $\mathcal{I}$ is finite and consists of n elements, such a product is the set of n–tuples $n = (x, y, z, \dots)$ of elements of the corresponding M_α. If $\mathcal{I}$ is infinite then an "n–tuple" n is a function which is defined on $\mathcal{I}$ and for any $\alpha \in \mathcal{I}$ it takes a value in M_α. The statement of the existence of at least one such function is equivalent to the axiom of choice. In a strong form this is the following.

Axiom of choice (strong form). *There exists a function $f : \mathcal{V} \to \mathcal{V}$ which associates to each non–empty set M an element $f(M)$ of this set.*

There is also a weaker form of this axiom (and that is the one we will use).

Axiom of choice (weak form). *For any set A there exists a function $f_A : 2^A \to A$ which associates to every non–empty subset $B \subset A$ an element $f_A(B)$ from this subset.*

An application of the strong form gives (the existence of) $\prod_{\alpha\in\mathcal{I}} M_\alpha$, where $\mathcal{I}$ is the class of all functions $f : \mathcal{I} \to \bigcup_{\alpha\in\mathcal{I}} M_\alpha$ such that $f(\alpha) \in M_\alpha$, for any $\alpha \in \mathcal{I}$. If $\mathcal{I}$ is a set then the direct product $\prod_{\alpha\in\mathcal{I}} M_\alpha$ is also a set (the definition is the same). Here the weak form of the axiom of choice is sufficient, because $\bigcup_{\alpha\in\mathcal{I}} M_\alpha$ in this case is a set.

5. Binary relations. Equivalence. It was already noted in subsection 1 that a correspondence $\Theta \subset \mathcal{A} \times \mathcal{A}$, where $\mathcal{A}$ is some class, is called a *relation*, or to be more precise, a *binary relation*, because there are two elements involved in such a relation. The relation $\Delta_\mathcal{A} = \{(x, x) | \, x \in \mathcal{A}\}$ is called a diagonal or the equality. It should be noted that the notation $(x, y) \in \Theta$ is equivalent to the $x\Theta y$ one. Usually, when considering binary relations the following properties are distinguished:

1) *Reflexivity*: for any $x \in \mathcal{A}$ we have $x\Theta x$.
2) *Symmetry*: if $x\Theta y$ then $y\Theta x$.
3) *Anti–symmetry*: if $x\Theta y$ and $y\Theta x$, then $x = y$.
4) *Transitivity*: if $x\Theta y$ and $y\Theta z$ then $x\Theta z$.

A reflexive, symmetric and transitive relation is called an *equivalence*. The equivalence relations on a class vary from the "smallest" relation, the one of equality, to the "largest" one where all elements from $\mathcal{A}$ are mutually equivalent. We should note also that an equivalence relation is often denoted by the sign $\sim$, i.e. $x\Theta y$ is the same as $x \sim y$. Because of the defining properties of an equivalence relation we obtain the formulas: $x \sim x$; if $x \sim y$ then $y \sim x$; if $x \sim y$, $y \sim z$ then $x \sim z$.

Having fixed an equivalence relation $\sim$ on the set M and having taken some $x \in M$ we can denote by $\bar{x}$ the set of all $y \in \mathcal{A}$ such that $x \sim y$. For any $x, y \in M$ either $\bar{x} = \bar{y}$, or $\bar{x} \cap \bar{y} = \emptyset$. Therefore the set M decomposes into a union of mutually non-intersecting subsets of the form $\bar{x}$, where $x \in M$. These subsets are called the equivalence classes of $\sim$. It should be remarked that if $M = \bigcup_{\alpha \in \mathcal{I}} M_\alpha$, and moreover, $M_\alpha \cap M_\beta = \emptyset$, for $\alpha \neq \beta$, then there is a unique equivalence relation $\sim$, for which the non-empty M_α are the equivalence classes. It is sufficient to set $x \sim y$, iff $x, y \in M_\alpha$, for some $\alpha \in \mathcal{I}$.

Thus we can assume that M is a union of mutually disjoint classes $\bar{x}$, $\bar{y}$, $\bar{z}, \ldots$ of an equivalence $\sim$. Then the set $\bar{M} = \{\bar{x}, \bar{y}, \bar{z}, \ldots\}$ the elements of which are the classes mentioned, is called the *quotient set* of the set M with respect to the relation $\sim$. We then write $\bar{M} = M/\sim$. The mapping $f : M \to M/\sim$ which associates to an element $x \in M$ the equivalence class $\bar{x}$ containing the element x, is called the natural (quotient) mapping.

Everything that was said above about equivalence classes and quotient sets can be extended to the case of classes if for any $x \in \mathcal{A}$ the collection of elements equivalent to x is a set.

6. Ordering. A reflexive, anti-symmetric and transitive relation $\Theta \subset \mathcal{A} \times \mathcal{A}$ on a class $\mathcal{A}$ is called a (partial) *ordering*. Usually, an ordering is denoted by the symbol $\leq$. We then write $x \leq y$ (or $y \geq x$), if $(x, y) \in \Theta$, and we say that x is smaller or equal to y (y is larger or equal to x). If $x \leq y$ and $x \neq y$, then we write $x < y$. By using this notations for the properties which define an ordering, the following formulas are obtained: $x \leq x$; if $x \leq y$ and $y \leq x$, then $x = y$; if $x \leq y$ and $y \leq z$, then $x \leq z$.

An example of an ordering on the class $\mathcal{V}$ of all sets is the inclusion of sets which was mentioned above. Various orderings can be considered on a set of natural numbers including the usual one under which $a \leq b$, if $b - a \in \mathbf{N}$, and another one under which $a \leq b$ if b is divisible by a without a remainder, etc..

An ordering $\leq$ on the class $\mathcal{A}$ is *linear (total)* if for any a, b we have $a \leq b$ or $b \leq a$ (i.e. any two elements are comparable). From the three examples of ordering presented above only the second one is linear.

The study of orderings in such "basic" fields of mathematics as the theory of sets is related to the natural procedures of calculation of elements in collections. Naturally, given a set of "numbers" of enumerated elements there should be a possibility to select the smallest of them. This property, when presented in a formal way, is called a property of minimality and is written in the following form. We call the element x_0 from the ordered class $\mathcal{A}$ *minimal* if from the condition $x \leq x_0$ follows $x = x_0$. The stronger restriction on the element x_0, of being the smallest in $\mathcal{A}$, consists of the fact that for any $x \in \mathcal{A}$ we have $x_0 \leq x$. We then say that the ordering $\leq$ on the class $\mathcal{A}$ satisfies the *minimality condition* if in any non–empty subclass $\mathcal{B}$ of the class $\mathcal{A}$ there is a minimal element. Analogously one defines a *maximal* element, the largest element and the maximality *condition.*

It should be noted that the minimality condition on $\mathcal{A}$ is equivalent to the following: any decreasing sequence $x_1 \geq x_2 \geq \ldots$, of elements from $\mathcal{A}$, has only a finite number of mutually different elements.

Finally, we say that the class $\mathcal{A}$ is *well–ordered* if it is ordered linearly and satisfies the condition of minimality. In this case any non–empty subclass contains a smallest element (as any two minimal elements are equal). For well–ordered classes the method of mathematical induction can be applied:

Theorem. *Let $\mathcal{B}$ be a non–empty subclass of a well–ordered class $\mathcal{A}$ which contains the smallest element of the class $\mathcal{A}$. We assume that $\alpha \in \mathcal{B}$ provided any β which is strictly smaller than α, lies in $\mathcal{B}$. Then $\mathcal{B} = \mathcal{A}$.*

The statement is evident from the fact that if $\mathcal{A} \neq \mathcal{B}$ then one can find the smallest $\alpha \in \mathcal{A} \setminus \mathcal{B}$. If $\beta \leq \alpha$ and $\beta \neq \alpha$, then $\beta \in \mathcal{B}$. Thus according to the condition $\alpha \in \mathcal{B}$.
□

A more often used form of this theorem occurs when the class $\mathcal{B}$ is given by some property inside $\mathcal{B}$. Moreover, since the interval $\sigma \leq \alpha < \tau$ of a well–ordered set is itself well–ordered, another version of the previous theorem can be formulated in the following way.

Theorem on transfinite induction. *Let $\mathcal{P}(\alpha)$ be a property of elements of a well–ordered class $\mathcal{A}$, and let $\sigma \leq \alpha < \tau$ be an interval of this class. Assume that $\mathcal{P}(\sigma)$ holds and that $\mathcal{P}(\alpha)$, $\alpha < \tau$, holds if $\mathcal{P}(\beta)$ is valid for all $\beta \geq \sigma$, which are strictly smaller than α. Then $\mathcal{P}(\alpha)$ holds for all α, $\sigma \leq \alpha < \tau$.*

$*$ 1) There are a number of orderings that are considered in algebra. One of these is the so–called *lexicographic* ordering (or to be more precise, there is a range of orderings presented below that are called thus). The lexicographical ordering is introduced on the set $W(X)$ of words in a linearly ordered alphabet X. We define $u \leq v$ by using induction with respect to the length of the word u. We assume the empty word to be the smallest element. Further, if $u = xu'$, $v = yv'$, $x, y \in X$, $u', v' \in W(X)$, then $u \leq v$, if either $x < y$, or $x = y$ and $u' \leq v'$. So if $X = \{a, b\}$, $a < b$, then, for instance, $a^n < b$, $ba^m b < b^2$ for $m, n > 0$ and etc..

2) The theory of the ordered sets contains the theory of *lattices*. In order to define a lattice first of all we shall define the *exact lower bound* of two elements a, b from

an ordered set L to be an element $c = \inf\{a, b\}$ such that $c \leq a$, $c \leq b$, and if $c' \leq a$, $c' \leq b$, for some $c' \in L$, then $c' \leq c$. Analogously, the *exact upper bound* $c = \sup\{a, b\}$ is defined. If in a partially ordered set each pair of elements possesses an exact lower and upper bound (it is easy to see that these bounds are unique) then such a set is called a lattice (earlier, in Russian literature, the term of *structure* was also used in this case). The operations $a+b$ and ab of taking exact upper and lower bounds of a, b have a number of properties

$$\begin{aligned} (a+b)+c &= a+(b+c), & (ab)c &= a(bc), \\ a+b &= b+a, & ab &= ba, \\ a+a &= a, & aa &= a, \\ (a+b)a &= a, & ab+a &= a. \end{aligned}$$

An algebra with operations + and $\cdot$ for which the axioms written above are valid, can be transformed into a lattice by setting that $a \leq b$, if $ab = a$ (or, equivalently, $a+b = b$).

An example of a *distributive lattice* (i.e. one with the additional property $a(b+c) = ab + ac$) is the set 2^M of all subsets of some set M which is ordered by inclusion of subsets. The collection of normal subgroups in a group, vector subspaces in a space, etc. is also an example of a lattice where the exact lower bound is the intersection in a set–theoretic sense, and the exact upper bound is the smallest, let us say, normal subgroup which contains the two given ones. As a rule, such a lattice is not distributive and satisfies a weaker relation which is called the *modular identity* ora *Dedekind law*: $(a+b)c = ac + bc = a + bc$ if $a \leq c$.

We can define also a *Boolean algebra* as a distributive lattice B with smallest element 0 and largest element 1 and with a unique complement a', for every $a \in L$, i.e. an element $a' \in B$ such that $a + a' = 1$ and $aa' = 0$. The lattice 2^M of all subsets of the set M is a Boolean algebra.

More details about lattices can be found in Birkhoff [1], Skorniakov [1]. *

7. Ordinal numbers. All well–ordered sets are obtained in a rather similar way. The exact meaning of this statement can be given with a help of the concept of isomorphism. We say that the ordered classes $\mathcal{A}$ and $\mathcal{B}$ are *isomorphic* if there exists the bijection $f : \mathcal{A} \to \mathcal{B}$ such that $a \leq b$ $(a, b \in \mathcal{A})$ iff $f(a) \leq f(b)$. We shall use the notation $\mathcal{O}$ to indicate the class of well–ordered sets, called ordinals, such that any well–ordered set is isomorphic to precisely one ordinal.

In order to construct such an $\mathcal{O}$ we call a class $\mathcal{A}$ *ordinal* if its elements are well–ordered with respect to the relation of belonging[2] $\in$ (i.e. $a \leq b$, if $a \in b$, or $a = b$), and each element from the class is a subset. An example of such class is the empty set and also the natural sequence $\mathbf{N} = \{0, 1, 2, \dots\}$ if we take into account a certain construction. The element 0 is an empty set, i.e. $0 = \emptyset$. The element 1 is $\{0\}$, i.e. it is the set of one element with the unique element 0. The element 2 is $\{0, 1\}$, 3

[2]The 'esti' relation (translator's note).

is $\{0, 1, 2\}$, etc.. It is clear that for $k < n$ we have $k \in n$ and also $k \subset n$, so our conditions are satisfied. An ordinal class which is also a set, is called an *ordinal number* or simply an *ordinal*. Let $\mathcal{O}$ be a class of all ordinal numbers. Then $\mathcal{O}$ is well–ordered with respect to the relation of belonging $\in$. In fact, it is easy to understand that the smallest element of each ordinal is $0 = \emptyset$. All further verifications are based on the following idea: any element γ of the ordinal α is a set of all elements from α which are contained in γ (strictly lower than γ, which is the same). In particular, if α and β are two ordinals then $\alpha \cap \beta \subset \alpha$, $\alpha \cap \beta \subset \beta$. The smallest element γ from α, which is not included in $\alpha \cap \beta$, is $\alpha \cap \beta$, and the same is valid for β. Thus, $\gamma \in \beta$. Therefore either $\alpha \cap \beta = \alpha$, or $\alpha \cap \beta = \beta$. If $\alpha \neq \beta$ then either $\alpha \in \beta$, or $\beta \in \alpha$. The relation $\alpha \leq \beta$ is anti–symmetric. Indeed, if $\alpha \in \beta$ and $\beta \in \alpha$ then $\alpha \subset \beta$ and $\beta \subset \alpha$, i.e. $\alpha = \beta$. If $\alpha \in \beta$, $\beta \in \gamma$ then $\alpha \in \beta \subset \gamma$, thus, $\alpha \in \gamma$. Let $\mathcal{B}$ be a non–empty subclass in $\mathcal{O}$ and $\alpha \in \mathcal{B}$; if α is smallest in $\mathcal{B}$ then the minimal element in $\mathcal{B}$ is found. Otherwise, $\alpha \cap \mathcal{B}$ is non–empty and the minimal element exists because of the fact that α is well–ordered: if γ is minimal in $\alpha \cap \mathcal{B}$ and for $\delta \in \mathcal{B}$ we have $\delta \leq \gamma$, then $\delta \in \gamma$ or $\delta = \gamma$ and $\delta \in \alpha \cap \mathcal{B}$, i.e. $\delta = \gamma$ and γ is minimal in $\mathcal{B}$. So the class of ordinals is well–ordered.

If α is an ordinal then $\alpha' = \alpha \cup \{\alpha\}$ is also an ordinal; moreover between α and α' there are no intermediate ordinals. We say that the ordinal β is *non–limit*, if $\beta = 0$ or $\beta = \alpha'$ for suitable ordinal α. Otherwise the ordinal is called a *limit* ordinal. A limit ordinal is equal to the union of all ordinals, contained in it.

Since the class of ordinals $\mathcal{O}$ is well–ordered it can be used for transfinite induction. For instance, relations, in particular, functions, mapping, etc. can be defined on $\mathcal{O}$ or on its intervals by giving their values on the smallest element and specifying how the relation is constructed for α if this relation is already given on the ordinals which are lower than α.

8. Order–types of sets. Returning to the concept of isomorphism it should be noted that the central result on the structure of well–ordered sets is the following

Theorem. *Every well–ordered set is isomorphic with precisely one ordinal.*

Proof. We shall construct a function $f : \mathcal{O} \to M$, where M is the given well–ordered set, by using transfinite induction. If m_0 is the smallest element in M then we set $f(0) = m_0$. If the function is defined for all $\beta < \alpha$ and the domain of its values P coincides with M then the definition is finished. Otherwise, we take the unique smallest $m \in M \setminus P$ and assume that $f\alpha = m$. Let D be the domain of definition and P the domain of values of the function $f : \mathcal{O} \to M$ thus obtained. Since $f : D \to P$ is bijective there exists an inverse function $f^{-1} : P \to D$. From the fact that P is a set, follows that D is a set, i.e. $D = \gamma = \bigcup_{\alpha<\gamma} \alpha$ is some ordinal. If $P \neq M$ and m is the smallest element from $M \setminus P$ then according to the definiton, $f(\gamma') = m \in P$. Thus, $P = M$ and we can construct an isomorphism of the set M and of the ordinal γ.

Let now $f : \alpha \to \beta$ be an isomorphism of two ordinals $\alpha \neq \beta$. Replacing f, if needed, by its inverse mapping we can assume that $\beta \in \alpha$. In this case we have an injection from α into α which retains the ordering, and such that $f(\alpha) \in \alpha$. We obtain a contradiction by noting that for any $\gamma \in \alpha$ we should have $\gamma = f(\gamma)$ or $\gamma \in f(\gamma)$. Indeed, suppose that for some $\gamma \in \alpha$ we have $f(\gamma) \in \gamma$. We denote by R the set of all such γ, and let γ_0 be the smallest element in R. Then $f(\gamma_0) \in \gamma_0$ and $f(f(\gamma_0)) \in f(\gamma_0)$. However $\delta = f(\gamma_0) \notin R$ and we have $f(\delta) \in \delta$. We arrived at a contradiction. The theorem is proved. □

If M is a well–ordered set then the unique ordinal which is determined by the theorem, is denoted by $\operatorname{ord} M$. The ordinal $\operatorname{ord} M$ is called the *order–type* of the set M. If M consists of n elements then $\operatorname{ord} M = n$ $(= \{n-1, \{n-1\}\}$. The set $\mathbf{N}$ of natural numbers with its natural ordering, the subset of even numbers, the set of squares of natural numbers all have the same order–type $\omega = \{0, 1, 2, \dots\}$. The set T of real numbers of the form $T = \{1 - \frac{1}{1}, 1 - \frac{1}{2}, 1 - \frac{1}{3}, \dots, 1\}$ has the type $\omega' = \{\omega, \{\omega\}\}$. We then write $\alpha' = \alpha + 1$, $\alpha'' = \alpha + 2, \dots$. Using this notation $\operatorname{ord} T = \omega + 1$.

9. Cardinality. Let $\mathcal{A}$ and $\mathcal{B}$ be two classes. We say that $\mathcal{A}$ and $\mathcal{B}$ are *of equal cardinality* if there exists a bijection $f : \mathcal{A} \to \mathcal{B}$. Then we say that the cardinality of the class $\mathcal{A}$ is equal to the cardinality of the class $\mathcal{B}$ and write $\operatorname{card} \mathcal{A} = \operatorname{card} \mathcal{B}$. One can define the cardinal numbers as the equivalence classes consisting of classes of equal cardinality, but one cannot speak of a class of cardinal numbers, as any such equivalence class with the exeption of $\{\emptyset\}$ is not a set. So the mapping f which assigns to each set M the one–element set $\{M\}$ is a bijection of the class $\mathcal{V}$ of all sets to a class of one–element sets. Consequently, the latter is not a set.

In order to compare classes with respect to cardinality the following is important

Cantor–Bernstein theorem. *If $\mathcal{A}$ is of equal cardinality with a subclass of $\mathcal{B}$, and $\mathcal{B}$ is of equal cardinality with a subclass of $\mathcal{A}$ then* $\operatorname{card} \mathcal{A} = \operatorname{card} \mathcal{B}$.

Lemma. *Let $F : \mathcal{X} \to \mathcal{X} \cup \mathcal{Y} \cup \mathcal{Z}$ be a bijection, and suppose moreover that $\mathcal{X} \cap \mathcal{Y} = \mathcal{X} \cap \mathcal{Z} = \mathcal{Y} \cap \mathcal{Z} = \emptyset$. Then there exists a bijection $G : \mathcal{X} \to \mathcal{X} \cup \mathcal{Y}$.*

Proof. We construct a (partial) function $H : \mathcal{X} \times \mathbf{N} \to \mathcal{X} \cup \mathcal{Y} \cup \mathcal{Z}$ by assuming with respect to induction that $H((x, 0)) = F(x)$ and $H((x, i+1)) = F(H(x, i))$ if $H(x, i) \in \mathcal{X}$. Let $\mathcal{X}^*$ be the subclass of those x for which there exists an i with the property $H(x, i) \in \mathcal{Z}$, and let $\mathcal{Y}^*$ be its complement in $\mathcal{X}$. We define $G(x) = x$ if $x \in \mathcal{X}^*$, and $G(x) = F(x)$, if $x \in \mathcal{Y}^*$. Then G is the desired bijection. □

Proof of the theorem. If $F : \mathcal{A} \to \mathcal{B}_1 \subset \mathcal{B}$, $G : \mathcal{B} \to \mathcal{A}_1 \subset \mathcal{A}$ are the given bijections and $\mathcal{C} = G(\mathcal{B}_1)$ then $\mathcal{C}$ is bijective to $\mathcal{X} = \mathcal{C} \cup (\mathcal{A}_1 \backslash \mathcal{C}) \cup (\mathcal{A} \backslash \mathcal{A}_1)$. According to the lemma $\mathcal{C}$ is bijective to $\mathcal{A}_1 = \mathcal{C} \cup (\mathcal{A}_1 \backslash \mathcal{C})$, i.e. $\mathcal{A}$ is also bijecitve to $\mathcal{A}_1$. But $\mathcal{A}_1$ is bijective to $\mathcal{B}$, so $\mathcal{A}$ also is bijective to $\mathcal{B}$. The theorem is proved. □

We write card $\mathcal{A} \leq$ card $\mathcal{B}$ if $\mathcal{A}$ is isomorphic to a subclass in $\mathcal{B}$ and card $\mathcal{A} <$ card $\mathcal{B}$, if, in addition to this, card $\mathcal{A} \neq$ card $\mathcal{B}$. The transitivity of the relation $\leq$ is evident and from the theorem follows that it is anti–symmetric.

Restricting ourselves to sets we note the following result:

Cantor theorem. *If M is a set, and $\mathcal{P}(M)$ is the set of all its subsets, then* card $M <$ card $\mathcal{P}(M)$.

Proof. It is evident that the mapping $F : x \to \{x\}$ defines a bijection of the set M and a subset in $\mathcal{P}(M)$. We shall prove that a bijection onto all of $\mathcal{P}(M)$ is impossible. In fact, let $G : M \to \mathcal{P}(M)$ be such a bijection. We consider the set N of all $x \in M$ such that $x \notin G(x)$. Then $N \in \mathcal{P}(M)$. Let $y \in M$ be such that $G(y) = N$. If $y \in N$ then $y \notin G(y) = N$, which is a contradiciton. If $y \notin N$ then $y \in G(y) = N$; again a contradiction. Thus the existence of the bijection G is impossible. □

Hartogs' theorem. *For any set M there exists an ordinal α which is not of equal cardinality with any subset from M.*

Proof. Let M be a counterexample for the theorem. For any $\alpha \in \mathcal{O}$ let N be the subset in M which is bijective to α. It is clear that the bijection with the well–ordered set α provides N with a structure of a well–ordered set which is isomorphic to α. We shall construct a mapping $F : \mathcal{O} \to Q = \mathcal{P}(\mathcal{P}(M \times M)) \times \mathcal{P}(M)$ by assigning to the ordinal α the set W of pairs (Θ, N), where N is a subset in M and which is well–ordered by the relation Θ and isomorphic to α. This is impossible, because $\mathcal{O}$ is not a set (see, subsection 1). □

In considering the concept of cardinality we have not used the axiom of choice (weak form). Omiting it the theory of cardinality would be incomplete: we are not able to compare cardinalities of two arbitrary sets.

10. Statements which are equivalent to the axiom of choice (weak form). In subsection 4 it has been noted that the axiom of choice is equivalent to the non–emptyness of Cartesian products of a non–empty collection of sets. We give some other statements with this property.

Zermelo theorem. *Any set can be well–ordered.*

Trichotomy.[3] *For any two sets M and N either M is in bijection with a subset in N, or N is in bijection with a subset in M, or* card N = card M.

In order to formulate the following statement we say that an element α of a partially ordered set P is an *upper bound* of the subset Q if for any $\beta \in Q$ we have $\beta \leq \alpha$. We call Q a *chain* if the ordering of the set P provides Q with a linear order.

Zorn's lemma. *If in the partially ordered set P any chain Q has an upper bound then there exists a maximal element in P.*

The equivalence of the Zermelo theorem and trichotomy (by using Hartogs theorem) is almost evident. In fact, if the Zermelo theorem is valid and M and N are two arbitrary sets then after well ordering them we can find the ordinals α, β which are of equal cardinality with M, N (subsection 8). However any two ordinals are comparable, so the cardinalitites of the sets M, N are also comparable. By using the Cantor–Bernstein theorem we get either card $M \leq$ card N, or card $N \leq$ card M. Inversily, let trichotomy hold and M be an arbitrary set. Let α be an ordinal which is not of equal cardinality with any such set from M. Then there is a bijection between M and a subset from α, what provides M with a total ordering.

We shall now show the equivalence between the Zermelo theorem, the axiom of choice and the Zorn lemma.

If the Zermelo theorem holds and M is some set then M can be well–ordered. The function f which assigns to every non–empty subset $N \subseteq M$ its smallest element, is the selection function as required in the axiom of choice.

Now let the axiom of choice be valid and assume that M is some partially ordered set of which is known that every chain of this set has an upper bound. Let b be an arbitrary element from M. According to the axiom of choice there exists a selection function f for the set M. With a help of transfinite induction we construct a partial function $F : \mathcal{O} \to M$, by setting that $F(0) = b$ and $F(\alpha) = f(u)$, where u is a set of those upper bounds of the set $Q_\alpha = \{F(\beta) \mid \beta < \alpha\}$, which do not lie in Q_α. There can be found γ such that all upper bounds of the set Q_γ lie in Q_γ. Indeed, otherwise, by extending the inductive construction we can get a bijection F between the class $\mathcal{O}$ and some subset from M, which is impossible. The set Q_γ is a chain by its very construction. Its upper bound a lies in Q_γ and, consequently, is unique. The element a is maximal in M, because if $a \leq c$, then c is an upper bound for Q_γ, i.e. $c = a$. Thus the Zorn lemma holds.

We assume that the Zorn lemma holds. We shall prove the validity of the Zermelo

[3] This word means a division into three disjoint possibilities. We shall use also the term "dichotomy" to denote a division of two disjoint possibilities.

theorem. Let M be a non-empty set. Consider the class $\mathcal{X}$ of all bijections which are defined on the class of ordinals $\mathcal{O}$ with values in M. According to the Hartogs theorem $\mathcal{X}$ is a set. Since the bijection is a subset of $\mathcal{O} \times M$, the set $\mathcal{X}$ is ordered with respect to the relation $\subset$. If C is a chain in $\mathcal{X}$ then one of two elements from C is a bijection which extends another one. Therefore the union of bijections from C is a new bijection, i.e. an element from M which is an upper bound for C. Since the conditions of the Zorn lemma are satisfied, $\mathcal{X}$ has a maximal element $g : \gamma \to N$, where $\gamma \in \mathcal{O}$, $N \subset M$. If $N \neq M$ and $b \in M \backslash N$ then the pair (γ, b) can be added to g, i.e. we extend g by putting $g(\gamma) = b$. The bijection thus obtained belongs to $\mathcal{X}$ and is strictly larger than g. We arrived at a contradiction. So g realizes a bijection of the ordinal γ and M. This allows to transfer the complete ordering to M with a help of γ. So M is well-ordered and the Zermelo theorem is valid. □

11. Cardinal numbers. Application of trichotomy shows that any set is of equal cardinality with some ordinal or, what is the same, with the set of ordinals that are smaller than the given one. Therefore the hierarchy of cardinalities of all sets is reduced to the hierarchy of cardinalities of ordinals. We call the smallest ordinal among all ordinals of equal cardinality a *cardinal number* (or a *cardinal*). Then the cardinal numbers (consequently, cardinalities) form a class $\mathcal{C} \subset \mathcal{O}$ and this class is totally ordered. The finite cardinals, ordinals and natural numbers can be identified among themselves in a rather obvious way. The cardinal of the cardinality of the first infinite ordinal $\omega = \{0, 1, 2, \dots\}$ is denoted by ω_0, its cardinality is that of the set of natural numbers and is called countable. For $\alpha \in \mathcal{O}$ the ordinal $\omega_{\alpha+1}$ is defined as the smallest in the set of ordinals the cardinality of which is strictly larger than the cardinality of the ordinal ω_α. If α is a limit ordinal then $\omega_\alpha = \bigcup_{\beta<\alpha} \omega_\beta$. From the Cantor–Bernstein theorem we can see that the cardinality of the ordinal ω_α is strictly larger than a cardinality of any ordinal ω_β, $\beta < \alpha$. The cardinality of this ordinal will be denoted by the Hebrew letter $\aleph_\alpha$. From the Cantor theorem follows that the cardinality $2^{\aleph_\alpha}$ of the set 2^{ω_α} is strictly larger than $\aleph_\alpha$ which is the cardinality of the set ω_α, i.e. $2^{\aleph_\alpha} \geq \aleph_{\alpha+1}$. The question about equality in the latter inequality is called the generalized continuum hypothesis. Investigations during several decades have shown that the continuum hypothesis not only does not contradict the system of axioms of the theory of sets (in the sense of the Zermelo–Frenkel system) but also does not depend on it.

Taking $\alpha = 0$ we have the usual continuum hypothesis. The reader knows of course, that the continuum (= the set $\mathbf{R}$ of real numbers) is of equal cardinality with the set $2^{\mathbf{N}}$ of subsets of the set $\mathbf{N}$ of natural numbers. The continuum hypothesis, consequently, is the assumption that the first uncountable cardinality $\aleph_1$ is the cardinality of the continuum $2^{\aleph_0}$.

∗ In order to introduce some more systematics in our ideas about ordinals it should be noted that any countable ordinal is of equal cardinality of a subset in the set $\mathbf{Q}$ of rational numbers with its natural ordering (see, Birkhoff [1]). ∗

The next subsection deals with the constructions of sets.

12. The direct and inverse limits of a family of sets. We call an ordering of the set $\mathcal{I}$ *directed* (upwards) if for any α, $\beta \in \mathcal{I}$ there exists $\gamma \in \mathcal{I}$ such that $\alpha \leq \gamma$ and $\beta \leq \gamma$. Let $(M_\alpha)_{\alpha \in \mathcal{I}}$ be a family of sets for which the set of indexes $\mathcal{I}$ is directed. We also assume that for any pair α, $\beta \in \mathcal{I}$ with the condition $\alpha \leq \beta$ there exists a mapping $\phi_{\alpha,\beta} : M_\alpha \to M_\beta$ such that, moreover, the following conditions are fulfilled: $\phi_{\alpha\alpha} = 1_{M_\alpha}$ and $\phi_{\beta\gamma}\phi_{\alpha\beta} = \phi_{\alpha\gamma}$ for any triple of indices satisfying $\alpha \leq \beta \leq \gamma$. We form the disjoint union $M = \bigsqcup_{\alpha \in \mathcal{I}} M_\alpha$ (also called the sum of the family considered). Introduce the following equivalence relation on M: $x \in M_\alpha$ and $y \in M_\beta$ are equivalent (we write $x \sim y$), if there can be found $\gamma \in \mathcal{I}$ such that $\alpha \leq \gamma$, $\beta \leq \gamma$ and $\phi_{\alpha\gamma}(x) = \phi_{\beta\gamma}(y)$. The quotient set $D = M/\sim$ with respect to this equivalence relation is called the *direct* (or *inductive*) limit of the family $(M_\alpha)_{\alpha \in \mathcal{I}}$ and is denoted by $D = \varinjlim M_\alpha$. Intuitively, the operation of direct limit corresponds to gluing the sets together according to the scheme given by the gluing functions $\phi_{\alpha\beta}$. The direct limit provides a precise gluing in the following sense. If a set E and a family of mappings $\psi_\alpha : M_\alpha \to E$ are given such that $\psi_\beta\phi_{\alpha\beta} = \psi_\alpha$, for any $\beta \geq \alpha$ (i.e. E is an "arbitrary gluing" according to the same scheme) then there exists a unique mapping $\psi : D \to E$ such that $\pi(\bar{x}) = \psi_\alpha(x)$, for any $x \in M_\alpha$.

In order to define the concept of an inverse (projective) limit we consider a family of sets $(N_\alpha)_{\alpha \in \mathcal{I}}$, where $\mathcal{I}$ is a directed set, and a set of mappings $\pi_{\alpha\beta} : N_\beta \to N_\alpha$, for any indices $\alpha \leq \beta$. Analogously to the above we require that $\pi_{\alpha\alpha} = 1_{N_\alpha}$, and if $\alpha \leq \beta \leq \gamma$, then $\pi_{\alpha\beta}\pi_{\beta\gamma} = \pi_{\alpha\gamma}$. Construct the direct product $N = \prod_{\alpha \in \mathcal{I}} N_\alpha$ and select in N the subset P which consists of those functions f such that for any $\alpha \leq \beta$ we have $f(\alpha) = \pi_{\alpha\beta}(f(\beta))$. The subset P is called the *inverse* (or *projective*) limit of the family $(N_\alpha)_{\alpha \in \mathcal{I}}$ and is denoted by $P = \varprojlim N_\alpha$. The inverse limit is the "smallest" set which maps into every N_α and is compatible with the mappings $\pi_{\alpha\beta} : N_\beta \to N_\alpha$. To be more exact, if Q is some set and $\Theta_\alpha : Q \to N_\alpha$, and moreover $\pi_{\alpha\beta}\Theta_\beta = \Theta_\alpha$, for any $\alpha \leq \beta$, then there exists a unique mapping $\Theta : Q \to P$ such that $(\Theta(x))(\alpha) = \Theta_\alpha(x)$, for any $x \in Q$. The intuitive meaning of the inverse limit is that of the most accurate "completion" of the family $(N_\alpha)_{\alpha \in \mathcal{I}}$. Note that there are cases where the inverse limit of a non–empty family is empty.

13. Categories. Combining classes of sets and collections of mappings which map each member from the selection into one another, leads to the concept of a category. A more detailed (and more abstract) formulation of what was said is the following. We say that a *category* $\mathcal{K}$ is given if there are given a class of *objects* $\mathrm{Ob}(\mathcal{K})$, classes of mutually non–intersecting sets of *morphisms* $\mathrm{Mor}(A,B)$ for all A, $B \in \mathrm{Ob}(\mathcal{K})$, such that for $\alpha \in \mathrm{Mor}(A,B)$, $\beta \in \mathrm{Mor}(B,C)$ there is defined a morphism $\gamma = \beta\alpha \in \mathrm{Mor}(A,C)$. In addition the following conditions are to be fulfilled:

(*i*) Any set $\mathrm{Mor}(A,A)$ contains a morphism 1_A such that $\alpha 1_A = \alpha$, $1_A\beta = \beta$, for any $\alpha \in \mathrm{Mor}(A,B)$, $\beta \in \mathrm{Mor}(B,A)$.

(*ii*) If $\alpha \in \mathrm{Mor}(A, B)$, $\beta \in \mathrm{Mor}(B, C)$, $\gamma \in \mathrm{Mor}(C, D)$ then associativity of the product holds: $(\gamma\beta)\alpha = \gamma(\beta\alpha)$.

Conditions (*i*), (*ii*) imitate well–known properties of the product of mappings of sets. The fact is that in many actual cases the sets $\mathrm{Mor}(A, B)$ indeed consist of mappings but in general this need not be the case.

Examples of categories. 1) The category of sets (objects) and its mappings (morphisms).

2) The category of groups and its homomorphisms.

3) The category of modules over a fixed ring; the morphisms are the homomorphisms of modules.

4) The category of homomorphisms of groups. Its objects are pictures (diagrams) of the form $A \xrightarrow{\phi} B$, where A, B are the groups, and ϕ is a homomorphism of groups. If $A' \xrightarrow{\phi} B'$ is another homomorphism of groups then a pair (α, β) of homomorphisms $\alpha : A \to A'$, $\beta : B \to B'$ is called a morphism if the diagram

$$\begin{array}{ccc} A & \xrightarrow{\phi} & B \\ {\scriptstyle\alpha}\downarrow & & \downarrow{\scriptstyle\beta} \\ A' & \xrightarrow{\phi'} & B' \end{array}$$

is commutative, i.e. $\beta\phi = \phi'\alpha$. If (α', β') is a morphism of the homomorphism $A' \xrightarrow{\phi'} B'$ in the homomorphism $A'' \xrightarrow{\phi''} B''$ then the product of morphisms $(\alpha', \beta')(\alpha, \beta)$ is the morphism $(\alpha'\alpha, \beta'\beta)$. All verifications are immediate.

In order to establish connections between two categories we use the concept of a functor. We say that F is a *covariant functor* from a category $\mathcal{K}$ into a category $\mathcal{L}$ if for any object A from $\mathcal{K}$ there is given an object $F(A)$ from $\mathcal{L}$ and for any morphism $\phi \in \mathrm{Mor}(A, B)$ there is given a morphism $F(\phi) \in \mathrm{Mor}(F(A), F(B))$, such that, moreover $F(1_A) = 1_{F(A)}$ and if $\phi \in \mathrm{Mor}(A, B)$, $\psi \in \mathrm{Mor}(B, C)$ then $F(\psi\phi) = F(\psi)F(\phi)$. An example of a covariant functor can be a so–called forgetful functor. For instance, if $\mathcal{G}$ is the category of groups, $\mathcal{S}$ is the category of sets then the forgetful functor assigns to any group to the set of its underlying elements, and to every homomorphism is assigned itself but now as a mapping of sets.

Analogously, we define a *contravariant functor* G from $\mathcal{K}$ into $\mathcal{L}$. Here it is required that to a morphism $\phi \in \mathrm{Mor}(A, B)$ there corresponds a morphism $G(\phi) \in \mathrm{Mor}(G(B), G(A))$; in addition the rule $F(\psi\phi) = F(\psi)F(\phi)$ is replaced by $G(\psi\phi) = G(\phi)G(\psi)$. For instance, having fixed in some category $\mathcal{K}$ an object A, by assigning to each object $X \in \mathcal{K}$ the set $\mathrm{Mor}(X, A)$, and to a morphism $\phi \in \mathrm{Mor}(X, Y)$ the mapping $F(\phi)$, which takes $f \in \mathrm{Mor}(Y, A)$ into $f\phi \in \mathrm{Mor}(X, A)$, we obtain a contravariant functor from $\mathcal{K}$ into the category of sets $\mathcal{S}$. This functor is called *representable*. (The assignement of from X to $\mathrm{Mor}(A, X)$, for a fixed A, gives an example of a covariant functor from $\mathcal{K}$ into $\mathcal{S}$, which is also called representable).

Two categories $\mathcal{K}$ and $\mathcal{K}'$ are called *equivalent* if there exists a covariant functor F from $\mathcal{K}$ into $\mathcal{K}'$ such that for any A, $B \in \mathrm{Ob}(\mathcal{K})$ the mapping $F : \mathrm{Mor}(A, B) \to \mathrm{Mor}(F(A), F(B))$ is bijective and such that for any object $\bar{A} \in \mathrm{Ob}(\mathcal{K}')$ there can be found an object $A \in \mathrm{Ob}(\mathcal{K})$ such that the objects $F(A)$ and $\bar{A}$ are isomorphic.

Under an isomorphism of the objects U, V of some category we understand a morphism $\phi \in \mathrm{Mor}(U, V)$ which has an inverse morphism, i.e. a $\psi \in \mathrm{Mor}(V, U)$, such that $\phi\psi = 1_V$ and $\psi\phi = 1_U$.

Finally, we should mention an important class of categories – the additive categories. A category $\mathcal{K}$ is called *additive* if for any A, $B \in \mathrm{Ob}(\mathcal{K})$ the set of morphisms $\mathrm{Mor}(A, B)$ is an abelian group (in this case the set is more often denoted as $\mathrm{Hom}(A, B)$) and if, moreover, the following natural requirement is fulfilled: for any objects A, B, C, D of the category $\mathcal{K}$ and for any morphisms ϕ, $\psi \in \mathrm{Hom}(A, B)$, $\chi \in \mathrm{Hom}(D, A)$ and $\omega \in \mathrm{Hom}(B, C)$ the following equalities hold

$$\begin{aligned} \chi(\phi + \psi) &= \chi\phi + \chi\psi, \\ (\phi + \psi)\omega &= \phi\omega + \psi\omega. \end{aligned}$$

For additive categories $\mathcal{K}$, $\mathcal{L}$ an additivity requirement is often included in the requirements for a functor $F : \mathcal{K} \to \mathcal{L}$

$$F(\phi + \psi) = F(\phi) + F(\psi).$$

So in the case of an equivalence of additive categories $F : \mathcal{K} \to \mathcal{K}'$ the groups $\mathrm{Hom}(A, B)$ and $\mathrm{Hom}(F(A), F(B))$ are isomorphic for any pair of objects A, $B \in \mathrm{Ob}(\mathcal{K})$.

It is clear that the category of modules over any ring is additive. If we impose some natural restrictions on an additive category then it is possible to show that this category is isomorphic to the category of modules over a suitable ring.

14. Category of topological spaces. We say that a *topology* $\mathcal{T}$ is given on a set M if there is specified a family of $\mathcal{T}$ subsets which are called *open* and which fulfills the following conditions:

$O1.$ $\emptyset, M \in \mathcal{T}$;
$O2.$ If $U_1, U_2 \in \mathcal{T}$, then $U_1 \cap U_2 \in \mathcal{T}$;
$O3.$ If $(U_\alpha)_{\alpha \in \mathcal{I}}$ is a family of open subsets then

$$\bigcup_{\alpha \in \mathcal{I}} U_\alpha \in \mathcal{T}.$$

From the axiom $O2$ it follows that the intersection of any finite family of open subsets is open. An open set U which contains some point p is called a *neighbourhood* of that point. Any family of open sets from which the remaining open sets can be obtained by taking union is called an open *basis* $\mathcal{B}$ of the topology $\mathcal{T}$. For instance, the set **R** of real numbers is a topological space with a countable open basis. Any subset U together with any point $p \in U$ contains some interval $(a, b) = \{x \mid a < x < b\}$, in which p is situated, is called open in **R**. A basis $\mathcal{B}$ consists of open intervals with rational endpoints.

A topology is called *discrete* if every subset from M is open.

A subset F of a topological space M with topology $\mathcal{T}$ is called *closed* if for some open subset U we have $F \cup U = M$, $F \cap U = \emptyset$ or, what is the same, $F = M \backslash U$.

The family of *closed* subsets $\mathcal{C}$ satisfies the conditions:

$F1$. $\emptyset, M \in \mathcal{C}$;
$F2$. If $F_1, F_2 \in \mathcal{C}$, then $F_1 \cup F_2 \in \mathcal{C}$;
$F3$. If $(F_\alpha)_{\alpha \in \mathcal{I}}$ is a family of closed subsets then

$$\bigcap_{\alpha \in \mathcal{I}} F_\alpha \in \mathcal{C}.$$

The properties $F1$–$F3$ can be used to define a topological space with the help of the system $\mathcal{C}$ of closed subsets.

A topology can also be defined by using the concept of closure. Namely, if $S \subset M$ and M is provided with the topology $\mathcal{T}$ which has the system of closed sets $\mathcal{C}$, then according to the property $F3$ the intersection of all closed subsets containing S is a closed subset $\bar{S}$ containing S. It is clear that $\bar{S}$ is the smallest closed subset which contains S. It is called the *closure* of the set S. It can be easily seen that the following properties are valid:

$P1$. $\bar{\emptyset} = \emptyset$;
$P2$. $S \subset \bar{S}$;
$P3$. $\bar{\bar{S}} = \bar{S}$;
$P4$. $\overline{S_1 \cup S_2} = \bar{S}_1 \cup \bar{S}_2$.

In the case of discrete topology $\bar{S} = S$, for any $S \subset M$.

If M, N are two topological spaces then a mapping $f : M \to N$ is called *continuous* if the preimage $f^{-1}(U)$ of any open set $U \subset N$ is open in M. This definition generalizes the definition of a continuous function of a real variable in a natural way. The condition that the preimage of a closed set be closed, or the condition $f(\bar{S}) \subset \overline{f(S)}$ are equivalent to continuity. It is clear that a composition of continuous mappings is continuous and that the identical mapping 1_M of any topological space M is also continuous. So the class of topological spaces as a class of objects together with the class of continuous mappings as the set of morphisms $\mathrm{Mor}(M, N)$ is a category (of topological spaces).

A bijective mapping $f : M \to N$ that is continuous in both directions is called a *homeomorphism*. From the topological point of view homeomorphic spaces are undistinguishable. We shall call a mapping $g : M \to N$ *open* if the image $g(U)$ of an open set U is open.

A subset M' of a topological space M turns into a topological space with a topology $\mathcal{T}'$, by setting $U' \in \mathcal{T}'$ iff $U' = M' \cap U$, for some $U \in \mathcal{T}$. In other words, the open subsets in M' are intersections of open subsets of the set M with M'. It can be easily seen that according to this definition F' is closed in M' iff there can be found the closed subset F in M such that $F' = M' \cap F$. Moreover, the closure $\tilde{S}$ of the set $S \subset M'$ in the topology $\mathcal{T}'$ is given by $\tilde{S} = M' \cap \bar{S}$ by means of the closure $\bar{S}$ of S in $\mathcal{T}$.

The natural imbedding $f : M' \to M$ is a continuous mapping of topological spaces.

If $f : M \to N$ is a continuous mapping, $N' \subset N$ and $f(M') \subset N'$ then the restriction $f' : M \to N'$ is also a continuous mapping. As an example of a subspace we consider the subset $\mathbf{R}_+$ of positive real numbers in the real line $\mathbf{R}$. The mappings $e^x : \mathbf{R} \to \mathbf{R}_+$ and $\ln x : \mathbf{R}_+ \to \mathbf{R}$ are mutually inverse homeomorphisms of these two spaces.

The *direct product* $M = M_1 \times M_2$ of two topological spaces M_1, M_2 with topologies $\mathcal{T}_1$, $\mathcal{T}_2$ can be made into a topological space with the topology $\mathcal{T}$ by taking as a basis of $\mathcal{T}$ the sets of the form $U = U_1 \times U_2$, where $U_1 \in \mathcal{T}_1$, $U_2 \in \mathcal{T}_2$. It is clear that this definition can be extended to any number of factors. For instance, if $\mathbf{T}^1$ is a unit circle in the plane $\mathbf{R}^2$ then $\mathbf{T}^n = \underbrace{\mathbf{T}^1 \times \ldots \times \mathbf{T}^1}_{n}$ is called a *n–dimensional torus.*

The analogue of a finite set (in a category of sets) in the category of topological spaces is a concept of a compact set. Indeed, a topological space M is called *compact* if from any covering $M = \bigcup_{\alpha \in \mathcal{I}} U_\alpha$ of the set M by open sets we can select a finite subcovering, i.e. $M = U_{\alpha_1} \cup \ldots \cup U_{\alpha_m}$, for suitable $\alpha_1, \ldots, \alpha_m \in \mathcal{I}$. A well–known (and easily proved) equivalent condition consists of the fact that if $(F_\alpha)_{\alpha \in \mathcal{I}}$ is an arbitrary system of closed subsets in M, such that every finite subsystem $F_{\alpha_1}, \ldots, F_{\alpha_m}$ has a non–empty intersection, then the system itself has a non–empty intersection. A space M is called *locally compact* if for any point $p \in M$ there exists a neighbourhood U which has compact closure $\bar{U}$. An example of a compact space is the circle $\mathbf{T}^1$. The space $\mathbf{R}^1$ is not compact but locally compact. The topological product of any finite number of compact (locally compact) spaces posesses the same property. So $\mathbf{T}^n$ is compact and $\mathbf{R}^n$ is locally compact. It is clear that the image of a compact space under a continuous mapping is compact.

A topological space M is called *connected* if it cannot be divided into two non–empty non–intersecting open subsets U and V. Evidently, this is equivalent to the condition that it is impossible to represent M as a non–intersecting union of non–empty closed subsets F and G. A connected subset M' is a connected topological space in a sense of the topology $\mathcal{T}'$ which is induced by the topology $\mathcal{T}$ of the space M. Let $(S_\alpha)_{\alpha \in \mathcal{I}}$ be some family of connected subsets in M; moreover let there be a point p in M

which appears in each S_α, $\alpha \in \mathcal{I}$. Then the union $S = \bigcup_{\alpha \in \mathcal{I}} S_\alpha$ of the given family is a connected set. Indeed, if $S = U \cup V$, where U, V are open, then p lies in one of U, V, let say U. Since V is non-empty, there can be found $q \in V \cap S_\alpha$. Then $S_\alpha = (U \cap S_\alpha) \cup (V \cap S_\alpha)$ is a disconnected set; contradiction.

From what was just said it is clear that each point p of the topological space M lies in a maximal connected subset $K(p)$. For two points p and q either $K(p) = K(q)$, or $K(p) \cap K(q) = \emptyset$. Therefore M splits into a disjoint union of maximal connected sets of the form $K(p)$ which are called the *connected components* of the set M. It is not difficult to see that $\overline{K(p)} = K(p)$, i.e. the connected components are always closed. If all components of the set M are the separate points then M is called a *totally disconnected* set. It is evident that the image of a connected space under a continuous mapping is connected. An example of a non–discrete totally disconnected space is the rational straight line $\mathbf{Q}^1 \subset \mathbf{R}^1$. Finally, it should be noted that a product of a finite number of connected spaces is connected.

The concept of topological space as presented at the beginning of this subsection is, as a rule, too general to obtain interesting theorems. Therefore, one puts additional conditions on the on topology in the form of *separation* axioms.

Axiom $\mathbf{T_0}$. *For any two different points p, q there can be found an open set U which contains only one of these points, i.e.* $|U \cap \{p, q\}| = 1$.

Axiom $\mathbf{T_1}$. *For any two different points p, q there can be found an open set U which contains the (given beforehand) point p but does not contain q.*

Axiom $\mathbf{T_2}$ (Hausdorff axiom). *For any two different points p, q there can be found open sets U, V such that $p \in U$, $q \in V$ and $U \cap V = \emptyset$.*

Axiom $\mathbf{T_3}$. This includes T_1 and the following property of *regularity*: *for any closed set F and a point $p \notin F$ there can be found open sets U and V such that $F \subset V$, $p \in V$ and $U \cap V = \emptyset$.*

Axiom $\mathbf{T_4}$. This includes T_1 and the following property of *normality*: *for any two non–intersecting closed sets F, G there exist open sets U, V such that $F \subset U$, $G \subset V$, $U \cap V = \emptyset$.*

15. Category of analytic manifolds. An open set U of the topological set M is called a n–dimensional *chart* if there exists a homeomorphism $\phi : U \to B$, where B is an open subset in $\mathbf{R}^n$. Two n–dimensional charts U, V with homeomorphisms $\phi : U \to B \subset \mathbf{R}^n$, $\psi : V \to C \subset \mathbf{R}^n$ are called *compatible* if the mappings $\psi\phi^{-1} : B \to C$ and $\phi\psi^{-1} : C \to B$ are analytic functions, i.e. they are defined by power series in a neighbourhood of any point from B (respectively, from C). A collection M of the pairwise compatible n–dimensional charts of a topological space M which cover this space is called an *atlas*. A topological space M which is provided with an n–dimensional atlas $\mathfrak{A}$ is called a *n–dimensional analytic manifold*. Usually, the atlas $\mathfrak{A}$ is supplemented with all possible charts which are compatible with the charts from $\mathfrak{A}$. In this way the *complete* atlas $\mathfrak{A}(M)$ of the manifold M is obtained.

If M and N are two analytic manifolds of dimensions m and n then a mapping $f : M \to N$ is called a *morphism* if f is continuous and for any chart $U \in \mathfrak{A}(M)$ with homeomorphism $\phi : U \to B \subset \mathbf{R}^n$ and any chart $V \in \mathfrak{A}(N)$ with homeomorphism $\psi : V \to C \subset \mathbf{R}^n$ the composition $\psi f \phi^{-1} : \phi(W) \to B$ is an analytic function; here $W = U \cap f^{-1}(V)$ (analyticity is within the limits of the charts). The fact that the identity mapping is a morphism follows from the requirement of compatability for the charts of an atlas. From the analyticity of composition of analytic functions it follows that the composition of morphisms is a morphism. Thus we have arrived at the category of real analytic manifolds.

If $f : M \to \mathbf{R}$ is a morphism then we simply say that f is an analytic function on M. The set of such functions is denoted by $\mathcal{A}(M)$.

Analogously the categories of complex, p–adic and other manifolds are introduced.

An analytic isomorphism (*diffeomorphism*) is a morphism $f : M \to N$ such that its inverse mapping is also a morphism.

An open subset N of a manifold M is a manifold if in the capacity of its atlas one takes the set of charts of the complete atlas $\mathfrak{A}(M)$ for which the domains of definition lie entirely in N. Since $\mathbf{R}^n$, evidently, is a n–dimensional variety then any of its open subsets M is an n–dimensional variety. Inversily, if some topological space M is a union of a family of its open subsets $(U_\alpha)_{\alpha \in \mathcal{I}}$ and each U_α has a structure of analytic variety, and in addition, the structure of variety for $U_\alpha \cap U_\beta$ is induced from U_α and coincides with the structure which is induced from U_β, then M has a unique structure of analytic variety which induces the existent structure on every U_α.

The product $M \times N$ of varieties of dimensions m and n is an $(m+n)$–dimensional variety. An atlas $\mathfrak{A}(M \times N)$ can be given in the form $\mathfrak{A}(M) \times \mathfrak{A}(N)$, i.e. every chart has the form $U \times V$, where $U \in \mathfrak{A}(M)$, $V \in \mathfrak{A}(N)$ with homeomorphism $\chi : U \times V \to B \times C \subset \mathbf{R}^{n+n}$, where $\phi : U \to B \subset \mathbf{R}^m$, $\psi : V \to C \subset \mathbf{R}^n$ are the homeomorphisms corresponding to the charts U, V. We can introduce a structure of n–dimensional analytic variety on the torus $\mathbf{T}^n$, $n > 1$, given such a structure on the circle $\mathbf{T}^1$. The structure on $\mathbf{T}^1 \subset \mathbf{C}^1$ can be defined, for instance, by the two charts U, V where U is obtained by removing from $\mathbf{T}^1$ the point $z = 1$, and V by removing the point $z = -1$. The function $\phi : U \to R$ is defined by the rule $\phi(z) = \arg z$, $\psi : V \to \mathbf{R}$ by the rule $\psi(z) = \arg(-z)$. We have $\phi(U) = (0, 2\pi) = \psi(V)$,

$$\psi\phi^{-1}(\lambda) = \begin{cases} \lambda + \pi, & 0 < \lambda < \pi \\ \lambda - \pi, & \pi < a < 2\pi \end{cases} = \phi\psi^{-1}(\lambda),$$

i.e. the relation between coordinates (charts) is analytic.

Another example of an (n^2–dimensional) analytic real variety is the set $G = GL_n(\mathbf{R})$ of non–singular square matrices of order n with the coefficients from the field $\mathbf{R}$. The set G is considered together with the topology of the space $\mathbf{R}^{n^2}$ induced on G by means of the natural imbedding $G \subset M_n(\mathbf{R})$. A chart $U \subset G$ is defined as the neighbourhood of the unit matrix E which consists of the matrices of the form $E+X$ such that $|x_{ij}| < \frac{1}{2n}$; the mapping ϕ is given by the formula $\phi(E+X) = X$; then $\phi(U) = \{X \mid |x_{ij}| < \frac{1}{2n}\}$ and ϕ is a homeomorphism. We define in the point A a chart U' as the set $U' = AU$, with homeomorphism $\phi' : U' \to \mathbf{R}^{n^2}$ $\phi'(Y) = \phi(A^{-1}Y)$. If $U' = AU$, $U'' = BU$, $V = AU \cap BU$, then

$$\phi''\phi^{-1}(Y) = \phi''(A(E+Y)) = \phi(B^{-1}A(E+Y)) = B^{-1}A - E + B^{-1}AY.$$

Thus $\phi''\phi'^{-1}$ is a linear function; the analogous fact is valid also for $\phi'\phi''^{-1}$.

The concept of coordinates (relative to a chart U of an atlas $\mathfrak{A}(M)$) is defined for the points of the n–dimensional analytic real variety M. Indeed, if $\phi : U \to B \subset \mathbf{R}^n$, $p \in U$, then $\phi(p) = p_1e_1 + \ldots + p_ne_n$, where $\{e_1, \ldots, e_n\}$ is the standard basis of the space $\mathbf{R}^n$; in this case the $(p_1, \ldots, p_n)$ are called the coordinates of the point p in the chart U. From the definition of variety follows that if $(p_1', \ldots, p_n')$ are the coordinates of the point p in the chart $U' \in \mathfrak{A}(M)$ then the relation between $(p_1, \ldots, p_n)$ and $(p_1', \ldots, p_n')$ is analytic and depends only on U, U'.

Now we introduce the concept of the *tangent space* $T_p(M)$ in a point p of the n–dimensional variety M. In fact, $T_p(M)$ is a copy of the space $\mathbf{R}^n$. Its connection with M is following. Let V be a chart containing p. We call a smooth mapping $x : [0,1] \to V$, satisfying $x(0) = p$, a curve $x(t)$, $0 \leq t \leq 1$, which passes through the point p. The vector

$$\xi = \frac{d\,x(t)}{dt}\Big|_{t=0} \in T_p(M) \tag{1}$$

is called the tangent of the curve $x(t)$ at the point $p \in M$. It is clear that for any $\xi \in T_p(M)$ equation (1) (in fact, it is a system of n equations with respect to the coordinates of the point $x(t)$) has a solution $x(t)$ with the condition $x(0) = a$. Therefore we can consider the space $T_p(M)$ as the collection of all tangent vectors to M at the point p. This identification suffers from non–invariance with respect to the choice of a chart. If V' and V'' are two charts which contain some point with the coordinates x' in V' and x'' in V'', and $x' \mapsto x''$ is the analytic mapping giving change in coordinates for these charts, then the Jacobian matrix $J = (a_{ij}) = \left(\frac{\partial x_i''}{\partial x_j'}\right)_{x'=a'}$ is non–singular. In this case the coordinates ξ'' of the vector which is assigned to some curve in the chart V'', will be connected to the coordinates ξ' of that vector with respect to V' by the formula

$$\xi'' = J\xi'.$$

In order to obtain real invariance the algebra $\mathcal{A}_p(M)$ of germs of analytic functions on M at a point p with values in $\mathbf{R}$ should be considered for each point $p \in M$.

The following is considered. Two analytic functions (f', U') and (f'', U'') which are defined in neighbourhoods of U' and U'' of a point p is called equivalent if they coincide on $U' \cap U''$. A class of equivalent functions is called a germ in the point p. Sums and products of classes are well defined by sums and products of their representatives on the intersection of their domains of definition. As a result the algebra $\mathcal{A}_p(M)$ is obtained. A linear mapping $\delta : \mathcal{A}_p(M) \to \mathbf{R}$ is called a *derivation* if

$$\delta(fg) = \delta(f)g(p) + f(p)\delta(g). \tag{3}$$

If a chart V around the point p is fixed then the coordinate functions $x_1, \ldots, x_n$ are defined. In this case the derivation is determined by its values on the coordinate functions, and the space of all derivations is an n–dimensional space which is identified with $T_p(M)$; for any vector $\xi = (\xi_1, \ldots, \xi_n)$ a derivation ∂_ξ is defined by means of the rule

$$\partial_\xi(f) = \xi_1 \frac{\partial f}{\partial x_1}\bigg|_{x=p} + \ldots \xi_n \frac{\partial f}{\partial x_n}\bigg|_{x=p}, \tag{4}$$

where in the right hand part occur the usual derivatives in the point p. The statement that a derivation is determined by its values on the coordinate functions can be proved by representing the function f in the form

$$f = f(p) + \sum_i \frac{\partial f}{\partial x_i}\bigg|_{x=p} (x_i - p_i) + \sum_{k,l} p_{kl}(x_k - p_k)(x_l - p_l), \tag{5}$$

where p_{kl} are functions analytic in the point p ($k, l = 1, \ldots, n$). Using property (3) from which is it evident that the derivation of a constant is zero, we obtain

$$\delta f = \sum_{i=1}^{n} \frac{\partial f}{\partial x_i}\bigg|_{x=p} \partial x_i,$$

or

$$\delta = (\delta x_1)\frac{\partial}{\partial x_1} + \ldots + (\delta x_n)\frac{\partial}{\partial x_n}. \tag{6}$$

The row vectors of derivations

$$\left(\frac{\partial}{\partial x_1'}, \ldots, \frac{\partial}{\partial x_n'}\right), \quad \left(\frac{\partial}{\partial x_1''}, \ldots, \frac{\partial}{\partial x_n''}\right)$$

are connected by means of the matrix J from equality (2). The notation $\partial_\xi(f)$ is shortened to $\xi(f)$.

Finally, for a morphism $\phi : M \to N$ of two analytic varieties and for any point p we can define the *tangent mapping* $T_p(\phi) : T_p(M) \to T_{\phi(p)}(N)$. If $\xi \in T_p(M)$, $f \in \mathcal{A}_{\phi(p)}(N)$ then we define $T_p(\phi)\xi \in T_{\phi(p)}(N)$ by setting

$$(T_p(\phi)\xi)(f) = \xi(f \circ \phi);$$

here $f \circ \phi \in \mathcal{A}_p(M)$ is the composition of a morphism and an analytic function, i.e. it is an analytic function. It should be noted that this definition is compatible with the two interpretations of a tangent vector given above. Indeed, if ξ is defined by formula (1) as a tangent vector to the curve $x(t)$ at the point $p = x(0)$ then $T_p(\phi)\xi$ is tangent to the curve $\phi(x(t))$ at the point $\phi(p)$. This result is obtained directly from the rule of differentiation of a composite function.

We shall illustrate what was said by the example of the n^2-dimensional variety $GL_n(\mathbf{R})$. In this case every tangent space $T_A(GL_n(\mathbf{R}))$ is naturally identified with $M_n(\mathbf{R})$. Equation (1) takes the form

$$B = \left.\frac{d\,A(t)}{dt}\right|_{t=0},$$

where $A(t)$, $0 \leq t \leq 1$ is a curve satisfying $A(0) = A$. We consider the left shift $l(C) : X \to CX$, where $C, X \in GL_n(\mathbf{R})$ as a morphism $\phi : GL_n(\mathbf{R}) \to GL_n(\mathbf{R})$. In this case

$$T_A(l(C))B = CB.$$

Indeed, according to what was noted above

$$T_A(l(C))B = \left.\frac{d\,l(C)(A(t))}{dt}\right|_{t=0} =$$
$$= \left.\frac{d\,CA(t)}{dt}\right|_{t=0} = C\left.\frac{d\,A(t)}{dt}\right|_{t=0} = CB.$$

References.

Birkhoff [1], Bourbaki [3,4], Vinberg [1], Kleene [1], Mal'cev [2], Mendelson [1], Pontryagin [1], Serre [1], Tsalenko–Shulgeifer [1], Sagle–Walde [1].

REFERENCES

S.I. Adian.

[1] *The Burnside problem and identities in groups,* Nauka, 1975.

V.A. Andrunakevich, J.M Riabukhin.

[1] *Radicals of algebras and structural theory,* Nauka, 1979.

E. Artin.

[1] *Geometric algebra,* Interscience, 1957.

Automorphisms of classical groups, Mir, 1976.

J.A. Bahturin.

[1] *Identities in Lie algebras,* VNU Science Press, Utrecht, 1985.

[2] *Lectures on Lie algebras,* Akademie–Verlag, 1978.

J.A Bahturin, A.J. Ol'shanskii.

[1] *Identities,* In Algebra II, Encycl. Math. Sci., 18, (1991) Springer–Verlag.

G. Bergman.

[1] *Everybody knows what Hopf algebra is,* In Contemp. Math., 43, (1985) 25–48.

G. Birkhoff.

[1] *Theory of lattices,* XXV, AMS Colloq. Publ., 1973.

L.A. Bokut', I.V. L'vov, V.K. Kharchenko.

[1] *Non–commutative rings,* In Algebra II, Encycl. Math. Sci., 18, (1991) Springer–Verlag.

Z. I. Borevich, I.R. Shafarevich.

[1] *The theory of numbers,* Nauka, 1964.

N. Bourbaki.

[1] *Algebra. Modules, rings, forms,* Nauka, 1966.

[2] *Groupes et algèbras de Lie,* Hermann, 1960.

[3] *General topology. Basic structures,* Nauka, 1968.

[5] *Algebra. Ch.X: Homological algebra,* Nauka, 1987.

J.W.S. Cassels, A. Fröhlich (eds.)

[1] *Algebraic number theory,* Academic Press, 1967.

A.W. Chatters, C.R. Hajarnavis.

[1] *Rings with chain condition,* Pitman, 1980.

A. Clifford, G. Preston.

[1] *Algebraic theory of semi–groups,* Mir, 1972.

P.M. Cohn.
[1] *Universal Algebra,* Harper & Row, 1965.
[2] *Algebra,* 1,2 John Wiley and Sons, 1974, 1977.
Ch. Curtis, I. Reiner.
[1] *Representation theory of finite groups and associative algebras,* Interscience, 1962.
J. Dixmier.
[1] *Algèbres enveloppantes,* Gauthier–Villars, 1974.
C. Faith.
[1] *Algebra I: Rings, modules and categories,* Springer–Verlag, Heidelberg, 1973.
[2] *Algebra II: Ring theory,* Springer–Verlag, Heidelberg, 1976.
N.I. Feldman.
[1] *The seventh Hilbert problem,* Moscow University, 1982.
D. Gorenstein.
[1] *Finite simple groups. Introduction to its classification,* Nauka, 1985.
R.I. Grigorchuk.
[1] *On the problem of Burnside periodic groups,* Functional analysis and its application, 14, No 1, (1980) 53–54.
M. Hall.
[1] *The theory of groups,* The Macmillan Co., 1959.
[2] *Combinatorial theory,* Blaisdell, 1967.
I. Herstein.
[1] *Carus Monograph,* 15, Wiley, 1962.
G. Hochschild.
[1] *Grad. text math.,* 75, Springer, 1981.
J. Humphreys.
[1] *Linear algebraic groups,* Nauka, 1980.
N. Jacobson.
[1] *Lie algebras,* Wiley–Interscience, 1962.
[2] *Lecture Notes in Math.,* Springer, 1975.
V.G. Kac.
[1] *Infinite–dimensional Lie algebras. An introduction,* Birkhäuser, 1983.
L.A. Kaluzhin.
[1] *Introduction into the general algebra,* Nauka, 1973.
M.I. Kargapolov, J.I. Merzliakov.
[1] *Grad. Texts Math.,* 62, Springer–Verlag, 1979.
A.A. Kirillov.
[1] *Elements of representation theory,* Springer–Verlag, 1976.
J.G. Kleiman.
[1] *On some problems of the theory of varieties of groups,* In Izv. Acad. Sci. USSR, Ser Mat. 47, No 1, 37–74.

S.C. Kleene.
[1] *Mathematical logic*, Mir, 1973.
A.I. Kostrikin.
[1] *Introduction to algebra*, Universitext, Springer–Verlag, 1982.
[2] *Around Burnside*, Springer–Verlag, 1990.
A.I. Kostrikin, Yu.I. Manin.
[1] *Linear algebra and geometry*, Nauka, 1986.
A.G. Kurosh.
[1] *Lectures on general algebra*, Pergamon Press, 1965.
[2] *General algebra. Lectures from 1969–1970 academic year*, Nauka, 1974.
J. Lambek.
[1] *Rings and modules*, Blaisdells, 1966.
S. Lang.
[1] *Algebra*, Addison–Wesley, 1965.
R. Lyndon, P. Shupp.
[1] *Combinatorial theory of groups*, Mir, 1980.
S. MacLane.
[1] *Homology*, Springer–Verlag, 1963.
V. Magnus, A. Carrass, D. Solitaire.
[1] *Combinatorial theory of groups*, Wiley, 1966.
A.I. Mal'cev.
[1] *Algebraic systems*, Nauka, 1970
[2] *Algorithms and recursive functions*, Springer–Verlag, 1973.
E. Mendelson.
[1] *Introduction to mathematical logic*, Mir, 1971.
R.V. Moody, A. Pianzola.
[1] *Algebras, groups and geometries*, 4, (1987),
H. Neumann.
[1] *Varieties of groups*, Springer–Verlag, 1967.
A.J. Ol'shanskii.
[1] *Geometry of determining relations in groups*, Kluwer AP, 1991.
A.J. Ol'shanskii, A.L. Shmelkin.
[1] *Infinite groups*, In Algebra 4, (1989) VINITI.
L.S. Pontryagin.
[1] *Continuous groups*, Nauka, 1984.
J.P. Razmyslov.
[1] *Identities of algebras and their representations*, Nauka, 1989.
A. Sagle, R. Walde.
[1] *Introduction to Lie groups and Lie algebras*, Academic Press, 1973.
G.P. Serre.
[1] *Lie algebras and Lie groups*, Benjamin, 1965.
[2] *Algèbres de Lie semi–simples complexes*, Benjamin, 1966.

I.R. Shafarevich.
[1] *Introduction to algebraic geometry,* Nauka, 1972.
[2] *Algebra I, Encyclopaedia of Mathematical Sciences,* 11 Springer, 1990.
L.A. Skorniakov.
[1] *Elements of general algebra,* Nauka, 1983.
[2] *Mathematics. News in science abroad,* 21, Mir, 1981.
M.Sh. Tsalenko, E.G. Shulgeifer.
[1] *Fundaments of the theory of categories,* Nauka, 1974.
Van der Waerden B.L.
[1] *Algebra II,* Springer–Verlag, 1982.
E.V. Vinberg.
[1] *Linear representations of groups,* Nauka, 1985.
E.V. Vinberg, A.L. Onishchik.
[1] *Fundaments of the theory of Lie groups,* In Lie groups and Lie algebras 1, 20, (1988) VINITI.
[2] *Seminar on Lie groups and algebraic groups,* Nauka, 1988.
M.V. Volkov, L.N. Shevrin.
[1] *Identities of semi–groups,* Sov. Math., 29(11), 1985 1–64.
K.A. Zhevlakov, A.M.Slinko, I.P. Shestakov, A.I. Shirshov.
[1] *Rings that are nearly associative,* Acad. Press, 1982.

SYMBOL INDEX

SUBJECT INDEX

Other *Mathematics and Its Applications* titles of interest:

A.Ya. Helemskii: *The Homology of Banach and Topological Algebras.* 1989, 356 pp. ISBN 0-7923-0217-6

J. Martinez (ed.): *Ordered Algebraic Structures.* 1989, 304 pp. ISBN 0-7923-0489-6

V.I. Varshavsky: *Self-Timed Control of Concurrent Processes. The Design of Aperiodic Logical Circuits in Computers and Discrete Systems.* 1989, 428 pp. ISBN 0-7923-0525-6

E. Goles and S. Martinez: *Neural and Automata Networks. Dynamical Behavior and Applications.* 1990, 264 pp. ISBN 0-7923-0632-5

A. Crumeyrolle: *Orthogonal and Symplectic Clifford Algebras. Spinor Structures.* 1990, 364 pp. ISBN 0-7923-0541-8

S. Albeverio, Ph. Blanchard and D. Testard (eds.): *Stochastics, Algebra and Analysis in Classical and Quantum Dynamics.* 1990, 264 pp. ISBN 0-7923-0637-6

G. Karpilovsky: *Symmetric and G-Algebras. With Applications to Group Representations.* 1990, 384 pp. ISBN 0-7923-0761-5

J. Bosak: *Decomposition of Graphs.* 1990, 268 pp. ISBN 0-7923-0747-X

J. Adamek and V. Trnkova: *Automata and Algebras in Categories.* 1990, 488 pp. ISBN 0-7923-0010-6

A.B. Venkov: *Spectral Theory of Automorphic Functions and Its Applications.* 1991, 280 pp. ISBN 0-7923-0487-X

M.A. Tsfasman and S.G. Vladuts: *Algebraic Geometric Codes.* 1991, 668 pp. ISBN 0-7923-0727-5

H.J. Voss: *Cycles and Bridges in Graphs.* 1991, 288 pp. ISBN 0-7923-0899-9

V.K. Kharchenko: *Automorphisms and Derivations of Associative Rings.* 1991, 386 pp. ISBN 0-7923-1382-8

A.Yu. Olshanskii: *Geometry of Defining Relations in Groups.* 1991, 513 pp. ISBN 0-7923-1394-1

F. Brackx and D. Constales: *Computer Algebra with LISP and REDUCE. An Introduction to Computer-Aided Pure Mathematics.* 1992, 286 pp. ISBN 0-7923-1441-7

N.M. Korobov: *Exponential Sums and their Applications.* 1992, 210 pp. ISBN 0-7923-1647-9

D.G. Skordev: *Computability in Combinatory Spaces. An Algebraic Generalization of Abstract First Order Computability.* 1992, 320 pp. ISBN 0-7923-1576-6

E. Goles and S. Martinez: *Statistical Physics, Automata Networks and Dynamical Systems.* 1992, 208 pp. ISBN 0-7923-1595-2

Other *Mathematics and Its Applications* titles of interest:

M.A. Frumkin: *Systolic Computations.* 1992, 320 pp. ISBN 0-7923-1708-4

J. Alajbegovic and J. Mockor: *Approximation Theorems in Commutative Algebra.* 1992, 330 pp. ISBN 0-7923-1948-6

I.A. Faradzev, A.A. Ivanov, M.M. Klin and A.J. Woldar: *Investigations in Algebraic Theory of Combinatorial Objects.* 1993, 516 pp. ISBN 0-7923-1927-3

I.E. Shparlinski: *Computational and Algorithmic Problems in Finite Fields.* 1992, 266 pp. ISBN 0-7923-2057-3

P. Feinsilver and R. Schott: *Algebraic Structures and Operator Calculus. Vol. I. Representations and Probability Theory.* 1993, 224 pp. ISBN 0-7923-2116-2

A.G. Pinus: *Boolean Constructions in Universal Algebras.* 1993, 350 pp. ISBN 0-7923-2117-0

V.V. Alexandrov and N.D. Gorsky: *Image Representation and Processing. A Recursive Approach.* 1993, 200 pp. ISBN 0-7923-2136-7

L.A. Bokut′ and G.P. Kukin: *Algorithmic and Combinatorial Algebra.* 1993, 469 pp. ISBN 0-7923-2313-0

Y. Bahturin: *Basic Structures of Modern Algebra.* 1993, 419 pp. ISBN 0-7923-2459-5

MIX
Papier aus verantwortungsvollen Quellen
Paper from responsible sources
FSC® C105338

If you have any concerns about our products,
you can contact us on
ProductSafety@springernature.com

In case Publisher is established outside the EU,
the EU authorized representative is:
Springer Nature Customer Service Center GmbH
Europaplatz 3, 69115 Heidelberg, Germany

Printed by Libri Plureos GmbH
in Hamburg, Germany